Applied Mathematical Sciences
Volume 174

For other titles published in this series, go to
www.springer.com/series/34

Weizhang Huang • Robert D. Russell

Adaptive Moving Mesh Methods

Springer

Weizhang Huang
Department of Mathematics
The University of Kansas
Lawrence, KS 66045
USA
huang@math.ku.edu

Robert D. Russell
Department of Mathematics
Simon Fraser University
Burnaby, BC V5A 1S6
Canada
rdr@cs.sfu.ca

ISSN 0066-5452
ISBN 978-1-4419-7915-5 e-ISBN 978-1-4419-7916-2
DOI 10.1007/978-1-4419-7916-2
Springer New York Dordrecht Heidelberg London

Mathematics Subject Classification (2010): 41A05, 41A15, 6500, 65D17, 65D18, 65K10, 65L10, 65L50, 65L60, 65M06, 65M08, 65M15, 65M20, 65M50, 65M60, 65N40, 65N50, 74S05, 74S10, 74S20, 80M10

Printed on acid-free paper

Springer is part of Springer Science+Business Media (www.springer.com)

To our families: Fang, Eileen, Andrew and Robin, Nathan, Claire, Maria, and Simon.

Preface

Solutions of partial differential equations (PDEs) arising in science and engineering frequently have large variations occurring over small portions of the physical domain, and a major challenge when solving such problems is to appropriately resolve the solution behavior there. When finite difference or finite element methods are employed, a fine mesh is required in those particular parts of the physical domain. Using a uniform mesh throughout the physical domain can become a substantial, indeed a formidable, computational expense, especially in multidimensions where the number of mesh points required can be prohibitively large. A practical and often indispensable alternative is to place a high proportion of mesh points in the regions of large solution variation and few points in the rest of the domain. With this basic idea of *mesh adaptivity*, the total number of mesh points required is much smaller than with a uniform mesh, and significant economies are achieved.

The purpose of this book is to present the theoretical and practical aspects of mesh adaptivity, with particular emphasis on its application to time-dependent PDEs. Given the ubiquitous need for mesh adaptivity in the various areas of science and engineering, a proliferation of methods have been developed in the past. While on one hand this makes the study of mesh adaptivity an exciting, multifaceted endeavor, it has also made it a daunting task for the potential user of adaptive mesh techniques to know where to begin looking for a method suitable for his or her particular needs. In this book a major effort is made to make the general topic of mesh adaptivity more accessible, both practically and theoretically, to a broad audience. The intent, however, is not to provide a review of all adaptive mesh methods, but rather a detailed discussion of one type of method, the r-adaptive mesh method or the moving mesh method, which reflects the authors' research interest in this field over the past 20 years. Fortuitously, the fundamental principles presented are applicable in a much wider range of contexts, and there are often close (often complementary) relationships between the moving mesh method considered here and other important types of adaptive mesh methods such as the h- and p-adaptive methods.

It is important to say at the outset that mesh adaptivity has its place and should not be viewed as a panacea. For problems with smooth solutions, a uniform mesh suffices and is usually preferred over a nonuniform one because it readily lends itself to efficient solution. Even when adaptivity is required, an isotropic mesh can be preferred for the same reason if it can resolve the solution without using an undue number of mesh points. Anisotropic meshes are favored when there is a need for better alignment of the mesh with certain solution directions, such as those arising due to boundary or interior layers and sharp interfaces. The distinction between isotropic and anisotropic meshes is an important theme in the chapters on multidimensional adaptation.

The type of adaptive mesh methods considered in this book for solving PDEs is characterized by the fact that a mesh moves continuously in time while adapting to the evolving structures in the solution. These are called adaptive moving mesh methods, or simply *moving mesh methods* for short. Since point locations are dynamically *relocated* during the course of numerical computation, a moving mesh method is also called an r-adaptive method in the finite element community. The analysis of adaptivity within the moving mesh context focuses on how to optimally choose mesh points, where the computational cost is normally correlated with the number of mesh points used. The main goals of this book are to carry out this adaptivity analysis, understand the existing methods, and develop new ones, by relying on two key tools, the so-called *equidistribution* and *alignment* conditions. While the concept of equidistribution has been well-known and used in the mesh adaptivity community for many years, an understanding of alignment in multidimensions is relatively recent. These tools are discussed in detail in Chapters 2 and 4.

It is useful to note that the formulations of the moving mesh strategies presented in this book are generally independent of specific types of physical PDEs being solved (although discretization of the physical PDEs themselves are obviously not). Throughout the book, moving mesh methods are described mainly for parabolic PDEs. When they are applied to other types of PDEs, discretization schemes suitable for the underlying physical PDEs should be used.

An outline of the book is as follows.

Chapter 1 provides an introduction to moving mesh methods in one spatial dimension (1D). Specifically, simple moving mesh finite difference and finite element methods for time-dependent PDEs are described and implemented in Matlab. (The Matlab codes are available online on the first author's homepage.) The basic components of moving mesh methods – the mesh movement strategy, the PDE discretization for moving meshes, and the overall solution procedure – are discussed at the end of the chapter.

Chapter 2 provides a more detailed treatment of these three components of moving mesh methods in 1D. The well-known equidistribution principle plays a fundamental role in the design of mesh movement strategies throughout. The optimality properties of the equidistributing meshes (discrete case) and coordinate transformations (continuous case) are discussed. Using equidistribution, a number of mesh equations and moving mesh equations (MMPDEs) – continuous forms of mesh movement strategies formulated in terms of coordinate transformations – are developed for steady-state and time-dependent problems, respectively. Practical implementation issues, including discretization of mesh equations and physical PDEs and the overall solution procedure, are addressed. In particular, the discretization of the physical PDEs for a moving mesh using finite differences or finite elements can be done using either the quasi-Lagrange approach or the rezoning approach, and the coupled system of mesh and physical PDEs can be solved either simultaneously or alternately.

The key to the success of moving mesh methods lies in a suitable choice of a mesh density function. This function controls mesh concentration through the equidistribution principle and typically measures the difficulty in the spatial numerical approximation of the underlying problem. Several sections in this chapter are devoted to the selection of the mesh density function based on an interpolation error estimate, on scaling invariance, or on an a posteriori error estimate, with the optimal bound for interpolation error or solution error also obtained for the corresponding equidistributing mesh. Numerical results for a number of nontrivial physical examples of time-dependent PDEs are included.

The remaining chapters discuss and expand upon mesh adaptivity issues in the considerably more challenging multidimensional contexts. For these higher spatial dimensions, one needs the basic tools of advanced calculus to transform PDEs between the physical space and the computational space, and these are reviewed in Chapter 3. For the situation where the time-varying mesh is assumed to be known, a detailed discussion of implementation issues for the finite difference and finite element methods are then given for both the quasi-Lagrange and rezoning approaches.

Chapters 4 and 5 for the most part generalize the steady-state mesh adaptation strategies in Chapter 2 to multidimensions. The situation with mesh adaptivity now becomes much more complicated. The equidistribution principle, specifying only the volume of mesh elements, is no longer sufficient for determining a multidimensional mesh. An additional condition is needed for specifying the shape and orientation of mesh elements. Chapter 4 presents the basic principles of multidimensional mesh adaptivity, including this needed alignment condition. The mesh adaptivity is driven by a solution-dependent monitor function (a symmetric positive definite matrix function related naturally to the 1D mesh density function), which defines a metric on the physical domain. A fundamental interpretation of multidimensional mesh adaptivity is as a technique to generate an M-uniform mesh (a uniform mesh in the metric space), which in turn provides a natural control of mesh equidistribution and alignment, the role of the latter being to ensure that the mesh is properly aligned with behavior of the physical solution. These equidistribution and alignment conditions are analyzed and related to the mesh quality in a mathematically precise way. Interpretations are given from both a discrete (mesh) and continuous (coordinate transformation) perspective.

Chapter 5 discusses how to choose an optimal monitor function for a given interpolation error bound or a posteriori error estimate, as well as a monitor function based upon other geometric and physical considerations. The rigorous theoretical treatment of interpolation error in general Sobolev spaces, given for both the isotropic and anisotropic cases, provides a strong result showing how the optimal choice of monitor function leads to an error bounded by an optimal solution-dependent factor times an optimal order of $1/N$, with N being the number of mesh elements. Interpolation error bounds associated with non-optimal monitor functions and optimal monitor functions for a posteriori error bounds are also addressed. Furthermore, various practical aspects of computing monitor functions are discussed.

The final two chapters are devoted to a discussion of the specific mesh adaptation strategies. For some, the development has been directly driven by the equidistribution and alignment conditions, while many others can be shown to be closely related. Even more generally, these two conditions can be used to facilitate an understanding of virtually all of them. The mesh adaptation strategies are separated into two loose categories, variational methods and velocity-based methods, which are addressed in Chapters 6 and 7, respectively. The variational methods determine the coordinate transformation needed for mesh generation as a minimizer of an adaptation functional typically designed to measure difficulty in the numerical simulation and to achieve desired mesh properties such as smoothness. Various theoretical and practical issues for the functionals are dealt with, including the existence and invertibility of minimizers, derivation of the corresponding Euler-Lagrange equations, and their finite difference and finite element discretization. A general MMPDE strategy is also developed from the steady-state adaptation functionals. A major effort in the chapter has been made to describe and analyze the large variety of variational methods which were originally developed based upon radically different motivations. Notably, a number of these methods, while designed in the context of mesh generation, can be very easily modified to perform mesh adaptation through proper insertion of a monitor function. Numerical examples are chosen to both illustrate strengths and weakness of the methods and to demonstrate their utility for solving nontrivial physical problems.

Chapter 7 discusses velocity-based methods, which target the mesh velocity directly and subsequently determine mesh point locations by integrating the velocity field. These are Lagrange type methods, being more or less motivated by Lagrangian methods in fluid dynamics for which the mesh coordinates are obtained as particle trajectories by integrating a flow velocity. The first method considered, the GCL (Geometric Conservation Law) method, forces the mesh velocity field to satisfy an equidistribution condition and a curl condition. It shares common features with a method in Chapter 6 based upon the Monge-Kantorovich optimal mass transform problem, where the mapping itself is forced to satisfy such conditions. The GCL method is shown to be closely related to Lagrangian methods in fluid dynamics and a moving mesh method based on deformation maps. Two finite element methods, one based upon GCL and the other the original version of the moving finite element method (MFE), are discussed. Finally, some other physically motivated velocity-based methods are described.

Given the comprehensive treatment of the topics, navigating the book in a way which best suits a reader's interests can be a difficult task, and with this in mind we provide some guidance. While the analysis on optimal error bounds for adaptive meshes found in §§2.4, 2.9, 5.1, 5.2, 5.4, and 6.2 is fairly technical and intimidating for those lacking expertise in the mathematical theory of finite element methods, we want to emphasize the importance of the fundamental results in those sections. Nevertheless, some first time readers may wish to skip the above mentioned sections and focus on the summaries given at the end of §2.4 and §5.2 to avoid being caught up in unfamiliar technical details. In general, the reader may often skip sections not directly related to their research interest. For example, a reader mainly interested in theoretical aspects of adaptivity could skip most of the sections devoted exclusively to implementation issues such as §§2.6, 3.2, 3.4, 5.3, and 6.3. As well, readers mainly interested in finite element methods may skip those sections on finite difference methods and discretizations.

The book can be used as a textbook for an advanced, semester-long course in the numerical solution of partial differential equations. Such a course could cover the basic principles of adaptive mesh movement in 1D, higher dimensional discretization for PDEs on moving meshes, and general principles of mesh adaptation in Chapters 1–4, discussion of monitor functions in §5.2.5 and §5.3, and explanation of variational methods in §6.1 and §6.3. The methods described in Chapters 6 and 7 can be selected based upon the particular interests of the students and/or the instructor, although we would encourage using some treatment of the equidistribution and alignment principles in §6.4.

Finally, it is fair to say that moving mesh methods as a whole are still in a relatively early phase of development. Many of them are at the experimental stage, and almost all require further mathematical justification. Rigorous analysis of moving mesh methods for solving time-dependent PDEs has only been carried out for some very simple model problems to date, and more ways to improve their efficiency and robustness will no doubt be developed. For example, more systematic numerical studies on how to reduce the costs in solving the overall system of mesh and physical PDEs are needed, not to mention how to balance time stepping with spatial mesh adaptation – a question that has received very little attention in the literature. It is the authors' hope that this book will serve as a springboard for the reader who wishes to learn and master basic methods for mesh adaptation in general, and mesh movement in particular, and that it will serve as a stepping stone toward a more complete understanding of and development of practicable moving mesh methods.

Acknowledgment. We are indebted to many colleagues and former graduate students for their invaluable discussion and comments. We are particularly grateful to Jens Lang, Weishi Liu, Chris Paige, and Xiangmin Xu for their careful reading of portions of the manuscript and to Chris J. Budd and Weiming Cao for their long term collaboration and support.

Lawrence, Kansas and Vancouver, British Columbia *Weizhang Huang*
July, 2010 *Robert D. Russell*

Contents

Chapter 1
Introduction

1.1 A model problem

In this first chapter, we introduce the basic principles of adaptivity and moving mesh methods for solving partial differential equations in one spatial dimension. In particular, two adaptive methods are described and used to solve a simple model problem – an initial-boundary value problem consisting of Burgers' equation

$$u_t = \varepsilon u_{xx} - \left(\frac{u^2}{2}\right)_x, \quad x \in (0,1),\ t > 0 \tag{1.1}$$

subject to the boundary conditions

$$u(0,t) = u(1,t) = 0 \tag{1.2}$$

and initial condition

$$u(x,0) = \sin(2\pi x) + \frac{1}{2}\sin(\pi x). \tag{1.3}$$

Here, $\varepsilon > 0$ is a physical parameter. For small ε, the solution has a smooth initial profile and develops a steep front. The front propagates toward the right end and eventually dies out due to the homogeneous Dirichlet boundary condition at $x = 1$. The difficulty with the numerical solution of the problem lies in the resolution of this propagating steep front.

W. Huang and R.D. Russell, *Adaptive Moving Mesh Methods*, Applied Mathematical Sciences 174, DOI 10.1007/978-1-4419-7196-2_1,

1.2 A moving finite difference method

1.2.1 Finite difference method on a fixed mesh

The first numerical method we consider is a standard method of lines for the partial differential equation (PDE) (1.1). Specifically, the PDE is first discretized in the spatial domain, and then the resulting system of ordinary differential equations (ODEs) is integrated using an ODE solver. The main advantage of the method of lines is the separate treatments of the spatial and temporal components of the PDE, so that attention can be focused on each of them in turn.

To illustrate, we consider the finite difference solution of the model problem on a uniform spatial mesh. Given a positive integer N, define the mesh

$$\mathscr{T}_h: \quad x_j = (j-1)h, \quad j = 1, ..., N \tag{1.4}$$

where $h = 1/(N-1)$. A semi-discretization of Burgers' equation (1.1) using central finite differences in space is given by

$$\frac{du_j}{dt} = \frac{\varepsilon}{h^2}\left(u_{j+1} - 2u_j + u_{j-1}\right) - \frac{1}{4h}\left(u_{j+1}^2 - u_{j-1}^2\right), \qquad j = 2, ..., N-1 \tag{1.5}$$

where $u_j(t)$ is an approximation to the solution $u = u(x,t)$ at $x = x_j$, i.e., $u_j(t) \approx u(x_j,t)$. The discrete boundary and initial conditions become

$$u_1(t) = 0, \quad u_N(t) = 0, \quad t > 0 \tag{1.6}$$

$$u_j(0) = \sin(2\pi x_j) + \frac{1}{2}\sin(\pi x_j), \quad j = 1, ..., N. \tag{1.7}$$

The boundary conditions (1.6) are replaced in the actual implementation by the ODE form

$$\frac{du_1}{dt} = 0, \quad \frac{du_N}{dt} = 0. \tag{1.8}$$

The equations (1.5) and (1.8), with initial conditions (1.7), constitute an initial value problem which can in principle be conveniently integrated using an ODE solver.

Figure 1.1 (a) and (b) show computed solutions for Burgers' equation with $\varepsilon = 10^{-2}$ using this approach. The results are obtained using the Matlab ODE solver "ode15i," which is based on the backward differentiation formulas (BDFs) of orders 1–5 [296, 297, 298]. Note that with a uniform mesh of 21 points, oscillations in the computed solution are visible, indicating that the steep front is not adequately resolved on the uniform mesh. These oscillations can be eliminated by using a finer mesh of 81 equidistant points, as shown in Figure 1.1(b).

When ε is smaller, a correspondingly finer mesh has to be used to resolve the steep front. This is illustrated in Figure 1.2, where oscillations are still visible in the

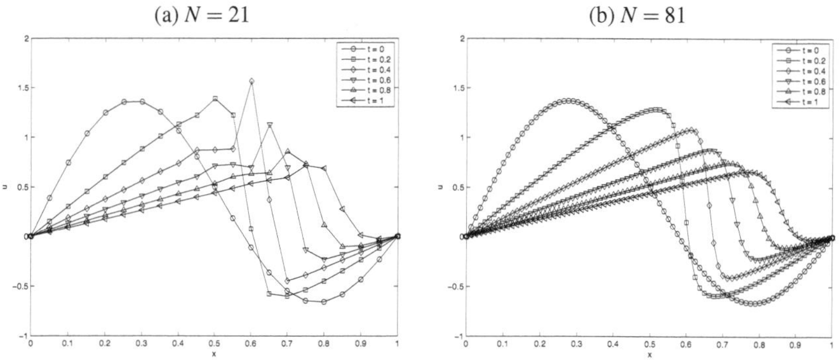

Fig. 1.1 Computed solutions obtained with a uniform mesh for Burgers' equation with $\varepsilon = 10^{-2}$ are shown at $t = 0$, 0.2, 0.4, 0.6, 0.8, and 1.0.

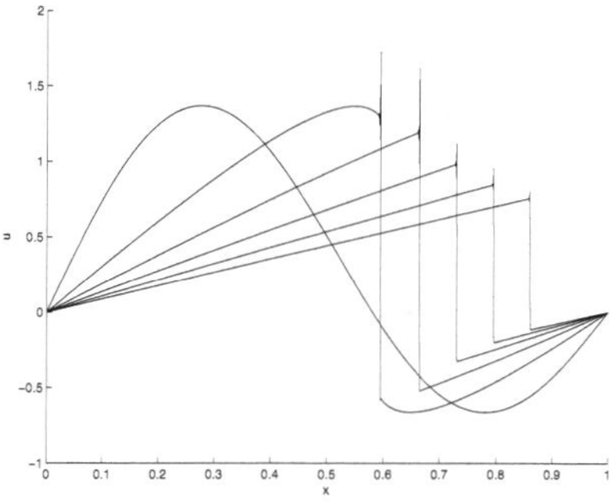

Fig. 1.2 Computed solutions at $t = 0$, 0.2, 0.4, 0.6, 0.8, and 1.0 obtained with a uniform mesh of 2001 points for Burgers' equation with $\varepsilon = 10^{-4}$.

solution computed with a uniform mesh of 2001 points for the case $\varepsilon = 10^{-4}$. Use of a very fine uniform mesh is expensive in terms of computer time and memory, and much more so for two- and three-dimensional problems, making mesh adaptation necessary.

1.2.2 Finite difference method on an adaptive moving mesh

The adaptive solution of the model problem requires that the mesh points be concentrated around the steep front and dynamically adjusted to follow the front as it propagates in time. Such a dynamically adjusting mesh is referred to as an *adaptive moving mesh*.

Adaptive mesh movement is often best understood by interpreting the problem in terms of a suitable coordinate transformation. Specifically, we assume for the moment that a time-dependent coordinate transformation $x = x(\xi,t) : \Omega_c \equiv [0,1] \to \Omega \equiv [0,1]$ is given, where Ω_c and Ω are the computational and physical domains, respectively. Generally speaking, this transformation is chosen such that the solution in the transformed spatial variable,

$$\hat{u}(\xi,t) = u(x(\xi,t),t),$$

is smooth and in principle economical to approximate using a uniform mesh. A corresponding moving mesh can be described as

$$\mathscr{T}_h(t): \qquad x_j(t) = x(\xi_j,t), \quad j = 1,...,N \tag{1.9}$$

for the fixed, uniform mesh on Ω_c,

$$\mathscr{T}_h^c: \quad \xi_j = \frac{j-1}{N-1}, \quad j = 1,...,N. \tag{1.10}$$

A finite difference discretization of Burgers' equation for $\hat{u}(\xi,t)$ on this moving mesh can be derived using the so-called *quasi-Lagrange approach*. Burgers' equation is transformed from the physical domain to the computational domain using the coordinate transformation as follows: By the chain rule,

$$\hat{u}_\xi = u_x x_\xi, \quad \hat{u}_t = u_t + u_x x_t, \tag{1.11}$$

where $x_t = \frac{\partial x}{\partial t}(\xi,t)$ determines the mesh speed. In the new coordinates (ξ,t) Burgers' equation (1.1) becomes

$$\hat{u}_t - \frac{\hat{u}_\xi}{x_\xi} x_t = \frac{\varepsilon}{x_\xi}\left(\frac{\hat{u}_\xi}{x_\xi}\right)_\xi - \frac{1}{x_\xi}\left(\frac{\hat{u}^2}{2}\right)_\xi. \tag{1.12}$$

Central finite differences can then be used to discretize (1.12) on the uniform computational mesh $\mathscr{T}_h^c$. This yields

$$\begin{aligned}\frac{du_j}{dt} - \frac{(u_{j+1}-u_{j-1})}{(x_{j+1}-x_{j-1})}\frac{dx_j}{dt} &= \frac{2\varepsilon}{(x_{j+1}-x_{j-1})}\left[\frac{(u_{j+1}-u_j)}{(x_{j+1}-x_j)} - \frac{(u_j-u_{j-1})}{(x_j-x_{j-1})}\right] \\ &\quad -\frac{1}{2}\frac{(u_{j+1}^2-u_{j-1}^2)}{(x_{j+1}-x_{j-1})}, \quad j = 2,...,N-1\end{aligned} \tag{1.13}$$

where $u_j(t) \approx \hat{u}(\xi_j,t) = u(x_j(t),t)$.

This begs the question of how one determines the coordinate transformation $x = x(\xi,t)$. We defer a detailed discussion of moving mesh methods until Chapter 2, but for now suppose that $x(\xi,t)$ is determined by solving the so-called *moving mesh PDE* (MMPDE)

$$x_t = \frac{1}{\rho\tau}\left(\rho x_\xi\right)_\xi, \tag{1.14}$$

supplemented with the boundary conditions

$$x(0,t) = 0, \quad x(1,t) = 1. \tag{1.15}$$

Here, $\rho = \rho(x,t)$ is called a *mesh density specification function*, or simply *mesh density function*, whose purpose is to control the concentration or density of the mesh, and $\tau > 0$ is a user-specified parameter for adjusting the response time of mesh movement to changes in $\rho(x,t)$. The smaller τ, the more quickly the mesh responds to changes in ρ. Likewise, the mesh moves slowly when a large value of τ is used.

The proper choice of a mesh density function is key to the success of the moving mesh method. One popular choice is

$$\rho = \left(1 + \frac{1}{\alpha} |u_{xx}|^2\right)^{\frac{1}{3}}, \tag{1.16}$$

where α is the intensity parameter given by

$$\alpha = \max\left\{1,\ \left[\int_0^1 |u_{xx}|^{\frac{2}{3}} dx\right]^3\right\}. \tag{1.17}$$

(Such strategies for choosing the mesh density function are explained in Chapter 2.)

Semi-discretization of MMPDE (1.14) on the uniform mesh $\mathscr{T}_h^c$ gives, for $j = 2, \ldots, N-1$,

$$\frac{dx_j}{dt} = \frac{1}{\rho_j \tau \Delta\xi^2}\left[\frac{\rho_{j+1}+\rho_j}{2}(x_{j+1}-x_j) - \frac{\rho_j+\rho_{j-1}}{2}(x_j - x_{j-1})\right], \tag{1.18}$$

and the boundary conditions (1.15) become

$$\frac{dx_1}{dt} = 0, \quad \frac{dx_N}{dt} = 0. \tag{1.19}$$

Here, $\Delta\xi = 1/(N-1)$, and

$$\rho_j = \left(1 + \frac{1}{\alpha_h}\left|u_{xx,j}\right|^2\right)^{\frac{1}{3}}, \quad j = 1, \ldots, N \tag{1.20}$$

$$\alpha_h = \max\left\{1,\ \left[\sum_{j=2}^{N} \frac{1}{2}(x_j - x_{j-1})\left(\left|u_{xx,j}\right|^{\frac{2}{3}} + \left|u_{xx,j-1}\right|^{\frac{2}{3}}\right)\right]^3\right\}, \tag{1.21}$$

where the second derivative is approximated by

$$
\begin{cases}
u_{xx,j} = \dfrac{2}{(x_{j+1}-x_{j-1})}\left[\dfrac{(u_{j+1}-u_j)}{(x_{j+1}-x_j)} - \dfrac{(u_j-u_{j-1})}{(x_j-x_{j-1})}\right], \\
\qquad\qquad j = 2,...,N-1 \\
u_{xx,1} = \dfrac{2\left[(x_2-x_1)(u_3-u_1)-(x_3-x_1)(u_2-u_1)\right]}{(x_3-x_1)(x_2-x_1)(x_3-x_2)}, \\
u_{xx,N} = \dfrac{2\left[(x_{N-1}-x_N)(u_{N-2}-u_N)-(x_{N-2}-x_N)(u_{N-1}-u_N)\right]}{(x_{N-2}-x_N)(x_{N-1}-x_N)(x_{N-2}-x_{N-1})}.
\end{cases}
\tag{1.22}
$$

If u is not smooth, the discrete mesh density function computed this way can often change abruptly and unnecessarily slow down the computation. To obtain a smoother mesh and also make the MMPDE easier to integrate, it is common practice in the context of moving mesh methods to smooth the mesh density function. A simple but effective smoothing scheme is weighted averaging, e.g.,

$$
\begin{cases}
\rho_j := \frac{1}{4}\rho_{j-1} + \frac{1}{2}\rho_j + \frac{1}{4}\rho_{j+1}, \quad j = 2,...,N-1 \\
\rho_1 := \frac{1}{2}\rho_1 + \frac{1}{2}\rho_2, \\
\rho_N := \frac{1}{2}\rho_{N-1} + \frac{1}{2}\rho_N,
\end{cases}
\tag{1.23}
$$

where the symbol ":=" stands for the operation in which the right-hand side terms are calculated and the final value is saved to the variable on the left-hand side. Several sweeps of the scheme may be applied each integration step. (Four sweeps are used for the numerical results presented in this chapter.)

Equations (1.13) and (1.18), supplemented with the boundary conditions (1.8) and (1.19), form a coupled system of $2N$ ordinary differential equations for the physical solution $u_1(t)$, ..., $u_N(t)$ and the mesh $x_1(t)$, ..., $x_N(t)$. Let

$$
\mathbf{y} = [u_1(t),...,u_N(t),x_1(t),...,x_N(t)]^T, \quad \mathbf{y}' = \frac{d\mathbf{y}}{dt}.
$$

Then the ODE system can be written in the implicit form

$$
\boldsymbol{f}(t,\mathbf{y},\mathbf{y}') = \mathbf{0}. \tag{1.24}
$$

Performance of an ODE solver can generally be improved by utilizing the nonzero structure of the Jacobian matrices $\frac{\partial \boldsymbol{f}}{\partial \mathbf{y}}$ and $\frac{\partial \boldsymbol{f}}{\partial \mathbf{y}'}$. For the current system, this structure has the form

$$
\frac{\partial \mathbf{f}}{\partial \mathbf{y}} = \left[\begin{array}{ccccc|ccccc}
* & * & & & & * & * & & & \\
* & * & * & & & * & * & * & & \\
 & \ddots & \ddots & \ddots & & & \ddots & \ddots & \ddots & \\
 & & * & * & * & & & * & * & * \\
 & & & * & * & & & & * & * \\
\hline
* & * & \cdots & * & * & * & * & \cdots & * & * \\
* & * & \cdots & * & * & * & * & \cdots & * & * \\
\vdots & \vdots & & \vdots & \vdots & \vdots & \vdots & & \vdots & \vdots \\
* & * & \cdots & * & * & * & * & \cdots & * & * \\
* & * & \cdots & * & * & * & * & \cdots & * & *
\end{array} \right] \tag{1.25}
$$

$$
\frac{\partial \mathbf{f}}{\partial \mathbf{y}'} = \left[\begin{array}{ccccc|ccccc}
* & & & & & * & & & & \\
 & * & & & & & * & & & \\
 & & \ddots & & & & & \ddots & & \\
 & & & * & & & & & * & \\
 & & & & * & & & & & * \\
\hline
 & & & & & * & & & & \\
 & & & & & & * & & & \\
 & & & & & & & \ddots & & \\
 & & & & & & & & * & \\
 & & & & & & & & & *
\end{array} \right]. \tag{1.26}
$$

Returning to the model problem, the solution and mesh trajectories computed for $\varepsilon = 10^{-4}$ are shown in Figure 1.3. In the computation, an initial uniform mesh with $N = 41$ and the value $\tau = 10^{-2}$ are used. Experience has shown that this value for τ works well for most problems. Oscillations are no longer visible in the computed solution, as the formation and propagation of the steep front are resolved on the adaptive moving mesh of 41 points. The mesh trajectories show how the mesh points respond quickly to the change in the solution.

1.3 A moving finite element method

1.3.1 Finite element method on a fixed mesh

The finite element discretization in space for the model problem is based on the Galerkin formulation of Burgers' equation (1.1). Let

$$
V = H_0^1(0,1) \equiv \{u \mid u \in H^1(0,1),\ u(0) = u(1) = 0\}.
$$

(a) Computed solution. (b) Mesh trajectories.

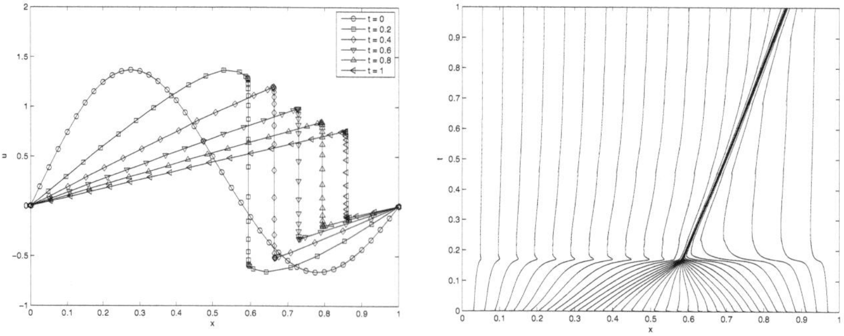

Fig. 1.3 (a) The computed solution for Burgers' equation with $\varepsilon = 10^{-4}$, obtained with an adaptive moving mesh of 41 points, is shown at $t = 0$, 0.2, 0.4, 0.6, 0.8, and 1.0. (b) The corresponding mesh trajectories.

(See Appendix A for the definition of Sobolev space $H^1(0,1)$.) For a given time $T > 0$, the Galerkin formulation of the model problem is the following: Find $u(\cdot,t) \in V$ for $0 < t \leq T$ such that

$$\int_0^1 u_t \phi dx = \int_0^1 \left(-\varepsilon u_x + \frac{1}{2} u^2 \right) \phi_x dx, \quad \forall \phi \in V, \ 0 < t \leq T \tag{1.27}$$

and the initial condition (1.3) holds. The equation (1.27) is obtained by multiplying (1.1) by ϕ, integrating both sides of the resulting equation over the interval $(0,1)$, and integrating by parts on the right-hand side.

We consider the linear finite element approximation on the uniform mesh $\mathscr{T}_h$ in (1.4). Specifically, define the basis functions (hat functions)

$$\phi_j(x) = \begin{cases} \dfrac{x - x_{j-1}}{x_j - x_{j-1}}, & \text{for } x \in [x_{j-1}, x_j] \\ \dfrac{x_{j+1} - x}{x_{j+1} - x_j}, & \text{for } x \in [x_j, x_{j+1}] \\ 0, & \text{otherwise} \end{cases} \qquad j = 1, ..., N \tag{1.28}$$

and let V^h be the $(N-2)$-dimensional subspace of V spanned by the basis functions $\phi_2, ..., \phi_{N-1}$, i.e.,

$$V^h = \text{span}\{\phi_2, ..., \phi_{N-1}\}.$$

A linear finite element approximation $u^h(\cdot,t) \in V^h$ for $0 < t \leq T$ to the exact solution u of the model problem is then required to satisfy

$$\int_0^1 u_t^h \phi dx = \int_0^1 \left(-\varepsilon u_x^h + \frac{1}{2} \left(u^h \right)^2 \right) \phi_x dx, \quad \forall \, \phi \in V^h, \, 0 < t \leq T \tag{1.29}$$

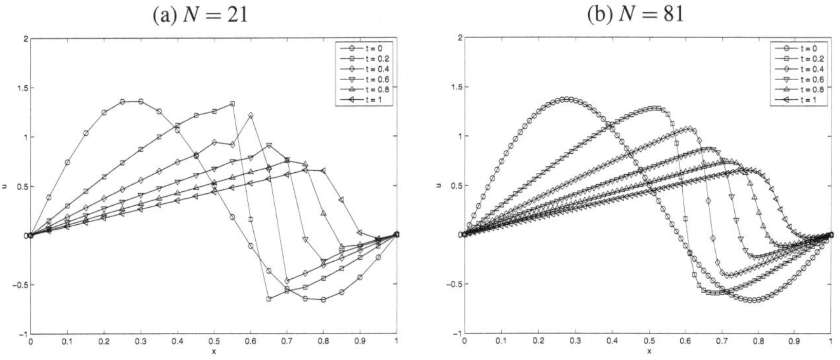

Fig. 1.4 Finite element solutions for Burgers' equation with $\varepsilon = 10^{-2}$, obtained with a uniform mesh, are shown at $t = 0,\ 0.2,\ 0.4,\ 0.6,\ 0.8$, and 1.0.

and

$$u^h(x_j,0) = u(x_j,0), \quad j = 1,...,N. \tag{1.30}$$

Writing

$$u^h(x,t) = \sum_{j=1}^{N} u_j(t)\phi_j(x), \tag{1.31}$$

where $u_j(t) \approx u(x_j,t)$, and taking $\phi = \phi_k(x)$, $k = 2,...,N-1$ in (1.29) leads to

$$\begin{aligned}\sum_{j=1}^{N} \frac{du_j}{dt} \int_0^1 \phi_j(x)\phi_k(x)dx = &-\varepsilon \sum_{j=1}^{N} u_j \int_0^1 \phi_j'(x)\phi_k'(x)dx \\ &+ \frac{1}{2}\int_0^1 \left(\sum_{j=1}^{N} u_j\phi_j(x)\right)^2 \phi_k'(x)dx, \quad k = 2,...,N-1. \end{aligned}\tag{1.32}$$

These equations and the boundary conditions (1.8) constitute a nonlinear system of N ordinary differential equations, which can be integrated with N initial conditions (1.30) for the unknown variables $u_1(t),...,u_N(t)$.

An implementation of the finite element method is described below. Figure 1.4 (a) and (b) show the numerical solutions for Burgers' equation with $\varepsilon = 10^{-2}$, obtained using the implementation. The results are comparable to those obtained using the finite difference method in §1.2.1. Once again, oscillations in the solution computed with a uniform mesh of 21 points are significant, unlike for the solution on a finer mesh of 81 points. Like the finite difference method, the finite element method requires a very large number of equidistant points to resolve the steep front when ε is small.

Implementation of finite element method. The initial value problem consisting of the ODE system (1.32) and (1.8), with initial conditions (1.30), can be solved using a standard ODE solver. The residual of the system, i.e., the function $\boldsymbol{f}(t,\mathbf{y},\mathbf{y}')$ in the implicit form (1.24) with the unknowns $\mathbf{y} = [u_1,...,u_N]^T$ (since the mesh points are fixed), can be easily calculated. From (1.32),

$$\begin{aligned} f_k(t,\mathbf{y},\mathbf{y}') &= \int_0^1 u_t^h \phi_k dx - \int_0^1 \left(-\varepsilon u_x^h + \frac{1}{2}\left(u^h\right)^2\right)(\phi_k)_x dx \\ &= \int_{x_{k-1}}^{x_k} \left(\frac{du_{k-1}}{dt}\phi_{k-1} + \frac{du_k}{dt}\phi_k\right)\phi_k dx \\ &\quad - \int_{x_{k-1}}^{x_k} \left(-\varepsilon\left(u_{k-1}\phi_{k-1} + u_k\phi_k\right)_x + \frac{1}{2}\left(u_{k-1}\phi_{k-1} + u_k\phi_k\right)^2\right)(\phi_k)_x dx \\ &\quad + \int_{x_k}^{x_{k+1}} \left(\frac{du_k}{dt}\phi_k + \frac{du_{k+1}}{dt}\phi_{k+1}\right)\phi_k dx \\ &\quad - \int_{x_k}^{x_{k+1}} \left(-\varepsilon\left(u_k\phi_k + u_{k+1}\phi_{k+1}\right)_x + \frac{1}{2}\left(u_k\phi_k + u_{k+1}\phi_{k+1}\right)^2\right)(\phi_k)_x dx \end{aligned}$$

for $k = 2,...,N-1$, and from the boundary conditions (1.8),

$$f_1(t,\mathbf{y},\mathbf{y}') = \frac{du_1}{dt}, \quad f_N(t,\mathbf{y},\mathbf{y}') = \frac{du_N}{dt}.$$

A procedure to compute the residual element by element is the following:

(i) Set $f_k = 0$ for $k = 1,...,N$.
(ii) For $k = 2,...,N$, compute

$$\begin{aligned} f_{k-1} &:= f_{k-1} + \int_{x_{k-1}}^{x_k} \left(\frac{du_{k-1}}{dt}\phi_{k-1} + \frac{du_k}{dt}\phi_k\right)\phi_{k-1} dx \\ &\quad - \int_{x_{k-1}}^{x_k} \left(-\varepsilon\left(u_{k-1}\phi_{k-1} + u_k\phi_k\right)_x + \frac{1}{2}\left(u_{k-1}\phi_{k-1} + u_k\phi_k\right)^2\right)(\phi_{k-1})_x dx, \\ f_k &:= f_k + \int_{x_{k-1}}^{x_k} \left(\frac{du_{k-1}}{dt}\phi_{k-1} + \frac{du_k}{dt}\phi_k\right)\phi_k dx \\ &\quad - \int_{x_{k-1}}^{x_k} \left(-\varepsilon\left(u_{k-1}\phi_{k-1} + u_k\phi_k\right)_x + \frac{1}{2}\left(u_{k-1}\phi_{k-1} + u_k\phi_k\right)^2\right)(\phi_k)_x dx. \end{aligned}$$

(iii) Modify f_1 and f_N according to the boundary conditions:

$$f_1 := \frac{du_1}{dt} - 0, \quad f_N := \frac{du_N}{dt} - 0.$$

Due to the simplicity of the model problem, the integrals in the above formulas are simple enough to evaluate exactly, but in general, numerical quadrature is needed. In such a case, it suffices to use the two-point Gaussian quadrature rule satisfying

$$\int_{x_{k-1}}^{x_k} f(x)dx = \frac{x_k - x_{k-1}}{2}\left(f(x_{k,1}) + f(x_{k,2})\right) + O\left((x_k - x_{k-1})^5\right),$$

where $x_{k,1} = x_{k-1} + s_1(x_k - x_{k-1})$, $x_{k,2} = x_{k-1} + s_2(x_k - x_{k-1})$ for the two Gaussian points

$$s_1 = \frac{1}{2}\left(1 - \frac{1}{\sqrt{3}}\right), \quad s_2 = \frac{1}{2}\left(1 + \frac{1}{\sqrt{3}}\right).$$

1.3.2 Finite element method on an adaptive moving mesh

A moving mesh strategy for the finite difference method as discussed in §1.2.2 (e.g., using the MMPDE (1.14)) applies in essentially the same way for the finite element method. For this reason, we focus our discussion in this subsection on the finite element semi-discretization of the model problem on the moving mesh $\mathscr{T}_h(t)$ in (1.9).

Since the mesh is moving, the basis functions and the approximation function space are now time-dependent, viz.,

$$\phi_j(x,t) = \begin{cases} \dfrac{x - x_{j-1}(t)}{x_j(t) - x_{j-1}(t)}, & \text{for } x \in [x_{j-1}(t), x_j(t)] \\ \dfrac{x_{j+1}(t) - x}{x_{j+1}(t) - x_j(t)}, & \text{for } x \in [x_j(t), x_{j+1}(t)] \\ 0, & \text{otherwise} \end{cases} \qquad j = 1, ..., N \tag{1.33}$$

and

$$V^h(t) = \text{ span}\{\phi_2(\cdot,t), ..., \phi_{N-1}(\cdot,t)\}.$$

The finite element approximation can be defined in a similar fashion as for a fixed mesh: Find $u^h(\cdot,t) \in V^h(t)$ for $0 < t \leq T$ such that

$$\int_0^1 u_t^h \phi dx = \int_0^1 \left(-\varepsilon u_x^h + \frac{1}{2}\left(u^h\right)^2\right)\phi_x dx \quad \forall \phi \in V^h(t),\ 0 < t \leq T, \tag{1.34}$$

and

$$u^h(x_j(0),0) = u(x_j(0),0), \quad j = 1, ..., N. \tag{1.35}$$

The time derivative of u^h now requires special attention. Writing

$$u^h(x,t) = \sum_{j=1}^{N} u_j(t)\phi_j(x,t), \tag{1.36}$$

where $u_j(t) \approx u(x_j(t),t)$, the time derivative becomes

$$u_t^h(x,t) = \sum_{j=1}^{N} \left(\frac{du_j}{dt}(t)\phi_j(x,t) + u_j(t)\frac{\partial \phi_j}{\partial t}(x,t) \right).$$

A direct calculation using the basis function ϕ_j in (1.33) shows that

$$\frac{\partial \phi_j}{\partial t}(x,t) = -\frac{\partial \phi_j}{\partial x}(x,t)X_t(x,t), \tag{1.37}$$

where $X_t(x,t)$ is the linear interpolant of the nodal mesh speeds, i.e.,

$$X_t(x,t) = \sum_{j=1}^{N} \frac{dx_j}{dt}(t)\phi_j(x,t). \tag{1.38}$$

Thus,

$$\begin{aligned} u_t^h(x,t) &= \sum_{j=1}^{N} \left(\frac{du_j}{dt}(t)\phi_j(x,t) - u_j(t)\frac{\partial \phi_j}{\partial x}(x,t)\, X_t(x,t) \right) \\ &= \sum_{j=1}^{N} \frac{du_j}{dt}(t)\phi_j(x,t) - \frac{\partial u^h}{\partial x}(x,t)\, X_t(x,t). \end{aligned}$$

Inserting this into (1.34) yields, for any $t \in (0,T]$ and $\phi \in V^h(t)$,

$$\int_0^1 \left(\sum_{j=1}^{N} \frac{du_j}{dt}\phi_j - \frac{\partial u^h}{\partial x}\, X_t \right) \phi dx = \int_0^1 \left(-\varepsilon u_x^h + \frac{1}{2}\left(u^h\right)^2 \right) \phi_x\, dx. \tag{1.39}$$

Observe that the mesh movement introduces an extra convection term,

$$-\frac{\partial u^h}{\partial x}\, X_t.$$

Interestingly, this term has a similar form to the convection term

$$-\frac{\hat{u}_\xi}{x_\xi}x_t = -\frac{\partial u}{\partial x}x_t$$

in (1.12), derived using the coordinate transformation $x(\xi,t)$.

The system of ODEs for $u_1,\ldots,u_N$ is obtained by substituting $\phi = \phi_k(x,t)$ into (1.39), viz., for $k = 2,\ldots,N-1$,

$$\int_0^1 \left(\sum_{j=1}^{N} \frac{du_j}{dt}\phi_j - \frac{\partial u^h}{\partial x}\, X_t \right) \phi_k dx = \int_0^1 \left(-\varepsilon u_x^h + \frac{1}{2}\left(u^h\right)^2 \right) (\phi_k)_x\, dx, \tag{1.40}$$

and these equations are supplemented with the boundary conditions (1.8).

Combining with the mesh movement conditions, we now have a coupled system consisting of the discrete physical equations (1.40), mesh equations (1.18), and cor-

(a) Computed solution. (b) Mesh trajectories.

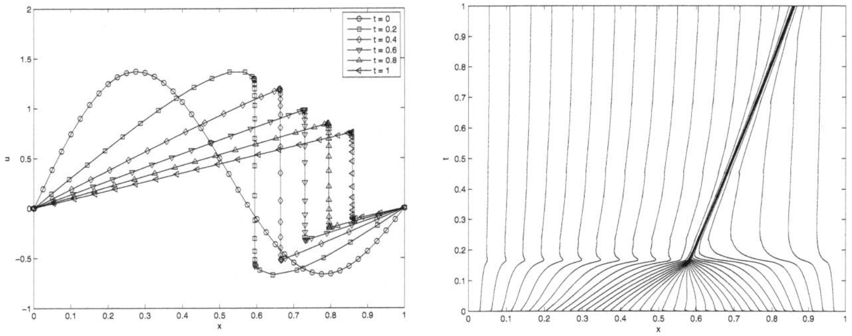

Fig. 1.5 (a) The finite element solution for Burgers' equation with $\varepsilon = 10^{-4}$, obtained with an adaptive moving mesh of 41 points, is shown at $t = 0,\ 0.2,\ 0.4,\ 0.6,\ 0.8$, and 1.0. (b) The corresponding mesh trajectories.

responding boundary conditions for variables $\mathbf{y} = [u_1(t), \ldots, u_N(t), x_1(t), \ldots, x_N(t)]^T$. It can be cast in the form (1.24), with $\frac{\partial f}{\partial \mathbf{y}}$ having a nonzero structure as in (1.25) and the Jacobian matrix $\frac{\partial f}{\partial \mathbf{y}'}$ having the slightly different structure

$$\frac{\partial f}{\partial \mathbf{y}'} = \left[\begin{array}{cccc|cccc} * & * & & & * & * & & \\ * & * & * & & * & * & * & \\ & \ddots & \ddots & \ddots & & \ddots & \ddots & \ddots \\ & & * & * \; * & & & * & * \; * \\ & & & * \; * & & & & * \; * \\ \hline & & & & * & & & \\ & & & & & * & & \\ & & & & & & \ddots & \\ & & & & & & & * \\ & & & & & & & \; * \end{array}\right]. \tag{1.41}$$

Figure 1.5 shows the solution and mesh trajectories obtained for Burgers' equation with $\varepsilon = 10^{-4}$. An ODE solver has been used to integrate the resulting system of ordinary differential equations in the implicit form (1.24). Four smoothing cycles with the weighted averaging (1.23) are applied to the monitor function each time it is computed. The results are almost identical to those shown in Figure 1.3 for the finite difference method. Once again, the steep front is resolved on an adaptive mesh of 41 points, and the mesh responds well to the change in the solution.

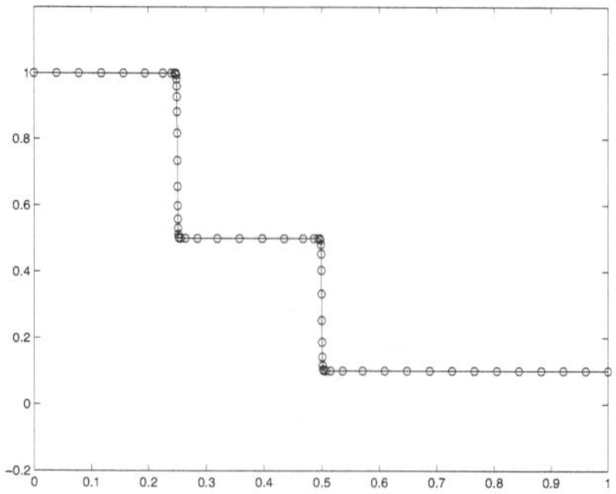

Fig. 1.6 An adaptive initial mesh of 61 points is shown on the graph of the initial solution (1.42).

1.4 Burgers' equation with an exact solution

As another example, an initial-boundary value problem for Burgers' equation with a known solution is solved with the moving finite difference method described in §1.2. (Results for the moving finite element method in §1.3 are similar, and not given.) The problem consists of Burgers' equation (1.1) together with Dirichlet boundary and initial conditions chosen such that the exact solution is

$$u(x,t) = \frac{0.1e^{\frac{-x+0.5-4.95t}{20\varepsilon}} + 0.5e^{\frac{-x+0.5-0.75t}{4\varepsilon}} + e^{\frac{-x+0.375}{2\varepsilon}}}{e^{\frac{-x+0.5-4.95t}{20\varepsilon}} + e^{\frac{-x+0.5-0.75t}{4\varepsilon}} + e^{\frac{-x+0.375}{2\varepsilon}}}, \tag{1.42}$$

where ε is taken as $\varepsilon = 10^{-4}$. Initially, the solution has two steep fronts (physically corresponding to shock waves in a fluid) traveling toward the right end. They merge around $t = 0.55$ and form a steeper shock wave.

Since the initial solution, $u(x,0)$, has two steep fronts, an adaptive initial mesh should be used to start the integration. Such a mesh is generated using a time-continuation method. Specifically, if we define $v(x,t) = tu(x,0)$, where t here denotes the continuation parameter, then v satisfies the differential equation

$$\frac{\partial v}{\partial t} = u(x,0), \quad x \in [0,1],\ 0 < t \le T, \tag{1.43}$$

subject to the initial condition $v(x,0) = 0$. Starting with a uniform mesh, the system consisting of (1.43) and a suitable MMPDE is integrated from $t = 0$ to $t = T$ to obtain an adaptive mesh for $v(x,T) = Tu(x,0)$, and for sufficiently large T, this mesh gives a suitable approximation for $u(x,0)$ (or the multiple $Tu(x,0)$). Here, using $T = 1$ and $N = 61$, the graph of $u(x,0)$ and the points on the graph corresponding to the adaptive mesh obtained this way are shown in Figure 1.6.

Once this adaptive initial mesh for $u(x,0)$ is obtained, Burgers' equation (1.1) with the corresponding boundary conditions is integrated using the moving finite difference method. Figure 1.7 shows a computed solution and the corresponding mesh trajectories obtained with a moving mesh of 61 points. The corresponding time step size used in the integration is shown in Figure 1.8. The convergence history in Figure 1.9 shows that when a moving mesh is used, the error at $t = 1$ in the H^1 semi-norm (see Appendix A) converges at the rate $O(N^{-1})$. In contrast, when a uniform mesh is used, the method does not converge for the range of values of N considered.

To give a sense of the efficiency or the cost-effectiveness of the adaptive moving mesh method, the H^1 semi-norm of the error at $t = 1$ is plotted in Figure 1.10 as a function of the scaled CPU time for three cases – one with uniform meshes, and the other two with adaptive meshes obtained using the intensity parameter $\alpha = \alpha(u)$ defined in (1.17) and using $\alpha = 1$. The error is smaller with adaptive meshes than with a uniform mesh for the same amount of the CPU time and, to reach the same level of error, more CPU time is required when a uniform mesh is used. In this sense, the adaptive moving mesh method is more efficient than a uniform mesh method.

No special effort has been made to optimize the performance of the methods used for Figure 1.10. Generally speaking, a uniform mesh method runs much faster than an adaptive mesh method for the same number of mesh points. This is because the linear algebraic systems resulting from implicit time discretization are tridiagonal and can be solved extremely fast when a uniform mesh is used. In contrast, for the moving mesh method with $\alpha = \alpha(u)$, the Jacobian matrix has a denser nonzero structure (cf. (1.25)), so computing its finite difference approximations (typically used in an ODE solver) and doing the inversion require more CPU time than for a tridiagonal system. The situation can be improved (cf. Figure 1.10) by using a constant intensity parameter such as $\alpha = 1$ in (1.16) which leads to a Jacobian matrix with a sparser nonzero structure

$$\frac{\partial f}{\partial \mathbf{y}} = \begin{bmatrix} A & A \\ B & B \end{bmatrix}, \tag{1.44}$$

where

$$A = \begin{bmatrix} * & * & & & \\ * & * & * & & \\ & \ddots & \ddots & \ddots & \\ & & * & * & * \\ & & & * & * \end{bmatrix}, \quad B = \begin{bmatrix} * & \cdots & * & & & & \\ \vdots & \ddots & & \ddots & & & \\ * & & & \ddots & & & \\ & \ddots & & \ddots & & \ddots & \\ & & & & \ddots & & * \\ & & & \ddots & & \ddots & \vdots \\ & & & & * & \cdots & * \end{bmatrix}.$$

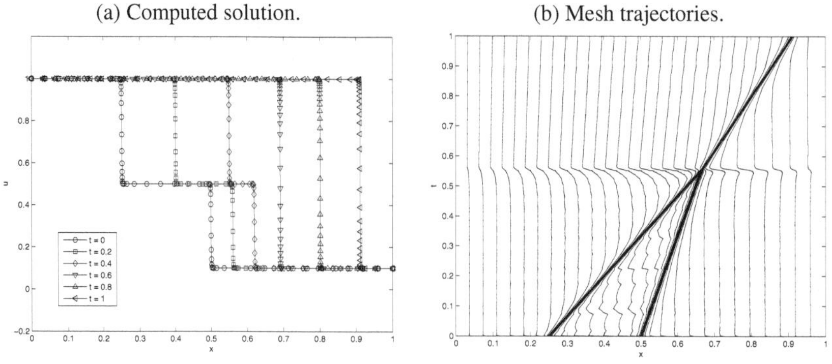

Fig. 1.7 (a) The finite difference solution for Burgers' equation with $\varepsilon = 10^{-4}$, obtained with an adaptive moving mesh of 61 points, is shown at $t = 0$, 0.2, 0.4, 0.6, 0.8, and 1.0. (b) The corresponding mesh trajectories.

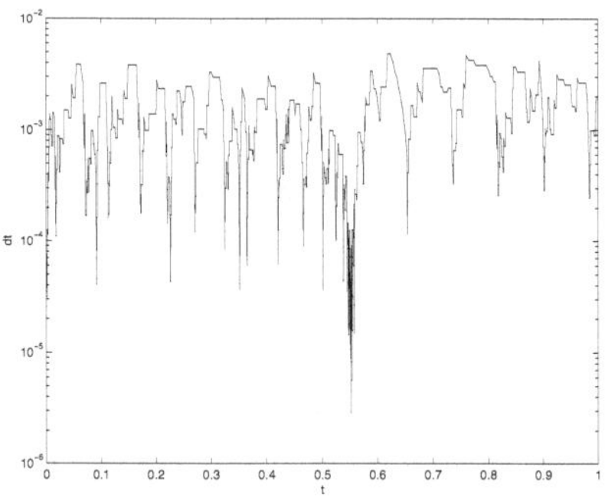

Fig. 1.8 The time step size used in the adaptive moving mesh solution of Burgers' equation with $\varepsilon = 10^{-4}$ and 61 points is plotted as function of time. The relative and absolute tolerances for the time step control are taken as $rtol = 10^{-6}$ and $atol = 10^{-4}$, respectively, for the Matlab ODE solver "ode15i" (using a backward differentiation formula of order 5).

The upper and lower bandwidths of B are equal to 2 plus the number of sweeps used in smoothing the mesh density function. Strategies for choosing the mesh density function and the intensity parameter are considered in detail in Chapter 2.

The difference in time required to solve algebraic systems for the uniform and adaptive mesh methods turns out to be less significant in two and three dimensions, where the algebraic systems for the uniform mesh method are generally much more expensive to solve.

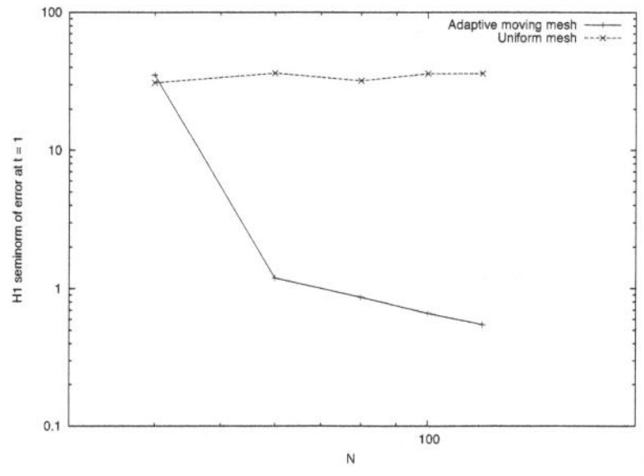

Fig. 1.9 The H^1 semi-norm of the error at $t = 1$ is plotted against the number of mesh points, N, for uniform and adaptive moving meshes.

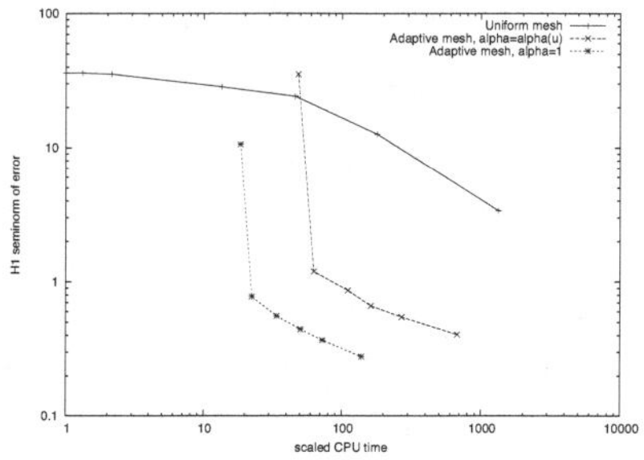

Fig. 1.10 The H^1 semi-norm of the error at $t = 1$ is plotted against the CPU time scaled by the time required for the computation of the case with a uniform mesh of 100 points.

1.5 Basic components of a moving mesh method

Thus far, we have seen that a moving mesh method has three major components: the strategy used to move the mesh, the method employed to discretize the physical PDE, and the approach used to solve the coupled system of physical and mesh equations. In particular, for the finite difference and finite element moving mesh methods described in this chapter, the mesh is moved using the moving mesh PDE (1.14), the physical PDE is discretized with the quasi-Lagrange approach using finite differences or finite elements, and the coupled system of physical and mesh PDEs is solved simultaneously with a general ODE solver. Motivated by this elementary understanding of the moving mesh method for solving Burgers' equation, we complete this chapter with a brief discussion of various options available when implementing these three components of a moving mesh method.

1.5.1 Mesh movement strategies

Mesh movement is usually performed either by solving an elliptic or parabolic system of PDEs involving the mesh coordinate transformation or by doing a direct error-based minimization. The derivation of the moving mesh system is typically motivated by the equidistribution principle. It makes use of an error density function (which is referred to as a mesh density function) which is required to be evenly distributed among all the mesh elements. In one dimension, the equidistribution condition, together with suitable boundary conditions, uniquely determines a mesh for a given mesh density function. However, as we shall see, this is not the case in multidimensions. Generally speaking, a condition regularizing the shape of mesh elements, in addition to equidistribution, is needed to determine a suitable multidimensional mesh. For instance, for isotropic mesh adaptation, mesh elements are required to be nearly equilateral, whereas for anisotropic mesh adaptation, a shape-alignment condition is often imposed on mesh elements. Studies of basic principles of mesh adaptation, including equidistribution and shape-alignment conditions, are formally addressed in Chapter 4. Note that while the formulation of a mesh movement strategy is independent of the type of the physical PDE to be solved, the discretization method and solution strategy are not.

A variational approach is perhaps the most natural one for formulating elliptic or parabolic PDE based mesh generators. (The moving mesh PDE (1.14), for instance, is derived using the variational approach.) For this approach, mesh equations are defined as the Euler-Lagrange equations of a functional specially designed for the purpose of mesh adaptation. A number of adaptation functionals have been developed in the past based on error estimates and geometric considerations. They are discussed in Chapter 6.

Variational strategies are special examples of the so-called *location-based* mesh movement strategies which control directly the location of mesh points. A different group of strategies is *velocity-based* since it targets directly the mesh velocity and obtains the location of mesh points by integrating the velocity field. A majority of methods of this type are motivated by the Lagrangian method in fluid dynamics where the mesh coordinates, defined to follow fluid particles, are obtained by integrating flow velocity. Velocity-based strategies are presented in Chapter 7.

1.5.2 Discretization of PDEs on a moving mesh

Finite differences and finite elements have been used in this chapter for spatial discretization of the physical PDE on a moving mesh. As we have seen, the physical PDE is discretized on the computational domain when finite differences are used and

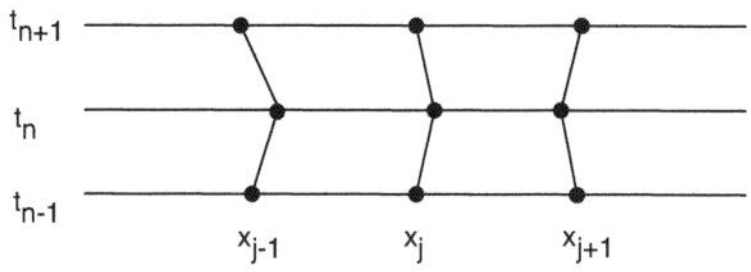

Fig. 1.11 The mesh points are considered to move continuously for the quasi-Lagrange approach of temporal discretization of physical PDEs.

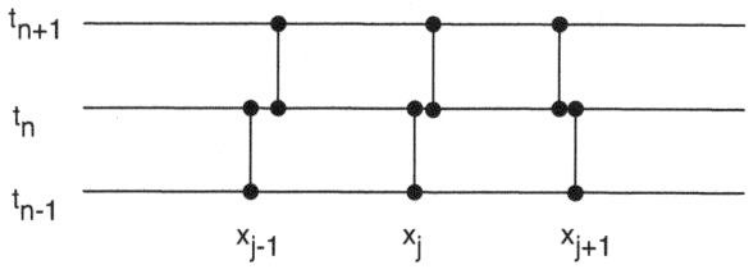

Fig. 1.12 The mesh is considered to vary only at time instants $t = t_n$, $n = 0, 1, \ldots$ for the rezoning approach of temporal discretization of physical PDEs.

on the physical domain when finite elements are used. This is because normally a finite difference discretization can only be conveniently carried out on a rectangular or a cubic mesh in the computational domain, whereas a finite element discretization can be used directly on a non-uniform mesh in the physical domain. Discretization in the computational domain requires that the physical PDE be transformed into the computational variables, often giving a complicated form. In contrast, discretization on the physical mesh can avoid such a complexity.

The effect of mesh movement in the time discretization of the physical PDE can be treated with either *the quasi-Lagrange approach* or *the rezoning approach*. With the quasi-Lagrange approach, the mesh points are considered to move continuously in time (cf. Figure 1.11), and physical time derivatives are transformed into time derivatives along mesh trajectories supplemented with a convective term reflecting mesh movement. The new time derivative and the extra convection term are typically treated in the same way as other terms in the physical PDE. Alternatively, with the rezoning approach, the mesh points are considered to move in an intermittent manner in time (cf. Figure 1.12). More precisely, the mesh is updated at each time level using certain mesh equations or generators, the physical solution is interpolated from the old mesh to the new one, and the physical PDE is then discretized on the new mesh, which is held fixed for the current time step. Interpolation of the physical solution is a crucial step for the success of this approach, and it is often necessary to use a conservative interpolation scheme which preserves some solution quantities. This issue is discussed in detail in Chapters 2 and 3.

1.5.3 Simultaneous or alternate solution

We have seen that for a moving mesh method, the discrete physical PDE and the mesh equation give a coupled system. Some basic features of the solution procedure for this system, along with the approach used to treat the mesh movement in the physical PDE, are briefly discussed here. By design the rezoning approach involves solving alternately for the physical solution and the mesh. For the quasi-Lagrange approach, the physical PDE and the mesh equation for the moving mesh method can be solved either *simultaneously* or *alternately*.

A simultaneous solution procedure is illustrated in Figure 1.13. For it, the mesh equation and physical PDE are treated as one large system which is solved simultaneously for the mesh and physical solution. This is illustrated in §1.2 and §1.3, where the mesh equation (1.18) is solved together with physical equation (1.13) or (1.32). Simultaneous solution is in principle relatively simple and has the advantage that standard, well-developed ODE solvers can be directly applied to the integration of the extended system arising from applying the method of lines to the mesh and physical equations. Moreover, the physical solution and the mesh are tightly coupled, with the mesh responding promptly to any change occurring in the physical solution. The main disadvantage of simultaneous solution is the highly nonlinear coupling between the physical solution and the mesh. Even a linear physical PDE can result in a highly nonlinear equation in the new variables (cf. (1.13)). The extended system also has a more complicated structure (cf. (1.25)) and often loses features which the physical PDE may have in the physical variables, such as symmetry and positive definiteness. These factors often make the extended system more difficult and expensive to solve.

The alternate solution procedure is illustrated in Figure 1.14. A mesh x^{n+1} at the new time level is first generated using the mesh and the physical solution (x^n, u^n) at the current time level, and the solution u^{n+1} is then obtained at the new time level. Note that this mesh x^{n+1} adapts only to the current solution u^n and thus lags in time. This will not generally cause much trouble if the time step is reasonably small or the solution does not have abrupt changes in time. If the lag of the mesh in time causes a serious problem, several iterations of solving for the mesh and the physical PDE at each new time level can be used (cf. Figure 1.14). The main advantages of the alternate solution procedure are (i) its flexibility (the mesh generation part can be coded separately as a module to incorporate into the PDE solver) and (ii) its potential efficiency at each time step (structures for each of the physical and mesh equations can be fully explored to improve efficiency). Since the mesh adaptation is not tied to the solution process for the physical PDE, the mesh generator does not have to take the form of a differential equation. A minimization-based mesh generator, for example, can suffice equally well. In addition to the above-mentioned disadvantage of the lag of the mesh in time, the alternate solution procedure runs the risk of

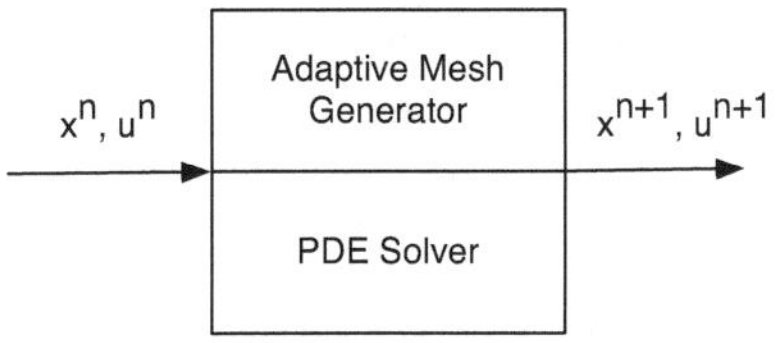

Fig. 1.13 Illustration of a simultaneous solution procedure.

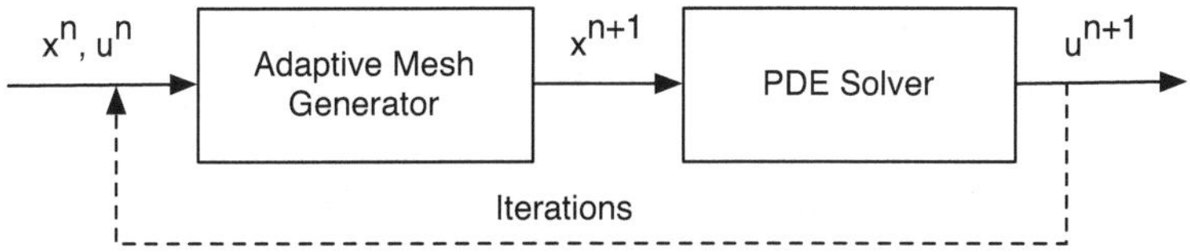

Fig. 1.14 Illustration of an alternate solution procedure.

causing instability in the integration because it does not have a mechanism built in to force the system back on track once the mesh is not generated accurately enough at one time step. Relatively speaking, this risk is smaller with the simultaneous solution method because the physical solution and the mesh are forced to satisfy the physical PDE and the mesh equation simultaneously at each time step.

The simultaneous solution procedure has been limited mainly to one-dimensional problems in space, and most of the existing moving mesh methods for multidimensional computation employ an alternate solution procedure.

1.6 Biographical notes

Roughly speaking, mesh movement algorithms can be classified into the two groups, *velocity-based* algorithms and *location-based* ones [85], cf. §1.5.1. Many of the velocity-based algorithms have been motivated by the Lagrangian method in fluid dynamics (e.g., see Batchelor [38]), and a major consideration in their development has been to avoid mesh tangling, an undesired property of the Lagrangian method. Examples include the method of Yanenko et al. [345] that is of Lagrange-type. In the work of Anderson and Rai [13], the mesh is moved according to attraction and repulsion pseudo-forces between nodes motivated by a spring model in mechanics. The moving finite element method (MFE) of Miller and Miller [258] and Miller [253] has aroused considerable interest. It computes the solution and the mesh simultaneously by minimizing the residual of the PDEs written in a finite element form. Penalty terms are added to avoid possible singularities in the mesh movement equations; see [88, 89]. A way of treating the singularities but without using penalty functions has been proposed by Wathen and Baines [339]. Petzold [274] obtains an equation for mesh velocity by minimizing the time variation of both the unknown variable and the spatial coordinate in computational coordinates and adding a diffusion-like term to the mesh equation. Liao and his coworkers [53, 226, 234, 236, 232, 241] employ a deformation map to move the mesh. In [84], Cao, Huang, and Russell develop the GCL method, which is based on the Geometric Conservation Law. Interestingly, the deformation map method can be viewed as a special example of the GCL method (see §7.1). A similar idea has been used by Baines et al. [31, 32, 33] for the development of the so-called moving mesh finite element method.

In contrast, location-based mesh movement algorithms provide a direct control of the location of mesh points. A natural and important approach for designing this type of algorithm is the variational approach for which the mesh point relocation and movement are determined by minimizing some functional formulated to measure error or difficulty in numerical simulation. Many location-based algorithms have been developed as variational ones, whereas some others have been based on elliptic PDEs or other considerations. For example, Winslow [341] and Thompson et al. [324] use a system of elliptic PDEs for generating boundary-fitted meshes. Winslow [342] proposes to generate adaptive meshes through a variable diffusion model. The idea is generalized by Brackbill and Saltzman [58], who combine functionals representing mesh adaptivity, smoothness, and orthogonality. This is further modified by Brackbill [57] to include directional control in mesh adaptation and to require the terms contained in the functional to be dimensionally homogeneous. The method of Dorfi and Drury [124] is linked to a functional associated with the well-known equidistribution principle [115, 186] while that of Dvinsky [129] is based on the energy of a harmonic mapping for mesh adaptation. Examples of other mesh adaptation functionals can be found in Jacquotte et al. [202, 203, 204] (based on mechanical models), Knupp [211] (using vector fields), Knupp et al. [212, 214] (using a weighted or reference Jacobian matrix), Huang and Russell [189] and Cao et al. [82] (using a generalized variable diffusion functional with a matrix-valued diffusion coefficient), and Huang [176] (based on the so-called equidistribution and isotropy (or alignment) conditions). These functionals are discussed in Chapter 6. The moving mesh PDE (MMPDE) method developed in [81, 185, 186, 189, 190, 282] moves the mesh through the gradient flow equation of an adaptation functional. Tang et al. [228, 229, 316] use the generalized variable diffusion functional (cf. [82, 189]) as their adaptation functional, but discretize the physical PDE in the rezoning approach. Budd and Williams [71] use a parabolic Monge-Ampère equation to move adaptive meshes. The methods of Ren and Wang [280] and Ceniceros and Hou [95] also deserve special attention.

There exist a number of review articles and books addressing (at least partially) moving mesh methods. Review articles include Russell and Christiansen [285], Thompson et al. [326], Thompson [323], Eiseman [132, 133], Hawken et al. [169],

Thompson and Weatherill [327], Huang and Russell [191], Cao et al. [85], Sloan [303], and more recently, Huang [181] and Budd et al. [68]. Hawken et al. [169] give a particularly extensive overview and list of references on moving mesh methods before 1990. Relevant books include Thompson et al. [325], Ascher et al. [16], Knupp and Steinberg [213], Baines [29], Zegeling [347], Carey [86], and Liseikin [238]. Relevant conference proceedings and edited books include Babuška et al. [23], Castillo [92], Shi et al. [299], and Tang and Xu [319].

This book is mainly concerned with the r-adaptive, or moving mesh, method; for other types of adaptive mesh methods, especially the h-adaptive mesh method and the Adaptive Mesh Refinement (AMR) method, the interested reader is referred to books such as Baden et al. [27], Carey [86], Ern and Guermond [135], Frey and George [150], George [154], Lang [222], Linß [237], Plewa et al. [275], and Sarris [293].

1.7 Exercises

1. Assume that the function $u = u(x)$ is sufficiently smooth around point x. Find the order of the truncation error for the following finite difference approximations:

$$\frac{du}{dx}(x) \approx \frac{u(x+h) - u(x)}{h},$$
$$\frac{du}{dx}(x) \approx \frac{u(x) - u(x-h)}{h},$$
$$\frac{du}{dx}(x) \approx \frac{u(x+h) - u(x-h)}{2h},$$
$$\frac{du}{dx}(x) \approx \frac{u(x+\frac{h}{2}) - u(x-\frac{h}{2})}{h},$$
$$\frac{d^2u}{dx^2}(x) \approx \frac{u(x+h) - 2u(x) + u(x-h)}{h^2},$$

where h is a small positive number.

2. Derive the three-point central finite difference approximation to the second derivative $\frac{d^2u}{dx^2}(x)$ on a non-uniform mesh. What are the leading terms in the truncation error?
3. Assume that functions $u(x)$ and $p(x)$ are sufficiently smooth around point x. Derive the approximation

$$\frac{d}{dx}\left(p(x)\frac{du}{dx}\right) \approx \frac{1}{h}\left(\frac{(p(x+h)+p(x))}{2}\frac{(u(x+h)-u(x))}{h} - \frac{(p(x-h)+p(x))}{2}\frac{(u(x)-u(x-h))}{h}\right),$$

where h is a small positive number. Find the leading terms in the truncation error of the approximation.

4. Consider a central finite difference approximation on a uniform mesh to the boundary value problem

$$-u'' + u' = 1, \quad \forall x \in (0,1)$$
$$u(0) = u(1) = 0.$$

(a) Derive the scheme; (b) find the local truncation error; and (c) write down the matrix form of the resulting algebraic system explicitly.

5. Prove (1.11) using the chain rule.
6. For sufficiently smooth functions $u = u(x)$ and $x = x(\xi)$, let $\hat{u} = u(x(\xi))$. Show that

$$\frac{d^2u}{dx^2} = \frac{1}{x_\xi}\frac{d}{d\xi}\left(\frac{1}{x_\xi}\frac{d\hat{u}}{d\xi}\right),$$

where $x_\xi = dx/d\xi$.

7. Derive (1.12).
8. Derive the semi-discrete scheme (1.13).
9. For $\rho = 1$, find the general solution of MMPDE (1.14) and boundary condition (1.15) for any initial coordinate transformation $x(\xi,0) = x_0(\xi)$. Discuss the monotonicity of the solution in space and its asymptotical behavior as $t \to \infty$. (Hint: Set $x = \xi + \phi$ and solve the equation for ϕ using the Fourier series method or the method of separation of variables.)
10. Evaluate the integrals

$$\int_{x_{k-1}}^{x_k} \phi_{k-1}\phi_k dx, \quad \int_{x_{k-1}}^{x_k} \phi_k\phi_k dx,$$

$$\int_{x_{k-1}}^{x_k} \phi_{k-1}'\phi_k' dx, \quad \int_{x_{k-1}}^{x_k} \phi_k'\phi_k' dx,$$

$$\int_{x_{k-1}}^{x_k} \phi_{k-1}'\phi_k dx, \quad \int_{x_{k-1}}^{x_k} \phi_k'\phi_k dx,$$

where ϕ_k and ϕ_{k-1} are the basis functions defined in (1.28).

11. Consider a linear finite element approximation on a uniform mesh to the boundary value problem

$$\begin{cases} -u'' + u' = 1, & \forall x \in (0,1) \\ u(0) = u(1) = 0. \end{cases}$$

(a) Using the results in Problem 10, derive the scheme and (b) write down the matrix form of the resulting algebraic system explicitly.

12. Implement on computer the finite difference and finite element schemes in Problems 4 and 11.

13. Use direct calculation to derive (1.37).

Chapter 2
Adaptive Mesh Movement in 1D

In this chapter we discuss more formally the principles of adaptive mesh movement in 1D. The underlying mesh selection problem itself is quite simple to state: If one wishes to approximate a given function $u(x)$ using its values at a finite number of mesh points, how should these points be chosen? The answer can usually be given as follows: one chooses a so-called *mesh density function* $\rho(x)$, which in some way indicates the error in the numerical approximation, and the mesh points are then placed in such a way that distances between them are smaller in regions where $\rho(x)$ is larger, and the distances are larger in regions where $\rho(x)$ is smaller. For the one-dimensional case, adaptivity is predicated on what is called *equidistribution*, which is considered in some detail here. The argument for choosing the mesh density function $\rho(x)$ will normally be motivated by the desire to minimize an error in interpolating a function or by solving a differential equation, although in special cases other arguments such as one based on scaling invariance are used.

Our overall goal in this chapter is to get the reader thinking about how to compute an adaptive mesh. Fundamental to our approach is to equate the problem of finding an adaptive mesh to finding a suitable coordinate transformation. Some implementations of equidistribution are presented, largely so as to motivate ideas which are central to the topics of later chapters. As well, a purpose is to give a basic error analysis which demonstrates the advantages of adaptive meshes over uniform ones for approximating non-smooth functions or for solving differential equations having non-smooth solutions. In doing so, we introduce specific tools which are also useful for the study of higher dimensional mesh adaptivity, which is considered in Chapter 4.

W. Huang and R.D. Russell, *Adaptive Moving Mesh Methods*, Applied Mathematical Sciences 174, DOI 10.1007/978-1-4419-7196-2_2,

2.1 The equidistribution principle

2.1.1 Equidistribution

The concept of equidistribution has played a fundamental role in mesh adaptation. Given an integer $N > 1$ and a continuous function $\rho = \rho(x) > 0$ on a bounded interval $[a,b]$, *equidistribution* entails finding a mesh $\mathscr{T}_h : x_1 = a < x_2 < \cdots < x_N = b$ which evenly distributes ρ among the subintervals determined by the mesh points, in the sense that

$$\int_{x_1}^{x_2} \rho(x)dx = \cdots = \int_{x_{N-1}}^{x_N} \rho(x)dx. \tag{2.1}$$

That is, the area under $\rho(x)$ is the same for every subinterval. A mesh $\mathscr{T}_h$ satisfying this relation is called an *equidistributing mesh* for $\rho = \rho(x)$. The function ρ is referred to as the *mesh density specification function*, or simply the *mesh density function*, and its square, $\rho(x)^2$, as the *monitor function*. We emphasize at the outset that this terminology differs from that found in much of the research literature to date, where ρ is typically called the monitor function. However, it is consistent with the notation used in the multi-dimensional context, where the monitor function refers to a matrix-valued function used for specifying the size, shape, and orientation of mesh elements, and the mesh density function refers to the square-root of its determinant, a scalar function specifying the size of mesh elements (see Chapter 4). Moreover, as we shall see in this chapter and Chapter 4, the mesh density function is proportional to the density of the mesh when the equidistribution principle is satisfied.

One often requires a mesh density function to only satisfy the weak condition that it be non-negative, implying that it can vanish locally. A consequence is that the equidistributing mesh may be non-unique, which complicates many theoretical proofs and the actual construction of equidistributing meshes. For these reasons, we assume henceforth that any mesh density function is by definition strictly positive, i.e., that there exists a constant γ such that

$$\rho(x) \geq \gamma > 0, \quad \forall x \in [a,b]. \tag{2.2}$$

As seen in §2.4, mesh density functions can be defined in this way without any practical loss of generality.

Proposition 2.1.1 *For a given integer $N > 0$, there exists a unique equidistributing mesh of N points satisfying (2.1) for any strictly positive mesh density function.*

Proof. Rewrite (2.1) as

$$\int_a^{x_j} \rho(x)dx = \frac{(j-1)}{(N-1)}\sigma, \qquad j = 1,...,N \tag{2.3}$$

where

$$\sigma = \int_a^b \rho(x)dx. \tag{2.4}$$

By (2.2), $\int_a^{\hat{x}} \rho(x)dx$ is a strictly monotone increasing function of $\hat{x}$, so each x_j is uniquely determined. □

Equation (2.1) can be rewritten as

$$(x_j - x_{j-1}) \langle \rho \rangle_{I_j} = \frac{\sigma}{(N-1)}, \qquad j = 2,...,N \tag{2.5}$$

where σ is defined in (2.4), $I_j = (x_{j-1}, x_j)$, and $\langle \rho \rangle_{I_j}$ is the integral average of $\rho(x)$ on the interval $[x_{j-1}, x_j]$,

$$\langle \rho \rangle_{I_j} = \frac{1}{|I_j|} \int_{I_j} \rho(x)dx = \frac{1}{x_j - x_{j-1}} \int_{x_{j-1}}^{x_j} \rho(x)dx. \tag{2.6}$$

While this discrete form is critical for actual computation, for mathematical understanding it can be more useful to consider a continuous form. Specifically, suppose that the mesh $\mathscr{T}_h$ is to be generated using a coordinate transformation $x = x(\xi) : [0,1] \to [a,b]$ in such a way that

$$x_j = x(\xi_j), \qquad j = 1,...,N$$

where

$$\xi_j = \frac{(j-1)}{(N-1)}, \qquad j = 1,...,N$$

is a uniform mesh on $[0,1]$. Then (2.3) becomes

$$\int_a^{x(\xi_j)} \rho(x)dx = \sigma\xi_j, \qquad j = 1,...,N.$$

More generally, a continuous mapping $x = x(\xi)$ is called an *equidistributing coordinate transformation* for $\rho(x)$ if it satisfies the condition

$$\int_a^{x(\xi)} \rho(x)dx = \sigma\xi, \qquad \forall \xi \in (0,1) \tag{2.7}$$

with σ defined in (2.4). Differentiating with respect to ξ, one sees that $x(\xi)$ also satisfies

$$\rho(x)\frac{dx}{d\xi} = \sigma. \tag{2.8}$$

We later see that it is useful to formulate the equidistribution relation in terms of the inverse coordinate transformation, $\xi = \xi(x) : [a,b] \to [0,1]$. From (2.8) we have

$$\frac{1}{\rho(x)}\frac{d\xi}{dx} = \frac{1}{\sigma}. \tag{2.9}$$

From (2.5) and (2.8) one sees that equidistribution requires the interval length $(x_j - x_{j-1})$ (respectively, $dx/d\xi$ in the continuous form) be small in places where ρ_j (respectively, $\rho(x)$) is large.

Example 2.1.1 Find the equidistributing coordinate transformation for the mesh density function

$$\rho(x) = 1 + R(1 - \tanh^2(Rx)), \qquad x \in [-1,1]$$

where $R = 100$.

To solve this problem, we rewrite (2.9) as

$$\frac{d\xi}{dx} = \frac{\rho(x)}{\sigma},$$

where for this mesh density function

$$\sigma = \int_{-1}^{1} \rho(x)dx = 2 + \tanh(R) - \tanh(-R).$$

Solving the differential equation together with the boundary conditions $\xi(-1) = 0$ and $\xi(1) = 1$ yields

$$\xi = \frac{1 + x + \tanh(Rx) - \tanh(-R)}{2 + \tanh(R) - \tanh(-R)}. \tag{2.10}$$

It is not obvious that (2.10) can even be solved analytically for $x(\xi)$. The coordinate transformation and its inverse are plotted in Figure 2.1. Note that $\xi = \xi(x)$ has a steep gradient near $x = 0$ where $\rho(x)$ attains its maximum, whereas $x = x(\xi)$ changes much more smoothly. One might conclude that it is better to use mesh equations formulated in terms of $x(\xi)$ than $\xi(x)$, but we see in §2.3 that for certain situations the formulations involving $\xi(x)$ can in fact have some computational advantages. □

2.1.2 *Optimality of equidistribution*

The popularity of equidistribution is due largely to its optimality properties. In the context of approximating a function $u = u(x)$, which is either a given function be-

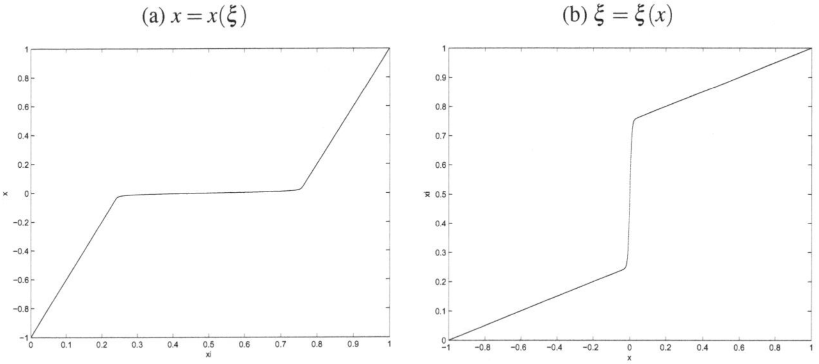

Fig. 2.1 Equidistributing coordinate transformation $x = x(\xi)$ and inverse coordinate transformation $\xi = \xi(x)$ for Example 2.1.1.

ing interpolated or the solution of a PDE being solved, the mesh- and solution-dependent factor in an error estimate typically has the general form

$$E(\mathscr{T}_h) \equiv (N-1)^s \sum_{j=2}^{N} (h_j f_j)^{s+1}. \tag{2.11}$$

See the truncation error in §2.4, for example. The factor $(N-1)^s$ is included so that $E(\mathscr{T}_h)$ is asymptotically independent of the mesh, as we shall see in (2.12) below. In (2.11), $h_j = |I_j| = x_j - x_{j-1}$ is the mesh spacing, $s > 0$ is a real number, and f_j denotes some average on $I_j = (x_{j-1}, x_j)$ of a positive function f which generally depends upon a derivative or derivatives of u. The mesh function value f_j can appear in a variety of forms, but in the limit it approximates the integral average of $f(x)$ on I_j, i.e.,

$$f_j \approx \langle f \rangle_{I_j} \equiv \frac{1}{x_j - x_{j-1}} \int_{x_{j-1}}^{x_j} f(x)dx.$$

We shall assume that this is always the case. Noticing that

$$h_j = \frac{(x_j - x_{j-1})}{(\xi_j - \xi_{j-1})}(\xi_j - \xi_{j-1}) \approx \left.\frac{dx}{d\xi}\right|_{\xi=\xi_j} (N-1)^{-1},$$

in the limit we can derive the following continuous form for the bound (2.11):

$$\begin{aligned} E(\mathscr{T}_h) &\equiv (N-1)^s \sum_{j=2}^{N} (h_j f_j)^{s+1} \\ &\longrightarrow E[\xi] \equiv \int_a^b \left(\left| \frac{dx}{d\xi} \right| f \right)^s f\, dx, \quad \text{as} \quad \max_j h_j \to 0. \end{aligned} \tag{2.12}$$

An *optimal coordinate transformation* is defined as a coordinate transformation $\xi = \xi(x)$ minimizing $E[\xi]$.

Theorem 2.1.1 (Optimality of the equidistributing coordinate transformation) *Suppose that an error bound is of form (2.11) for a real number $s > 0$ and a continuous, strictly positive function $f = f(x)$ on $[a,b]$. Letting $\sigma = \int_a^b f(x)dx$, then the inequality*

$$E[\xi] \equiv \int_a^b \left(\left| \frac{dx}{d\xi} \right| f \right)^s f\, dx \geq \sigma^{s+1} \tag{2.13}$$

holds for all invertible coordinate transformations $x(\xi)$ from $[0,1]$ to $[a,b]$.

In addition, the lower bound is attained for a coordinate transformation satisfying the equidistribution relation (2.8) for the mesh density function $\rho = f(x)$. Thus, any equidistributing coordinate transformation for $\rho = f(x)$ is an optimal coordinate transformation with respect to the asymptotic error bound $E[\xi]$ defined in (2.12).

Proof. We first prove that the functional $E[\xi]$ has a constant lower bound for all invertible coordinate transformations. By Theorem A.0.3 (with $r := -1$, $s := s$, $w := f/\sigma$, and $f := \left(f \left|\frac{dx}{d\xi}\right|\right)^s$) in Appendix A, we get

$$\left[\int_a^b \left(\left| \frac{dx}{d\xi} \right| f \right)^s \frac{f}{\sigma} dx \right]^{\frac{1}{s}} \geq \left[\int_a^b \left(\left| \frac{dx}{d\xi} \right| f \right)^{-1} \frac{f}{\sigma} dx \right]^{-1},$$

which implies (2.13).

Note that the lower bound σ^{s+1} is a constant independent of the coordinate transformation. It is easy to see that this lower bound is attained for any coordinate transformation satisfying the equidistribution relation (2.8) for $\rho = f(x)$. □

While (2.12) and Theorem 2.1.1 are concerned with the continuous form of the error bound, the actual discrete error (2.11) is described in the following theorem.

Theorem 2.1.2 (Optimality of the equidistributing mesh) *Suppose that an error bound is of the form (2.11) for a real number $s > 0$ and a continuous, strictly positive function $f = f(x)$ on $[a,b]$. Then*

$$E(\tilde{\mathscr{T}}_h) \equiv (N-1)^s \sum_{j=2}^{N} \left(\tilde{h}_j \tilde{f}_j\right)^{s+1} \geq \tilde{\sigma}_h^{s+1} \longrightarrow \sigma^{s+1}, \qquad \forall \tilde{\mathscr{T}}_h \in \mathscr{T}_N. \tag{2.14}$$

Here, $\mathscr{T}_N \equiv \{\tilde{\mathscr{T}}_h : y_1 = a < y_2 < \ldots < y_N = b\}$ is the set of all partitions of N points, $\tilde{I}_j = (y_{j-1}, y_j)$, $\tilde{h}_j = y_j - y_{j-1}$, $\tilde{\sigma}_h = \sum_j \tilde{h}_j \tilde{f}_j$, where $\tilde{f}_j$ denotes a certain average of f on $\tilde{I}_j$, and the limit is taken as $\max_j \tilde{h}_j \to 0$.

In addition,

$$E(\mathscr{T}_h) = \sigma_h^{s+1} \longrightarrow \sigma^{s+1} \qquad as \qquad \max_j (x_j - x_{j-1}) \to 0 \tag{2.15}$$

for any equidistributing mesh $\mathscr{T}_h$ satisfying

$$(x_j - x_{j-1})\rho_j = \frac{\sigma_h}{N-1}, \quad j = 2, ..., N \tag{2.16}$$

where $\sigma_h = \sum_j h_j \rho_j$ and $\rho_j = f_j$.

Proof. From Theorem B.0.11 in Appendix B,

$$\left[\frac{1}{N-1}\sum_j \left(\tilde{h}_j \tilde{f}_j\right)^{s+1}\right]^{\frac{1}{s+1}} \geq \frac{1}{N-1}\sum_j \tilde{h}_j \tilde{f}_j,$$

and inequality (2.14) follows. The equality in (2.15) is obtained by simply inserting (2.16) into the bound (2.11), and the asymptotic result follows assuming that f_j approximates the integral average of f on I_j in the limit. □

Note that the quantities $\tilde{\sigma}_h$ and σ_h in Theorem 2.1.2 are mesh-dependent. As a consequence, an equidistributing mesh does not necessarily give the lowest error bound among all partitions of N points and is therefore not necessarily the optimal mesh. Nevertheless, an equidistributing mesh is *asymptotically optimal* in the sense that its error bound $E(\mathscr{T}_h)$ converges to the constant lower bound in (2.14) as long as

$$\max_j (x_j - x_{j-1}) \to 0. \tag{2.17}$$

Fortuitously, condition (2.17), a necessary condition for ensuring this asymptotic behavior of the error bound, is shown in §2.4 to hold for the equidistributing mesh as $N \to \infty$. In particular, we show that

$$\max_j (x_j - x_{j-1}) \leq \frac{2(b-a)}{N-1},$$

provided the mesh density function is suitably chosen.

Linearly varying error measures. The error bound in Theorem 2.1.2 is proportional to a power of σ_h for the equidistributing mesh, so σ_h may be viewed as a measure of the total "error" over the physical domain and the mesh density function ρ_j as an "error" density.[1] Notice that ρ_j $(= f_j)$ depends upon the mesh, but not in a crucial way, and under the condition (2.17) it will converge to $\rho(x_j)$ as $N \to \infty$. As

---

[1] The mesh density functions defined in §2.4 and §2.9 are mostly scaled by a regularization parameter $\alpha_h > 0$. In those cases, it is more appropriate to view $\alpha_h \rho_j$ as an "error" density and $\alpha_h \sigma_h$ as the total "error" over the physical domain.

a consequence, the corresponding "error" measure,

$$h_j \rho_j, \tag{2.18}$$

varies linearly with h_j (the size of I_j). Finding a linearly varying error measure to use to define the mesh density function for an equidistributing mesh is done in early mesh adaptivity research; e.g., see Ascher et al. [16] (Chapter 9), Pereyra and Sewell [273], and Lentini and Pereyra [227]. As we shall see in §2.5.4, it provides a natural tool for error estimation and control.

2.1.3 Equidistributing meshes as uniform meshes in a metric space

Thus far, we have considered adaptivity primarily from the point of view of the mesh generation problem, and we have derived the equidistribution principle from the desire to properly control the size of mesh elements. It is also useful, especially in multi-dimensions, to view an equidistributing mesh as a uniform mesh in a metric space (cf. Chapter 4). The main advantage of this approach is that a uniform mesh, even in a metric space, can be described both geometrically and analytically in a relatively simple manner.

In the current 1D situation, we can define the monitor function as $M = \rho(x)^2$, as we discuss in detail in Chapter 4. The distance between any two points c and d in the metric specified by M is given by

$$\int_c^d \sqrt{\det(M)}dx = \int_c^d \rho(x)dx,$$

where $\det(\cdot)$ denotes the determinant of a matrix. It is easy to see that a uniform mesh in the metric specified by M, or *an M-uniform mesh* for short, satisfies the equidistribution relation (2.1). Thus, generating an equidistributing mesh for a given ρ is equivalent to generating an M-uniform mesh.

2.1.4 Another view of equidistribution

Another closely related but somewhat different approach is to consider the problem of function approximation. Specifically, given a function $u = u(x)$ defined on $[a,b]$, we seek a coordinate transformation $x = x(\xi) : [0,1] \to [a,b]$ for which $u(x(\xi))$ is sufficiently well-behaved that it can be approximated efficiently on a uniform mesh in this new coordinate.

Take piecewise constant interpolation as an example. For a given uniform mesh,

$$\xi_j = \frac{(j-1)}{(N-1)}, \quad j = 1,\ldots,N$$

a piecewise constant approximation can be defined as

$$u(x(\xi)) \approx u(x(\xi_{j-\frac{1}{2}})), \quad \forall \xi \in [\xi_{j-1}, \xi_j],\ j = 2,\ldots,N$$

where $\xi_{j-\frac{1}{2}}$ is the midpoint of the interval $[\xi_{j-1}, \xi_j]$. The approximation error can be expressed as

$$u(x(\xi)) - u(x(\xi_{j-\frac{1}{2}})) = (\xi - \xi_{j-\frac{1}{2}}) \int_0^1 \frac{d}{d\xi} u(x(\xi_{j-\frac{1}{2}} + s(\xi - \xi_{j-\frac{1}{2}})))ds.$$

If a coordinate transformation can be chosen such that

$$\left| \frac{d}{d\xi} u(x(\xi)) \right| = c \quad \forall \xi \in [0,1] \tag{2.19}$$

for a fixed positive constant c, then

$$\max_{\xi \in [\xi_{j-1}, \xi_j]} \left| u(x(\xi)) - u(x(\xi_{j-\frac{1}{2}})) \right| = \frac{c}{2(N-1)}, \tag{2.20}$$

so the size of the error is roughly the same on all mesh intervals. As a consequence, a uniform mesh will be efficient for resolving $u(x(\xi))$ with piecewise constant interpolation.

The condition (2.19) can be rewritten as

$$\left| \frac{du}{dx} \right| \frac{dx}{d\xi} = c.$$

If $|du/dx|$ is strictly positive on $[a,b]$, then defining $\rho(x) = |du/dx|$ we have

$$\rho(x) \frac{dx}{d\xi} = c. \tag{2.21}$$

The constant c must satisfy a compatibility condition, viz., integrating (2.19) over $[0,1]$ we see that

$$c = \int_0^1 \rho(x) \frac{dx}{d\xi} d\xi = \int_a^b \rho(x) dx \equiv \sigma.$$

Inserting this into (2.21) gives precisely the equidistribution relation (2.8).

If on the other hand $|du/dx|$ is not strictly positive on $[a,b]$, then we could instead require

$$\sqrt{\left(\frac{du}{d\xi}\right)^2 + \alpha \left(\frac{dx}{d\xi}\right)^2} = c,$$

where $\alpha > 0$ is a regularization parameter. But this again leads to the equidistribution relation (2.8), with the mesh density function now being given by

$$\rho = \sqrt{1 + \frac{1}{\alpha}\left(\frac{du}{dx}\right)^2}.$$

Thus, in both cases we get the same adaptivity relations as derived previously from a different (mesh generation) viewpoint. The distinction between these approaches is rather subtle in one dimension, but in higher dimensions it is more interesting, and provides insight into different adaptivity strategies, as we see in Chapter 4.

2.2 Computation of equidistributing meshes

Although its existence and uniqueness are guaranteed theoretically, in practice the equidistributing mesh can rarely be found exactly because the integrals in (2.1) must normally be approximated. Thus, one has to rely on numerical methods for finding approximations to the equidistributing mesh even when $u(x)$ is given explicitly.

2.2.1 De Boor's algorithm

A simple yet useful approximation method for finding an equidistributing mesh is de Boor's algorithm [115], which is described below.

Assume that the mesh density function is known on an arbitrary background mesh $\mathscr{T}_b : y_1 = a < y_2 < \cdots < y_K = b$, which can be thought of as a prescribed mesh or as the current approximate mesh in an iterative process. The idea behind de Boor's algorithm is to approximate $\rho = \rho(x)$ on this background mesh by a piecewise constant function of the form

$$p(x) = \begin{cases} \frac{1}{2}(\rho(y_1) + \rho(y_2)), & \text{for } x \in [y_1, y_2] \\ \frac{1}{2}(\rho(y_2) + \rho(y_3)), & \text{for } x \in (y_2, y_3] \\ \cdots & \\ \frac{1}{2}(\rho(y_{K-1}) + \rho(y_K)), & \text{for } x \in (y_{K-1}, y_K] \end{cases} \tag{2.22}$$

and to then find the equidistributing mesh for this piecewise constant function. Denoting

$$P(x) = \int_a^x p(x)dx,$$

then

$$P(y_j) = \sum_{i=2}^{j} (y_i - y_{i-1}) \frac{\rho(y_i) + \rho(y_{i-1})}{2}, \qquad j = 2, \ldots, K$$

and the equidistribution relation (2.3) now reads as

$$P(x_j) = \xi_j P(b), \qquad j = 2, \ldots, N-1$$

where $P(b) = P(y_K)$. To find x_j, $2 \le j \le N-1$, one first determines the index k such that

$$P(y_{k-1}) < \xi_j P(b) \le P(y_k).$$

Then, since $P(x)$ is piecewise linear, x_j can be directly calculated from

$$(x_j - y_{k-1}) \frac{\rho(y_{k-1}) + \rho(y_k)}{2} = \xi_j P(b) - P(y_{k-1})$$

or

$$x_j = y_{k-1} + \frac{2(\xi_j P(b) - P(y_{k-1}))}{\rho(y_{k-1}) + \rho(y_k)}.$$

Note that one call to de Boor's algorithm results in an equidistributing mesh only for the piecewise constant function $p(x)$ defined in (2.22). Iteration is normally required to obtain a good approximation to the equidistributing mesh associated with the underlying mesh density function $\rho = \rho(x)$. A simple iteration procedure is to let the current approximation now be the background mesh and to use the nodal values of the mesh density function on it to generate a new mesh. A sequence of meshes of N points, $\{\hat{\mathscr{T}}_h^{(n)}\}_{n=0}^{\infty}$, can be generated this way. When it converges, $\hat{\mathscr{T}}_h \equiv \lim_{n\to\infty} \hat{\mathscr{T}}_h^{(n)}$ satisfies

$$(\hat{x}_j - \hat{x}_{j-1}) \frac{\rho(\hat{x}_{j-1}) + \rho(\hat{x}_j)}{2} = (\xi_j - \xi_{j-1}) \hat{\sigma}_h, \qquad j = 2, \ldots, N \tag{2.23}$$

where

$$\hat{\sigma}_h = \sum_{j=2}^{N} (\hat{x}_j - \hat{x}_{j-1}) \frac{\rho(\hat{x}_{j-1}) + \rho(\hat{x}_j)}{2}. \tag{2.24}$$

Pryce [277] proves that the mesh sequence produced by the corresponding iterative procedure with a piecewise *linear* approximation to the mesh density function converges to a limit mesh which satisfies the relation (2.23) provided that the mesh density function is sufficiently smooth and N is sufficiently large. He also shows that the convergence rate depends upon the number of mesh points N, where the larger N, the faster the iteration converges. Subsequent theoretical results of Xu et al. [343] show convergence for the above case of piecewise constant interpolation.

For a given approximation to an equidistributing coordinate transformation or mesh, it is useful to know how closely it satisfies the equidistribution principle. In the continuous case, the *equidistribution quality measure* is defined as

$$Q_{eq}(x) = \frac{\rho x_\xi}{\sigma}, \tag{2.25}$$

where $x_\xi = \frac{dx}{d\xi}$. This definition is motivated from (2.8). It is a simple but instructive exercise to show that $\|Q_{eq}\|_\infty \equiv \max_x Q_{eq}(x) \geq 1$ and that the mesh satisfies (2.8) if and only if $\max_x Q_{eq}(x) = 1$. A *discrete* version of the measure can be defined as

$$Q_{eq,j} = \frac{(\rho(x_j)+\rho(x_{j-1}))}{2\sigma_h} \cdot \frac{(x_j - x_{j-1})}{(\xi_j - \xi_{j-1})}, \qquad j = 2, ..., N \tag{2.26}$$

where

$$\sigma_h = \sum_{j=2}^{N} \frac{(x_j - x_{j-1})}{(\xi_j - \xi_{j-1})} \cdot \frac{(\rho(x_j)+\rho(x_{j-1}))}{2}.$$

Like the continuous equidistribution quality measure, it satisfies $\max_j Q_{eq,j} \geq 1$.

Example 2.2.1 Consider the mesh density function

$$\begin{aligned} \rho(x) &= 1 + 20(1 - \tanh^2(20(x - 0.25))) + 30(1 - \tanh^2(30(x - 0.5))) \\ &\quad + 10(1 - \tanh^2(10(x - 0.75))) \quad \text{for} \quad x \in [0, 1]. \end{aligned} \tag{2.27}$$

As for Example 2.1.1, an analytical formula can be found for the inverse coordinate transformation $x(\xi)$ for this mesh density function.

Starting with a uniform mesh, a sequence of meshes is generated with de Boor's algorithm to obtain a good approximation to the equidistributing mesh. The number of iterations, $Iter$, required for the difference between two consecutive iterates, $\|\mathscr{T}_h^{(n+1)} - \mathscr{T}_h^{(n)}\|_\infty \equiv \max_j \|x_j^{(n+1)} - x_j^{(n)}\|$, to reach 10^{-8} is listed for various values of N in Table 2.1. For $N = 11$, the algorithm fails to converge (denoted by N.Cnvg). The results confirm the theory of Pryce [277] and Xu et al. [343] that the iteration converges faster for larger N. The table also lists the difference between the exact equidistributing mesh $\mathscr{T}_h$ for $\rho = \rho(x)$ and the convergent mesh $\hat{\mathscr{T}}_h$, measured as $\max_j |\hat{x}_j - x_j|$, and the ratio $\max_j \{\frac{|\hat{x}_j - x_j|}{x_{j+1} - x_j}, \frac{|\hat{x}_j - x_j|}{x_j - x_{j-1}}\}$. The latter shows how far a computed mesh point is from the corresponding one in the (exact) equidistributing mesh relative to the size of the neighboring subintervals. The mesh differences decrease at roughly the rate $O(\frac{1}{N^2})$ (as predicted theoretically) while the mesh ratios converge at $O(\frac{1}{N})$.

Numerical results obtained with $N = 81$ and $tol = 10^{-8}$ are plotted in Figure 2.2 (a) – (d). It can be seen in Figure 2.2(b) that the *measure of the mesh density* $d(x)$, with its nodal values being defined as $d(x_j) = \frac{1}{x_{j+1} - x_j}$ ($\approx \frac{N}{x_\xi}$), is roughly proportional to the mesh density (specification) function $\rho(x)$ in Figure 2.2(a). This indicates that the mesh is approximately equidistributing (see (2.5)). Note that the mesh points are concentrated near the points $x = 0.25$, 0.5, and 0.75, where the mesh

Table 2.1 Results for de Boor's algorithm for Example 2.2.1. *Iter* is the number of iterations required for the difference between two consecutive iterates to reach 10^{-8}.

N	11	21	41	81	161	321	641
Iter	N.Cnvg	66	13	10	9	7	6
$\max_j \lvert\hat{x}_j - x_j\rvert$		3.58e-2	1.68e-2	6.77e-3	2.12e-3	5.69e-4	1.48e-4
$\max_j\{\frac{\lvert\hat{x}_j-x_j\rvert}{x_{j+1}-x_j}, \frac{\lvert\hat{x}_j-x_j\rvert}{x_j-x_{j-1}}\}$		9.00e-1	6.33e-1	3.72e-1	1.88e-1	9.22e-2	4.53e-2

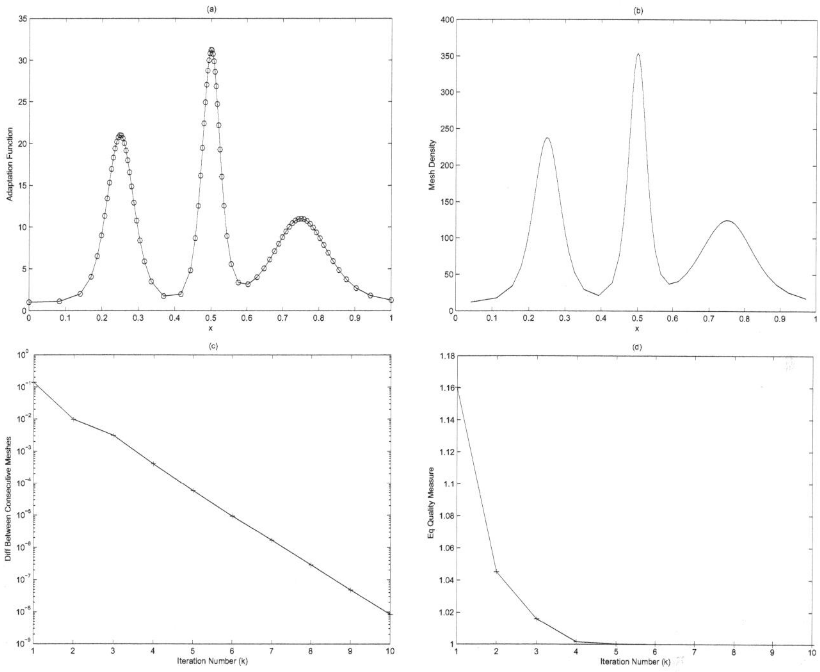

Fig. 2.2 Numerical results for Example 2.2.1 using de Boor's algorithm and $N = 81$. (a) Points corresponding to the convergent mesh shown on graph of $\rho = \rho(x)$. (b) Density of the mesh $d(x)$ plotted against x. (c) Difference between two consecutive iterates, $\|\mathscr{T}_h^{(n+1)} - \mathscr{T}_h^{(n)}\|_\infty \equiv \max_j \|x_j^{(n+1)} - x_j^{(n)}\|$, plotted against iteration number. (d) Equidistribution quality measure, $\|Q_{eq}\|_\infty$, plotted against iteration number.

density function has relative maxima. The convergence history is shown in Figure 2.2(c). The last figure shows that $\|Q_{eq}\|_\infty$ rapidly approaches 1, meaning that the equidistribution relation (2.5) or (2.8) is nearly satisfied after a few iterations. □

2.2.2 BVP method

While de Boor's algorithm is simple and reliable, it unfortunately does not extend easily to higher dimensions. We consider here a method for computing an equidistributing mesh which is based on a boundary-value-problem (BVP) formulation of equidistribution. Higher dimensional generalizations are considered in later chapters.

There are several ways to derive such a formulation. First, differentiating (2.8) with respect to ξ, one sees that $x(\xi)$ satisfies the quasi-linear second-order differential equation

$$\frac{d}{d\xi}\left(\rho(x)\frac{dx}{d\xi}\right) = 0, \tag{2.28}$$

subject to the boundary conditions

$$x(0) = a, \quad x(1) = b. \tag{2.29}$$

A variational formulation of the coordinate transformation problem which is particularly useful and fundamental when considering the multi-dimensional case is the following: Find a transformation $x = x(\xi)$ which satisfies (2.29) and minimizes the functional

$$I[x] = \frac{1}{2}\int_0^1 \left(\rho(x)\frac{dx}{d\xi}\right)^2 d\xi. \tag{2.30}$$

From basic calculus of variations (e.g., see [106] and [153]), a minimizer satisfies the Euler-Lagrange equation of the functional.

Proposition 2.2.1 *The Euler-Lagrange equation of (2.30) is given by (2.28).*

Proof. Let $x = x(\xi)$ be a minimizer and $\delta x(\xi)$ an arbitrary perturbation satisfying the homogeneous boundary conditions $\delta x(0) = \delta x(1) = 0$. Define the function

$$g(\varepsilon) \equiv I[x+\varepsilon\delta x] = \frac{1}{2}\int_0^1 \left(\rho(x+\varepsilon\delta x)\frac{d(x+\varepsilon\delta x)}{d\xi}\right)^2 d\xi.$$

Then the variation of $I[x]$ can be found as

$$
\begin{aligned}
\delta I &\equiv \left. \frac{dg}{d\varepsilon} \right|_{\varepsilon=0} \\
&= \frac{1}{2} \int_0^1 2\rho(x) \frac{dx}{d\xi} \left(\rho'(x) \frac{dx}{d\xi} \delta x + \rho(x) \frac{d\delta x}{d\xi} \right) d\xi \\
&= \int_0^1 \left[\rho(x)\rho'(x) \left(\frac{dx}{d\xi} \right)^2 - \frac{d}{d\xi} \left(\rho^2(x) \frac{dx}{d\xi} \right) \right] \delta x \, d\xi + \rho^2 \frac{dx}{d\xi} \delta x \Big|_0^1 \\
&= - \int_0^1 \rho(x) \frac{d}{d\xi} \left(\rho(x) \frac{dx}{d\xi} \right) \delta x \, d\xi + \rho^2 \frac{dx}{d\xi} \delta x \Big|_0^1 .
\end{aligned}
$$

Setting $\delta I = 0$ gives

$$
- \int_0^1 \rho(x) \frac{d}{d\xi} \left(\rho(x) \frac{dx}{d\xi} \right) \delta x \, d\xi + \rho^2 \frac{dx}{d\xi} \delta x \Big|_0^1 = 0 \qquad \forall\, \delta x : \, \delta x(0) = \delta x(1) = 0.
$$

Since this holds for all perturbations δx, by the vanishing theorem in calculus [153] we obtained the Euler-Lagrange equation (2.28) for $I[x]$. □

One can alternatively formulate the optimization problem in terms of the inverse coordinate transformation $\xi = \xi(x) : \; [a,b] \to [0,1]$. Differentiating (2.9) with respect to x, one obtains the linear second-order differential equation

$$
\frac{d}{dx} \left(\frac{1}{\rho(x)} \frac{d\xi}{dx} \right) = 0 \tag{2.31}
$$

for $\xi(x)$, subject to the boundary conditions

$$
\xi(a) = 0, \quad \xi(b) = 1. \tag{2.32}
$$

Equation (2.31) is the Euler-Lagrange equation for the functional

$$
I[\xi] = \frac{1}{2} \int_a^b \frac{1}{\rho(x)} \left(\frac{d\xi}{dx} \right)^2 dx, \tag{2.33}
$$

where for notational convenience we again write the functional simply as I. A minimum of $I[\xi]$ thus satisfies (2.31).

There are two major advantages in formulating the equidistribution relation in terms of the inverse coordinate transformation $\xi = \xi(x)$ instead of the coordinate transformation $x = x(\xi)$. The first one is that the functional I in (2.33) is quadratic and its Euler-Lagrange equation (2.31) is linear. In contrast, the functional in (2.30) is generally not quadratic and (2.28) is nonlinear. The other, while not a relevant consideration in the one-dimensional case, is that the linearity of (2.31) makes it easier to ensure existence, uniqueness, and well-posedness of the (inverse) coordinate transformation in multi-dimensions (e.g., see [129]).

Unfortunately, neither (2.31) nor (2.33) is in a form amenable to direct computation of an equidistributing mesh because the inverse coordinate transformation does not directly give the node locations on the physical domain. (See §2.3.2 for the use of the inverse coordinate transformation in the computation of equidistributing meshes.) A common practice is to interchange the roles of the dependent and independent variables in the Euler-Lagrange equation for $I[\xi]$. In the one-dimensional case, such an interchange simply transforms (2.31) into (2.28), but in higher dimensions this interchange bears a considerable cost, as we see in Chapter 6.

We are now in the position to introduce the BVP method. Suppose that an approximation $\mathscr{T}_h^{(n)}$ to $\mathscr{T}_h$ and a mesh density function $\rho^{(n)}$ defined on it are given. Discretizing (2.28) on a computational mesh $\mathscr{T}_h^c : \xi_j,\ j = 1,...,N$ using central finite differences, we get, for $j = 2,...,N-1$,

$$\frac{2}{\xi_{j+1}-\xi_{j-1}} \left(\frac{\rho(x_{j+1}^{(n)}) + \rho(x_j^{(n)})}{2} \cdot \frac{(x_{j+1}^{(n+1)} - x_j^{(n+1)})}{(\xi_{j+1}-\xi_j)} \right.$$
$$\left. - \frac{\rho(x_j^{(n)}) + \rho(x_{j-1}^{(n)})}{2} \cdot \frac{(x_j^{(n+1)} - x_{j-1}^{(n+1)})}{(\xi_j - \xi_{j-1})} \right) = 0. \tag{2.34}$$

Keeping ρ fixed for the current iteration, this system together with the boundary conditions

$$x_1^{(n+1)} = a, \qquad x_N^{(n+1)} = b, \tag{2.35}$$

is solved for the new approximation $\mathscr{T}_h^{(n+1)}$. Since the mesh density function ρ is chosen to reflect large solution variations and is often highly nonlinear, the BVP (2.28), (2.29) is also highly nonlinear. When freezing ρ on the current mesh, the iteration scheme (2.34) may fail to converge, especially when N is small and/or $\rho(x)$ changes abruptly. This is seen in the next example. As for many iteration schemes, convergence can be improved by using relaxation or a quasi-time approach, where (2.28) is embedded into a time-dependent mesh movement PDE having a steady state solution as its solution (cf. §**??**).

Example 2.2.2 The mesh density function is the same as in Example 2.2.1, i.e.,

$$\begin{aligned}\rho = {} & 1 + 20(1-\tanh^2(20(x-0.25))) + 30(1-\tanh^2(30(x-0.5))) \\ & + 10(1-\tanh^2(10(x-0.75))) \quad \text{for} \quad x \in [0,1].\end{aligned}$$

The BVP algorithm is used for computing the equidistributing mesh with the iteration convergence tolerance set at $tol = 10^{-8}$. This produces an adaptive mesh almost identical to that for de Boor's algorithm (cf. Figure 2.2), which is not surprising since the mesh sequences converge to the same limit mesh. The convergence

Table 2.2 The number of iterations required to achieve $\|\mathscr{T}_h^{(n+1)} - \mathscr{T}_h^{(n)}\|_\infty < 10^{-8}$ for the BVP method for Example 2.2.2.

N	11	21	41	81	161	321	641
Iter	N.Cnvg	N.Cnvg	N.Cnvg	77	39	40	40

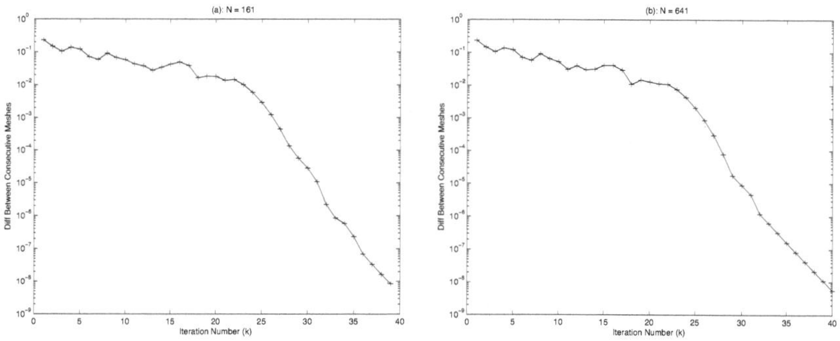

Fig. 2.3 Difference between two consecutive iterates, $\|\mathscr{T}_h^{(n+1)} - \mathscr{T}_h^{(n)}\|_\infty \equiv \max_j \|x_j^{(n+1)} - x_j^{(n)}\|$, is plotted against iteration number for (a) $N = 161$ and (b) $N = 641$ for Example 2.2.2 with the BVP algorithm.

properties of the BVP algorithm are shown in Table 2.2. The algorithm does not converge when N is small. Interestingly, for larger N the required number of iterations first decreases with N and then stays around 40 for large N. This convergence behavior is shown for large values of N in Figure 2.3. One possible interpretation for this phenomenon is that when N is large, the convergence of the algorithm is not merely influenced by the approximation error in the mesh density function, but is also largely determined by the particular linearization employed in (2.34). □

2.3 Moving mesh PDEs

2.3.1 MMPDEs in terms of coordinate transformation

Up to this point, we have studied equidistribution for time-independent problems. For the numerical solution of time-dependent problems, the mesh density function will generally depend upon the solution and hence on time, so it is necessary to employ a time-dependent mesh or coordinate transformation as well. In principle we can still use the boundary value problem corresponding to the time-independent problem (2.28) and (2.29) to determine such a coordinate transformation $x = x(\xi, t)$, viz.,

$$\frac{\partial}{\partial \xi}\left(\rho(x,t)\frac{\partial x}{\partial \xi}\right) = 0, \tag{2.36}$$

$$x(0,t) = a, \quad x(1,t) = b. \tag{2.37}$$

However, there are several advantages to employing a PDE that explicitly involves the mesh speed. First, a semi-discretization of such a mesh equation using finite differences or finite elements gives a system of ODEs, whereas a semi-discretization of (2.36) produces a system of algebraic equations. When applied to the numerical solution of PDEs, as seen in §1.2 and §1.3, the former leads to a system of ODEs and the latter to a system of DAEs (differential-algebraic equations). It is known that an ODE system is often easier to integrate than a system of differential-algebraic equations. Moreover, introduction of mesh speed into the mesh equation provides a degree of temporal smoothing for mesh movement, which is necessary for accurate integration of many physical PDEs. Furthermore, as mentioned in §2.2, a mesh equation involving mesh speed can be used as a quasi-time approach for obtaining an equidistributing mesh for a time-independent mesh density function, i.e., for computing $x(\xi) = \lim_{t\to\infty} x(\xi,t)$, where t is used as a continuation parameter.

A mesh equation involving mesh speed is referred to as a *moving mesh PDE* (MMPDE). There are numerous ways of formulating MMPDEs (e.g., see [185, 186]), and indeed, it can be unclear which MMPDEs to prefer without careful analysis and computational comparison. In this section we derive a few of the more popular MMPDEs using the approach of [189, 190], where an MMPDE is chosen as the gradient flow equation of an adaptation functional. This approach has the important advantage that it can be straightforwardly extended to multi-dimensions.

For a given mesh density function $\rho = \rho(x,t)$, the functional corresponding to (2.33) is now

$$I[\xi] = \frac{1}{2}\int_a^b \frac{1}{\rho(x,t)}\left(\frac{\partial \xi}{\partial x}\right)^2 dx. \tag{2.38}$$

The direction for ξ which reduces the value of $I[\xi]$ is given by the gradient or heat flow equation for this functional (hereafter, simply called the *gradient flow equation*), which has the form

$$\frac{\partial \xi}{\partial t} = -\frac{P}{\tau}\frac{\delta I}{\delta \xi}, \tag{2.39}$$

where $\delta I/\delta \xi$ is the functional derivative of $I[\xi]$, $\tau > 0$ is a user specified parameter for adjusting the response time of mesh movement to changes in $\rho(x,t)$, and P is a positive-definite differential operator which can be chosen with considerable flexibility, as we discuss below. From standard calculus of variations, this functional derivative is defined (cf. §2.2) in terms of the first variation through the relation

$$\delta I = \int_a^b \frac{\delta I}{\delta \xi} \delta \xi dx, \quad \forall \delta \xi : \delta \xi(a) = \delta \xi(b) = 0.$$

Arguing as for the steady state case for $I[\xi]$, it can be shown that

$$\frac{\delta I}{\delta \xi} = -\frac{\partial}{\partial x}\left(\frac{1}{\rho}\frac{\partial \xi}{\partial x}\right),$$

so (2.39) becomes

$$\frac{\partial \xi}{\partial t} = \frac{P}{\tau}\frac{\partial}{\partial x}\left(\frac{1}{\rho}\frac{\partial \xi}{\partial x}\right). \tag{2.40}$$

The use of the gradient flow equation can be largely motivated by the following stability result.

Theorem 2.3.1 *Let $\xi = \xi^*(x,t)$ be a minimizer of functional (2.38). Assume that the mesh density function is chosen such that, for a given $T > 0$,*

$$0 < \rho_{min} \le \rho(x,t) \le \rho_{max}, \quad \left|\frac{\partial \xi^*}{\partial t}\right| \le S, \quad \forall\, (x,t) \in [a,b] \times [0,T] \tag{2.41}$$

where ρ_{min}, ρ_{max}, and S are positive constants. Then, the difference between the solution $\xi(x,t)$ to (2.40) (with $P = 1$) and the minimizer $\xi^*(x,t)$, $w(x,t) \equiv \xi(x,t) - \xi^*(x,t)$, is bounded by

$$\begin{aligned}\left(\int_a^b w(x,t)^2 dx\right)^{1/2} &\le e^{-\frac{2t}{\tau\rho_{max}}}\left(\int_a^b w(x,0)^2 dx\right)^{1/2} \\ &+ \frac{\tau \rho_{max} S \sqrt{b-a}}{2}\left(1 - e^{-\frac{2t}{\tau\rho_{max}}}\right), \quad \forall t \in [0,T].\end{aligned} \tag{2.42}$$

Proof. We first notice that the minimizer $\xi = \xi^*(x,t)$ of the functional (2.38) satisfies the equation (cf. (2.31))

$$\frac{\partial}{\partial x}\left(\frac{1}{\rho(x,t)}\frac{\partial \xi^*}{\partial x}\right) = 0 \tag{2.43}$$

and the boundary conditions (2.32). It is not difficult to show that

$$\xi^*(x,t) = \frac{\int_a^x \rho(\tilde{x},t)d\tilde{x}}{\int_a^b \rho(\tilde{x},t)d\tilde{x}}, \tag{2.44}$$

from which the time derivative can be found as

$$\frac{\partial \xi^*}{\partial t} = \frac{\int_a^x \frac{\partial \rho(\tilde{x},t)}{\partial t} d\tilde{x}}{\int_a^b \rho(\tilde{x},t)d\tilde{x}} - \frac{\int_a^b \frac{\partial \rho(\tilde{x},t)}{\partial t} d\tilde{x}}{\int_a^b \rho(\tilde{x},t)d\tilde{x}} \xi^*(x,t). \tag{2.45}$$

The difference between the solution $\xi(x,t)$ to (2.40) (with $P=1$) and the minimizer $\xi^*(x,t)$ satisfies the homogeneous boundary conditions $w(a,t) = w(b,t) = 0$ and differential equation

$$\frac{\partial w}{\partial t} = \frac{1}{\tau}\frac{\partial}{\partial x}\left(\frac{1}{\rho}\frac{\partial w}{\partial x}\right) - \frac{\partial \xi^*}{\partial t}.$$

Multiplying this equation by w, integrating with respect to x over $[a,b]$, and using integration by parts on the diffusion term, we have

$$\frac{1}{2}\frac{d}{dt}\int_a^b w^2 dx = -\frac{1}{\tau}\int_a^b \frac{1}{\rho}\left(\frac{\partial w}{\partial x}\right)^2 dx - \int_a^b \frac{\partial \xi^*}{\partial t} w dx.$$

Applying Schwarz's inequality to the last term gives

$$\frac{1}{2}\frac{d}{dt}\int_a^b w^2 dx \le -\frac{1}{\tau}\int_a^b \frac{1}{\rho}\left(\frac{\partial w}{\partial x}\right)^2 dx + \left(\int_a^b w^2 dx\right)^{1/2}\left(\int_a^b \left(\frac{\partial \xi^*}{\partial t}\right)^2 dx\right)^{1/2}.$$

Using the assumptions on $\rho(x,t)$ and applying Poincaré's inequality (cf. Theorem A.0.7 in Appendix A) to the diffusion term, we then have

$$\frac{1}{2}\frac{d}{dt}\int_a^b w^2 dx \le -\frac{2}{\tau\rho_{max}}\int_a^b w^2 dx + \left(\int_a^b w^2 dx\right)^{1/2} S\sqrt{b-a}.$$

Rewriting the left-hand side as

$$\frac{1}{2}\frac{d}{dt}\int_a^b w^2 dx = \left(\int_a^b w^2 dx\right)^{1/2}\frac{d}{dt}\left(\int_a^b w^2 dx\right)^{1/2}$$

and dividing both sides of the inequality by the factor $\left(\int_a^b w^2 dx\right)^{1/2}$ gives

$$\frac{d}{dt}\left(\int_a^b w^2 dx\right)^{1/2} \le -\frac{2}{\tau\rho_{max}}\left(\int_a^b w^2 dx\right)^{1/2} + S\sqrt{b-a}.$$

Integrating both sides in time we obtain (2.42). □

It is remarked that the assumption (2.41) is only made for theoretical purposes. The restrictions simply require that $\rho(x,t)$ be bounded away from zero and from above and not change too fast in time (cf. (2.45)). Moreover, for the time-independent case where $\rho = \rho(x)$, it is easy to see that $S=0$, and (2.42) implies that

$$\left(\int_a^b w(x,t)^2 dx\right)^{1/2} \le e^{-\frac{2t}{\tau\rho_{max}}}\left(\int_a^b w(x,0)^2 dx\right)^{1/2}, \qquad (2.46)$$

i.e., $\xi(x,t)$ tends to $\xi^*(x)$ exponentially as $t \to \infty$. When $\rho = \rho(x,t)$ depends upon time, the theorem implies that

$$\left(\int_a^b (\xi(x,t) - \xi^*(x,t))^2 dx \right)^{1/2} = O(\tau). \tag{2.47}$$

In other words, $\xi(x,t)$ stays of the order $O(\tau)$ close to the minimizer $\xi^*(x,t)$ in the L^2 sense.

Returning to MMPDE (2.40), note that it is formulated in terms of the inverse coordinate transformation $\xi = \xi(x,t)$. As previously mentioned, this formulation is common in the context of variational mesh adaptation since it gives a mapping which in higher dimensions is less likely to be singular – e.g., see Dvinsky [129] and Chapter 6. But (2.40) is not convenient to use in actual computation since $\xi = \xi(x,t)$ does not explicitly specify the node location on the physical domain. A mesh equation for the coordinate transformation $x = x(\xi,t)$ instead can be obtained by interchanging the roles of dependent and independent variables in (2.40). This interchange can be done as follows: Differentiating both sides of the identity

$$\xi = \xi(x(\xi,t),t) \tag{2.48}$$

with respect to ξ while holding t fixed gives

$$1 = \frac{\partial \xi}{\partial x}\frac{\partial x}{\partial \xi} \quad \text{or} \quad \frac{\partial x}{\partial \xi} = \left(\frac{\partial \xi}{\partial x}\right)^{-1}. \tag{2.49}$$

Differentiating (2.48) with respect to t while holding ξ fixed gives

$$0 = \frac{\partial \xi}{\partial t} + \frac{\partial \xi}{\partial x}\frac{\partial x}{\partial t}$$

or

$$\frac{\partial x}{\partial t} = -\frac{\partial x}{\partial \xi}\frac{\partial \xi}{\partial t}. \tag{2.50}$$

From (2.49) and (2.50), the gradient flow equation (2.40) becomes

$$\frac{\partial x}{\partial t} = \frac{1}{\tau}\frac{\partial x}{\partial \xi} P \left(\rho \frac{\partial x}{\partial \xi}\right)^{-2} \left(\frac{\partial x}{\partial \xi}\right)^{-1} \frac{\partial}{\partial \xi}\left(\rho \frac{\partial x}{\partial \xi}\right). \tag{2.51}$$

By choosing $P = (\rho x_\xi)^2$ in (2.51), we obtain the so-called MMPDE5 [186]:

$$\text{(MMPDE5):} \qquad \frac{\partial x}{\partial t} = \frac{1}{\tau}\frac{\partial}{\partial \xi}\left(\rho \frac{\partial x}{\partial \xi}\right). \tag{2.52}$$

A slightly different choice $P = (\rho x_\xi)^2/\rho$ is suggested in [175] to make the mesh equation more spatially balanced throughout the physical domain. This results in the modified MMPDE5

$$\text{(modified MMPDE5):} \qquad \frac{\partial x}{\partial t} = \frac{1}{\tau \rho} \frac{\partial}{\partial \xi}\left(\rho \frac{\partial x}{\partial \xi}\right). \tag{2.53}$$

Moreover, the choices

$$P = -\frac{\partial x}{\partial \xi}\left(\frac{\partial}{\partial \xi}\rho \frac{\partial}{\partial \xi}\right)^{-1}\left(\rho \frac{\partial x}{\partial \xi}\right)^2 \frac{\partial x}{\partial \xi}$$

and

$$P = -\frac{\partial x}{\partial \xi}\left(\frac{\partial^2}{\partial \xi^2}\right)^{-1}\left(\rho \frac{\partial x}{\partial \xi}\right)^2 \frac{\partial x}{\partial \xi}$$

lead to the so-called MMPDE4 and MMPDE6

$$\text{(MMPDE4):} \qquad \frac{\partial}{\partial \xi}\left(\rho \frac{\partial x_t}{\partial \xi}\right) = -\frac{1}{\tau} \frac{\partial}{\partial \xi}\left(\rho \frac{\partial x}{\partial \xi}\right), \tag{2.54}$$

$$\text{(MMPDE6):} \qquad \frac{\partial^2 x_t}{\partial \xi^2} = -\frac{1}{\tau} \frac{\partial}{\partial \xi}\left(\rho \frac{\partial x}{\partial \xi}\right). \tag{2.55}$$

Note that these MMPDEs all have in common the fact that they contain the left-hand-side term of the equidistribution relation (2.36), i.e.,

$$\frac{1}{\tau} \frac{\partial}{\partial \xi}\left(\rho \frac{\partial x}{\partial \xi}\right). \tag{2.56}$$

This term plays the role of a driving force for mesh movement and provides the mechanism to pull the mesh back toward equidistribution of the mesh density function ρ when it drifts away from equidistribution. The term vanishes when the equidistribution relation is satisfied, giving no mesh movement at that point. Moreover, when ρ is time independent and $x(\xi,t) = x^*(\xi)$ holds initially, where $x^*(\xi)$ is the corresponding equidistributing coordinate transformation (see (2.28)), then $x(\xi,t) = x^*(\xi)$ holds for all time. In this case, $x^*(\xi)$ is an exact solution to these MMPDEs.

The above MMPDEs can be discretized in a standard way. Consider the modified MMPDE5 as an example. A spatial discretization of (2.53) (or (1.14)) with central finite differences on a uniform computational mesh is given in (1.18) and for convenience repeated here: for $j = 2, \ldots, N-1$,

$$\frac{dx_j}{dt} = \frac{1}{\rho_j \tau \Delta \xi^2}\left[\frac{\rho_{j+1} + \rho_j}{2}(x_{j+1} - x_j) - \frac{\rho_j + \rho_{j-1}}{2}(x_j - x_{j-1})\right], \tag{2.57}$$

where $x_j \approx x(\xi_j,t)$ and $\rho_j = \rho(x_j,t)$. The dependence of ρ on x and t often occurs only through its dependence on the physical solution $u = u(x,t)$. In such a case, we can write $\rho_j = \rho(u(x_j,t))$. Use of (2.57) can be viewed as the method-of-lines ap-

Table 2.3 The number of time steps required to achieve $\max_j |x_j^{n+1} - x_j^n| \le 10^{-8}$ for modified MMPDE5 (2.58) for Example 2.3.1.

$\Delta t \setminus N$	11	21	41	81	161	321	641
1	N.Cnvg	N.Cnvg	N.Cnvg	236	105	91	88
0.1	N.Cnvg	N.Cnvg	102	85	78	76	74
0.01	117	121	135	140	145	146	146

proach, either for computing the mesh for a given mesh density function $\rho = \rho(x,t)$ or for the simultaneous solution of a coupled system of mesh and physical equations (cf. §1.2). Discretizing further in time and using the backward Euler discretization, e.g., but calculating the mesh density function at the previous time step, we obtain

$$\frac{x_j^{n+1} - x_j^n}{\Delta t} = \frac{1}{\rho_j^n \tau \Delta \xi^2} \left[\frac{\rho_{j+1}^n + \rho_j^n}{2}(x_{j+1}^{n+1} - x_j^{n+1}) - \frac{\rho_j^n + \rho_{j-1}^n}{2}(x_j^{n+1} - x_{j-1}^{n+1}) \right], \tag{2.58}$$

where $x_j^n \approx x(\xi_j^n, t_n)$ and $\rho_j^n = \rho(x_j^n, t_n)$. Note that (2.58) is a parabolic version of (2.34).

Example 2.3.1 The mesh density function is the same one as for Example 2.2.1, i.e.,

$$\rho = 1 + 20(1 - \tanh^2(20(x - 0.25))) + 30(1 - \tanh^2(30(x - 0.5))) + 10(1 - \tanh^2(10(x - 0.75))) \quad \text{for} \quad x \in [0,1].$$

Starting with an initial uniform mesh, the scheme (2.58) with $\tau = 1$ is used to compute an approximation to the equidistributing mesh. The number of time steps (NTS) required to achieve $\max_j |x_j^{n+1} - x_j^n| \le 10^{-8}$ is listed in Table 2.3 for three choices of Δt. One can see that for a fixed Δt, the convergence of the algorithm improves as N increases and NTS stays almost constant for large N. NTS is plotted against Δt in Figure 2.4 for the case $N = 641$. The result shows $NTS = O(\Delta t^{-0.9})$ for small Δt, meaning that the convergence of the computation occurs around the time instant $t_c = NTS \times \Delta t^{0.9}$. Interestingly, for relatively large Δt, NTS changes very little and is not necessarily even monotone decreasing. □

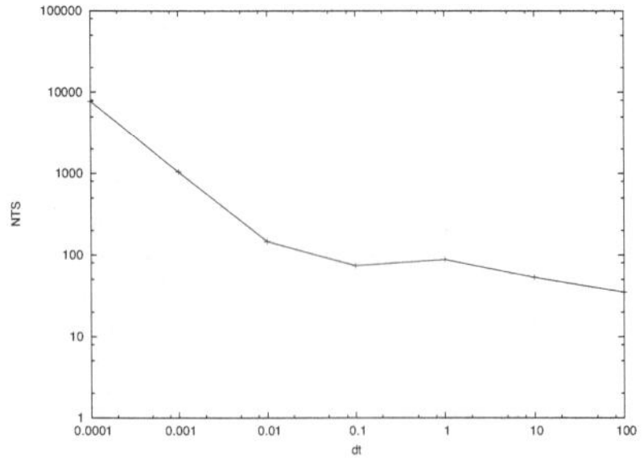

Fig. 2.4 The number of time steps required to achieve $\max_j |x_j^{n+1} - x_j^n| \leq 10^{-8}$ for the scheme (2.58) with $N = 641$ is plotted against Δt for Example 2.3.1.

2.3.2 MMPDEs in terms of inverse coordinate transformation

Mesh equations (2.31) and (2.40) have not normally been used for the computation of adaptive meshes. This is mainly because (i) the inverse coordinate transformation does not directly give the node locations on the physical domain and (ii) the inverse coordinate transformation has a steep gradient in the regions where $\rho(x)$ is large (cf. Figure 2.1) so that obtaining a reasonably accurate approximation requires a priori having a good adaptive mesh (which is precisely what we want to compute). Nevertheless, there are some situations where these difficulties can be fairly easily overcome. For instance, when solving a time-dependent problem the mesh at the previous time step can be used for solving (2.31) or (2.40). Moreover, the new physical mesh can be obtained from a new ξ mesh using either an iteration procedure (see Hagmeijer [163]) or linear interpolation, which is what is described in this section.

Once again, we take the simple case of the MMPDE (2.40) with $P = 1$. Thus, consider

$$\frac{\partial \xi}{\partial t} = \frac{1}{\tau} \frac{\partial}{\partial x} \left(\frac{1}{\rho} \frac{\partial \xi}{\partial x} \right), \tag{2.59}$$

which is referred to as MMPDE5xi due to its similarity to (2.52). Assume that a physical mesh

$$\mathscr{T}_h^n : \quad x_j^n \approx x(\xi_j, t_n), \; j = 1, \ldots, N$$

is available at time t_n, where

$$\mathscr{T}_h^c : \quad \xi_j = \frac{j-1}{N-1}, \; j = 1, \ldots, N$$

is the uniform computational mesh. Using a backward Euler-type time discretization of (2.59), we obtain

$$\frac{\xi_j^{n+1}-\xi_j^n}{\Delta t} = \frac{2}{\tau(x_{j+1}^n - x_{j-1}^n)}\left(\frac{\xi_{j+1}^{n+1}-\xi_j^{n+1}}{x_{j+1}^n - x_j^n}\frac{2}{\rho_{j+1}^n+\rho_j^n} - \frac{\xi_j^{n+1}-\xi_{j-1}^{n+1}}{x_j^n - x_{j-1}^n}\frac{2}{\rho_j^n+\rho_{j-1}^n}\right), \tag{2.60}$$

which combined with the boundary conditions $\xi_1^{n+1}=0$ and $\xi_N^{n+1}=1$ gives a system of equations for the new computational mesh

$$\mathscr{T}_h^{c,n+1}: \quad \xi_j^{n+1}, \; j=1,\dots,N.$$

Since the mesh density function is computed at $t=t_n$, i.e., $\rho_j^n = \rho(x_j^n, t_n)$, the system is linear, but as a consequence there is a time lagging problem with this scheme (see discussion on time lagging in §1.5.3 and §2.6).

By construction, the new computational mesh $\mathscr{T}_h^{c,n+1}$ is related to the physical mesh $\mathscr{T}_h^n$ by

$$\xi_j^{n+1} \approx \xi(x_j^n, t_{n+1}), \quad j=1,\dots,N$$

or

$$x_j^n \approx x(\xi_j^{n+1}, t_{n+1}), \quad j=1,\dots,N \tag{2.61}$$

where $\xi = \xi(x,t)$ denotes the inverse of the coordinate transformation $x=x(\xi,t)$. From this relation, one can see that the function $x=x(\xi,t_{n+1})$ can be approximated by the piecewise linear polynomial interpolating the paired points (ξ_j^{n+1}, x_j^n) formed by the nodes of meshes $\mathscr{T}_h^{c,n+1}$ and $\mathscr{T}_h^n$, viz.,

$$x(\xi,t_{n+1}) \approx x(\xi) = \Pi_1(\mathscr{T}_h^{c,n+1}, \mathscr{T}_h^n; \xi), \tag{2.62}$$

where Π_1 denotes the piecewise linear polynomial. Recalling that the desired new physical mesh $\mathscr{T}_h^{n+1}$ is related to the uniform computational mesh $\mathscr{T}_h^c$ by

$$x_j^{n+1} \approx x(\xi_j, t_{n+1}), \quad j=1,\dots,N \tag{2.63}$$

from (2.62) one can obtain the new physical mesh as

$$x_j^{n+1} = \Pi_1(\mathscr{T}_h^{c,n+1}, \mathscr{T}_h^n; \xi_j), \quad j=1,\dots,N. \tag{2.64}$$

Linear interpolation is employed because it is relatively easy to implement and preserves the monotonicity of the coordinate transformation, preventing mesh-point crossover.

One advantage of having used the mesh equation (2.31) or (2.40) is that the discretization (2.60) is linear in the ξ mesh so no nonlinear iteration was needed in computing it. The x mesh is simply computed using linear interpolation in (2.64) instead of having to consider an iterative scheme like those in §2.2. This can work

Table 2.4 The number of time steps NTS required to achieve $\max_j |x_j^{n+1} - x_j^n| \leq 10^{-8}$ using discretization (2.60) for MMPDE5xi for Example 2.3.2.

$\Delta t \setminus N$	11	21	41	81	161	321	641
1	N.Cnvg	35	19	18	17	16	16
0.1	112	90	84	83	82	82	81
0.01	799	701	652	627	627	625	625

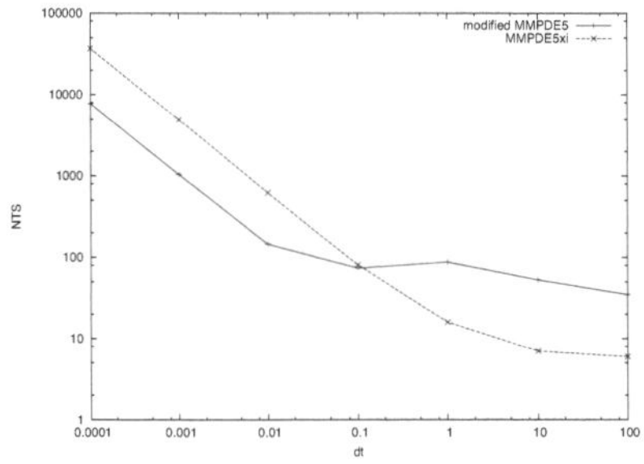

Fig. 2.5 The number of time steps required to achieve $\max_j |x_j^{n+1} - x_j^n| \leq 10^{-8}$ for MMPDEs (2.59) and (2.53) with $N = 641$ is plotted against Δt for Example 2.3.2.

well since one has a good approximation for the mesh from the previous time step. Experience indicates that these mesh equations tend to produce a mesh closer to the equidistributing one than the algorithms like (2.28), which involve a mesh equation formulation in terms of $x(\xi)$, largely because the nonlinear nature of the equidistribution process has been taken into consideration in such a simple way.

Example 2.3.2 The computation in Example 2.3.1 is repeated using the above-described scheme with $\tau = 1$ and an initial uniform mesh. The number of time steps (NTS) required to achieve $\max_j |x_j^{n+1} - x_j^n| \leq 10^{-8}$ is listed in Table 2.4 for three choices of Δt and plotted as a function of Δt in Figure 2.5. Once again, NTS is of the order $O(\Delta t^{-0.9})$ for relatively small Δt and decreases slowly as N increases for a fixed Δt. Interestingly, the results also show that modified MMPDE5 works better for small Δt whereas MMPDE5xi is better for large Δt. Moreover, as can be seen in Figure 2.6, for large Δt the difference between two consecutive meshes, $\max_j |x_j^{n+1} - x_j^n|$, oscillates more significantly for modified MMPDE5 than MMPDE5xi. □

To conclude this section, we remark that MMPDEs involving $x = x(\xi, t)$ are more convenient to use when the simultaneous solution procedure for the coupled system of mesh and physical equations is desired (cf. §1.2 and §1.3). On the other hand,

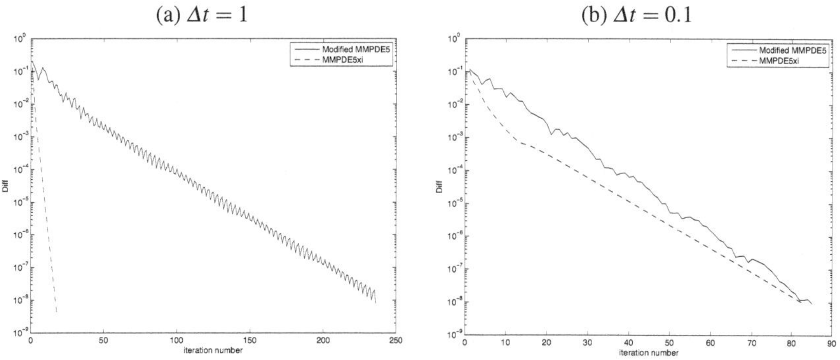

Fig. 2.6 Example 2.3.1. The difference between two consecutive meshes, $\max_j |x_j^{n+1} - x_j^n|$, is plotted as function of n (iteration number) for discretization of the modified MMPDE5 (2.53) and MMPDE5xi (2.59) with $N = 81$.

MMPDEs in terms of either $x = x(\xi, t)$ or $\xi = \xi(x,t)$ can be used with an alternate solution procedure – see §2.6 for further discussion.

2.4 Mesh density functions based on interpolation error

A key to the success of mesh equidistribution for the adaptive numerical solution of differential equations lies in the selection of an appropriate mesh density function. Here we give an error analysis for the basic adaptive equidistribution problem introduced in the previous sections. For the novice, the functional analysis tools required for this analysis may seem technical and the task of interpreting the results daunting. Nevertheless, such a study is invaluable: it provides an understanding of the *practical principles* needed to design and actually implement algorithms on computers, as well as a knowledge of the inherent advantages and limitations of different mesh adaptation strategies.

Our error analysis for the problem is based on error estimates for polynomial interpolation. There are several reasons why this is the case. First, interpolation error estimates are simple to use, economical to compute, and most of all, problem independent. As a consequence, adaptive mesh generation codes based on interpolation error are used for a large class of problems. Second, the use of interpolation error can be theoretically justified for the important case of finite element approximation of elliptic partial differential equations. This is because such an error often dominates the finite element error. Third, estimates of other errors, such as the truncation error for a finite difference discretization of a differential equation, can often be written in a similar form to that of interpolation error estimates. When this is the case, the procedure based on interpolation error for defining the mesh density function can be straightforwardly applied to these errors. Finally, it is often hard to obtain reliable estimates for the error for complicated and/or highly nonlinear problems. Fortuitously, interpolation error estimates, used almost out of necessity, turn out to frequently be the most reliable ones.

2.4.1 Interpolation error estimates

Here we consider estimates of interpolation error in the Sobolev space $H^{k+1}(a,b)$, where $k \geq 0$ will throughout denote the degree of interpolating piecewise polynomials. A purpose here is to familiarize the reader with and motivate these estimates in a simple setting. Results in multi-dimensions and in a more general Sobolev space $W^{l,p}(\Omega)$ are given in Chapter 4.

Given a mesh $\mathscr{T}_h: \; x_1 = a < x_2 < \ldots < x_N = b$ on $[a,b]$, let

$$h_j = x_j - x_{j-1}, \quad I_j = (x_{j-1}, x_j), \quad j = 2, \ldots, n; \quad h = \max_j h_j.$$

Denote by Π_k an interpolation or approximation operator preserving piecewise polynomials of degree $\leq k$ on $\mathscr{T}_h$. Then a standard interpolation error bound (e.g., see Ciarlet [104]) is the following: for $0 \leq m \leq k$,

$$|u - \Pi_k u|^2_{H^m(a,b)} \leq C \sum_{j=2}^{N} h_j^{1+2(k-m+1)} \langle u \rangle^2_{H^{k+1}(I_j)}, \qquad \forall u \in H^{k+1}(a,b) \qquad (2.65)$$

where C is a constant independent of the mesh $\mathscr{T}_h$ and the function u, and the scaled H^{k+1} semi-norm of u on I_j is

$$\langle u \rangle_{H^{k+1}(I_j)} = \left[\frac{1}{h_j} \int_{x_{j-1}}^{x_j} |u^{(k+1)}|^2 dx \right]^{\frac{1}{2}}.$$

On a uniform mesh, (2.65) reduces to the classic result

$$|u - \Pi_k u|_{H^m(a,b)} \leq C N^{-(k-m+1)} |u|_{H^{k+1}(a,b)} \qquad \text{for } 0 \leq m \leq k. \qquad (2.66)$$

To help familiarize the reader with (2.65), a proof for the special case where interpolation is done with Taylor polynomials is given below. It uses Taylor's basic expansion theorem. (Another special case is given in Exercise Problem 9.)

Theorem 2.4.1 (Taylor's Theorem) *Suppose that* $u \in C^n[a,b]$ *and* $u^{(n+1)}$ *exists on* $[a,b]$ *for some integer* $n \geq 0$. *Then for* $x, x_0 \in [a,b]$, $u(x)$ *can be expanded about* x_0 *as*

$$u(x) = \sum_{j=0}^{n} \frac{(x-x_0)^j}{j!} u^{(j)}(x_0) + R_n(x) \equiv T_n(u) + R_n(x),$$

where the remainder term for the Taylor polynomial $T_n(u)$ is given by

$$R_n(x) = \frac{1}{n!} \int_{x_0}^{x} (x-t)^n u^{(n+1)}(t)dt = \frac{(x-x_0)^{n+1}}{(n+1)!} u^{(n+1)}(\xi)$$

for some ξ between x_0 and x.

Proof of (2.65). For an arbitrary function $u \in H^{k+1}(a,b)$, consider the interpolation error from the kth Taylor polynomial on the subinterval I_j ($2 \leq j \leq N$). Taking $x_0 := x_{j-1}$ and $n := k$ in Theorem 2.4.1 and differentiating the remainder term m times ($m \leq k$) gives

$$R_k^{(m)}(x) = \frac{1}{(k-m)!} \int_{x_{j-1}}^{x} (x-t)^{k-m} u^{(k+1)}(t)dt.$$

From Schwarz's inequality,

$$\begin{aligned}
&\int_{x_{j-1}}^{x_j} |R_k^{(m)}(x)|^2 dx \\
&= \frac{1}{(k-m)!^2} \int_{x_{j-1}}^{x_j} \left| \int_{x_{j-1}}^{x} (x-t)^{k-m} u^{(k+1)}(t)dt \right|^2 dx \\
&\leq \frac{1}{(k-m)!^2(2(k-m)+1)} \int_{x_{j-1}}^{x_j} (x-x_{j-1})^{2(k-m)+1} dx \int_{x_{j-1}}^{x_j} |u^{(k+1)}(t)|^2 dt \\
&= C h_j^{2(k-m+1)} \int_{x_{j-1}}^{x_j} |u^{(k+1)}(t)|^2 dt,
\end{aligned}$$

where $C = ((k-m)!^2(2(k-m)+1)2(k-m+1))^{-1}$. Thus,

$$|u - T_k(u)|^2_{H^m(I_j)} \leq C h_j^{1+2(k-m+1)} \langle u \rangle^2_{H^{k+1}(I_j)}. \tag{2.67}$$

Summing (2.67) from $j = 2$ to N yields

$$\sum_{j=2}^{N} |u - T_k(u)|^2_{H^m(I_j)} \leq C \sum_{j=2}^{N} h_j^{1+2(k-m+1)} \langle u \rangle^2_{H^{k+1}(I_j)},$$

which gives (2.65). □

In practical computation, piecewise constant interpolation ($k = 0$) and linear interpolation ($k = 1$) are commonly used. For the latter case, the error can be measured in either the L^2 norm ($m = 0$) or the H^1 semi-norm ($m = 1$).

2.4.2 Optimal mesh density functions

Based on the interpolation error inequality (2.65), the best mesh would be the one minimizing the bound on the right-hand side. Minimizing this directly is usually impractical since the highly nonlinear and non-convex nature of the error bound gives a nasty minimization problem. Instead, an indirect approach is considered, namely, an equidistribution approach where an optimal mesh which equidistributes an appropriately defined mesh density function is sought.

The optimal mesh density function can be obtained using Theorem 2.1.2 where the optimality property for the equidistributing mesh is given. To this end, we rewrite (2.65) as

$$|u - \Pi_k u|^2_{H^m(a,b)} \le C \sum_{j=2}^{N} \left(h_j \langle u \rangle_{H^{k+1}(I_j)}^{\frac{2}{1+2(k-m+1)}} \right)^{1+2(k-m+1)} . \tag{2.68}$$

Since $\langle u \rangle_{H^{k+1}(I_j)}$ may vanish locally, we regularize (2.68) using a positive constant α_h (to be specified) to define a strictly positive mesh density function, i.e.,

$$\begin{aligned} |u - \Pi_k u|^2_{H^m(a,b)} &\le C \sum_{j=2}^{N} \left(h_j \left[\alpha_h + \langle u \rangle^2_{H^{k+1}(I_j)} \right]^{\frac{1}{1+2(k-m+1)}} \right)^{1+2(k-m+1)} \\ &= C\, \alpha_h \sum_{j=2}^{N} \left(h_j \left[1 + \frac{1}{\alpha_h} \langle u \rangle^2_{H^{k+1}(I_j)} \right]^{\frac{1}{1+2(k-m+1)}} \right)^{1+2(k-m+1)} \end{aligned} \tag{2.69}$$

The mesh- and solution-dependent factor in this bound is

$$\begin{aligned} E(\mathscr{T}_h) &= (N-1)^{2(k-m+1)} \alpha_h \\ &\times \sum_{j=2}^{N} \left(h_j \left[1 + \frac{1}{\alpha_h} \langle u \rangle^2_{H^{k+1}(I_j)} \right]^{\frac{1}{1+2(k-m+1)}} \right)^{1+2(k-m+1)} . \end{aligned} \tag{2.70}$$

From Theorem 2.1.2, if the (optimal) mesh density function is defined as

$$\rho_j = \left[1 + \frac{1}{\alpha_h} \langle u \rangle^2_{H^{k+1}(I_j)} \right]^{\frac{1}{1+2(k-m+1)}}, \quad j = 2, \ldots, N \tag{2.71}$$

then the right-hand side of (2.69) attains its minimum asymptotically for the equidistributing mesh. Thus,

$$|u - \Pi_k u|^2_{H^m(a,b)} \le C\, N^{-2(k-m+1)}\, \alpha_h\, \sigma_h^{1+2(k-m+1)}, \tag{2.72}$$

where

$$\sigma_h = \sum_j h_j \rho_j .$$

The optimal mesh density function can also be obtained with a slightly more direct optimization approach. Notice that the error bound (2.70) can be regarded as a function of the mesh density function since an equidistributing mesh of N elements is determined uniquely by ρ through proper boundary conditions and the equidistribution relation

$$h_j \rho_j = \frac{\sigma_h}{N-1}, \quad j = 2, ..., N. \tag{2.73}$$

Then the optimal mesh density function is obtained by minimizing the error bound among all possible ρ, viz.,

$$\min_{\text{admissible } \rho} E(\mathscr{T}_h(\rho)).$$

Once again, direct solution of this minimization problem is impractical. Following the proof of Theorem 2.1.2, we can first find a lower bound of $E(\mathscr{T}_h)$ and then show that the lower bound can be attained with an equidistributing mesh for an appropriately chosen (optimal) mesh density function. Indeed, from the arithmetic-mean geometric-mean inequality (cf. Theorem B.0.11) we have

$$E(\mathscr{T}_h) \geq \alpha_h \left(\sum_{j=2}^{N} h_j \left[1 + \frac{1}{\alpha_h} \langle u \rangle^2_{H^{k+1}(I_j)} \right]^{\frac{1}{1+2(k-m+1)}} \right)^{1+2(k-m+1)}. \tag{2.74}$$

One can easily see that when the mesh density function is chosen as in (2.71), equality in (2.74) holds for any mesh equidistributing ρ and

$$E(\mathscr{T}_h) = \alpha_h \sigma_h^{1+2(k-m+1)},$$

which, combined with (2.69), gives (2.72).

Clearly, the size of α_h determines the level of impact the derivatives of u have on the adaptation, and for this reason α_h is often referred to as the *(adaptation) intensity parameter*. From the above derivation, we see that α_h plays at least two important roles:

(i) to define a strictly positive mesh density function ρ (a necessity to guarantee the asymptotic optimality of the equidistributing mesh), and
(ii) appropriately chosen, to ensure that the equidistributing mesh has the property

$$h = \max_j h_j \to 0 \text{ as } N \to \infty. \tag{2.75}$$

Following [175, 193, 195], there are two additional criteria for choosing α_h:

(a) to ensure that the mesh density function defined in (2.71) is invariant under a scaling transformation of u (i.e., multiplying u by a constant scaling parameter should not change the mesh distribution), and
(b) to ensure that $\sigma_h \equiv \sum_j h_j \rho_j \leq 2(b-a)$.

The bound $2(b-a)$ in Criterion (b) can be changed to a different value, which in turn affects the distribution of mesh points between regions of large and small ρ. To see this, we first show that the ratio

$$\frac{\sum_j h_j(\rho_j - 1)}{\sum_j h_j \rho_j} \tag{2.76}$$

is a reasonable indicator of the percentage of the mesh points concentrated in the regions with relatively large ρ. Denote by Ω' the union of intervals where $\langle u \rangle_{H^{k+1}(I_j)} \ll \alpha_h$ or from (2.71), $\rho_j \approx 1$. Then, the complement of Ω', $\Omega'' = [a,b] \backslash \Omega'$, represents the union of intervals where $\langle u \rangle_{H^{k+1}(I_j)} > \alpha_h$ or $\langle u \rangle_{H^{k+1}(I_j)} \approx \alpha_h$ (i.e., $\rho_j > 2$ or $\rho_j \approx 2$), or the regions with relatively large ρ. From the equidistribution condition (2.73), $\rho_j \propto 1/h_j$; in other words, the mesh density function provides a *measure of the mesh density* $d(x_j) = 1/h_j$. Thus, the number of the mesh points contained in Ω'' is proportional to

$$\begin{aligned}
\sum_{I_j \in \Omega''} \rho_j h_j &= \sum_j \rho_j h_j - \sum_{I_j \in \Omega'} \rho_j h_j \\
&\approx \sum_j \rho_j h_j - \sum_{I_j \in \Omega'} h_j \\
&= \sum_j \rho_j h_j - |\Omega'| \\
&\approx \sum_j \rho_j h_j - (|\Omega'| + |\Omega''|) \\
&= \sum_j \rho_j h_j - \sum_j h_j \\
&= \sum_j h_j(\rho_j - 1).
\end{aligned}$$

Here, we have assumed $|\Omega''| \approx 0$, i.e., the regions of large ρ are small. This is true in many practical problems where Ω'' corresponds to layer regions. Since the total number of mesh points is proportional to $\sum_j \rho_j h_j$, we see that the ratio (2.76) is a reasonable indicator for the percentage of the mesh points contained in the region with relatively large ρ.

The reason for choosing α_h such that $\sum_j h_j \rho_j \le 2(b-a)$ is that then

$$\frac{\sum_j h_j(\rho_j - 1)}{\sum_j h_j \rho_j} = 1 - \frac{(b-a)}{\sum_j h_j \rho_j} \ge 0.5,$$

so that at least 50% of the mesh points are concentrated in the regions with relatively large ρ (where $\rho_j > 2$ or $\rho_j \approx 2$).

Defining α_h to satisfy Criterion (b) is straightforward: From Jensen's inequality (Theorem B.0.12),

$$\sigma_h \equiv \sum_j h_j \rho_j \le \sum_j h_j \left[1 + \left(\frac{1}{\alpha_h} \langle u \rangle^2_{H^{k+1}(I_j)} \right)^{\frac{1}{1+2(k-m+1)}} \right]$$
$$\le (b-a) + \alpha_h^{-\frac{1}{1+2(k-m+1)}} \sum_j h_j \langle u \rangle_{H^{k+1}(I_j)}^{\frac{2}{1+2(k-m+1)}},$$

so Criterion (b) is satisfied by simply setting

$$\alpha_h = \left[\frac{1}{b-a} \sum_j h_j \langle u \rangle_{H^{k+1}(I_j)}^{\frac{2}{1+2(k-m+1)}} \right]^{1+2(k-m+1)}. \qquad (2.77)$$

Clearly, Criterion (a) is also satisfied with this choice.

Inserting (2.77) into (2.72), we have proved the following theorem:

Theorem 2.4.2 *Suppose that $u \in H^{k+1}(a,b)$. If the mesh density function and intensity parameter are chosen by (2.71) and (2.77), respectively, then for $0 \le m \le k$ the interpolation error for the corresponding equidistributing mesh satisfies*

$$|u - \Pi_k u|_{H^m(a,b)} \le C N^{-(k-m+1)} \left[\sum_j h_j \langle u \rangle_{H^{k+1}(I_j)}^{\frac{2}{1+2(k-m+1)}} \right]^{\frac{1+2(k-m+1)}{2}}$$
$$= C N^{-(k-m+1)} (b-a)^{\frac{1+2(k-m+1)}{2}} \alpha_h^{\frac{1}{2}}. \qquad (2.78)$$

We shall see that using the term $N^{-(k-m+1)}(b-a)^{\frac{1+2(k-m+1)}{2}} \alpha_h^{\frac{1}{2}}$ allows one to get a computable *error estimate* on an equidistributing mesh. The reliability of this estimate is seen for a number of examples in §2.5.4.

Since the equidistributing mesh satisfies (2.73) for the mesh density function (2.71), the facts that $\sigma_h \le 2(b-a)$ and $\rho_j \ge 1$ imply

$$h_j \le \frac{2(b-a)}{N-1}, \quad j = 2, \ldots, N. \qquad (2.79)$$

Thus, the desired condition (2.75) is satisfied as well. It follows from the L^2 integrability of $u^{(k+1)}$ and Theorem 2.4.2 that

$$\lim_{N \to \infty} \alpha_h = \left[\frac{1}{b-a} \int_a^b |u^{(k+1)}|^{\frac{2}{1+2(k-m+1)}} dx \right]^{1+2(k-m+1)} \qquad (2.80)$$

and

$$\lim_{N \to \infty} N^{(k-m+1)} |u - \Pi_k u|_{H^m(a,b)} \le C \left\| u^{(k+1)} \right\|_{L^{\frac{2}{1+2(k-m+1)}}(a,b)}. \qquad (2.81)$$

The convergence rate of the limits (2.80) and (2.81) can be obtained under some additional conditions on u. To this end, we need the following two lemmas first proven in He and Huang [170].

Lemma 2.4.1 *For any real number $0 < \gamma \le 1$ and any mesh $\mathscr{T}_h$ for $\Omega \equiv (a,b)$,*

$$\|v\|^{2\gamma}_{L^{2\gamma}(\Omega)} \le \sum_j h_j^{1-\gamma} \|v\|^{2\gamma}_{L^2(I_j)} \le |\Omega|^{1-\gamma} \|v\|^{2\gamma}_{L^2(\Omega)} \qquad \forall v \in L^2(\Omega). \tag{2.82}$$

Proof. Using Theorems A.0.3 and B.0.11, the estimates follow from

$$\sum_j h_j \langle v \rangle^{2\gamma}_{L^2(I_j)} = \sum_j h_j \left(\frac{1}{h_j} \int_{I_j} |v|^2 dx \right)^{\gamma} \ge \sum_j h_j \left(\frac{1}{h_j} \int_{I_j} |v|^{2\gamma} dx \right) = \|v\|^{2\gamma}_{L^{2\gamma}(\Omega)},$$

$$\left(\frac{1}{|\Omega|} \sum_j h_j \left(\frac{1}{h_j} \|v\|^2_{L^2(I_j)} \right)^{\gamma} \right)^{\frac{1}{\gamma}} \le \frac{1}{|\Omega|} \sum_j h_j \left(\frac{1}{h_j} \|v\|^2_{L^2(I_j)} \right) = \frac{1}{|\Omega|} \|v\|^2_{\Omega}.$$

□

Lemma 2.4.2 *For any real number $0 < \gamma \le \frac{1}{2}$ and any mesh $\mathscr{T}_h$ for $\Omega \equiv (a,b)$, there holds, for any $v \in L^2(\Omega) \cap W^{1,1}(\Omega)$,*

$$\|v\|^{2\gamma}_{L^{2\gamma}(\Omega)} \le \sum_j h_j^{1-\gamma} \|v\|^{2\gamma}_{L^2(I_j)} \le \|v\|^{2\gamma}_{L^{2\gamma}(\Omega)} + 2h^{2\gamma} |\Omega|^{1-2\gamma} \|v'\|^{2\gamma}_{L^1(\Omega)}, \tag{2.83}$$

where $h = \max_j h_j$.

Proof. The left inequality is a consequence of Lemma 2.4.1.

To prove the right inequality, define the element-wise average of v as

$$v_{I_j} = \frac{1}{h_j} \int_{I_j} v dx.$$

From Corollary B.0.1 and Theorem A.0.3 it follows that

$$
\begin{aligned}
&\sum_j h_j^{1-\gamma} \|v\|_{L^2(I_j)}^{2\gamma} - \|v\|_{L^{2\gamma}(\Omega)}^{2\gamma} \\
&= \sum_j h_j^{1-\gamma} \|v - v_{I_j} + v_{I_j}\|_{L^2(I_j)}^{2\gamma} - \sum_j \int_{I_j} |v|^{2\gamma} dx \\
&\le \sum_j h_j^{1-\gamma} \|v - v_{I_j}\|_{L^2(I_j)}^{2\gamma} + \sum_j h_j^{1-\gamma} \|v_{I_j}\|_{L^2(I_j)}^{2\gamma} - \sum_j \int_{I_j} |v|^{2\gamma} dx \\
&= \sum_j h_j^{1-\gamma} \|v - v_{I_j}\|_{L^2(I_j)}^{2\gamma} + \sum_j \int_{I_j} (|v_{I_j}|^{2\gamma} - |v|^{2\gamma}) dx \\
&\le \sum_j h_j^{1-\gamma} \|v - v_{I_j}\|_{L^2(I_j)}^{2\gamma} + \sum_j \int_{I_j} |v_{I_j} - v|^{2\gamma} dx \\
&\le \sum_j h_j^{1-\gamma} \|v - v_{I_j}\|_{L^2(I_j)}^{2\gamma} + \sum_j h_j \left(\frac{1}{h_j} \int_{I_j} |v_{I_j} - v|^2 dx \right)^{\gamma} \\
&\le 2 \sum_j h_j^{1-\gamma} \|v - v_{I_j}\|_{L^2(I_j)}^{2\gamma}. \qquad (2.84)
\end{aligned}
$$

Generally speaking, the term $\|v - v_{I_j}\|_{L^2(I_j)}$ can be estimated using Poincaré's inequality (cf. Theorem A.0.9). However, for the current 1D situation a sharper bound can be obtained. Indeed, from the assumption $v \in W^{1,1}(\Omega)$ we have

$$
\begin{aligned}
\|v - v_{I_j}\|_{L^2(I_j)}^2 &= \int_{I_j} \left| v(x) - \frac{1}{h_j} \int_{I_j} v(t) dt \right|^2 dx \\
&= \frac{1}{h_j^2} \int_{I_j} \left| \int_{I_j} (v(x) - v(t)) dt \right|^2 dx \\
&= \frac{1}{h_j^2} \int_{I_j} \left| \int_{I_j} \int_t^x v'(s) ds dt \right|^2 dx \\
&\le h_j \|v'\|_{L^1(I_j)}^2. \qquad (2.85)
\end{aligned}
$$

Combining (2.85) with (2.84) and using Hölder's inequality we obtain

$$
\begin{aligned}
\sum_j h_j^{1-\gamma} \|v\|_{L^2(I_j)}^{2\gamma} - \|v\|_{L^{2\gamma}(\Omega)}^{2\gamma} &\le 2 \sum_j h_j \|v'\|_{L^1(I_j)}^{2\gamma} \\
&\le 2h^{2\gamma} \sum_j h_j^{1-2\gamma} \|v'\|_{L^1(I_j)}^{2\gamma} \\
&\le 2h^{2\gamma} \left(\sum_j h_j^{(1-2\gamma)\cdot \frac{1}{1-2\gamma}} \right)^{1-2\gamma} \cdot \left(\sum_j \|v'\|_{L^1(I_j)}^{2\gamma \cdot \frac{1}{2\gamma}} \right)^{2\gamma} \\
&= 2h^{2\gamma} |\Omega|^{1-2\gamma} \|v'\|_{L^1(\Omega)}^{2\gamma},
\end{aligned}
$$

which gives the right inequality of (2.83). □

Using Lemma 2.4.2 we can obtain in following theorem giving the convergence rate of α_h to α as $N \to \infty$.

Theorem 2.4.3 *Suppose that $u \in H^{k+1}(a,b)$. If the mesh density function and intensity parameter are chosen by (2.71) and (2.77), respectively, then for $0 \le m \le k$ the interpolation error for the corresponding equidistributing mesh satisfies*

$$\lim_{N\to\infty} N^{(k-m+1)} \, |u - \Pi_k u|_{H^m(a,b)} \le C \left\| u^{(k+1)} \right\|_{L^{\frac{2}{1+2(k-m+1)}}(a,b)}. \tag{2.86}$$

If u further satisfies $u^{(k+2)} \in L^1(a,b)$, then

$$|u - \Pi_k u|_{H^m(a,b)} \le C N^{-(k-m+1)} \left(\left\| u^{(k+1)} \right\|_{L^{\frac{2}{1+2(k-m+1)}}(a,b)} + \frac{1}{N} \| u^{(k+2)} \|_{L^1(a,b)} \right). \tag{2.87}$$

Proof. (2.86) is simply (2.81). (Inequality (2.86) can also be proven using (2.87) and the fact that functions satisfying $u^{(k+2)} \in L^1(a,b)$ are dense in $H^{k+1}(a,b)$; e.g., see [170].)

Applying Lemma 2.4.2 to (2.77) and using (2.79) we get

$$\begin{aligned} \left\| u^{(k+1)} \right\|^{\frac{2}{1+2(k-m+1)}}_{L^{\frac{2}{1+2(k-m+1)}}(a,b)} &\le \alpha_h^{\frac{1}{1+2(k-m+1)}} (b-a) \\ &\le \left\| u^{(k+1)} \right\|^{\frac{2}{1+2(k-m+1)}}_{L^{\frac{2}{1+2(k-m+1)}}(a,b)} + 2(b-a) \left(\frac{2 \| u^{(k+2)} \|_{L^1(a,b)}}{N-1} \right)^{\frac{2}{1+2(k-m+1)}}. \end{aligned}$$

Inequality (2.87) now follows by combining the above inequality, the definition for α_h (in (2.77)), and (2.78). □

As discussed in §2.2, only approximate equidistributing meshes are found in practical computation because the integrals involved in the equidistribution relationship must generally be calculated numerically. As a result, we wish to extend the global interpolation error bounds in Theorems 2.4.2 and 2.4.3 to more general meshes. A natural generalization of the equidistributing mesh is *a quasi-equidistributing mesh* satisfying

$$\frac{(N-1)\, h_j \rho_j}{\sigma_h} \le \kappa_{eq} \quad j = 2, \ldots, N \tag{2.88}$$

for some constant $\kappa_{eq} \ge 1$ independent of j and N. Here, $\sigma_h = \sum_j h_j \rho_j$. Note that when $\kappa_{eq} = 1$, one has an exact equidistributing mesh. Moreover, if we define a discrete analog to the continuous equidistribution quality measure (2.25) by

$$Q_{eq}(I_j) = \frac{(N-1)\, h_j \rho_j}{\sigma_h}, \quad j = 2, \ldots, N, \tag{2.89}$$

equation (2.88) gives

$$\max_j Q_{eq}(I_j) \le \kappa_{eq}.$$

Furthermore, when ρ and α_h are chosen as in (2.71) and (2.77), respectively, (2.88) implies that a quasi-equidistributing mesh has the property

$$h_j \le \frac{2(b-a)\kappa_{eq}}{N-1}, \quad j = 2, \ldots, N. \tag{2.90}$$

With these properties of quasi-equidistributing meshes, we can prove the following analogue to Theorems 2.4.2 and 2.4.3 about equidistributing meshes using similar techniques.

Theorem 2.4.4 *Suppose that $u \in H^{k+1}(a,b)$. If the mesh density function and intensity parameter are chosen by (2.71) and (2.77), respectively, then for $0 \le m \le k$ the interpolation error for a quasi-equidistributing mesh satisfying (2.88) is bounded by*

$$|u - \Pi_k u|_{H^m(a,b)} \le C N^{-(k-m+1)} \kappa_{eq}^{k-m+1} \left[\sum_j h_j \langle u \rangle_{H^{k+1}(I_j)}^{\frac{2}{1+2(k-m+1)}} \right]^{\frac{1+2(k-m+1)}{2}} \tag{2.91}$$

and

$$\lim_{N\to\infty} N^{(k-m+1)} |u - \Pi_k u|_{H^m(a,b)} \le C \kappa_{eq}^{k-m+1} \left\| u^{(k+1)} \right\|_{L^{\frac{2}{1+2(k-m+1)}}(a,b)}. \tag{2.92}$$

Furthermore, if u satisfies $u^{(k+2)} \in L^1(a,b)$, then

$$|u - \Pi_k u|_{H^m(a,b)} \le C N^{-(k-m+1)} \kappa_{eq}^{k-m+1} \times \left(\left\| u^{(k+1)} \right\|_{L^{\frac{2}{1+2(k-m+1)}}(a,b)} + \frac{\kappa_{eq}}{N} \| u^{(k+2)} \|_{L^1(a,b)} \right). \tag{2.93}$$

This brings us to the heart of the matter: What is the advantage of an adaptive mesh over a uniform one? The answer can be seen by comparing the respective error bounds (2.86) and (2.66) for equidistributing and uniform meshes. Their main difference lies in the solution dependent factors $\|u^{(k+1)}\|_{L^{\frac{2}{1+2(k-m+1)}}(a,b)}$ in (2.86) and $|u|_{H^{k+1}(a,b)} \equiv \|u^{(k+1)}\|_{L^2(a,b)}$ in (2.66). They are of the same size when $u^{(k+1)}$ is smooth. However, the former is much smaller than the latter when $u^{(k+1)}$ is not smooth. Thus, the main advantage of an adaptive mesh is that *when the solution is not smooth, the error bound on an adaptive mesh contains a much smaller solution-*

dependent factor and overall is much smaller than the bound on a uniform mesh with the same number of elements. This also suggests that a *smoothness indicator* of the solution be defined as [178]

$$
\begin{aligned}
Q_{soln}(u) &\equiv \frac{\left\langle u^{(k+1)} \right\rangle_{L^2(a,b)}}{\left\langle u^{(k+1)} \right\rangle_{L^{\frac{2}{1+2(k-m+1)}}(a,b)}} \\
&= \frac{\left(\frac{1}{b-a} \int_a^b |u^{(k+1)}|^2 dx \right)^{\frac{1}{2}}}{\left(\frac{1}{b-a} \int_a^b |u^{(k+1)}|^{\frac{2}{1+2(k-m+1)}} dx \right)^{\frac{1+2(k-m+1)}{2}}}.
\end{aligned} \tag{2.94}
$$

Roughly speaking, $Q_{soln}(u) = O(1)$ indicates that the solution is smooth, while $Q_{soln}(u) \gg 1$ implies that it is not smooth.

Finally, it is useful to note that ρ_j and α_h have the continuous forms

$$
\rho(x) = \left[1 + \frac{1}{\alpha} |u^{(k+1)}(x)|^2 \right]^{\frac{1}{1+2(k-m+1)}}, \tag{2.95}
$$

$$
\alpha = \left[\frac{1}{b-a} \int_a^b |u^{(k+1)}|^{\frac{2}{1+2(k-m+1)}} dx \right]^{1+2(k-m+1)}. \tag{2.96}
$$

2.4.3 Error bounds for commonly used non-optimal mesh density functions

Here we consider interpolation error bounds for meshes equidistributing some commonly used non-optimal mesh density functions. (See §2.4.4 for comparison of these bounds with those for optimal mesh density functions.) Recall the equidistribution relation

$$
h_j \rho_j = \frac{\sigma_h}{N-1}, \quad j = 2, \ldots, N
$$

where $\sigma_h = \sum h_j \rho_j$. From (2.65), an interpolation error bound for such a mesh is

$$
\begin{aligned}
|u - \Pi_k u|^2_{H^m(a,b)} &\le C \sum_{j=2}^{N} h_j \left(\frac{\sigma_h}{N \rho_j} \right)^{2(k-m+1)} \langle u \rangle^2_{H^{k+1}(I_j)} \\
&= \frac{C \sigma_h^{2(k-m+1)}}{N^{2(k-m+1)}} \sum_{j=2}^{N} h_j \frac{\langle u \rangle^2_{H^{k+1}(I_j)}}{\rho_j^{2(k-m+1)}}.
\end{aligned}
$$

Assuming that $\max_j h_j \to 0$ as $N \to \infty$, we can take the limit in the above inequality and obtain

$$\lim_{N\to\infty} N^{k-m+1}\,|u-\Pi_k u|_{H^m(a,b)}$$

$$\leq C\left[\int_a^b \frac{|u^{(k+1)}(x)|^2}{\rho(x)^{2(k-m+1)}}dx\right]^{\frac{1}{2}}\left[\int_a^b \rho(x)dx\right]^{k-m+1}, \qquad (2.97)$$

where we have used $\sigma_h \approx \sigma = \int_a^b \rho(x)dx$.

It is instructive to look at several special cases.

(i) The first choice is $\rho = 1$. For this case, a uniform mesh results and (2.97) becomes

$$\lim_{N\to\infty} N^{k-m+1}\,|u-\Pi_k u|_{H^m(a,b)} \leq C\left[\int_a^b |u^{(k+1)}|^2 dx\right]^{\frac{1}{2}},$$

which is essentially the classical bound (2.66) for interpolation error on a uniform mesh.

(ii) The second choice is the optimal mesh density function (2.95). It is not difficult to show that (2.97) reduces to (2.86).

(iii) The third choice is the arc-length mesh density function

$$\rho = \sqrt{1+|u'|^2}, \qquad (2.98)$$

which is aimed at equidistributing the arc-length of the solution curve over the mesh points. This has proven to be fairly popular in the moving mesh community for several reasons. First, a discrete approximation to the first-order derivative of the solution is often smoother than those for higher-order derivatives, and the corresponding mesh equation is thus easier to solve. Second, obtaining a discrete approximation is easier for the first-order derivative than higher-order ones. Finally, and more importantly, the arc-length mesh density function has proven successful for certain classes of problems. (Comparison of the performances of this and some second-order derivative based mesh density functions can be seen in §2.4.5 and 2.5.1 in this and the next sections, as well as in the work of Blom and Verwer [52].) For a mesh equidistributing (2.98), the error bound is

$$\lim_{N\to\infty} N^{k-m+1}\,|u-\Pi_k u|_{H^m(a,b)}$$

$$\leq C\left[\int_a^b \frac{|u^{(k+1)}|^2}{(1+|u'|^2)^{(k-m+1)}}dx\right]^{\frac{1}{2}}\left[\int_a^b \left(1+|u'|^2\right)^{\frac{1}{2}} dx\right]^{k-m+1}. \qquad (2.99)$$

If the first integral on the right-hand side is bounded, then the interpolation error can basically be bounded by $\|u'\|_{L^2(a,b)}$. This may explain why the arc-length mesh density function works well for certain applications.

(iv) The final choice is the curvature mesh density function

$$\rho = \left(1 + |u''|^2\right)^{\frac{1}{4}}, \tag{2.100}$$

which has also been often used in practice. It can be motivated by the fact that the second-order derivative should be used for mesh adaptation related to linear interpolation. For a mesh equidistributing (2.100), the bound (2.97) reduces to

$$\lim_{N\to\infty} N^{k-m+1} \, |u - \Pi_k u|_{H^m(a,b)} \\ \leq C \left[\int_a^b \frac{|u^{(k+1)}|^2}{(1+|u''|^2)^{(k-m+1)/2}} dx \right]^{\frac{1}{2}} \left[\int_a^b \left(1+|u''|^2\right)^{\frac{1}{4}} dx \right]^{k-m+1}. \tag{2.101}$$

2.4.4 Summary of mesh density functions and error bounds

For easy reference as well as comparison purposes, we list here the mesh density functions described in the preceding two subsections and their corresponding error bounds for equidistributing meshes in the commonly used cases of piecewise constant and piecewise linear interpolation.

These error bounds are examined for an analytical example in §2.4.5 and several numerical examples in §2.5.4.

(a) Piecewise constant interpolation ($k = 0$ and $m = 0$).

- Uniform mesh (2.66):

$$N \, \|u - \Pi_0 u\|_{L^2(a,b)} \leq C \left[\int_a^b |u'|^2 dx \right]^{\frac{1}{2}}. \tag{2.102}$$

- Optimal mesh density function (2.95), (2.96), and (2.86):

$$\rho = \left(1 + \frac{1}{\alpha}|u'|^2\right)^{\frac{1}{3}}, \qquad \alpha = \left[\frac{1}{b-a}\int_a^b |u'|^{\frac{2}{3}} dx\right]^3, \tag{2.103}$$

$$\lim_{N\to\infty} N \, \|u - \Pi_0 u\|_{L^2(a,b)} \leq C \left[\int_a^b |u'|^{\frac{2}{3}} dx \right]^{\frac{3}{2}}. \tag{2.104}$$

- Arc-length mesh density function (2.98) and (2.99):

$$\rho = \left(1 + |u'|^2\right)^{\frac{1}{2}},$$

$$\lim_{N\to\infty} N \, \|u - \Pi_0 u\|_{L^2(a,b)} \leq C \left[\int_a^b \frac{|u'|^2}{1+|u'|^2} dx \right]^{\frac{1}{2}} \left[\int_a^b \left(1+|u'|^2\right)^{\frac{1}{2}} dx \right]. \tag{2.105}$$

From Hölder's inequality, it can be shown (see Problem 12) that

$$\left[\int_a^b |u'|^{\frac{2}{3}} dx\right]^{\frac{3}{2}} \leq \left[\int_a^b \frac{|u'|^2}{1+|u'|^2} dx\right]^{\frac{1}{2}} \left[\int_a^b (1+|u'|^2)^{\frac{1}{2}} dx\right]. \tag{2.106}$$

Therefore, ignoring the generic constants in the bounds, the error bound in (2.104) is smaller than that in (2.105).

(b) Piecewise linear interpolation with error measured in the L^2 norm ($k = 1$ and $m = 0$).

- Uniform mesh (2.66):

$$N^2 \|u - \Pi_1 u\|_{L^2(a,b)} \leq C \left[\int_a^b |u''|^2 dx\right]^{\frac{1}{2}}. \tag{2.107}$$

- Optimal mesh density function (2.95), (2.96), and (2.86):

$$\rho = \left(1 + \frac{1}{\alpha}|u''|^2\right)^{\frac{1}{5}}, \qquad \alpha = \left[\frac{1}{b-a}\int_a^b |u''|^{\frac{2}{5}} dx\right]^5, \tag{2.108}$$

$$\lim_{N\to\infty} N^2 \|u - \Pi_1 u\|_{L^2(a,b)} \leq C \left[\int_a^b |u''|^{\frac{2}{5}} dx\right]^{\frac{5}{2}}. \tag{2.109}$$

- Arc-length mesh density function (2.98) and (2.99):

$$\begin{aligned} &\rho = (1+|u'|^2)^{\frac{1}{2}}, \\ &\lim_{N\to\infty} N^2 \|u - \Pi_1 u\|_{L^2(a,b)} \\ &\qquad \leq C \left[\int_a^b \frac{|u''|^2}{(1+|u'|^2)^2} dx\right]^{\frac{1}{2}} \left[\int_a^b (1+|u'|^2)^{\frac{1}{2}} dx\right]^2. \end{aligned} \tag{2.110}$$

- Curvature mesh density function (2.100) and (2.101):

$$\begin{aligned} &\rho = (1+|u''|^2)^{\frac{1}{4}}, \\ &\lim_{N\to\infty} N^2 \|u - \Pi_1 u\|_{L^2(a,b)} \\ &\qquad \leq C \left[\int_a^b \frac{|u''|^2}{1+|u''|^2} dx\right]^{\frac{1}{2}} \left[\int_a^b (1+|u''|^2)^{\frac{1}{4}} dx\right]^2. \end{aligned} \tag{2.111}$$

From Hölder's inequality we have

$$\left[\int_a^b |u''|^{\frac{2}{5}} dx\right]^{\frac{5}{2}} \leq \left[\int_a^b \frac{|u''|^2}{1+|u''|^2} dx\right]^{\frac{1}{2}} \left[\int_a^b (1+|u''|^2)^{\frac{1}{4}} dx\right]^2, \tag{2.112}$$

so the error bound (2.111) is larger than (2.109) ignoring the generic constants.

(c) Piecewise linear interpolation with error measured in the H^1 semi-norm ($k = 1$ and $m = 1$).

- Uniform mesh (2.66):

$$N \, \|(u - \Pi_1 u)'\|_{L^2(a,b)} \le C \left[\int_a^b |u''|^2 dx \right]^{\frac{1}{2}}. \tag{2.113}$$

- Optimal mesh density function (2.95), (2.96), and (2.86):

$$\rho = \left(1 + \frac{1}{\alpha}|u''|^2\right)^{\frac{1}{3}}, \qquad \alpha = \left[\frac{1}{b-a}\int_a^b |u''|^{\frac{2}{3}} dx\right]^3, \tag{2.114}$$

$$\lim_{N\to\infty} N \, \|(u - \Pi_1 u)'\|_{L^2(a,b)} \le C \left[\int_a^b |u''|^{\frac{2}{3}} dx \right]^{\frac{3}{2}}. \tag{2.115}$$

- Arc-length mesh density function (2.98) and (2.99):

$$\begin{aligned} &\rho = (1 + |u'|^2)^{\frac{1}{2}}, \\ &\lim_{N\to\infty} N \, \|(u - \Pi_1 u)'\|_{L^2(a,b)} \\ &\qquad \le C \left[\int_a^b \frac{|u''|^2}{1+|u'|^2} dx \right]^{\frac{1}{2}} \left[\int_a^b (1+|u'|^2)^{\frac{1}{2}} dx \right]. \end{aligned} \tag{2.116}$$

- Curvature mesh density function (2.100) and (2.101):

$$\begin{aligned} &\rho = (1 + |u''|^2)^{\frac{1}{4}}, \\ &\lim_{N\to\infty} N \, \|(u - \Pi_1 u)'\|_{L^2(a,b)} \\ &\qquad \le C \left[\int_a^b \frac{|u''|^2}{(1+|u''|^2)^{1/2}} dx \right]^{\frac{1}{2}} \left[\int_a^b (1+|u''|^2)^{\frac{1}{4}} dx \right]. \end{aligned} \tag{2.117}$$

Once again, from Hölder's inequality we have

$$\left[\int_a^b |u''|^{\frac{2}{3}} dx \right]^{\frac{3}{2}} \le \left[\int_a^b \frac{|u''|^2}{(1+|u''|^2)^{1/2}} dx \right]^{\frac{1}{2}} \left[\int_a^b (1+|u''|^2)^{\frac{1}{4}} dx \right], \tag{2.118}$$

implying that error bound (2.117) is larger than (2.115) when ignoring generic constants.

Note that assuming $\alpha = 1$, the curvature mesh density function $\rho \sim |u''|^{\frac{1}{2}}$ lies between those for the optimal mesh density functions (2.108) ($\rho \sim |u''|^{\frac{2}{5}}$) and (2.114) ($\rho \sim |u''|^{\frac{2}{3}}$). Thus, it is not unreasonable to expect these mesh density functions to perform comparably to each other for most applications.

2.4.5 Error bounds for a function with boundary layer

In this subsection we examine error bounds for piecewise linear interpolation of the function

$$u(x) = \tanh(Rx), \qquad x \in (-1,1) \tag{2.119}$$

where R is a parameter. This function, having a sharp layer near $x = 0$ for large R, is representative of solutions having sharp layers with exponential growth or decay.

First consider the case where the interpolation error is measured in the H^1 semi-norm. For $k = 1$ and $m = 1$, the optimal mesh density function is calculated using (2.114). Bounds on the interpolation error for uniform and various equidistributing meshes are then found from (2.113), (2.115), (2.116), and (2.117). Since

$$u' = R(1-\tanh^2(Rx)), \quad u'' = -2R^2\tanh(Rx)\,(1-\tanh^2(Rx)), \tag{2.120}$$

for large R the derivatives u' and u'' are essentially zero in $(-1,1)$ except for an interval (denoted by I) where

$$u' = O(R), \quad u'' = O(R^2).$$

It is not difficult to see that I is centered at $x = 0$ and has a length of $O(1/R)$. It follows that

$$\begin{aligned}
\int_{-1}^{1} |u''|^2 dx &= \int_I O(R^4) dx = O(R^3),\\
\int_{-1}^{1} |u''|^{\frac{2}{3}} dx &= \int_I O(R^{\frac{4}{3}}) dx = O(R^{\frac{1}{3}}),\\
\int_{-1}^{1} (1+|u'|^2)^{\frac{1}{2}}\, dx &= 2 + \int_I O(R) dx = O(1),\\
\int_{-1}^{1} \frac{|u''|^2}{(1+|u''|^2)^{1/2}} dx &= \int_I O(R^2) dx = O(R),\\
\int_{-1}^{1} (1+|u''|^2)^{\frac{1}{4}}\, dx &= 2 + \int_I O(R) dx = O(1).
\end{aligned}$$

The first integral on the right-hand side of (2.116) must be estimated in a different way because u' and u'' reach their maxima at different locations. Define the function $\gamma(x) \geq 0$ implicitly through the equation

$$1-\tanh^2(Rx) = R^{-\gamma(x)}.$$

Since from (2.120)

$$u'(x) = O(R^{1-\gamma(x)}), \quad u''(x) = O(R^{2-\gamma(x)}),$$

Table 2.5 Function (2.119). Error bounds on uniform and equidistributing meshes for piecewise linear interpolation with $N \to \infty$ and R held fixed.

Mesh	$\|u - \Pi_1 u\|_{L^2}$	$\|(u - \Pi_1 u)'\|_{L^2}$
Uniform mesh	$O(R^{1.5}N^{-2})$	$O(R^{1.5}N^{-1})$
Mesh for optimal ρ (2.108) ($k = 1$, $m = 0$)	$O(R^{-0.5}N^{-2})$	
Mesh for optimal ρ (2.114) ($k = 1$, $m = 1$)		$O(R^{0.5}N^{-1})$
Mesh for arc-length ρ (2.98)	$O(R^{0.5}N^{-2})$	$O(R^{0.5}N^{-1})$
Mesh for curvature ρ (2.100)	$O(R^{-0.5}N^{-2})$	$O(R^{0.5}N^{-1})$

we have

$$\frac{|u''|^2}{1+|u'|^2}(x) = \frac{O(R^{2(2-\gamma(x))})}{1+O(R^{2(1-\gamma(x))})} = \begin{cases} O(R^2), & \text{for } 0 \le \gamma(x) \le 1 \\ O(R^{4-2\gamma(x)}), & \text{for } 1 < \gamma(x) \end{cases}$$

implying that the maximum of the integrand is $O(R^2)$. Since the length of the interval where the integrand reaches this magnitude is $O(1/R)$,

$$\int_{-1}^{1} \frac{|u''|^2}{1+|u'|^2} dx = O(R).$$

Inserting the above estimates of the integrals into (2.113), (2.115), (2.116), and (2.117), we find the error bounds on uniform and equidistributing meshes. These are summarized in the right column of Table 2.5. The middle column shows the corresponding results for the case $k = 1$ and $m = 0$. One can see that the error bounds for adaptive meshes have a much weaker R-dependence ($R^{0.5}$ for H^1 semi-norm and $R^{-0.5}$ for L^2 norm) than that for a uniform mesh ($R^{1.5}$). The R-dependence is the same for all mesh density functions except for the case $(k,m) = (1,0)$, where the arc-length mesh density function yields a stronger dependence ($R^{0.5}$). This indicates that the arc-length mesh density function may produce an adaptive mesh that is better than a uniform mesh but inferior to those generated using the optimal and curvature mesh density functions.

Some caution should be taken when interpreting the R-dependence of the error bounds because they involve the two parameters N and R, and they can have different limits for R large but fixed and $N \to \infty$ versus for N fixed and $R \to \infty$. The error bounds in Table 2.5 are valid in the limit as $N \to \infty$ and thus characterize the error behavior correctly only for sufficiently large N, say $N \ge N_0(R)$ for some number N_0 depending upon R. This can be seen in Figure 2.7 where the L^2 norm of the actual error of linear interpolation on a uniform mesh shows the second order convergence only after $N \ge 81$ for $R = 100$ and $N \ge 161$ for $R = 200$. A typical region in the N-R plane where an asymptotic error bound might give a correct characterization of the

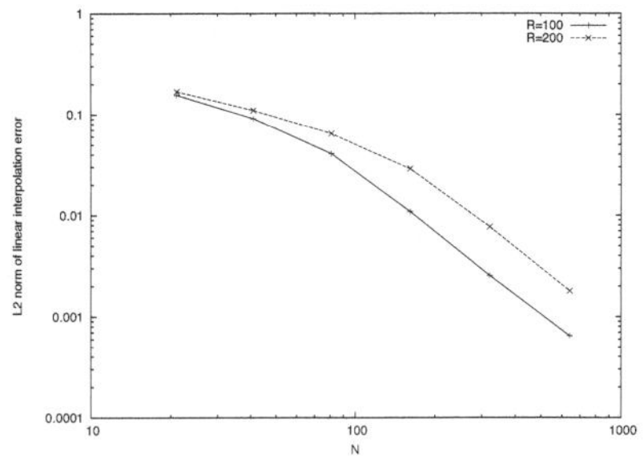

Fig. 2.7 Function (2.151). The L^2 norm of linear interpolation error on uniform meshes is plotted as function of N for $R = 100$ and $R = 200$.

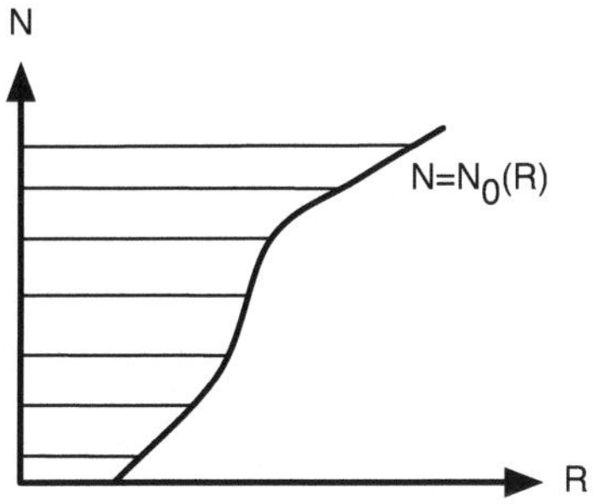

Fig. 2.8 Function (2.151). A sketch for the (shadowed) region in the N-R plane where an asymptotic bound characterizes error behavior correctly.

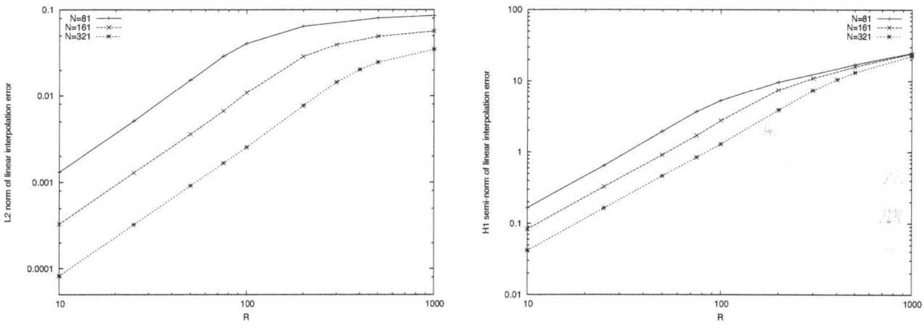

Fig. 2.9 Function (2.151). The L^2 norm and H^1 semi-norm of linear interpolation error on uniform meshes are plotted as function of R for $N = 81$, 161, and 321.

error behavior is sketched in Figure 2.8. A different interpretation of this region is that for a given N, the error bound has the correct R-dependence for small R, viz., $R < R_0(N)$ for some value R_0 depending upon N. This is illustrated in Figure 2.9, where both the L^2 norm and H^1 semi-norm of the actual error on a uniform mesh show the $O(R^{1.5})$ behavior only for relatively small R.

The actual error behavior with N fixed and $R \to \infty$ can be estimated as follows: For large R, both the exact solution u and u', and the piecewise linear interpolant $\Pi_1 u$ and $(\Pi_1 u)'$, are essentially zero away from $x = 0$. When the mesh is uniform, assume without loss of generality that there is only one mesh interval I_{j^*} where these functions are not all zero. Also assume that R is sufficiently large that I, the interval where $u' = O(R)$, is contained in I_{j^*}. The length of I_{j^*} is constant since N is fixed, and it is not difficult to show that

$$u = O(1), \quad u' = O(R), \quad \Pi_1 u = O(1), \quad (\Pi_1 u)' = O(1) \qquad \text{on } I_{j^*}.$$

It follows that

$$\int_{-1}^{1} |u - \Pi_1 u|^2 dx = \int_{I_{j^*}} O(1) dx = O(1)$$

and

$$\int_{-1}^{1} |(u - \Pi_1 u)'|^2 dx = \int_{I_{j^*}} |u' - O(1)|^2 dx = O(1) + \int_{I_{j^*}} |u'|^2 dx = O(R).$$

Thus, on a uniform mesh we have

$$\|u - \Pi_1 u\|_{L^2(-1,1)} = O(1), \qquad \text{as } R \to \infty \tag{2.121}$$

$$\|(u - \Pi_1 u)'\|_{L^2(-1,1)} = O(R^{0.5}), \qquad \text{as } R \to \infty. \tag{2.122}$$

In the case of an equidistributing mesh for a mesh density function, we need to estimate the length of I_{j^*}, now the union of the subintervals I_k where $u' = O(R)$. Consider the arc-length mesh density function as an example. The equidistribution relation implies

$$\sum_k \rho_k h_k \le \sum_j \rho_j h_j = \sigma_h = O(1).$$

Since $\rho_k = \sqrt{1 + (u')^2} = O(R)$ implies $h_k = O(\frac{1}{R})$ for each subinterval I_{j^*}, its length H_{j^*} satisfies

$$H_{j^*} = O(\frac{1}{R}). \tag{2.123}$$

Similar results can be obtained for the other mesh density functions considered here.

It is now straightforward to estimate the interpolation error. Since

$$u = O(1), \quad u' = O(R), \quad \Pi_1 u = O(1), \quad (\Pi_1 u)' = O(R)$$

on I_{j^*}, it follows that

$$\int_{-1}^{1} |u - \Pi_1 u|^2 dx = \int_{I_{j^*}} O(1) dx = O(R^{-1})$$

Table 2.6 Function (2.119). Error bounds on uniform and equidistributing meshes for piecewise linear interpolation as $R \to \infty$ with N held fixed.

Mesh	$\|u-\Pi_1 u\|_{L^2}$	$\|(u-\Pi_1 u)'\|_{L^2}$
Uniform mesh	$O(1)$	$O(R^{0.5})$
Mesh for optimal ρ (2.108) ($k=1$, $m=0$)	$O(R^{-0.5})$	
Mesh for optimal ρ (2.114) ($k=1$, $m=1$)		$O(R^{0.5})$
Mesh for arc-length ρ (2.98)	$O(R^{-0.5})$	$O(R^{0.5})$
Mesh for curvature ρ (2.100)	$O(R^{-0.5})$	$O(R^{0.5})$

and

$$\int_{-1}^{1} |(u-\Pi_1 u)'|^2 dx = \int_{I_{j^*}} |(u-\Pi_1 u)'|^2 dx = O(R).$$

Thus, on a mesh equidistributing the optimal, arc-length, or curvature mesh density function we have

$$\|u-\Pi_1 u\|_{L^2(-1,1)} = O(R^{-0.5}), \qquad \text{as } R \to \infty \tag{2.124}$$

$$\|(u-\Pi_1 u)'\|_{L^2(-1,1)} = O(R^{0.5}), \qquad \text{as } R \to \infty. \tag{2.125}$$

These error estimates for $R \to \infty$ are summarized in Table 2.6. They are confirmed by the actual error shown in Figure 2.9 for a uniform mesh and Figure 2.13 (of Example 2.5.1) for adaptive meshes.

It is interesting to note that the error bounds as $N \to \infty$ and as $R \to \infty$ (cf. Tables 2.5 and 2.6) exhibit the same R-dependence for the optimal and the curvature mesh density functions. It leads one to conjecture that this is always the case when a proper adaptive mesh is used to approximate a function with sharp layers like that for $\tanh(Rx)$.

These various cases illustrate the subtlety required to properly interpret the error bounds. One of the remarkable properties of adaptive meshes is that for fixed N, the L^2 norm of the solution actually *decreases* as $R \to \infty$ (i.e., as the problem becomes more difficult). In contrast, note that when R is not large, all of the error bounds have the same asymptotic form, and a uniform mesh strategy is as good as any other as $N \to \infty$. This is not surprising, since it is a case where adaptivity is not needed, as would be seen from examining the smoothness measure or indicator (2.94). However, interpreting its size as "large" in turn varies from one situation to the next, depending upon what values of N are being used in practice.

2.5 Computation of mesh density functions and examples

2.5.1 *Recovery of solution derivatives*

As we have seen in the preceding section, the discrete forms of the optimal mesh density function (2.71) and intensity parameter (2.77) cannot generally be computed exactly because they involve integrals which must be approximated numerically. On the other hand, the continuous forms, (2.95) and (2.96), are simpler, and they give the same asymptotic error bound for the corresponding equidistributing mesh. For this reason, these continuous forms for ρ and α are recommended for use in practical computation.

Another issue for the computation of the mesh density functions is how to compute the requisite derivatives of u. This is not a problem when an analytical expression for u is available, as has been assumed thus far. However, in most practical applications only approximations to the nodal values of the solution are known, so the problem arises of how to compute approximate derivative values in terms of these nodal values. A simple way is to use finite difference approximations, as shown in (1.22) of Chapter 1. Other methods include gradient (and Hessian) recovery techniques, e.g., the ones developed by Zienkiewicz and Zhu [354, 355] and Zhang and Naga [353]. These techniques generally produce fairly reliable and accurate results, although theoretical proofs of convergence have usually been given only for quasi-uniform meshes. indexmesh!quasi-uniform

A gradient recovery technique adopted from [353] is considered here. Suppose that the solution values u_j, $j = 1, ..., N$, defined on a mesh $x_1 = a < x_2 < \cdots < x_N = b$ are known. Given an integer $p \geq 1$, let x_{j_i}, $i = 1, ..., 2p+1$ be the $2p+1$ neighboring mesh points closest to x_j (including x_j itself), and let their center be $\hat{x}_j$. Define

$$H_j = \max_{i=1,...,2p+1} |x_{j_i} - \hat{x}_j|.$$

Using the first three orthogonal Legendre polynomials $P_0(x) = 1$, $P_1(x) = x$, and $P_2(x) = (3x^2 - 1)/2$, one determines the quadratic polynomial

$$q(x) = \sum_{k=0}^{2} a_k P_k\left(\frac{x - \hat{x}_j}{H_j}\right)$$

which is a least squares fit to this data, i.e., one solves

$$\min_{a_0,a_1,a_2} \sum_{i=1,...,2p+1} (q(x_{j_i}) - u_{j_i})^2 .$$

The value of the first derivative of this quadratic polynomial at x_j is then used as the approximation to $u'(x_j)$, i.e., $u'(x_j) \approx q'(x_j)$.

The approximation to the second derivative can be calculated by differentiating the quadratic polynomial $q(x)$ twice and using $u''(x_j) \approx q''(x_j)$. It can also be calculated using the same procedure for the first derivative but based on the nodal approximations to the values $u'(x_j)$. The same procedure can be repeated until the $(k+1)$st derivative is computed.

Having computed the $(k+1)$st derivative (for convenience, denoted simply by $u^{(k+1)}$), the parameter α defined in (2.96) can be approximated using a quadrature rule such as the trapezoidal rule,

$$\alpha \approx \left[\frac{1}{b-a} \sum_{j=2}^{N} \frac{h_j}{2} \left(|u^{(k+1)}(x_{j-1})|^{\frac{2}{1+2(k-m+1)}} + |u^{(k+1)}(x_j)|^{\frac{2}{1+2(k-m+1)}} \right) \right]^{1+2(k-m+1)}, \qquad (2.126)$$

and the mesh density function in (2.95) computed at the nodes in the obvious way.

The choice of α from the definition (2.96) can be modified as in (1.17) of Chapter 1 to ensure that $\alpha \geq 1$. Indeed, the cases of arc-length and curvature mesh density functions, (2.98) and (2.100), can be considered as similar instances of this. The integral definition (2.96) generally results in a more balanced distribution of mesh points between the smooth and rough regions of the physical solution than simply choosing $\alpha = 1$; e.g., see Figure 2.10, where mesh trajectories obtained with two different choices of α are shown. It also eliminates the need for fine tuning the parameter for each application. On the other hand, there are disadvantages with the integral definition of α. One is that it leads to a denser Jacobian matrix for the mesh equation (cf. (1.25)), which is often more expensive to approximate and invert. The other is that it may cause ρ (and thus the mesh adaptation) to be extremely sensitive to the changes in $u^{(k+1)}$. This is especially true when α is small. In that situation a small error introduced in approximating $u^{(k+1)}$ will be amplified and cause wrong mesh concentration. The above compromise of bounding α away from zero helps to avoid this difficulty. More generally, one can choose

$$\alpha = \max\left\{ \gamma, \left[\frac{1}{b-a} \int_a^b |u^{(k+1)}|^{\frac{2}{1+2(k-m+1)}} dx \right]^{1+2(k-m+1)} \right\}, \qquad (2.127)$$

where γ is a positive number. Intuitively, the idea is to concentrate mesh points only in regions where $|u^{(k+1)}| \gg \gamma$. There is once again the question of how to best choose γ, but in practice the simple choice $\gamma = 1$ often suffices.

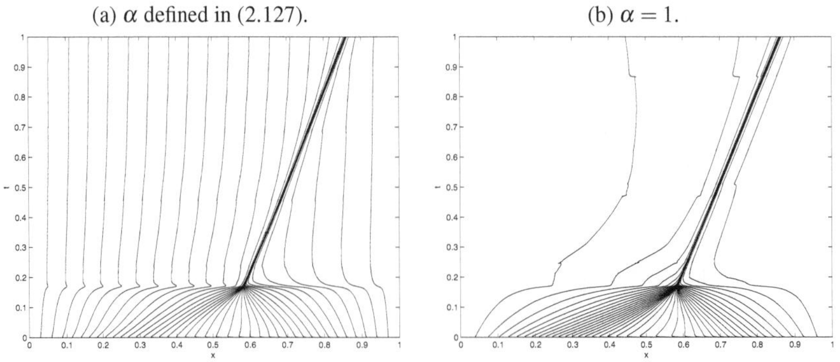

Fig. 2.10 Mesh trajectories obtained for model problem (1.1)–(1.3) using the mesh density function (1.16) with $\alpha = 1$ and with (2.127) ($a = 0$, $b = 1$, and $k = m = \gamma = 1$) or (1.17).

2.5.2 Smoothing of mesh density functions and smoothed MMPDEs

The approximate derivatives computed as above are often non-smooth, especially the higher order ones. As shown by Pryce [277], lack of smoothness may affect the convergence of the iteration for computing an equidistributing mesh. A smoother mesh density function can be obtained using a large value of p for the gradient recovery technique (as outlined in the preceding subsection) or by simply using a direct smoothing technique such as the averaging scheme given in (1.23). In practice, a successful strategy for most problems is to consecutively repeat the smoothing procedure three or four times.

More precisely, direct smoothing of the mesh density function is commonly based on use of an elliptic differential operator (especially the Laplace operator) or an approximation to it. For example, for a given ρ (viewed as a function of ξ through some coordinate transformation) a mesh density function having higher regularity or a smoother mesh density function, $\tilde{\rho}$, can be obtained as the solution of the BVP

$$\begin{cases} (I - \beta^{-2} \frac{d^2}{d\xi^2})\tilde{\rho} = \rho, & \forall \xi \in (0,1) \\ \frac{d\tilde{\rho}}{d\xi}(0) = \frac{d\tilde{\rho}}{d\xi}(1) = 0, \end{cases} \tag{2.128}$$

where $\beta > 0$ is a parameter and I is the identity operator. Indeed, we have the following theorem.

Theorem 2.5.1 *Suppose that ρ, when viewed as a function of ξ, is continuous on $[0,1]$ and satisfies*

$$0 < \underline{\rho} \le \rho(\xi) \le \overline{\rho} < \infty, \quad \forall \xi \in [0,1] \tag{2.129}$$

for some constants $\underline{\rho}$ and $\overline{\rho}$. Then, for any $\beta > 0$ the solution $\tilde{\rho}$ to BVP (2.128) exists and has the following properties:

$$0 < \underline{\rho} \le \tilde{\rho}(\xi) \le \overline{\rho} < \infty, \quad \forall \xi \in [0,1] \tag{2.130}$$

$$\frac{1}{\tilde{\rho}}\left|\frac{d\tilde{\rho}}{d\xi}\right| \le \beta, \quad \forall \xi \in [0,1]. \tag{2.131}$$

Proof. The differential equation in (2.128) is a second-order, constant coefficient, non-homogeneous ODE, and its solution can readily be found as

$$\tilde{\rho}(\xi) = \frac{\beta e^{\beta\xi}}{2}\left(c - \int_0^{\xi} e^{-\beta s}\rho(s)ds\right) + \frac{\beta e^{-\beta\xi}}{2}\left(c + \int_0^{\xi} e^{\beta s}\rho(s)ds\right),$$

where

$$c = \frac{1}{e^{\beta} - e^{-\beta}}\left(e^{\beta}\int_0^1 e^{-\beta s}\rho(s)ds + e^{-\beta}\int_0^1 e^{\beta s}\rho(s)ds\right).$$

Then (2.130) and (2.131) can be verified by observing that $c > 0$ and

$$\begin{aligned} c - \int_0^{\xi} e^{-\beta s}\rho(s)ds = \frac{1}{e^{\beta} - e^{-\beta}}\Big(& e^{\beta}\int_{\xi}^1 e^{-\beta s}\rho(s)ds \\ & + e^{-\beta}\int_0^1 e^{\beta s}\rho(s)ds + e^{-\beta}\int_0^{\xi} e^{\beta s}\rho(s)ds \Big) \\ > 0. & \end{aligned}$$

□

The property (2.131) indicates that the relative rate of change in $\tilde{\rho}$ is bounded by β. This can be seen more clearly in a discrete form (as we see in (2.134) below). Consider a central finite difference discretization of BVP (2.128) on a uniform mesh of N points,

$$\begin{cases} \tilde{\rho}_j - \dfrac{1}{\beta^2 \Delta\xi^2}\left(\tilde{\rho}_{j+1} - 2\tilde{\rho}_j + \tilde{\rho}_{j-1}\right) = \rho_j, \quad j = 2,\ldots,N-1 \\ \tilde{\rho}_1 = \tilde{\rho}_2, \quad \tilde{\rho}_{N-1} = \tilde{\rho}_N, \end{cases} \tag{2.132}$$

where $\Delta\xi = 1/(N-1)$. The following theorem is a discrete analog to Theorem 2.5.1 and can be proved similarly. (The interested reader is referred to [188] for the proof.)

Theorem 2.5.2 *Suppose that the assumption of Theorem 2.5.1 is satisfied. Then the solution to the difference equation (2.132) satisfies*

$$0 < \underline{\rho} \le \min_k \rho_k \le \tilde{\rho}_j \le \max_k \rho_k \le \overline{\rho} < \infty, \quad j = 1,\ldots,N-1 \tag{2.133}$$

$$\nu \le \frac{\tilde{\rho}_j}{\tilde{\rho}_{j-1}} \le \nu^{-1}, \quad j = 2, \ldots, N \tag{2.134}$$

where

$$\nu = \frac{\sqrt{1 + \frac{4}{\beta^2 \Delta \xi^2}} - 1}{\sqrt{1 + \frac{4}{\beta^2 \Delta \xi^2}} + 1}. \tag{2.135}$$

The smoothing defined by (2.128) or (2.132) is global in the sense that a differential equation or a linear system of algebraic equations must be solved for $\tilde{\rho}$. For efficiency, however, local smoothing is often sufficient. To develop local smoothing schemes, we can use the formal expansion

$$\begin{aligned} \tilde{\rho} &= (I - \beta^{-2} \frac{d^2}{d\xi^2})^{-1} \rho \\ &= \left[I + (\beta^{-2} \frac{d^2}{d\xi^2}) + (\beta^{-2} \frac{d^2}{d\xi^2})^2 + \cdots \right] \rho \end{aligned} \tag{2.136}$$

for large β. Truncating the expansion and approximating $(d^2)/(d\xi^2)$ with a central finite difference we obtain

$$\tilde{\rho}_j = \frac{1}{\beta^2 \Delta \xi^2} \rho_{j+1} + (1 - \frac{2}{\beta^2 \Delta \xi^2}) \rho_j + \frac{1}{\beta^2 \Delta \xi^2} \rho_{j-1}, \quad j = 2, \ldots, N-1 \tag{2.137}$$

which is an averaging if β is chosen such that $\beta^2 \Delta \xi^2 \ge 2$. In particular, when $\beta^2 \Delta \xi^2 = 4$ and the boundary condition is taken properly, (2.137) reduces to the averaging scheme (1.23).

A generalization is to use more terms in the expansion (2.136) or involve more neighboring points. For a given integer $p > 0$ and parameter $\gamma \in (0,1)$, we can use

$$\tilde{\rho}_j = \frac{\sum\limits_{k=\max(1,j-p)}^{\min(N,j+p)} \gamma^{|j-k|} \rho_k}{\sum\limits_{k=\max(1,j-p)}^{\min(N,j+p)} \gamma^{|j-k|}}, \quad j = 1, \ldots, N. \tag{2.138}$$

Interestingly, the smoothing of the mesh density function can be incorporated directly into the equidistribution principle to give "smoothed" MMPDEs. Notice that the equidistributing coordinate transformation for $\tilde{\rho}$ satisfies

$$\frac{d}{d\xi} \left(\tilde{\rho} \frac{dx}{d\xi} \right) = 0$$

or

$$\frac{d}{d\xi}\left((S^{-1}\rho)\frac{dx}{d\xi}\right) = 0,$$

where S is the smoothing operator, i.e., $S = (I - \beta^{-2}\frac{d^2}{d\xi^2})$. Integrating the above equation yields

$$(S^{-1}\rho)\frac{dx}{d\xi} = \theta,$$

where θ is a constant. Dividing both sides by $(dx)/(d\xi)$ and applying S, we obtain

$$\rho = \theta S\left(\frac{1}{\frac{dx}{d\xi}}\right),$$

which in turn leads to

$$\frac{1}{\theta} = \frac{1}{\rho}S\left(\frac{1}{\frac{dx}{d\xi}}\right).$$

Differentiating with respect to ξ yields the smoothed equidistribution equation

$$\frac{d}{d\xi}\left(\frac{1}{\rho}S\left(\frac{1}{\frac{dx}{d\xi}}\right)\right) = 0, \tag{2.139}$$

or

$$\frac{d}{d\xi}\left(\frac{1}{\rho}(I - \beta^{-2}\frac{d^2}{d\xi^2})\left(\frac{1}{\frac{dx}{d\xi}}\right)\right) = 0. \tag{2.140}$$

Similarly, the smoothed version of (2.31) is given by

$$\frac{d}{dx}\left(\frac{1}{\rho}S\frac{d\xi}{dx}\right) = 0. \tag{2.141}$$

The same procedure can be used for MMPDEs. With the *continuous measure of mesh density* defined as $d(\xi,t) = 1/((\partial x)/(\partial \xi))$, the smoothed MMPDE

$$\frac{\partial}{\partial \xi}\left(\frac{1}{\rho}(I - \beta^{-2}\frac{\partial^2}{\partial \xi^2})d_t\right) = -\frac{1}{\tau}\frac{\partial}{\partial \xi}\left(\frac{1}{\rho}(I - \beta^{-2}\frac{\partial^2}{\partial \xi^2})d\right) \tag{2.142}$$

is studied in [188]. This fourth-order PDE is augmented with the standard boundary conditions

$$x(0,t) = a, \quad x(1,t) = b \tag{2.143}$$

and the additional ones

$$\frac{\partial d}{\partial \xi}(0,t) = \frac{\partial d}{\partial \xi}(1,t) = 0. \tag{2.144}$$

Using a central finite difference semi-discretization of (2.142) on a uniform mesh with N points and denoting

$$y_{j+\frac{1}{2}} = -\frac{\Delta\xi\left(\frac{dx_{j+1}}{dt} - \frac{dx_j}{dt}\right)}{(x_{j+1}-x_j)^2} + \frac{1}{\tau}\frac{\Delta\xi}{x_{j+1}-x_j},$$

discretization of (2.142) gives, for $j = 1,...,N-1$,

$$\frac{1}{\rho_{j+\frac{1}{2}}}\left[y_{j+\frac{1}{2}} - \frac{1}{\beta^2\Delta\xi^2}\left(y_{j+\frac{3}{2}} - 2y_{j+\frac{1}{2}} + y_{j-\frac{1}{2}}\right)\right]$$
$$-\frac{1}{\rho_{j-\frac{1}{2}}}\left[y_{j-\frac{1}{2}} - \frac{1}{\beta^2\Delta\xi^2}\left(y_{j+\frac{1}{2}} - 2y_{j-\frac{1}{2}} + y_{j-\frac{3}{2}}\right)\right] = 0, \quad (2.145)$$

with the discrete boundary conditions

$$x_0 = a, \quad x_N = b, \quad y_{-\frac{1}{2}} = y_{\frac{1}{2}}, \quad y_{N-\frac{1}{2}} = y_{N+\frac{1}{2}}, \quad (2.146)$$

which approximate (2.143) and (2.144). The following theorem is proven in [188].

Theorem 2.5.3 *Suppose that the assumptions of Theorem 2.5.1 are satisfied.*

(i) The solution to (2.142), (2.143), and (2.144) satisfies

$$\left|\left(\frac{\partial x}{\partial \xi}(\xi,t)\right)^{-1}\frac{\partial^2 x}{\partial \xi^2}(\xi,t)\right| \le \beta, \quad \forall \xi \in (0,1),\ t > 0 \quad (2.147)$$

if the inequality holds for $t = 0$.

(ii) The solution to (2.145) and (2.146) satisfies

$$\nu \le \frac{x_{j+1}(t) - x_j(t)}{x_j(t) - x_{j-1}(t)} \le \nu^{-1}, \quad j = 2,...,N-1,\ t > 0 \quad (2.148)$$

if the inequality holds initially. Here ν is given in (2.135).

Inequality (2.148) can be shown to be a discrete analog of (2.147). A mesh satisfying (2.148) is said to be *locally quasi-uniform*, and such meshes normally lead to an approximation error of the same (asymptotic) order as a uniform one – e.g., see Kautský and Nichols [208, 209].

A different way of deriving smoothed MMPDEs is to combine MPPDE5 (2.52) and MMPDE6 (2.55) in such a way as to give

$$(I - \beta^{-2}\frac{\partial^2}{\partial \xi^2})x_t = \frac{1}{\tau}\frac{\partial}{\partial \xi}\left(\rho\frac{\partial x}{\partial \xi}\right). \quad (2.149)$$

This method of smoothing has also been used in the parabolic Monge-Ampère method proposed by Budd and Williams [71]; see Chapter 6 for discussion of the method.

2.5.3 Mesh density functions for solutions with multicomponents

The mesh density function has thus far been considered for the case where the solution $u(x)$ has only a single component. When the solution has multi-components u_l, $l = 1, ..., L$, a simple way to define $\rho(x)$ is to replace $|u^{(k+1)}|^2$ with $\sum_{l=1}^{L} |u_l^{(k+1)}|^2$ in (2.95), i.e., simply use

$$\rho(x) = \left[1 + \frac{1}{\alpha}\sum_{l=1}^{L} |u_l^{(k+1)}|^2\right]^{\frac{1}{1+2(k-m+1)}} \tag{2.150}$$

for a suitable constant α. This technique, which treats all components equally, can be too limiting. A more sophisticated alternative would be to use $\sum_l w_l |u_l^{(k+1)}|^2$ with user selected weights $w_l > 0$, $l = 1, ..., L$. Yet another option would be to compute the mesh density function for each individual component with its own regularization parameter, and then to suitably combine these mesh density functions (for example, by taking their maximum). The interested reader is referred to [93, 332] for discussion on this issue.

2.5.4 Examples with analytical functions

We now present some illustrative examples for which the adaptive meshes are generated from known functions and MMPDE5xi (2.59) is then integrated for a relatively small Δt. The derivatives in the mesh density functions are approximated using the quadratic least squares fit described earlier in this section. No smoothing is applied to the mesh density functions here because we want to understand the impact of linear interpolation on the accuracy. (A smoothed mesh density function can often lead not only to a more efficient computation, but also to a more accurate computed solution.) Also, the definition (2.96) for α is utilized without any modification such as setting a floor, or minimum value.

Example 2.5.1 Consider the function defined in (2.119), i.e.,

$$u(x) = \tanh(Rx), \qquad x \in (-1, 1) \tag{2.151}$$

where R is a parameter. For $R = 100$, a typical adaptive mesh and its density $d(x)$ are shown in Figure 2.11. Figure 2.12 shows the L^2 norm and H^1 semi-norm of the linear interpolation error as a function of N for a uniform mesh and for the meshes equidistributing the optimal, arc-length, and curvature mesh density functions. The numerical results confirm that the convergence orders are $O(N^{-2})$ for the L^2 norm and $O(N^{-1})$ for the H^1 semi-norm as $N \to \infty$ for all but the arc-length mesh density function, where the error has not yet reached the full convergence rates (given in Table 2.5) for the range of values of N considered. The results also show that all adaptive meshes produce smaller error than a uniform mesh does, while the optimal and curvature mesh density functions perform comparably to each other and significantly better than arc-length.

The linear interpolation error is shown in Figure 2.13 as a function of R for $N = 321$. The results agree with the theoretical predictions given in Tables 2.5 and 2.6. In particular, the interpolation error shows different R-dependence behaviors for small and large R for the uniform mesh case and the case with the arc-length mesh density function (with the L^2 error norm). For the other cases, the error behaves the same for both small and large R.

Figure 2.14 shows the minimal spacing as a function of R for meshes equidistributing the optimal, arc-length, and curvature mesh density functions. It is easy to see that $\min_j h_j = O(1/R)$, once again confirming the theoretical prediction (2.123).

Recall from Theorem 2.4.2 that the interpolation error is bounded by

$$|u - \Pi_k u|_{H^m(a,b)} \le C N^{-(k-m+1)} \sqrt{\alpha_h},$$

where α_h can be estimated during the process of computing the mesh density function from the computed solution. To study the relation between the bound $N^{-(k-m+1)}\sqrt{\alpha_h}$ and the actual error, we show the quantities

$$\frac{\|u - \Pi_1 u\|_{L^2(\Omega)}}{N^{-2}\sqrt{\alpha_h}} \quad \text{and} \quad \frac{|u - \Pi_1 u|_{H^1(\Omega)}}{N^{-1}\sqrt{\alpha_h}}$$

as functions N in Figure 2.15 (a) and (b), respectively. These quantities correspond to the optimal mesh density functions with $(k,m) = (1,0)$ and $(1,1)$. From the figures we can see that both ratios quickly tend to constant values. This suggests that the bound $N^{-(k-m+1)}\sqrt{\alpha_h}$ can serve as a reliable error indicator in error monitoring, or in error control using mesh refinement with addition or deletion of mesh points.
□

Example 2.5.2 Consider the function

$$u(x) = x^{\frac{11}{10}}, \quad x \in [0,1].$$

Since $u(x)$ is not in $H^2(0,1)$, the classical estimate (2.66) on a uniform mesh is not valid, whereas the estimate (2.86) is bounded.

A typical adaptive mesh for $N = 21$ and its mesh density function $d(x)$ are shown in Figure 2.16. Convergence histories of the error on uniform and adaptive meshes are shown in Figure 2.17. It can be seen that the error on an adaptive mesh is not only smaller than that on a uniform mesh, but also has a better convergence rate. Indeed, the L^2 norm and H^1 semi-norm of the error converge of the order $O(N^{-2})$

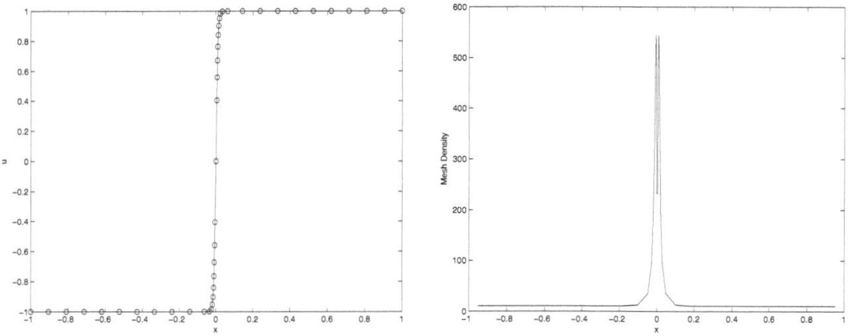

Fig. 2.11 Example 2.5.1. Using $N = 41$, $R = 100$, and the optimal mesh density function with $k = 1$ and $m = 1$, the function u and a converged adaptive mesh and its density $d(x)$ are shown.

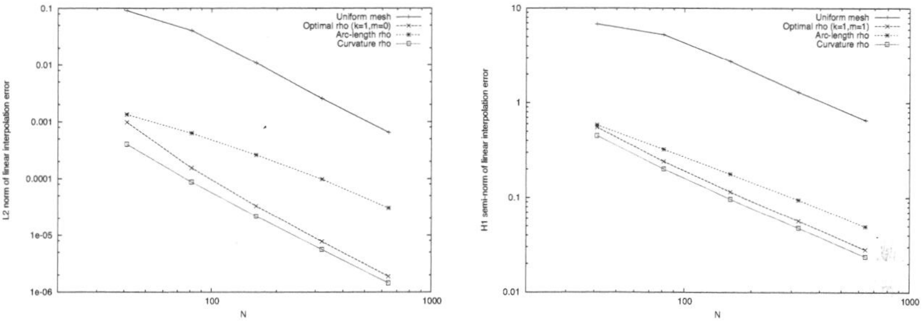

Fig. 2.12 Example 2.5.1. Using $R = 100$, the L^2 norm and H^1 semi-norm of linear interpolation error are plotted as functions of N for various mesh density functions.

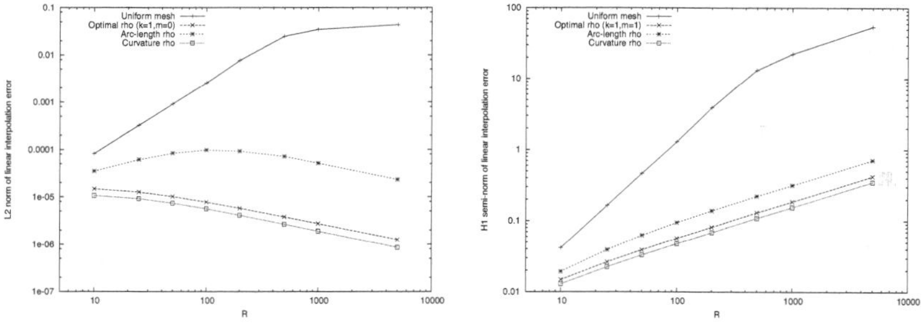

Fig. 2.13 Example 2.5.1. Using $N = 321$, the L^2 norm and H^1 semi-norm of linear interpolation error are plotted as functions of R for various mesh density functions.

and $O(N^{-1})$, respectively. This is consistent with what is predicted in the analysis in the preceding section. The error for a uniform mesh only converges of the order $O(N^{-1.6})$ for the L^2 norm and $O(N^{-0.6})$ for the H^1 semi-norm, although reasons for this particular rate of convergence remain to be investigated.

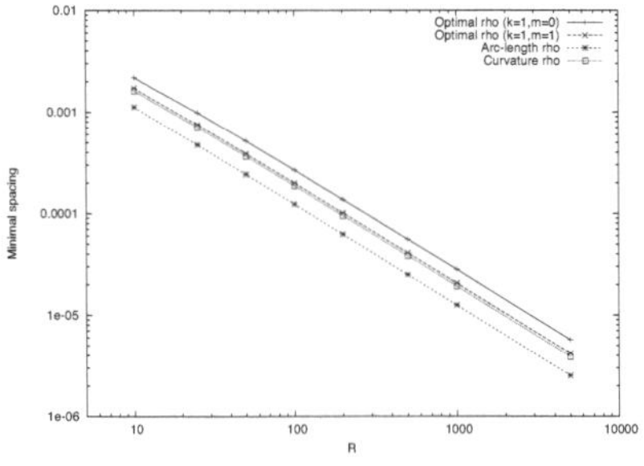

Fig. 2.14 Example 2.5.1. Using $N = 321$, the minimal spacing is plotted as function of R for various mesh density functions.

(a) $\frac{\|u-\Pi_1 u\|_{L^2(\Omega)}}{N^{-2}\sqrt{\alpha_h}}$ for $(k,m) = (1,0)$ (b) $\frac{|u-\Pi_1 u|_{H^1(\Omega)}}{N^{-1}\sqrt{\alpha_h}}$ for $(k,m) = (1,1)$

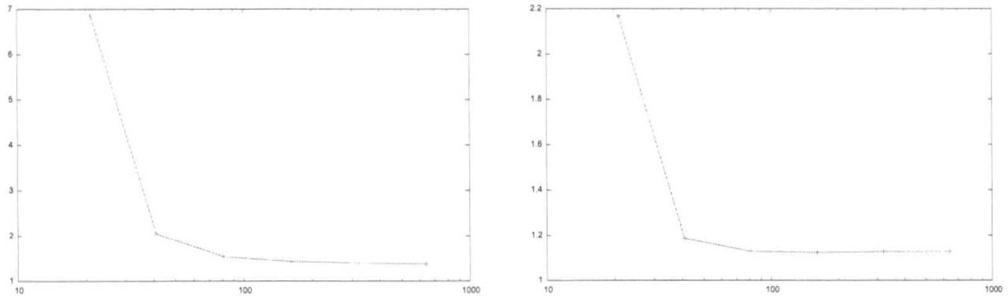

Fig. 2.15 Example 2.5.1. Ratio of the error estimate in (2.78) to the actual error for the optimal mesh density functions with $(k,m) = (1,0)$ and $(1,1)$ is plotted as function of N.

This difference in convergence rate becomes greater when the singularity of the function becomes stronger. For example, Figure 2.18 shows that the L^2 norm of the linear interpolation error for the function $u = x^{1/10}$ is $O(N^{-2})$ on an adaptive mesh but only $O(N^{-0.6})$ on a uniform mesh. The reliability of the error estimate in (2.78) is shown in Figure 2.19 (by comparing this error estimate with the actual error). □

Example 2.5.3 Finally, we take an example with the solution having two components,

$$u_1(x) = \tanh(100(x-0.3)), \quad u_2(x) = -\tanh(150(x-0.6)) \quad x \in [0,1]. \quad (2.152)$$

The numerical results are given in Figure 2.20 and 2.21. The benefit from using an adaptive mesh for this multicomponent problem is more striking. The converged equidistributed mesh is able to resolve the steep fronts, as seen in Figure 2.20, which results in an order of magnitude improvement over the uniform mesh results. Practically speaking, there are substantial savings in computation time to achieve a prescribed accuracy for the adaptive solution approximation. The reliability of the error estimate in (2.78) is shown in Figure 2.22. □

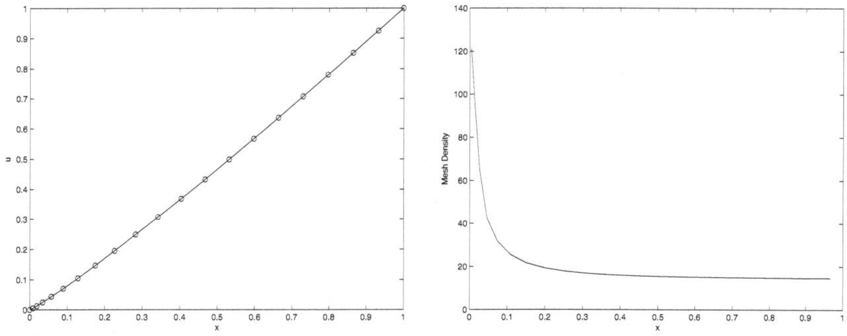

Fig. 2.16 Example 2.5.2. Using $N = 21$ and the optimal mesh density function with $k = 1$ and $m = 1$, the function u and a converged adaptive mesh and its density function $d(x)$ are shown.

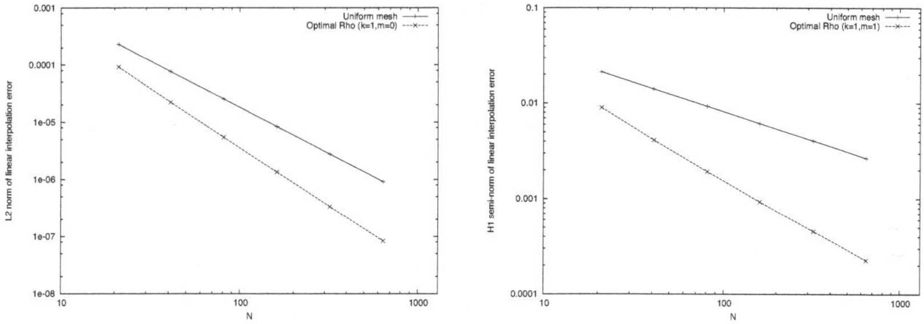

Fig. 2.17 Example 2.5.2. Convergence histories for the L^2 norm and H^1 semi-norm of the linear interpolation error is shown for uniform and adaptive meshes.

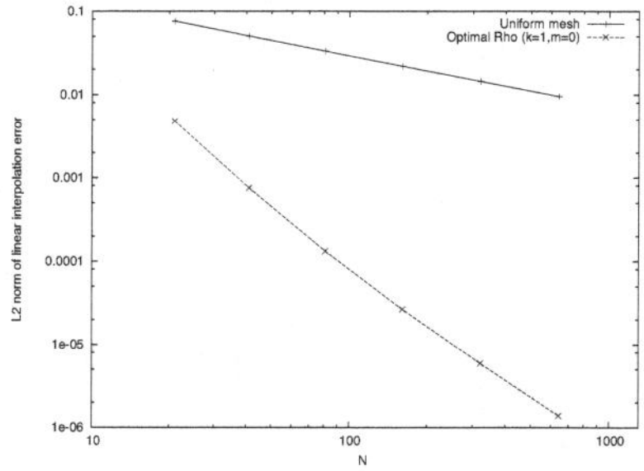

Fig. 2.18 Example 2.5.2. Convergence history for the L^2 norm of the linear interpolation error for function $u = x^{1/10}$ is shown for uniform and adaptive meshes.

2.6 Alternate solution procedures

In this chapter we have until now dealt mainly with strategies for mesh adaptation and movement. We turn next to the problem of solving a physical PDE. Recall that a moving mesh method has three major components: the moving mesh strategy, the

(a) $\frac{\|u-\Pi_1 u\|_{L^2(\Omega)}}{N^{-2}\sqrt{\alpha_h}}$ for $(k,m)=(1,0)$ (b) $\frac{|u-\Pi_1 u|_{H^1(\Omega)}}{N^{-1}\sqrt{\alpha_h}}$ for $(k,m)=(1,1)$

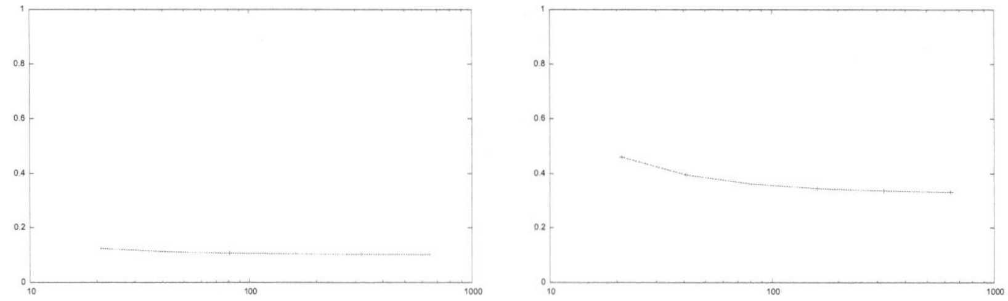

Fig. 2.19 Example 2.5.2. Ratio of the error estimate in (2.78) to the true error for the optimal mesh density functions with $(k,m)=(1,0)$ and $(1,1)$ is plotted as function of N.

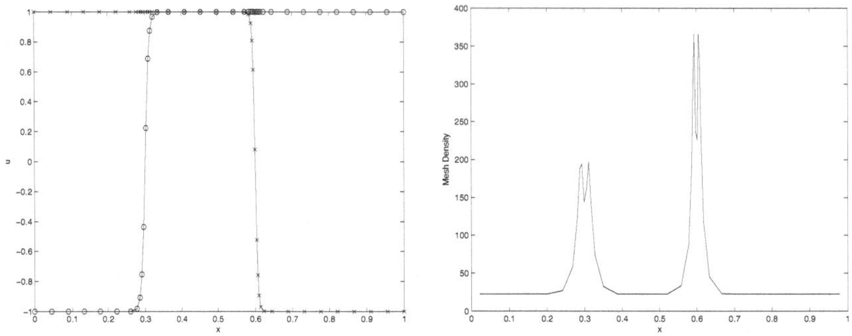

Fig. 2.20 Example 2.5.3. Using $N=41$ and the mesh density function with $k=1$ and $m=1$, the components u_1 and u_2, a converged adaptive mesh, and its density function $d(x)$ are shown.

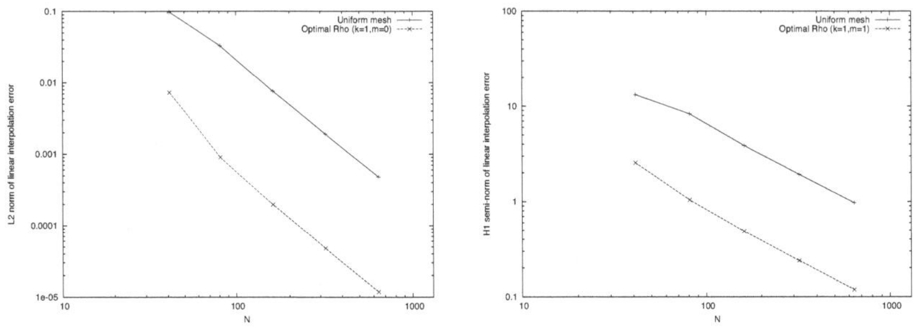

Fig. 2.21 Example 2.5.3. Convergence histories for the L^2 norm and H^1 semi-norm of the linear interpolation error are shown for uniform and adaptive meshes.

method used to discretize the physical PDE, and the procedure employed to solve the coupled system of physical and mesh equations. A brief discussion of these components has been given in §1.5. In particular, finite difference and finite element methods are used there for spatial discretization of PDEs on a moving mesh (see also §1.2 and §1.3). The effect caused by mesh movement needs special treatment

(a) $\frac{\|u-\Pi_1 u\|_{L^2(\Omega)}}{N^{-2}\sqrt{\alpha_h}}$ for $(k,m)=(1,0)$ (b) $\frac{|u-\Pi_1 u|_{H^1(\Omega)}}{N^{-1}\sqrt{\alpha_h}}$ for $(k,m)=(1,1)$

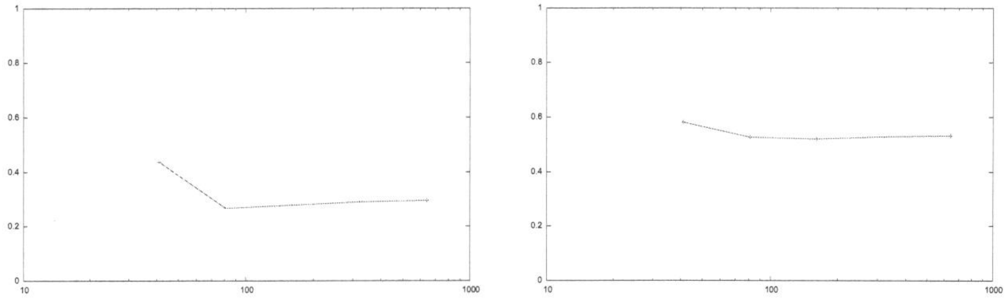

Fig. 2.22 Example 2.5.3. Ratio of the error estimate in (2.78) to the actual error for the optimal mesh density functions with $(k,m)=(1,0)$ and $(1,1)$ is plotted as function of N.

in the temporal discretization, and this typically involves taking either the quasi-Lagrange approach or the rezoning approach. With the quasi-Lagrange approach, the mesh is considered to move continuously in time (see Figure 1.11), and physical time derivatives are often transformed to time derivatives along mesh trajectories during the discretization process. The approach can be used along with simultaneous or alternate solution of the coupled system of the physical and mesh equations. On the other hand, with the rezoning approach the mesh is considered to move in an intermittent manner in time (cf. Figure 1.12), and the physical solution is interpolated from the old mesh to the new one. As a consequence, this approach is employed only with an alternate solution procedure.

The simultaneous solution procedure, along with the quasi-Lagrange treatment of mesh movement, has been used and explained in detail in Chapter 1. We focus our discussion in this section on alternate solution procedures with the quasi-Lagrange and the rezoning approaches for mesh movement in the temporal discretization of PDEs. While our interest here is in investigating these procedures in their own right for solving 1D problems, the understanding gained proves useful for solving higher dimensional problems, whose increased complexity normally necessitates resorting to some sort of alternate solution approach.

2.6.1 Alternate solution with quasi-Lagrange treatment of mesh movement

Consider first the quasi-Lagrange approach for mesh movement, along with an alternate solution procedure to solve the coupled system of physical and mesh equations. We illustrate this using the finite difference discretizations (1.13) of Burgers' equation and the modified MMPDE5 (1.18). One possible alternate temporal discretization procedure is the following:

The MP alternate solution procedure:

$$\frac{x_j^{n+1}-x_j^n}{\Delta t_n} = \frac{1}{\rho_j^n \tau \Delta\xi^2}\left[\frac{\rho_{j+1}^n+\rho_j^n}{2}(x_{j+1}^{n+1}-x_j^{n+1}) - \frac{\rho_j^n+\rho_{j-1}^n}{2}(x_j^{n+1}-x_{j-1}^{n+1})\right], \quad j=2,\ldots,N-1 \tag{2.153}$$

$$\begin{aligned}\frac{u_j^{n+1}-u_j^n}{\Delta t_n} &- \frac{1}{2}\left[\frac{(u_{j+1}^{n+1}-u_{j-1}^{n+1})}{(x_{j+1}^{n+1}-x_{j-1}^{n+1})}+\frac{(u_{j+1}^n-u_{j-1}^n)}{(x_{j+1}^n-x_{j-1}^n)}\right]\dot{x}_j^{n+1/2}\\ &= \frac{\varepsilon}{(x_{j+1}^{n+1}-x_{j-1}^{n+1})}\left[\frac{(u_{j+1}^{n+1}-u_j^{n+1})}{(x_{j+1}^{n+1}-x_j^{n+1})}-\frac{(u_j^{n+1}-u_{j-1}^{n+1})}{(x_j^{n+1}-x_{j-1}^{n+1})}\right]\\ &+ \frac{\varepsilon}{(x_{j+1}^n-x_{j-1}^n)}\left[\frac{(u_{j+1}^n-u_j^n)}{(x_{j+1}^n-x_j^n)}-\frac{(u_j^n-u_{j-1}^n)}{(x_j^n-x_{j-1}^n)}\right]\\ &- \frac{1}{4}\frac{(u_{j+1}^{n+1})^2-(u_{j-1}^{n+1})^2}{(x_{j+1}^{n+1}-x_{j-1}^{n+1})}-\frac{1}{4}\frac{(u_{j+1}^n)^2-(u_{j-1}^n)^2}{(x_{j+1}^n-x_{j-1}^n)},\\ &\qquad j=2,\ldots,N-1\end{aligned} \tag{2.154}$$

where $x_j^n \approx x(\xi_j,t_n)$, $u_j^n \approx u(x(\xi_j,t_n),t_n)$, $\rho_j^n = \rho(u_j^n)$, and $\dot{x}_j^{n+1/2} = (x_j^{n+1}-x_j^n)/\Delta t_n$. This is referred to as the *MP* procedure because the mesh equation (M) is integrated for one time step, followed by a one-step integration of the physical PDE (P).

Note that with this procedure the mesh density function is calculated using the solution approximation at $t=t_n$, which decouples the mesh equation from the physical one at the price of introducing a time lag in mesh movement. Moreover, the mesh equation is linearized by freezing the mesh density function at time $t=t_n$. As we see below, the linearization has a significant impact on the choice of the integration time step size. The M step of this MP procedure serves to generate the new mesh, $\mathscr{T}_h^{n+1}$, and non-parabolic-type strategies such as de Boor's algorithm described in §2.2, an algorithm based on MMPDE5xi (2.59), or one based on direct optimization of some error bound can be used for the same purpose. The backward Euler and midpoint discretizations are used in (2.153) and (2.154), respectively. A more accurate scheme may not be worthwhile for integrating the mesh equation because in general the location of mesh points does not need to be determined highly accurately. On the other hand, a higher order scheme is useful for improving the accuracy and efficiency of the integration of the physical PDE. The mesh and mesh speed needed in this computation can be calculated using linear interpolation in time, viz.,

$$x_j(t) = \frac{t-t_n}{\Delta t_n}x_j^{n+1} + \frac{t_{n+1}-t}{\Delta t_n}x_j^n, \quad j=1,\ldots,N \tag{2.155}$$

$$\dot{x}_j(t) = \frac{x_j^{n+1}-x_j^n}{\Delta t_n}, \quad j=1,\ldots,N \tag{2.156}$$

Table 2.7 The H^1 semi-norm of the error at $t = 1$ is listed for the Burgers' equation example in §1.4. The optimal mesh density function (2.114) for $k = 1$ and $m = 1$ is used for adaptive mesh movement.

	MP		$M^{10}P$		$M^{v}P$	
N	$\Delta t = 10^{-4}$	$\Delta t = 10^{-5}$	$\Delta t = 10^{-4}$	$\Delta t = 10^{-5}$	$\Delta t = 10^{-3}$	$\Delta t = 10^{-4}$
41	25.19	34.35	35.90	35.87	36.00	35.64
81	25.05	0.845	0.856	0.845	1.533	0.858
161	23.95	0.384	0.714	0.390	0.830	0.401
321	24.95	9.929	18.45	0.189	0.471	0.195
	PMP		$PM^{10}P$		$PM^{v}P$	
N	$\Delta t = 10^{-4}$	$\Delta t = 10^{-5}$	$\Delta t = 10^{-4}$	$\Delta t = 10^{-5}$	$\Delta t = 10^{-3}$	$\Delta t = 10^{-4}$
41	25.75	34.35	36.22	35.88	34.90	35.79
81	25.49	0.846	0.841	0.845	0.905	0.844
161	23.75	0.384	0.753	0.390	0.402	0.391
321	23.81	22.90	21.28	0.189	0.195	0.189

where $\Delta t_n = t_{n+1} - t_n$.

The main advantage of an alternate solution procedure is that by decoupling the solution of the physical equation and the mesh equation, each can be solved efficiently. Here, (2.153) is simply a tridiagonal system of linear equations for the unknown mesh locations x_j^{n+1}, while the equation (2.154) for the solution approximations u_j^{n+1} is similar to the one resulting from discretization on a fixed mesh. Unfortunately, this advantage comes with a cost: a relatively small time step size has to be used with the *MP* procedure to avoid sacrificing accuracy. This can be seen in the upper left portion of Table 2.7 where the H^1 semi-norm of the solution error for this *MP* procedure using $\Delta t = 10^{-4}$ and $\Delta t = 10^{-5}$ and various values of N is listed for the Burgers' equation example in §1.4. The *MP* and several other alternate solution procedures described below are employed for the integration of the system of mesh and physical equations. The physical PDE is integrated using a third-order SDIRK (Singly Diagonally Implicit Runge-Kutta) scheme (e.g., see [164, 165]) with a fixed time step.

The small time step size is attributable to the linearization of the mesh equation and the decoupling of the mesh movement from the physical PDE. The former results in a mesh $\mathscr{T}_h^{n+1}$ not closely satisfying the (nonlinear) equidistribution principle if Δt is large, while the latter introduces a time lag in the mesh movement.

One way to obtain a mesh more closely satisfying the equidistribution principle is to integrate the MMPDE from t_n to $t_n + \Delta t_n$ over a number of substeps. A mesh movement step with K fixed substeps of size $\Delta t_{n,k} = \Delta t_n / K$ is given as follows.

The mesh movement step of the $M^K P$ procedure:

(i) Let $\mathscr{T}_h^{n+1,1} = \mathscr{T}_h^n$.
(ii) For $k = 1,...,K$ do:

(a) Compute $\rho^{n+1,k}$ using piecewise linear interpolation for the mesh density function ρ^n on mesh $\mathscr{T}_h^n$, i.e.,

$$\rho_j^{n+1,k} = \Pi_1(\mathscr{T}_h^n, \rho^n; x_j^{n+1,k}), \qquad j = 1,....,N.$$

(b) Smooth the computed mesh density function by applying, e.g., the weighted averaging (1.23), several times.
(c) Compute $\mathscr{T}_h^{n+1,k+1}$ by solving the system consisting of

$$\frac{x_j^{n+1,k+1} - x_j^{n+1,k}}{\Delta t_{n,k}} = \frac{1}{\rho_j^{n+1,k}\tau\Delta\xi^2}\left[\frac{\rho_{j+1}^{n+1,k} + \rho_j^{n+1,k}}{2}(x_{j+1}^{n+1,k+1} - x_j^{n+1,k+1}) - \frac{\rho_j^{n+1,k} + \rho_{j-1}^{n+1,k}}{2}(x_j^{n+1,k+1} - x_{j-1}^{n+1,k+1})\right],$$
$$j = 2,...,N-1 \tag{2.157}$$

and the corresponding boundary conditions.

(iii) Set $\mathscr{T}_h^{n+1} = \mathscr{T}_h^{n+1,K+1}$.

Note that interpolation of the mesh density function from the mesh $\mathscr{T}_h^n$ to the mesh $\mathscr{T}_h^{n+1,k}$ is needed in each substep of this procedure. Piecewise linear interpolation is chosen because it preserves the monotonicity and creates no extra extrema for the mesh density function. Incidentally, while one may be tempted to interpolate ρ from the latest mesh $\mathscr{T}_h^{n+1,k-1}$ to the current one $\mathscr{T}_h^{n+1,k}$, experience has shown that this is a much less stable process than the one implemented above.

Numerical results obtained using this solution procedure with $K = 10$ are listed in Table 2.7. The improvement of the $M^{10}P$ procedure over the MP procedure is clear: the former now works well with $\Delta t = 10^{-5}$ for all N considered (aside from the fact that $N = 41$ is an insufficient number of mesh points) whereas the latter fails with the same time step size for $N = 321$.

The situation improves further when a variable substep size is used for integrating the mesh equation. Strategies for the substep size selection can be derived from local error control and/or stability considerations. However, a natural heuristic strategy is simply to require that the new mesh points $x_j^{n+1,k+1}$ move not too far from their original locations $x_j^{n+1,k}$ in each substep. For example, we may require, for $j = 2,...,N-1$,

$$\frac{1}{2}(x_j^{n+1,k} + x_{j-1}^{n+1,k}) \le x_j^{n+1,k+1} \le \frac{1}{2}(x_{j+1}^{n+1,k} + x_j^{n+1,k}),$$

or

$$x_j^{n+1,k} - \frac{1}{2}(x_j^{n+1,k} - x_{j-1}^{n+1,k}) \le x_j^{n+1,k+1} \le x_j^{n+1,k} + \frac{1}{2}(x_{j+1}^{n+1,k} - x_j^{n+1,k}). \qquad (2.158)$$

This is equivalent to

$$\Delta t_{n,k} \le \begin{cases} \frac{1}{2} \frac{(x_{j+1}^{n+1,k} - x_j^{n+1,k})}{\left|x_j^{n+1,k+1} - x_j^{n+1,k}\right| / \Delta t_{n,k}}, & \text{if } (x_j^{n+1,k+1} - x_j^{n+1,k}) > 0 \\ \frac{1}{2} \frac{(x_j^{n+1,k} - x_{j-1}^{n+1,k})}{\left|x_j^{n+1,k+1} - x_j^{n+1,k}\right| / \Delta t_{n,k}}, & \text{if } (x_j^{n+1,k+1} - x_j^{n+1,k}) < 0 \end{cases} \qquad (2.159)$$

for $j = 2, ..., N-1$. Letting

$$\overline{\Delta t_{n,k}} = \min_j \left\{ \begin{array}{ll} \frac{1}{2} \frac{(x_{j+1}^{n+1,k} - x_j^{n+1,k})}{\left|x_j^{n+1,k+1} - x_j^{n+1,k}\right| / \Delta t_{n,k}}, & \text{if } (x_j^{n+1,k+1} - x_j^{n+1,k}) > 0 \\ \frac{1}{2} \frac{(x_j^{n+1,k} - x_{j-1}^{n+1,k})}{\left|x_j^{n+1,k+1} - x_j^{n+1,k}\right| / \Delta t_{n,k}}, & \text{if } (x_j^{n+1,k+1} - x_j^{n+1,k}) < 0 \end{array} \right\}, \qquad (2.160)$$

then the constraint on the time step size (2.159) can be written as

$$\Delta t_{n,k} \le \overline{\Delta t_{n,k}}. \qquad (2.161)$$

The mesh movement step of this variable substepping control procedure (denoted by M^vP) is given in the following.

The mesh movement step of the M^vP procedure.

(i) Let $\mathscr{T}_h^{n+1,1} = \mathscr{T}_h^n$. Set $t = t_n$ and $\Delta t_{n,1} = \Delta t_n$.

(ii) For $k = 1, 2, ...$ do:

(a) Compute $\rho^{n+1,k}$ using piecewise linear interpolation for the mesh density function ρ^n on mesh $\mathscr{T}_h^n$, i.e.,

$$\rho_j^{n+1,k} = \Pi_1(\mathscr{T}_h^n, \rho^n; x_j^{n+1,k}), \qquad j = 1, ..., N.$$

(b) Smooth the computed mesh density function by applying, e.g., the weighted averaging (1.23), several times.

(c) Compute $\mathscr{T}_h^{n+1,k+1}$ by solving the system consisting of (2.157) and the corresponding boundary conditions.

(d) Compute $\overline{\Delta t_{n,k}}$ according to (2.160).

(e) If $\Delta t_{n,k} > \overline{\Delta t_{n,k}}$, set $\Delta t_{n,k} = \Delta t_{n,k}/2$, reject the current step, and go to (c). Otherwise, accept the current step, set $t = t + \Delta t_{n,k}$ and $\Delta t_{n,k+1} = \overline{\Delta t_{n,k}}$.

(f) If $t \ge t_n + \Delta t_n$, go to Step (iii). Otherwise, continue.

(iii) Set $\mathscr{T}_h^{n+1} = \mathscr{T}_h^{n+1,k+1}$.

Numerical results in Table 2.7 show significant improvement with this procedure, particularly for a relatively large fixed step size $\Delta t = 10^{-3}$, where first order convergence of the error in the H^1 semi-norm is observed when $N \geq 81$.

The time lagging problem can be overcome by iterating the alternate solutions of the mesh and physical equations several times. Such a strategy is motivated by the fact that if the iteration is continued until convergence (assuming that it is convergent), the procedure (called $(MP)^\infty$) will ultimately become a simultaneous solution procedure. Indeed, Beckett et al. [45, 46] have shown that for an initial-boundary value problem for Burgers' equation, the $(MP)^2$ and $(MP)^4$ procedures allow a larger time step size to be used in the integration of the coupled system without sacrificing accuracy. The goal is to find a balance between too many iterations, which sacrifice efficiency since each iteration requires about the same amount of work for the *MP* procedure, and too few iterations, which will not gain sufficient improvement. Unfortunately, the optimal number of iterations is very much application-dependent, and it is unclear how to choose it in practical computation.

To explain the basic iteration strategy, we consider a simple procedure called *PMP*: the physical equation is integrated on the mesh $\mathscr{T}_h^n$ (which is held fixed over the time step), followed by the *MP* procedure. The physical solution obtained on $\mathscr{T}_h^n$ can be regarded as a prediction to the solution at $t = t_{n+1}$, which hopefully produces a better new mesh $\mathscr{T}_h^{n+1}$ in the mesh movement step. For completeness, the details of this *PMP* procedure are given below.

The PMP alternate solution procedure:

$$
\begin{aligned}
\frac{\tilde{u}_j^{n+1} - u_j^n}{\Delta t_n} = {} & \frac{\varepsilon}{(x_{j+1}^n - x_{j-1}^n)} \left[\frac{(\tilde{u}_{j+1}^{n+1} - \tilde{u}_j^{n+1})}{(x_{j+1}^n - x_j^n)} - \frac{(\tilde{u}_j^{n+1} - \tilde{u}_{j-1}^{n+1})}{(x_j^n - x_{j-1}^n)} \right] \\
& + \frac{\varepsilon}{(x_{j+1}^n - x_{j-1}^n)} \left[\frac{(u_{j+1}^n - u_j^n)}{(x_{j+1}^n - x_j^n)} - \frac{(u_j^n - u_{j-1}^n)}{(x_j^n - x_{j-1}^n)} \right] \\
& - \frac{1}{4} \frac{(\tilde{u}_{j+1}^{n+1})^2 - (\tilde{u}_{j-1}^{n+1})^2}{(x_{j+1}^n - x_{j-1}^n)} - \frac{1}{4} \frac{(u_{j+1}^n)^2 - (u_{j-1}^n)^2}{(x_{j+1}^n - x_{j-1}^n)}, \\
& \qquad j = 2, \ldots, N-1
\end{aligned}
\tag{2.162}
$$

$$
\begin{aligned}
\frac{x_j^{n+1} - x_j^n}{\Delta t_n} = {} & \frac{1}{\tilde{\rho}_j^{n+1} \tau \Delta \xi^2} \left[\frac{\tilde{\rho}_{j+1}^{n+1} + \tilde{\rho}_j^{n+1}}{2} (x_{j+1}^{n+1} - x_j^{n+1}) \right. \\
& \left. - \frac{\tilde{\rho}_j^{n+1} + \tilde{\rho}_{j-1}^{n+1}}{2} (x_j^{n+1} - x_{j-1}^{n+1}) \right], \quad j = 2, \ldots, N-1
\end{aligned}
\tag{2.163}
$$

$$
\begin{aligned}
\frac{u_j^{n+1}-u_j^n}{\Delta t_n} &- \frac{1}{2}\left[\frac{(u_{j+1}^{n+1}-u_{j-1}^{n+1})}{(x_{j+1}^{n+1}-x_{j-1}^{n+1})}+\frac{(u_{j+1}^{n}-u_{j-1}^{n})}{(x_{j+1}^{n}-x_{j-1}^{n})}\right]\dot{x}_j^{n+1/2} \\
&= \frac{\varepsilon}{(x_{j+1}^{n+1}-x_{j-1}^{n+1})}\left[\frac{(u_{j+1}^{n+1}-u_{j}^{n+1})}{(x_{j+1}^{n+1}-x_{j}^{n+1})}-\frac{(u_{j}^{n+1}-u_{j-1}^{n+1})}{(x_{j}^{n+1}-x_{j-1}^{n+1})}\right] \\
&+ \frac{\varepsilon}{(x_{j+1}^{n}-x_{j-1}^{n})}\left[\frac{(u_{j+1}^{n}-u_{j}^{n})}{(x_{j+1}^{n}-x_{j}^{n})}-\frac{(u_{j}^{n}-u_{j-1}^{n})}{(x_{j}^{n}-x_{j-1}^{n})}\right] \\
&- \frac{1}{4}\frac{(u_{j+1}^{n+1})^2-(u_{j-1}^{n+1})^2}{(x_{j+1}^{n+1}-x_{j-1}^{n+1})}-\frac{1}{4}\frac{(u_{j+1}^{n})^2-(u_{j-1}^{n})^2}{(x_{j+1}^{n}-x_{j-1}^{n})}, \\
&\qquad j=2,\ldots,N-1 \qquad (2.164)
\end{aligned}
$$

where $\tilde{u}_j^{n+1} \approx u(x_j^n, t_{n+1})$ and $\tilde{\rho}_j^{n+1} = \rho(\tilde{u}_j^{n+1})$. The equations (2.163) and (2.164) are basically the same as (2.153) and (2.154) except that the mesh density function is now calculated using the solution $\tilde{u}_j^{n+1}$. Moreover, (2.162) can be obtained by setting $x_j^{n+1} := x_j^n$ in (2.164), so a code for solving (2.162) can also be used for solving (2.164).

The prediction step can also be combined with other alternate solution procedures, such as $M^K P$ and $M^\nu P$. Numerical results obtained with these procedures are listed in Table 2.7. Interestingly, this prediction step fails to allow larger time steps with PMP and $PM^{10}P$, but improvements are observed for $PM^\nu P$, where a mesh more closely satisfying the equidistribution principle is obtained.

Another issue the alternate solution procedure faces is the automatic selection of time integration steps for the the coupled system. A typical practice is to select the step size based on the accuracy of the physical solution. (See [164, 165] for time-control in Runge-Kutta schemes.) This is largely motivated by the observations that the ultimate goal of adaptive computation is to obtain an accurate physical solution instead of an accurate mesh, and the belief that it is generally unnecessary to determine the mesh at the same level of accuracy as the physical solution. Still, a major drawback of this time-step control mechanism is the complete lack of accuracy considerations for the mesh point locations. Modifications to remedy this lack of control of mesh accuracy are proposed by Blom et al. [51] and Beckett et al. [45, 46].

The time-step history is shown in Figure 2.23 for several alternate solution procedures combined with a time-step control mechanism based solely on the accuracy of the physical solution. One can see that $M^\nu P$ yields a much larger time step than MP, indicating the importance of obtaining a mesh at each time step that is close to the equidistributing mesh. Moreover, the prediction step shows significant improvements only for the $M^\nu P$ procedure.

Table 2.8 The H^1 semi-norm of the error at $t = 1$ is listed for the Burgers' equation example in §1.4. The methods are the same as for Table 2.7 except that MMPDE5xi is used for mesh adaptation.

	MP		M^2P		M^vP	
N	$\Delta t = 10^{-3}$	$\Delta t = 10^{-4}$	$\Delta t = 10^{-3}$	$\Delta t = 10^{-4}$	$\Delta t = 10^{-3}$	$\Delta t = 10^{-4}$
41	32.10	34.45	35.48	34.69	21.26	34.43
81	34.51	0.869	3.198	1.074	1.505	0.852
161	23.83	0.394	0.903	0.437	0.815	0.397
321	5.612	0.195	0.457	0.200	0.444	0.194
	PMP		PM^2P		PM^vP	
N	$\Delta t = 10^{-3}$	$\Delta t = 10^{-4}$	$\Delta t = 10^{-3}$	$\Delta t = 10^{-4}$	$\Delta t = 10^{-3}$	$\Delta t = 10^{-4}$
41	29.58	34.53	31.31	34.78	29.75	33.99
81	35.57	0.852	34.54	1.055	0.913	0.833
161	33.70	0.381	0.766	0.425	0.400	0.386
321	2.400	0.186	0.200	0.192	0.194	0.186

An interesting comparison (although not rigorous, since different integration schemes are used) can be made between the simultaneous solution in Figure 1.8 and the alternate solution in Figure 2.23(f). One can see that the time step associated with the alternate solution procedure is slightly smaller but much more erratic than that associated with the simultaneous solution procedure. This erratic time stepping behavior is partially attributed to the lack of control over mesh accuracy with the control mechanism used in the alternate solution procedure. It also causes erratic mesh movement, as seen in Figure 2.24. Relatively speaking, the mesh points within the layers move rather steadily (Figure 2.24(b)), and the error is not affected significantly by the rapid mesh oscillation (Figure 2.25 and Figure 2.26).

Recall that any algorithm based on MMPDE5xi (2.59) instead of MMPDE5 can also be used for meshing for an alternate solution procedure. Table 2.8 shows the results obtained using the algorithm described in §2.3.2. MMPDE5xi generally leads to larger time steps than MMPDE5 for the integration of the coupled system of mesh and physical PDEs. This can be partly explained by the fact that the algorithm based on MMPDE5xi produces a mesh that more accurately approximates the equidistribution relation.

To demonstrate the efficiency or cost-effectiveness of alternate solution procedures, we plot the H^1 semi-norm of the solution error at $t = 1$ as a function of the scaled CPU time in Figure 2.27. The PM^vP procedures with the modified MMPDE5 and MMPDE5xi are much more efficient than a uniform mesh method in terms of accuracy of the computed solution for a given amount of CPU time or in terms of the amount of CPU time required to reach the same level of error. In this sense, the disadvantages of smaller and more erratic time steps for alternate solution are more

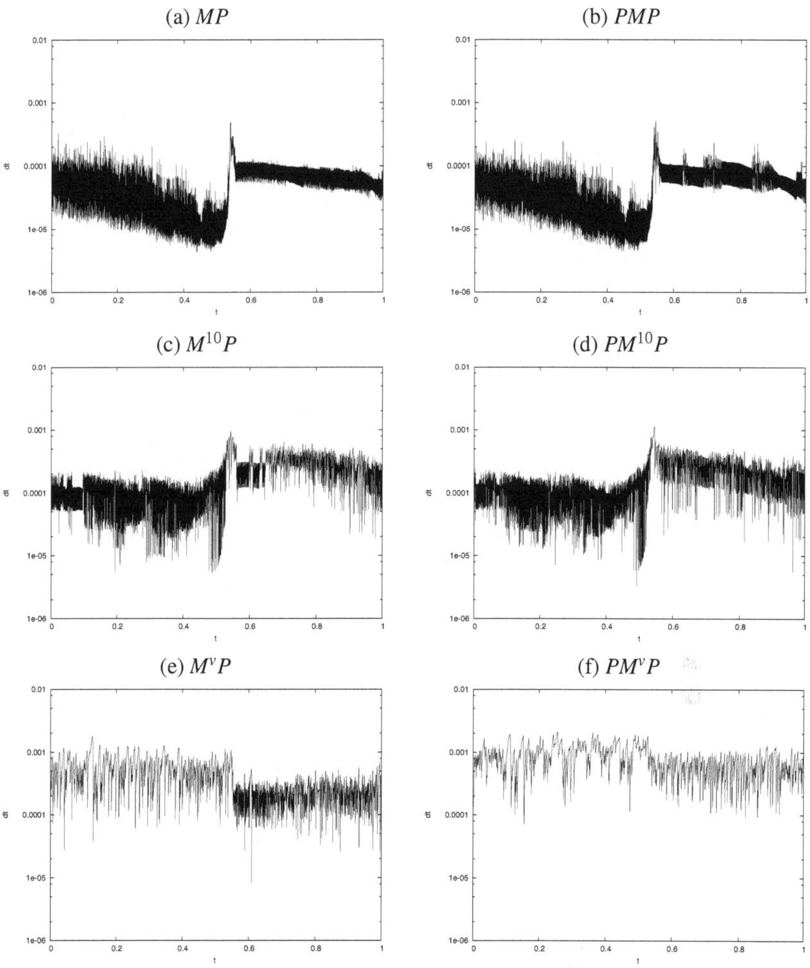

Fig. 2.23 The time step size is shown as function of time for several alternate solution procedures for the Burgers' equation example in §1.4. Absolute and relative tolerances $atol = 10^{-4}$ and $rtol = 10^{-6}$ are used in the automatic selection of time step size for the SDIRK (order 3) integration of Burgers' equation. The number of mesh points is $N = 61$.

than compensated for by the increased accuracy in the mesh and physical solution approximations at each time step. The curves in Figure 2.27 are similar to those in Figure 1.10 obtained with a simultaneous solution procedure.

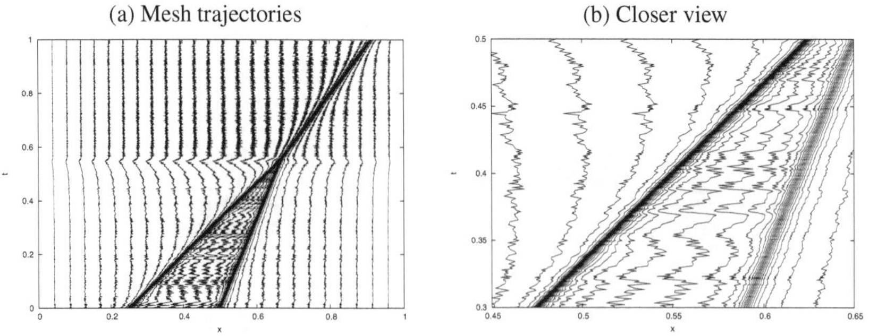

Fig. 2.24 Mesh trajectories with $N = 61$ are obtained with the PM^vP alternate solution procedure.

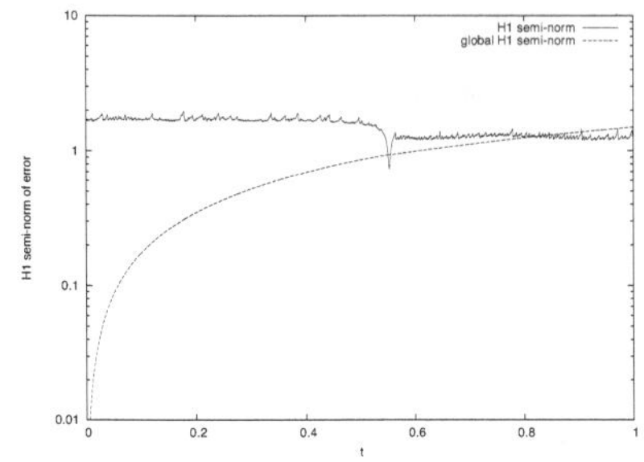

Fig. 2.25 The local and global H^1 semi-norms of the error are plotted as functions of time. The global H^1 semi-norm of error is defined as $\int_0^t |e|_{H^1(0,1)}(s)ds$.

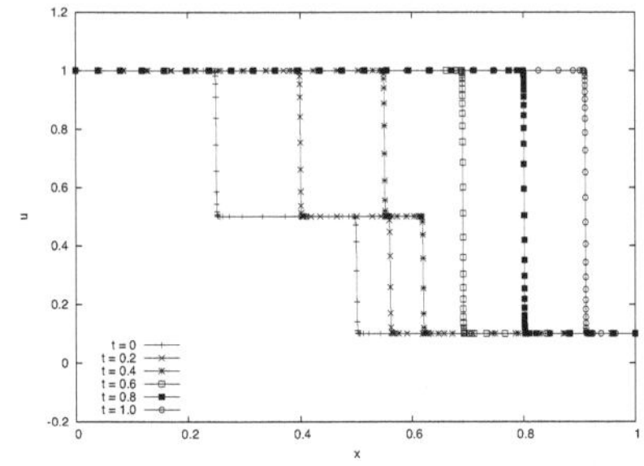

Fig. 2.26 Solutions at various time instants are computed with the PM^vP alternate solution procedure.

2.6.2 Rezoning treatment of mesh movement

We now turn our attention to the rezoning approach for temporal discretization. With this approach, the mesh moves in an *intermittent* manner. As a consequence, it is typically employed with an alternating solution procedure, and both parabolic and non-parabolic-type mesh movement strategies can be used for the adaptive mesh

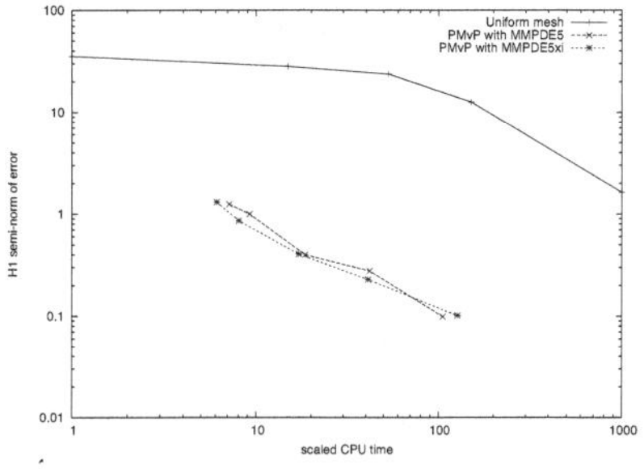

Fig. 2.27 H^1 semi-norm of error at $t = 1$ is plotted against CPU time scaled by time required for the computation in the case of 100 uniform mesh points. Alternate solution procedures are used to solve the coupled system of the mesh equation and Burgers' equation with $\varepsilon = 10^{-4}$.

computation. Once a new mesh is generated, the physical PDE is integrated for the current step with that mesh held fixed. For Burgers' equation, we solve

$$\frac{du_j}{dt} = \frac{2\varepsilon}{(x_{j+1}^{n+1} - x_{j-1}^{n+1})}\left[\frac{(u_{j+1} - u_j)}{(x_{j+1}^{n+1} - x_j^{n+1})} - \frac{(u_j - u_{j-1})}{(x_j^{n+1} - x_{j-1}^{n+1})}\right]$$
$$- \frac{1}{2}\frac{(u_{j+1}^2 - u_{j-1}^2)}{(x_{j+1}^{n+1} - x_{j-1}^{n+1})}, \qquad j = 2, \ldots, N-1 \qquad (2.165)$$

where $u_j(t) \approx u(x_j^{n+1}, t)$ for $t \in (t_n, t_{n+1}]$. Integration of this equation requires the initial data $u_j(t_n) = \tilde{u}_j^n \approx u(x_j^{n+1}, t_n)$. Since the solution approximation at $t = t_n$ is available only on the old mesh $\{x_j^n\}$, it is necessary to interpolate the solution approximation from the old mesh to the new one. Interpolation between moving meshes is discussed in the next subsection.

2.6.3 Interpolation on moving meshes

Denote the old and new meshes by $\mathscr{T}_h^0 = \{x_j^0\}$ and $\mathscr{T}_h^1 = \{x_j^1\}$, respectively. Assume that the physical solution $u = u(x)$ is available at the mesh points of $\mathscr{T}_h^0$, i.e., $u_j^0 = u(x_j^0)$. The task is to find approximations $u_j^1 \approx u(x_j^1)$ using interpolation.

A natural choice for interpolation is to use a piecewise linear or higher-degree polynomial interpolation scheme. For example, if for any mesh point x_j^1 an integer k can be found such that $x_j^1 \in [x_k^0, x_{k+1}^0]$, then the linear interpolant of u can be expressed as

$$u_j^1 = \frac{x_j^1 - x_k^0}{x_{k+1}^0 - x_k^0} u_{k+1}^0 + \frac{x_{k+1}^0 - x_j^1}{x_{k+1}^0 - x_k^0} u_k^0. \qquad (2.166)$$

On the other hand, the interpolation problem on a moving mesh can be formulated as an equivalent problem of solving a differential equation. Define a time continuation from the old mesh to the new one by

$$\mathscr{T}_h(t): \quad x_j(t) = (1-t)x_j^0 + t x_j^1, \quad t \in [0,1]. \tag{2.167}$$

Then the interpolation of $u = u(x)$ from $\mathscr{T}_h^0$ to $\mathscr{T}_h^1$ is equivalent to finding on the moving mesh $\mathscr{T}_h(t)$ the solution $v = v(x,t)$ of the differential equation

$$\frac{\partial v}{\partial t} = 0, \quad t \in (0,1] \tag{2.168}$$

subject to the initial condition $v(x_j(0),0) = u_j^0$, $j = 1,\dots,N$. The sought values u_j^1 ($j = 1,\dots,N$) are related to v by $u_j^1 = v_j^1 \approx v(x_j(1),1)$.

For the solution of (2.168) on moving mesh $\mathscr{T}_h(t)$, it is often convenient to transform it into the computational coordinate. From (1.12) we have [2]

$$\dot{v} - \frac{\dot{x}}{x_\xi} v_\xi = 0, \tag{2.169}$$

where $\dot{v} = \frac{\partial}{\partial t} v(x(\xi,t),t)\Big|_{\xi \text{ fixed}}$ and from (2.167) the mesh speed $\dot{x}$ is given by

$$\dot{x}_j = x_j^1 - x_j^0, \quad j = 1,\dots,N. \tag{2.170}$$

It can be verified that (2.169) can also be cast in the conservative form

$$\dot{\overline{(x_\xi v)}} - (\dot{x} v)_\xi = 0. \tag{2.171}$$

Thus, (2.168) becomes a hyperbolic equation on the computational domain where a fixed, uniform mesh is normally used. These equations can be discretized using finite differences, finite elements, finite volumes, or other methods. However, caution should be taken for the solution of these equations. While the total effect of the convection term in moving mesh methods is not well understood, a use of upwinding is often recommended to avoid a possible stability problem. It is well-known that an explicit, central finite difference scheme (or a finite-element equivalent) applied to a convection equation such as (2.169) or (2.171) is unconditionally unstable.

Moreover, numerical experience shows that a conservative interpolation scheme which preserves some physical solution quantity is often necessary, especially when the physical problem exhibits a strong hyperbolic feature; e.g., see Tang and Tang [316]. Consequently, a conservative scheme should generally be used for solving

[2] The composite function $\hat{v}(\xi,t) = v(x(\xi,t),t)$ is for convenience also denoted simply by v without causing confusion.

the interpolation PDE (2.168), (2.169), or (2.171). Such a scheme can more easily be derived from the conservative form (2.171) (for a more general form, see (3.25)). Indeed, integrating (2.171) over computational interval $[\xi_j, \xi_{j+1}]$ and changing variables we get

$$\frac{d}{dt}\int_{x_j(t)}^{x_{j+1}(t)} v dx = \dot{x}v|_{x_{j+1}(t)} - \dot{x}v|_{x_j(t)}.$$

Using the forward temporal discretization and approximation

$$\int_{x_j(t)}^{x_{j+1}(t)} v dx \approx v_{j+\frac{1}{2}}(t)(x_{j+1}(t) - x_j(t)),$$

we have

$$(x_{j+1}^{n+1} - x_j^{n+1})v_{j+\frac{1}{2}}^{n+1} = (x_{j+1}^n - x_j^n)v_{j+\frac{1}{2}}^n + \Delta t_n((\dot{x}v)_{j+1}^n - (\dot{x}v)_j^n), \tag{2.172}$$

where n indicates the n-th step for the continuation parameter time, and $(\dot{x}v)_j$ and $(\dot{x}v)_{j+1}$ are approximations of the "flux" $\dot{x}v$ at the cell faces $x = x_j$ and x_{j+1}, respectively. The first order upwind approximation of the flux takes the form

$$(\dot{x}v)_j = \frac{\dot{x}_j}{2}(v_{j+\frac{1}{2}} + v_{j-\frac{1}{2}}) + \frac{|\dot{x}_j|}{2}(v_{j+\frac{1}{2}} - v_{j-\frac{1}{2}}). \tag{2.173}$$

Higher order upwind approximations can also be used. The time step Δt_n should be chosen such that the Courant-Friedrichs-Lewy (CFL) condition [148] is satisfied:

$$\Delta t_n \max\left\{\frac{|\dot{x}_j^n|}{(x_{j+1}^n - x_j^n)}, \frac{|\dot{x}_j^n|}{(x_{j+1}^{n+1} - x_j^{n+1})}\right\} \le 1, \quad \forall j = 1, \ldots, N-1. \tag{2.174}$$

Scheme (2.172) defines an update for the cell centered variable $v_{j+\frac{1}{2}}$. It preserves the mass, i.e.,

$$\sum_{j=1}^{N-1}(x_{j+1}^{n+1} - x_j^{n+1})v_{j+\frac{1}{2}}^{n+1} = \sum_{j=1}^{N-1}(x_{j+1}^n - x_j^n)v_{j+\frac{1}{2}}^n.$$

This conservative interpolation scheme was proposed and used by Tang and Tang [316] for finite volume computation of hyperbolic PDEs. The current derivation is slightly different from the original one.

2.7 Examples of applications

In this section numerical results are presented for a selection of time-dependent PDEs which are standard test problems for adaptive mesh methods. These problems are solved using a moving finite difference method with the PM^vP alternate solution procedure described in the previous section. The modified MMPDE5 (2.53) is used with $\tau = 10^{-3}$ and the optimal mesh density function (2.114). The time integration is carried out using the SDIRK (order 3) scheme (see [164, 165]) with a time step selection procedure based only on the accuracy of the physical solution. The absolute and relative tolerances are chosen as $atol = rtol = 10^{-6}$.

Example 2.7.1 The first example is the advection-diffusion equation

$$\frac{\partial u}{\partial t} + V\frac{\partial u}{\partial x} = \varepsilon\frac{\partial^2 u}{\partial x^2}, \quad x \in (0,1) \tag{2.175}$$

where $\varepsilon > 0$ is the diffusion coefficient or diffusivity and V is the flow velocity, taken here as $V = 1$. Dirichlet boundary conditions at the endpoints $x = 0$ and $x = 1$ are chosen such that the exact solution of the problem is a traveling front given by

$$u(x,t) = \frac{1}{2}\text{erfc}\left(\frac{x-t}{\sqrt{4\varepsilon t}}\right) + \frac{1}{2}\exp\left(\frac{x}{\varepsilon}\right)\text{erfc}\left(\frac{x+t}{\sqrt{4\varepsilon t}}\right), \tag{2.176}$$

where $\text{erfc}(x)$ is the complementary error function. The smaller ε, the steeper the traveling front and the more difficult the problem is to solve numerically. Moreover, the exact solution is singular at $t = 0$. The integration starts at $t = 10^{-4}$ to avoid this singularity and stops at $t = 1$, when the steep front reaches the right endpoint. A solution for $\varepsilon = 10^{-5}$, computed with $N = 61$, is shown in Figure 2.28.

A closely related model is the advection-dispersion-reaction equation [195]

$$R\frac{\partial u}{\partial t} + V\frac{\partial u}{\partial x} = \varepsilon\frac{\partial^2 u}{\partial x^2} - \lambda R u, \tag{2.177}$$

where u is the concentration of a substance, ε the dispersivity, V the Darcy velocity, R the retardation factor, and λ the reaction factor. This equation models mass in the form of molecules or solid particles undergoing multiple processes in the subsurface, including advection, dispersion, and reaction. The physical parameters are taken as $\varepsilon = 10^{-5}$, $V = 1$, $R = 1.1$, and $\lambda = 1.1$ and the spatial domain is $[0,1]$. The initial and Dirichlet boundary conditions are chosen such that the exact solution is

$$\begin{aligned} u(x,t) = \frac{1}{2}\exp\left(\frac{Vx}{2\varepsilon}\right)\Bigg[&\exp\left(-\frac{x\sqrt{V^2+4\varepsilon\lambda R}}{2\varepsilon}\right)\text{erfc}\left(\frac{x-(t/R)\sqrt{V^2+4\varepsilon\lambda R}}{\sqrt{4\varepsilon t/R}}\right) \\ &+ \exp\left(\frac{x\sqrt{V^2+4\varepsilon\lambda R}}{2\varepsilon}\right)\text{erfc}\left(\frac{x+(t/R)\sqrt{V^2+4\varepsilon\lambda R}}{\sqrt{4\varepsilon t/R}}\right)\Bigg]. \end{aligned} \tag{2.178}$$

Results for the problem integrated from $t = 10^{-4}$ to $t = 1$ with $N = 61$ are shown in Figure 2.29. □

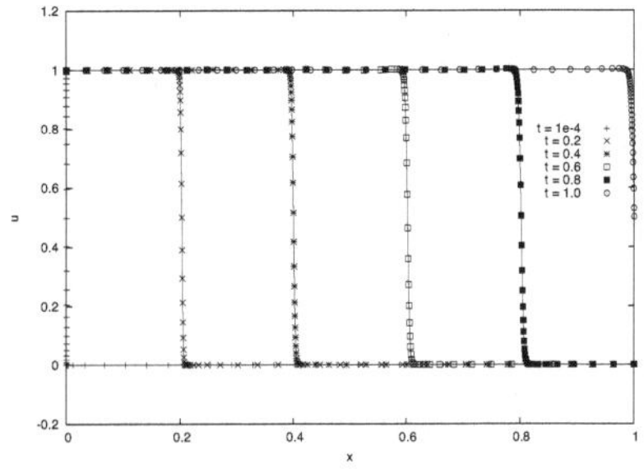

Fig. 2.28 Example 2.7.1. Computed solution of (2.175) (marked with symbols) is shown against graph of exact solution at several time instants.

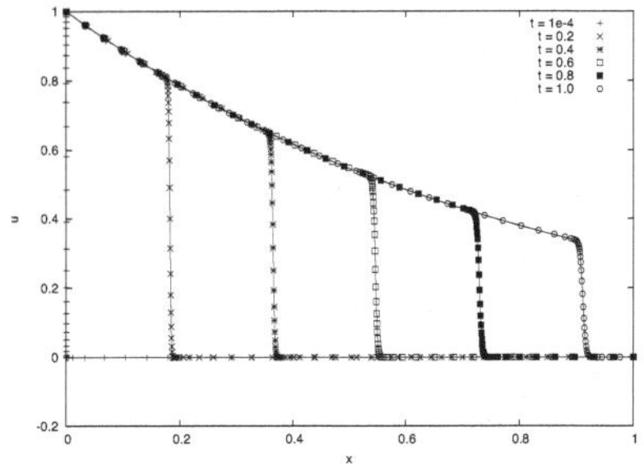

Fig. 2.29 Example 2.7.1. Computed solution of (2.177) (marked with symbols) is shown against graph of exact solution at several time instants.

Example 2.7.2 Richards' equation models the movement of water through unsaturated soil and has a dimensionless form in one dimension (e.g., see Cox and Payne [107]) as

$$\frac{\partial \theta}{\partial t} = \frac{\partial}{\partial z}\left(D(\theta)\frac{\partial \theta}{\partial z}\right) - \frac{\partial K(\theta)}{\partial z}, \tag{2.179}$$

where z is the vertical coordinate, θ is the water content, $D = D(\theta)$ is the soil water diffusivity, and $K = K(\theta)$ is the unsaturated hydraulic conductivity. We consider a case studied in [107] where $z \in [0,1]$, $t \in [0,1]$, $D = 2\theta$, and $K = \theta^3$. The initial and boundary conditions are chosen as

$$\begin{cases} \theta(z,0) = 0.15 + (0.3805 - 0.15)(1 - \tanh(1000z)), \\ \theta(0,t) = 0.3805, \quad \frac{\partial \theta}{\partial z}(1,t) = 0. \end{cases} \tag{2.180}$$

The analytical exact solution for this problem is unavailable. A solution obtained with $N = 61$ is shown in Figure 2.30.

A more difficult case where $D = 2\theta \times 10^{-4}$, $K = \theta^3 \times 2.5$, and there is a time-dependent boundary condition $\theta(0,t) = (0.3805 - 0.15)e^{-t} + 0.15$, is considered in Huang et al. [195]. A solution computed with $N = 61$ is shown in Figure 2.31. □

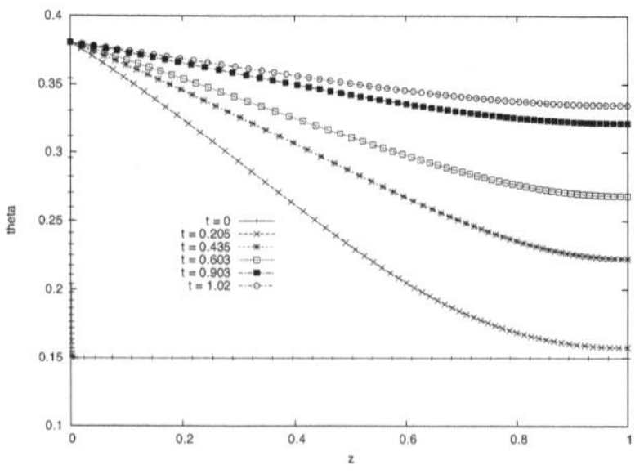

Fig. 2.30 Example 2.7.2. Solution for $D = 2\theta$, and $K = \theta^3$ is shown at several time instants.

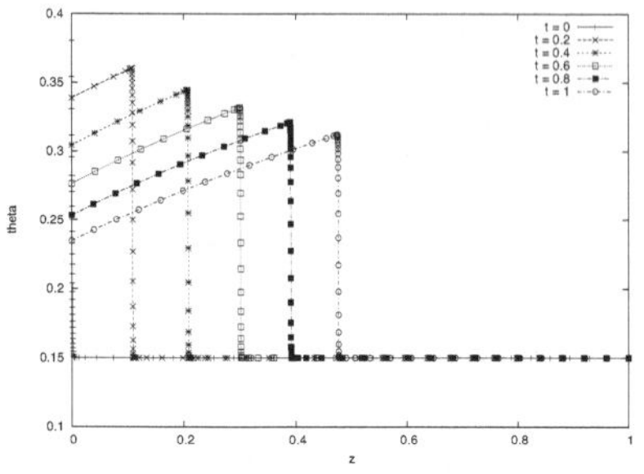

Fig. 2.31 Example 2.7.2. Solution for $D = 2\theta \times 10^{-4}$, and $K = \theta^3 \times 2.5$ is shown at several time instants.

Example 2.7.3 In this example we consider a coupled system for flow and brine transport in porous media. It models the interaction between flow and brine concentration, with a high salt concentration affecting the fluid density, and the fluid density having an impact on the fluid flow and brine transport. The governing equations for the isothermal, single-phase, two-component saturated flow model in one dimension are given by

$$\begin{aligned} \omega\rho\beta\frac{\partial P}{\partial t} + \omega\rho\gamma\frac{\partial C}{\partial t} &= -\frac{\partial(\rho v)}{\partial x}, \\ \omega\rho\frac{\partial C}{\partial t} &= -\rho v\frac{\partial C}{\partial x} + \frac{\partial}{\partial x}\left(\rho\lambda|v|\frac{\partial C}{\partial x}\right), \end{aligned} \tag{2.181}$$

where P, ρ, and v are the flow pressure, density, and velocity, respectively, and C is the salt concentration. The velocity is related to pressure through $v = -\frac{\kappa}{\mu}(\frac{\partial P}{\partial x} + \rho g)$, and the equation of state is $\rho = \rho_0 \exp(\beta(P - P_0) + \gamma C)$. Here, ρ_0 is the constant reference density, P_0 the constant reference pressure, ω the porosity, β the constant compressibility coefficient, γ the constant salt coefficient, κ the permeability, g the gravity, μ the viscosity, and λ the dispersion length. This problem has been studied by Zegeling et al. [352] and Huang et al. [195] using moving mesh techniques.

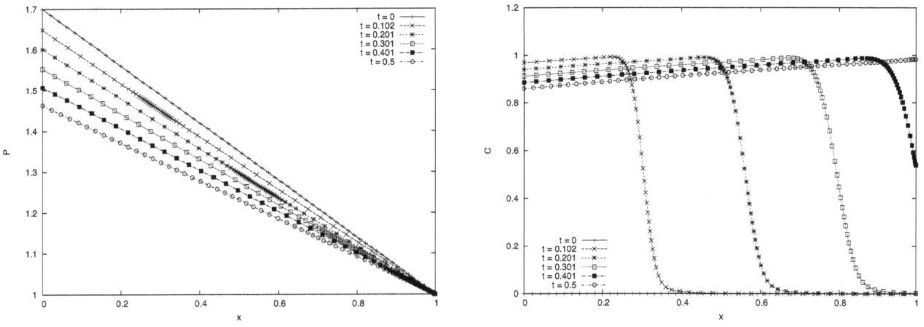

Fig. 2.32 Example 2.7.3. Computed solutions for P and C are shown at several time instants.

We consider the case where $\kappa = 1$, $\mu = 1$, $\omega = 0.2$, $\beta = 10^{-5}$, $\gamma = 0.1794$, $\lambda = 10^{-3}$, $g = 0.098$, $\rho_0 = 1$, and $P_0 = 1$. The time and space domains are taken as $[0, 0.5]$ and $[0, 1]$, respectively. The initial and boundary conditions are

$$P(x,0) = 1.7(1-x)+x, \quad C(x,0) = 0$$

and

$$P(0,t) = 1.7e^{-0.3t}, \quad P(1,t) = 1.0, \quad C(0,t) = e^{-0.3t}, \quad \frac{\partial C}{\partial x}(1,t) = 0.$$

Figure 2.32 shows the computed solution obtained with $N = 61$. □

Example 2.7.4 This example, arising in the modeling of flame propagation, involves the system of PDEs

$$\begin{aligned} \frac{\partial u}{\partial t} + f(u,v) &= \frac{\partial^2 u}{\partial x^2}, \\ \frac{\partial v}{\partial t} - f(u,v) &= \frac{\partial^2 v}{\partial x^2}, \end{aligned} \tag{2.182}$$

where $f(u,v) = 3.52 \times 10^6 \, ue^{-4/v}$ and $0 \le x \le 1$. The boundary conditions are

$$\begin{aligned} u_x(0,t) &= 0, \quad & u_x(1,t) &= 0, \\ v_x(0,t) &= 0, \quad & v(1,t) &= 1.2, \end{aligned}$$

and the initial conditions are

$$u(x,0) = 1, \quad v(x,0) = 1.2 + \tanh(1000(x-1)).$$

This model is proposed by Dwyer and Sanders [130], where $u(x,t)$ and $v(x,t)$ correspond to mass density and temperature, respectively. A constant value for the temperature at the right boundary models a heat source which generates a steep flame

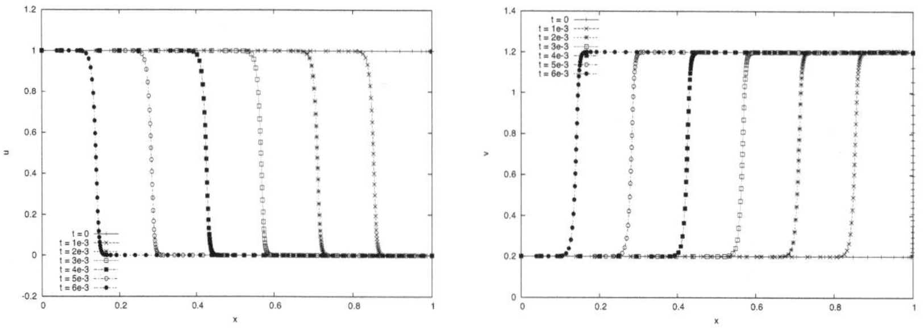

Fig. 2.33 Example 2.7.4. Computed solutions for u and v are shown at several time instants.

front. The front propagates from right to left at a relatively high velocity and reaches the left boundary slightly after $t = 0.006$. The problem is integrated up to this time, and a solution obtained with $N = 61$ is shown in Figure 2.33.

Example 2.7.5 Sod's shocktube problem [304] involves the one-dimensional Euler equations of gas dynamics in conservation form

$$\frac{\partial}{\partial t}\begin{bmatrix}\rho \\ (\rho u) \\ e\end{bmatrix} + \frac{\partial}{\partial x}\begin{bmatrix}(\rho u) \\ p+(\rho u)^2/\rho \\ (e+p)(\rho u)/\rho\end{bmatrix} = 0, \tag{2.183}$$

subject to the initial conditions

$$\begin{bmatrix}\rho \\ (\rho u) \\ e\end{bmatrix}(x,0) = \begin{cases}(1.0, 0.0, 2.5)^T, & \text{if } x \le 0.5 \\ (0.125, 0.0, 0.25)^T, & \text{if } x > 0.5\end{cases}$$

where ρ is the gas density, u velocity, e total internal energy per unit volume, and p pressure. The ideal gas equation of state is used,

$$p = 0.4(e - (\rho u)^2/(2\rho)). \tag{2.184}$$

To solve this problem with a moving (centered) finite difference method, we introduce some artificial viscosity by adding a diffusion term to each equation in (2.183), giving

$$\frac{\partial}{\partial t}\begin{bmatrix}\rho \\ (\rho u) \\ e\end{bmatrix} + \frac{\partial}{\partial x}\begin{bmatrix}(\rho u) \\ p+(\rho u)^2/\rho \\ (e+p)(\rho u)/\rho\end{bmatrix} = \varepsilon \frac{\partial^2}{\partial x^2}\begin{bmatrix}\rho \\ (\rho u) \\ e\end{bmatrix}, \tag{2.185}$$

where the small diffusion coefficient $\varepsilon > 0$ is chosen as $\varepsilon = 10^{-3}$. The initial condition is also smoothed so that

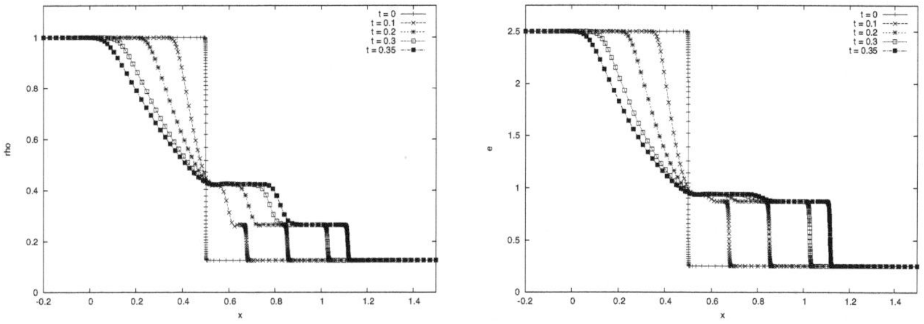

Fig. 2.34 Example 2.7.5. Computed solutions for ρ and e are shown at several time instants.

$$\begin{aligned}
\rho(x,0) &= -\frac{7}{16}\tanh\left(1000(x-0.5)\right)+\frac{9}{16},\\
(\rho u)(x,0) &= 0,\\
e(x,0) &= -\frac{9}{8}\tanh\left(1000(x-0.5)\right)+\frac{11}{8}.
\end{aligned} \tag{2.186}$$

These equations are solved over the time interval $0 \le t \le 0.35$, using artificial boundaries at $x=-0.2$ and $x=1.5$. The boundary conditions are chosen as

$$\rho_x(-0.2,t)=\rho_x(1.5,t)=0, \quad (\rho u)(-0.2,t)=(\rho u)(1.5,t)=0,$$

$$e_x(-0.2,t)=e_x(1.5,t)=0.$$

A computed solution with $N=121$ is shown in Figure 2.34. □

Example 2.7.6 Radiation diffusion equations are used to model problems in a variety of astrophysical and laboratory settings; e.g., see Mihalas and Mihalas [250] and Bowers and Wilson [56]. The equations coupling radiative diffusion and material temperature for matter at rest in the diffusion limit take a dimensionless form [56, 210] in one dimension as

$$\frac{\partial E}{\partial t} = \varepsilon\frac{\partial}{\partial x}\left(D\frac{\partial E}{\partial x}\right)+\frac{\sigma_a}{\delta}(T^4-E), \tag{2.187}$$

$$\frac{\partial T}{\partial t} = -\frac{\sigma_a}{\delta}(T^4-E), \tag{2.188}$$

where E is the dimensionless gray radiation energy density, T the dimensionless material temperature, $D=1/(3\sigma_a)$ the dimensionless diffusion coefficient, $\sigma_a = T^{-3}$, and $\varepsilon>0$ and $\delta>0$ are two dimensionless parameters. (Equations (2.187) and (2.188) are often referred to in the literature as non-equilibrium radiation diffusion equations.) A diffusion flux limiting is applied to the diffusion coefficient to avoid unphysical speed of radiation propagation. For example, Larsen's form for a flux-limited diffusion coefficient is given by

$$D_L = \left((3\sigma_a)^2 + \varepsilon\delta \left(\frac{1}{E}\frac{\partial E}{\partial x} \right)^2 \right)^{-\frac{1}{2}}. \tag{2.189}$$

Note that the sum of (2.187) and (2.188) gives the conservation equation

$$\frac{\partial (T+E)}{\partial t} = \varepsilon \frac{\partial}{\partial x}\left(D \frac{\partial E}{\partial x} \right). \tag{2.190}$$

We consider a case where the physical domain is $[0,1]$, $\varepsilon = 1$, and $\delta = 0.01$. The initial and boundary conditions are chosen as

$$E(x,0) = (1-\tanh(1000x))(1-10^{-4}) + 10^{-4}, \quad T(x,0) = E(x,0)^{\frac{1}{4}},$$

$$E(0,t) = 1, \quad E_x(1,t) = 0.$$

A similar setting has been considered by Knoll et al. [210], where the solution represents a thermal wave driven by a fixed value of E on the left boundary. A computed solution with $N = 61$ is shown in Figure 2.35.

An equilibrium situation has been studied by Lapenta and Chacón [225]. In this case, the radiation diffusion equation that assumes equilibrium between the material temperature T and radiation energy density $E = T^4$ takes the form

$$\frac{\partial}{\partial t}((1-\alpha)T + \alpha\eta E) = \eta \frac{\partial}{\partial x}\left(D \frac{\partial E}{\partial x} \right), \tag{2.191}$$

where $\alpha = 0$ and $\eta = 1$ are chosen for a matter-dominated system. The flux-limited diffusion coefficient is taken as

$$D_L = \left(3\sigma_a + \frac{1}{E}\left| \frac{\partial E}{\partial x} \right| \right)^{-1}. \tag{2.192}$$

A computed solution with $N = 61$ is shown in Figure 2.36, where the initial and boundary conditions are chosen as

$$E(x,0) = (1-0.8x)^4, \quad E(0,t) = 1, \quad E(1,t) = 0.2^4.$$

□

Example 2.7.7 Nagumo's equation [265] is used as a modeling tool in the physical sciences for problems varying from the movement of solidification fronts in material science to the propagation of action potentials in neuroscience. A dimensionless form of the equation with a cubic nonlinear reaction term in one dimension is given by

$$\frac{\partial u}{\partial t} = \varepsilon \frac{\partial^2 u}{\partial x^2} + \frac{1}{\delta} u(1-u)(u-a), \tag{2.193}$$

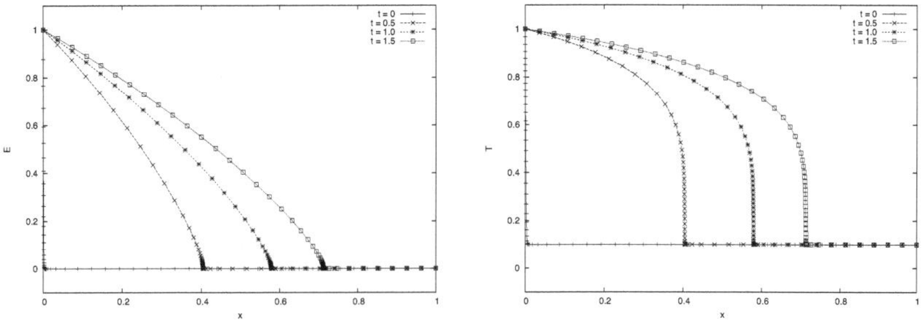

Fig. 2.35 Example 2.7.6. Computed solutions for E and T for non-equilibrium radiation diffusion are shown at several time instants.

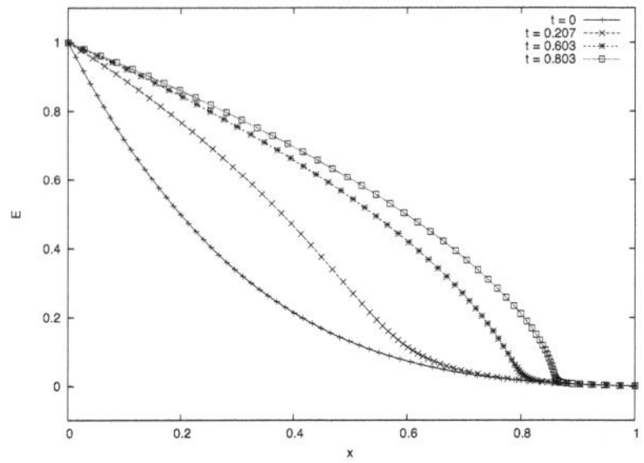

Fig. 2.36 The computed solutions obtained with $N = 61$ are shown at several time instants for equilibrium radiation diffusion in Example 2.7.6.

where $\varepsilon > 0$, $\delta > 0$, and $a \in [0,1]$ are dimensionless parameters. A major research interest for (2.193) has been to study its traveling wave solutions (e.g., see Evans [136]). One such solution is

$$u(x,t) = \frac{1}{2}\left[1 - \tanh\left(\frac{x-ct}{\sqrt{8\varepsilon\delta}}\right)\right], \tag{2.194}$$

where

$$c = (1-2a)\sqrt{\frac{\varepsilon}{2\delta}} \tag{2.195}$$

is the wave speed. Figure 2.37 shows a solution computed with $N = 61$, $\varepsilon = 10^{-3}$, $\delta = 10^{-3}$, and $a = 0$. The Dirichlet boundary conditions at $x = 0$ and $x = 1$ and the initial condition are chosen from the exact solution given in (2.194).

A closely related system is the FitzHugh-Nagumo equations [146, 265] where Nagumo's equation is coupled with an ordinary differential equation. The equations are a simplification of the Hodgkin-Huxley equation modeling the control of the electrical potential across a cell membrane. A dimensionless form of the FitzHugh-Nagumo equations in one dimension is

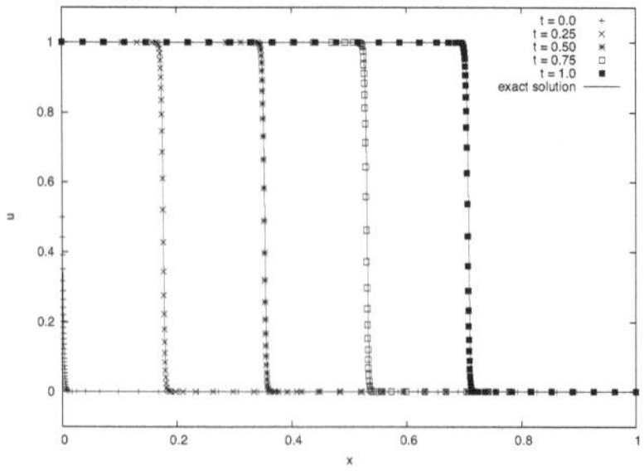

Fig. 2.37 Example 2.7.7. Computed solution of Nagumo's equation is shown at several time instants.

$$\frac{\partial u}{\partial t} = \varepsilon \frac{\partial^2 u}{\partial x^2} + \frac{1}{\delta}\left(u(1-u)(u-a) - w\right),$$
$$\frac{\partial w}{\partial t} = \beta(u - \gamma w), \tag{2.196}$$

where the dimensionless parameters $a \in [0,1]$, ε, δ, β, and γ are positive. We consider two cases. In the first case, the parameters are taken as $\varepsilon = 10^{-3}$, $\delta = 10^{-3}$, $a = 0.4$, $\beta = 0.1$, and $\gamma = 1.0$. The boundary conditions for u are

$$u_x(0,t) = u_x(1,t) = 0, \tag{2.197}$$

and the initial conditions are

$$\begin{cases} u(x,0) = 10(x-0.3)\exp(-100(x-0.35)^2), \\ w(x,0) = 0.1\exp(-100(x-0.3)^2). \end{cases} \tag{2.198}$$

Figure 2.38 shows a solution computed with $N = 321$ that simulates a traveling pulse solution.

In the second case, the parameters are taken as $\varepsilon = 10^{-3}$, $\delta = 10^{-3}$, $a = 0.1$, $\beta = 1$, and $\gamma = 0.7$. The initial conditions are

$$u(x,0) = \frac{1}{2}\tanh(1000(x-0.48)) - \frac{1}{2}\tanh(1000(x-0.52)), \quad w(x,0) = 0. \tag{2.199}$$

Figure 2.39 shows a bi-directional pulse wave solution computed with $N = 181$. □

Example 2.7.8 Stefan problems for phase change in matter are examples of moving boundary problems where a phase boundary moves with time (e.g., see Crank [109]). One scenario is icing (solidification) where the water at a uniform temperature $T_i > 0$ is confined to a half space $x > 0$ and the boundary surface at $x = 0$ is cooled down initially and then maintained at a temperature T_0 below the freezing level. The solidification process starts at the surface $x = 0$, and the liquid-solid in-

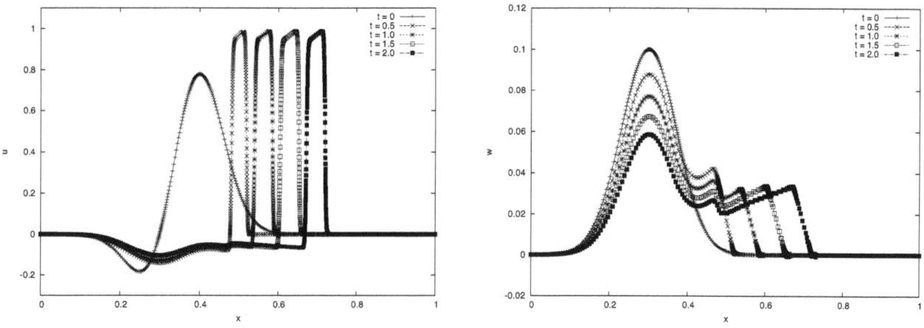

Fig. 2.38 Example 2.7.7. Computed solutions of FitzHugh-Nagumo equations are shown for boundary and initial conditions (2.197) and (2.198).

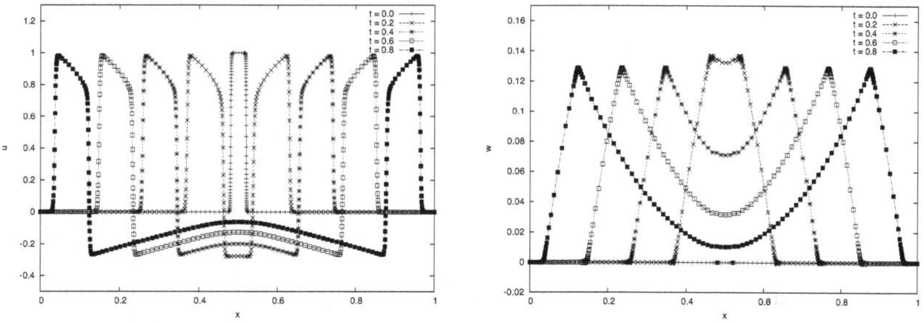

Fig. 2.39 Example 2.7.7. Computed solutions of FitzHugh-Nagumo equations are shown for boundary and initial conditions (2.197) and (2.199).

terface $x = s(t)$ moves in the positive x-direction. The one-dimensional governing equations for the solidification process in dimensionless form are

$$\frac{\partial T}{\partial t} = \frac{\partial^2 T}{\partial x^2}, \qquad 0 < x < \infty, \quad x \neq s(t) \tag{2.200}$$

$$T(s(t),t) = 0, \tag{2.201}$$

$$\frac{ds}{dt} = \frac{\partial T}{\partial x}(s(t)^-,t) - \frac{\partial T}{\partial x}(s(t)^+,t), \tag{2.202}$$

where T is the dimensionless temperature. The equations are supplemented with the constant initial condition and (initially inconsistent) boundary conditions

$$T(x,0) = T_i; \quad T(0,t) = T_0, \quad T(\infty,t) = T_i. \tag{2.203}$$

The exact solution, often referred to as Neumann's solution, is

$$T(x,t) = \begin{cases} \frac{T_0}{\operatorname{erf}(\lambda)} \left[\operatorname{erf}(\lambda) - \operatorname{erf}\left(\frac{x}{\sqrt{4t}}\right) \right], & \text{for } x \leq s(t) \\ -\frac{T_i}{1-\operatorname{erf}(\lambda)} \left[\operatorname{erf}(\lambda) - \operatorname{erf}\left(\frac{x}{\sqrt{4t}}\right) \right], & \text{for } x > s(t), \end{cases} \tag{2.204}$$

where λ is a constant satisfying

$$\frac{e^{-\lambda^2}}{\sqrt{\pi}}\left[\frac{T_0}{\operatorname{erf}(\lambda)}+\frac{T_i}{1-\operatorname{erf}(\lambda)}\right]+\lambda=0. \tag{2.205}$$

The location of the interface is given by

$$s(t)=\lambda\sqrt{4t}. \tag{2.206}$$

Direct numerical solution of (2.200)–(2.202) requires some form of front tracking to determine the location of the liquid-solid interface. Alternative approaches have been developed to avoid the need to explicitly track the moving front. We consider two of these, one based on the enthalpy formulation and the other using the phase-field equations.

The enthalpy formulation utilizes the enthalpy of the material, whose dimensionless form is related to the temperature by

$$H=\begin{cases} T, & \text{for } T<0\\ T+1, & \text{for } T>0.\end{cases} \tag{2.207}$$

The governing equation

$$\frac{\partial H}{\partial t}=\frac{\partial^2 T}{\partial x^2}, \qquad 0<x<\infty \tag{2.208}$$

is supplemented by the initial and boundary conditions (2.203). With this formulation, there is no need to explicitly impose (2.202) since the interface condition, and thus the location of the interface, is implicitly defined by (2.207) and (2.208). Indeed, (2.202) can be derived using the weak formulation of (2.208) and the fact that the solution $T(x,t)$ is smooth on the whole space-time domain except at the interface, where T is continuous but not smooth. The derivation procedure is similar to that used for deriving the Rankine-Hugoniot condition for a hyperbolic equation (e.g., see Thomas [321]).

Artificial diffusion is needed for the numerical solution of (2.208). This can be applied by regularizing the enthalpy function. An example of regularization given by Egolf and Manz [131] leads to

$$H_\varepsilon=\begin{cases} T+\frac{1}{2}e^{\frac{T}{\varepsilon}}, & \text{for } T<0\\ T+1-\frac{1}{2}e^{-\frac{T}{\varepsilon}}, & \text{for } T>0\end{cases} \tag{2.209}$$

where $\varepsilon>0$ is a small parameter. Figure 2.40 shows a solution computed with $N=81$ for $\varepsilon=10^{-4}$, $T_0=-0.06587$, $T_i=0.06587$, the spatial domain $[0,4]$, and the initial and Dirichlet boundary conditions consistent with Neumann's solution

(2.204). The integration is started from $t = 10^{-4}$ to avoid the singularity of the exact solution at $t = 0$.

For the phase-field approach, the governing equation (2.200) is coupled with a so-called phase field equation for a phase order parameter p. When the Caginalp free energy density [77] is used, the coupled system has the form

$$\frac{\partial T}{\partial t} + \frac{1}{2}\frac{\partial p}{\partial t} = \frac{\partial^2 T}{\partial x^2}, \tag{2.210}$$

$$\alpha l^2 \frac{\partial p}{\partial t} = l^2 \frac{\partial^2 p}{\partial x^2} - \frac{1}{2a}(p^3 - p) + 2T, \tag{2.211}$$

where α, l, and a are positive parameters. The solution for the phase-field model has been shown to converge to the solution of the Stefan problem as $a \to 0$, $l \to 0$, and $\sigma \equiv \frac{2l}{3a} \to 0$ while α is fixed (see Caginalp [76]). The thickness of the smeared interface is related to a and l by $\varepsilon = l\sqrt{a}$.

We consider a case studied by Mackenzie and Robertson [247], who also use a moving mesh technique. The parameters are taken as $a = 0.0625$, $l = 0.002$, and $\alpha = 1$, which give $\varepsilon = 5 \times 10^{-4}$ and $\sigma = 5.33 \times 10^{-3}$. The spatial domain is again $[0, 4]$, and the initial (at $t_0 = 10^{-4}$) and Dirichlet boundary conditions for T are chosen consistent with the exact solution (2.204) with $T_0 = -0.06587$ and $T_i = 0.06587$. The boundary conditions for p are chosen as

$$p(0,t) = \min_p \left(\frac{1}{8a}(p^2 - 1)^2 - 2pT_0 \right) \approx -1.0081, \quad \text{(closest to -1)}$$

$$p(4,t) = \min_p \left(\frac{1}{8a}(p^2 - 1)^2 - 2pT_i \right) \approx 1.0081, \quad \text{(closest to +1)}. \tag{2.212}$$

The initial condition is chosen as

$$p(x,t_0) = \begin{cases} p(0,t_0)\tanh(\frac{s(t_0)-x}{2\varepsilon}), & \text{for } x \le s(t_0) \\ p(4,t_0)\tanh(\frac{x-s(t_0)}{2\varepsilon}), & \text{for } x > s(t_0). \end{cases} \tag{2.213}$$

The computed solution using $N = 81$ and a fixed time step $\Delta t = 10^{-5}$ is shown in Figure 2.41.

□

2.8 Mesh density functions based on scaling invariance

The mesh density function in §2.4 is defined to minimize an interpolation error bound, but situations occur where it is useful to define the function motivated by

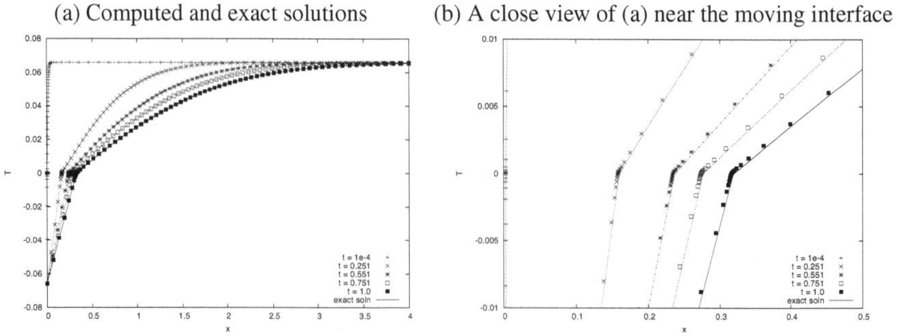

Fig. 2.40 Example 2.7.8. Computed solution of Stephan problem is shown for the enthalpy formulation.

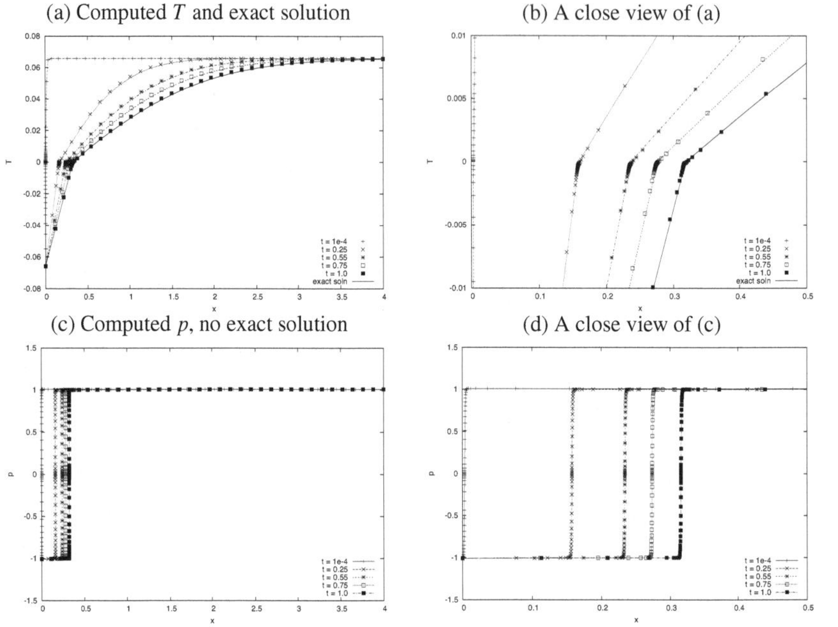

Fig. 2.41 Example 2.7.8. Computed solution of Stephan problem is shown for the phase-field formulation.

other considerations. A case in point is when solving physical PDEs having scaling invariance, which we consider here.

Scaling invariance is an important property of a broad class of PDEs having solutions that become unbounded or blow up in finite time. These PDEs arise from mathematical idealizations of models, e.g., describing combustion in chemicals, chemotaxis in cellular aggregates, and the formation of shocks in the inviscid Burgers' equation and the space-charge equations; e.g., see Pao [270]. A blowup in the solution often represents an important change in the properties of the model, such as the ignition of a heated gas mixture, and it is important that it is reproduced accurately in a numerical computation. Since a blowup typically occurs on increasingly small length scales as well as time scales, it is essential to use an adaptive mesh for its numerical simulation.

The study in this section focuses on the use of MMPDE5 in (2.52) for computing a blowup solution for the classic model problem

$$u_t = u_{xx} + u^p, \tag{2.214}$$

$$u(0,t) = u(1,t) = 0, \tag{2.215}$$

$$u(x,0) = u_0(x) > 0, \tag{2.216}$$

where $p > 1$ is a physical parameter and $u_0(x)$ is a given initial solution. However, the procedure itself is very general and has been used with a variety of MMPDEs for studying a number of physical PDEs with blowup solutions; e.g., see [67, 66, 184].

When the initial solution is sufficiently large, the solution of this initial-boundary value problem tends to infinity at a point $x^* \in (0,1)$ as $t \to T$, for some finite time $T > 0$. The quantities x^* and T are referred to as the blowup point and time, respectively.

A more precise description of the blowup profile of the solution is given in the following theorem [41].

Theorem 2.8.1 *Let $\beta = \frac{1}{p-1}$. If the initial solution is sufficiently large, then the solution to the initial-boundary value problem (2.214), (2.215), and (2.216) satisfies*

$$\lim_{t\to T}(T-t)^{\beta}u(x^* + \mu\left[(T-t)(\alpha - \log(T-t))\right]^{1/2}, t) = \beta^{\beta}\left[1 + \frac{\mu^2}{4p\beta}\right]^{-\beta} \tag{2.217}$$

uniformly for all $|\mu| \le C$ for a given constant $C > 0$, where α is a constant depending only upon the initial solution.

The theorem shows how both the time and length scales of blowup become increasingly small as $t \to T$. It also implies that the blowup profile of the solution can best be shown in the so-called kernel coordinate $\mu = \mu(x,t)$, which is fixed as $t \to T$ and defined as

$$\mu = (x - x^*)\left[(T-t)(\alpha - \log(T-t))\right]^{-1/2}. \tag{2.218}$$

The primary purpose here is finding suitable conditions under which MMPDE5 works satisfactorily. An MMPDE is judged to be satisfactory if it generates a coordinate transformation of the form

$$x(\xi,t) = x^* + (T-t)^{\frac{1}{2}}\left[\alpha - \log(T-t)\right]^{\frac{1}{2}} z(\xi,t), \tag{2.219}$$

with the property

$$z(\xi,t) = z_0(\xi) + o(1) \tag{2.220}$$

or

$$z(\xi,t) = z_0(\xi) + o(1) + O(\tau), \tag{2.221}$$

where $o(1)$ denotes terms tending to zero as $t \to T$ and $z_0(\xi)$ is a function depending only upon ξ. This is because, when the coordinate transformation is in the form (2.219) with property (2.220) or (2.221), the computational coordinate ξ is a function of μ and from Theorem 2.8.1, the solution profile in the peak region of blowup can be properly resolved in ξ.

2.8.1 Dimensional analysis, scaling invariance, and dominance of equidistribution

The main tool used in the study is dimensional analysis (e.g., see Barenblatt [37]). To begin with, we denote the dimensions of variables u, t, and x by $[u]$, $[t]$, and $[x]$, respectively. The dimensions of the terms u_t, u_{xx}, and u^p in the physical PDE (2.214) are then given by

$$[u_t] = \frac{[u]}{[t]}, \quad [u_{xx}] = \frac{[u]}{[x]^2}, \quad [u^p] = [u]^p.$$

The fact that all terms in the physical PDE are *dimensionally homogeneous* (i.e., all terms in the equation are of the same order of magnitude) implies

$$\frac{[u]}{[t]} = \frac{[u]}{[x]^2} = [u]^p.$$

This yields the dimension relations

$$[x] = [t]^{\frac{1}{2}}, \quad [u] = [t]^{-\frac{1}{p-1}} = [t]^{-\beta}. \tag{2.222}$$

Thus, if the dimension of t is changed by a factor $\lambda > 0$, the dimensions of x and u must change by factors $\lambda^{1/2}$ and $\lambda^{-\beta}$, respectively, to keep the physical equation dimensionally balanced. This suggests, and it is easy to verify, that the PDE (2.214) be invariant under the scaling transformation

$$\begin{cases} t \longrightarrow \lambda t, \\ x \longrightarrow \lambda^{\frac{1}{2}} x, \\ u \longrightarrow \lambda^{-\beta} u, \end{cases} \quad \forall \lambda > 0. \tag{2.223}$$

A common feature of scaling invariant PDEs is that they often admit so-called self-similar solutions. In the current situation, this type of solution satisfies the relation

$$u(x-x^*,T-t) = \lambda^{\beta} u(\lambda^{1/2}(x-x^*),\lambda(T-t)), \quad \forall\, \lambda > 0. \tag{2.224}$$

An example of a self-similar form is

$$u(x,t) = (T-t)^{-\beta} f\left(\frac{x-x^*}{\sqrt{T-t}}\right), \tag{2.225}$$

for a sufficiently smooth function f. The blowup profile given in (2.217) can be shown to be self-similar by using the self-similar form (2.225) and substituting the similarity variables

$$s = -log(T-t),\ y = (x-x^*)(T-t)^{-1/2},\ w(s,y) = (T-t)^{\beta} u(x,t)$$

into the PDE (2.214) (see Berger and Kohn [49]).

We now analyze the dimensions of MMPDE5 in (2.52). The computational domain for dimensional analysis can always be chosen as the unit interval, so the computational coordinate is dimensionless. Denoting the dimension of τ by $[\tau]$, the dimension equation for MMPDE5 is

$$\frac{[\tau]\,[x]}{[t]} = [\rho][x],$$

or simply

$$\frac{[\tau]}{[t]} = [\rho]. \tag{2.226}$$

For the situation where τ is taken to be constant, $[\tau] = 1$ and from (2.222),

$$[\rho] = [t]^{-1} = [u]^{\frac{1}{\beta}}. \tag{2.227}$$

This implies that MMPDE5 is invariant under the scaling transformation (2.223) if the mesh density function is chosen to be

$$\rho = u^{\frac{1}{\beta}}. \tag{2.228}$$

This type of scaling invariance based mesh density function is first used and analyzed in [67].

A generalization of (2.228) is to consider a mesh density function of the form

$$\rho = u^{\gamma}, \tag{2.229}$$

where $\gamma > 0$ is a parameter. This form satisfies the objective of concentrating more mesh points in the blowup region (where u is large) than in the rest of the domain, and we shall investigate how much flexibility there is in the choice of γ. This is motivated in part by the fact that for more complex problems, the choice of a mesh density function which gives scaling invariance is often far from straightforward. From (2.222) and (2.226), the dimension equation for (2.229) reduces to

$$[\tau]\ [t]^{\beta\gamma-1} = 1, \tag{2.230}$$

which indicates that the magnitude of the left-hand-side term of MMPDE5 in (2.52) is of order $[\tau]\ [t]^{\beta\gamma-1}$ relative to that of the right-hand-side term.

In the situation where τ is taken to be constant ($[\tau] = 1$), since the time scale of the underlying physical problem (2.214)–(2.220) can reasonably be taken as $[t] = T - t$, the left-hand side term of (2.230) vanishes as $t \to T$ when $\beta\gamma > 1$. In this case, MMPDE5 has *the dominance of equidistribution*, i.e., the equidistribution term (the right-hand side of MMPDE5) dominates the other term(s) in the equation. Obviously, this will not happen when $\beta\gamma < 1$. The critical case is $\beta\gamma = 1$, for which (2.230) is balanced, so MMPDE5 is dimensionally homogeneous and invariant under the scaling transformation (2.223). Thus, the MMPDE can be made to be equidistribution dominant by choosing constant τ sufficiently small.

The situation where τ is solution-dependent is sightly more complicated. It is discussed in §2.8.3.

2.8.2 MMPDE5 with constant τ

The dimensional analysis in the previous subsection shows that MMPDE5 has dominance of equidistribution when τ is a sufficiently small constant and the mesh density function is chosen in the form (2.229) with $\gamma\beta > 1$ or $\gamma\beta = 1$. We consider now whether or not this property is indeed sufficient to guarantee that MMPDE5 in (2.52) performs satisfactorily in practice.

When an MMPDE has the dominance of equidistribution, compared to the equidistribution term any other term is small or even vanishing as $t \to T$. It is thus reasonable when investigating blowup to only consider the equidistribution term when analyzing the MMPDE. For MMPDE5, this gives

$$\frac{\partial}{\partial \xi}\left(\rho \frac{\partial x}{\partial \xi}\right) \approx 0, \tag{2.231}$$

which is essentially the equidistribution relation (2.28). The approach of [67] is adopted here: treated as an equality, (2.231) is solved analytically using the exact form (2.217) for the solution $u(x,t)$. Since in practice the physical solution is what is sought by the computation, this approach is used for theoretical analysis only. Nevertheless, it determines what the "optimal" mesh is for the underlying initial-boundary value problem. In actual computation the mesh density function is approximated using the computed solution and the resulting mesh is thus truly adaptive. Later in this section, numerical results obtained this way are used to verify the theoretical findings.

Expanding the derivative on the left-hand side of (2.231) gives

$$\rho \frac{\partial^2 x}{\partial \xi^2} + \frac{\partial \rho}{\partial \xi} \frac{\partial x}{\partial \xi} \approx 0. \tag{2.232}$$

We seek a coordinate transformation in the form (2.219). Differentiating it with respect to ξ gives

$$x_\xi = (T-t)^{\frac{1}{2}} \left[\alpha - \log(T-t)\right]^{\frac{1}{2}} z_\xi, \tag{2.233}$$

$$x_{\xi\xi} = (T-t)^{\frac{1}{2}} \left[\alpha - \log(T-t)\right]^{\frac{1}{2}} z_{\xi\xi}. \tag{2.234}$$

Using the exact form (2.217) for the solution $u(x,t)$, from (2.219) and (2.229) we have

$$u(x,t) = (T-t)^{-\beta} \beta^\beta \left[1 + \frac{z^2}{4p\beta}\right]^{-\beta} + o\left((T-t)^{-\beta}\right), \tag{2.235}$$

$$\rho = (T-t)^{-\beta\gamma} \beta^{\beta\gamma} \left[1 + \frac{z^2}{4p\beta}\right]^{-\beta\gamma} + o\left((T-t)^{-\beta\gamma}\right), \tag{2.236}$$

$$\rho_\xi = -\frac{\gamma}{2p}(T-t)^{-\beta\gamma} \beta^{\beta\gamma} \left[1 + \frac{z^2}{4p\beta}\right]^{-\beta\gamma-1} z\, z_\xi + o\left((T-t)^{-\beta\gamma}\right). \tag{2.237}$$

Inserting these into (2.232) yields

$$\frac{d^2 z}{d\xi^2} \approx \frac{\gamma}{2p} \frac{z}{1 + \frac{z^2}{4p\beta}} \left(\frac{dz}{d\xi}\right)^2. \tag{2.238}$$

To determine suitable boundary conditions for $z(\xi)$, observe that from Theorem 2.8.1 and the form of the coordinate transformation (2.219), for any given constant $C > 0$, the mesh points $(x(\xi,t),t)$ with $|z(\xi,t)|$ and $|\dot{z}(\xi,t)| \le C$ will eventually fall in the blowup peak of the solution as t approaches T. This implies that the boundary conditions for the coordinate transformation, $x(0,t) = 0$ and $x(1,t) = 1$, must correspond to the limits of large $|z|$. Consequently, it is reasonable to choose the boundary conditions as

$$z(\xi,t) \to -\infty \text{ as } \xi \to 0; \quad z(\xi,t) \to +\infty \text{ as } \xi \to 1. \tag{2.239}$$

The boundary value problem (2.238) and (2.239) can be solved by viewing $dz/d\xi$ and z as new dependent and independent variables. The solution is

$$\frac{z}{\sqrt{4p\beta}} F\left(\frac{1}{2}, \beta\gamma; \frac{3}{2}; -\frac{z^2}{4p\beta}\right) \approx \frac{\sqrt{\pi}\,\Gamma(\beta\gamma - \frac{1}{2})}{\Gamma(\beta\gamma)} \left(\xi - \frac{1}{2}\right), \tag{2.240}$$

where $F(a,b;c;z)$ is the Gauss hypergeometric function with scalar parameters a, b, c, and z. It has the properties

$$\int_0^z \left(1+\frac{s^2}{4p\beta}\right)^{-\beta\gamma} ds = z\,F(\frac{1}{2},\beta\gamma;\frac{3}{2};-\frac{z^2}{4p\beta}),$$

$$\lim_{z\to\infty} \frac{z}{\sqrt{4p\beta}}\,F(\frac{1}{2},\beta\gamma;\frac{3}{2};-\frac{z^2}{4p\beta}) = \frac{\sqrt{\pi}}{2}\frac{\Gamma(\beta\gamma-\frac{1}{2})}{\Gamma(\beta\gamma)},$$

where Γ is the Gamma function.

For the special case $\beta\gamma = \frac{3}{2}$,

$$F(\frac{1}{2},\frac{3}{2};\frac{3}{2};-\frac{z^2}{4p\beta}) = \frac{1}{\sqrt{1+\frac{z^2}{4p\beta}}} \quad \text{and} \quad \Gamma(\frac{3}{2}) = \frac{\sqrt{\pi}}{2}.$$

From (2.240) we get

$$\frac{z}{\sqrt{1+\frac{z^2}{4p\beta}}} \approx 2\sqrt{4p\beta}(\xi-\frac{1}{2}),$$

or

$$z(\xi) \approx \frac{4\sqrt{p\beta}(\xi-\frac{1}{2})}{\sqrt{1-4(\xi-\frac{1}{2})^2}}. \tag{2.241}$$

From (2.219), we obtain the form

$$x(\xi,t) = x^* + (T-t)^{\frac{1}{2}}[\alpha - \log(T-t)]^{\frac{1}{2}} \left(\frac{4\sqrt{p\beta}(\xi-\frac{1}{2})}{\sqrt{1-4(\xi-\frac{1}{2})^2}} + o(1)\right) \tag{2.242}$$

for the coordinate transformation in the peak region of blowup. The solution in this region can be expressed in terms of the computational coordinate by

$$u(x(\xi,t),t) = (T-t)^{-\beta}\beta^{\beta}\left[\left(1-4(\xi-\frac{1}{2})^2\right)^{\beta} + o(1)\right]. \tag{2.243}$$

Arguing similarly, for the scaling invariance case $\gamma\beta = 1$ (and τ sufficiently small), the coordinate transformation and the physical solution in the peak region of blowup have the expressions

$$x(\xi,t) = x^* + (T-t)^{\frac{1}{2}}[\alpha - \log(T-t)]^{\frac{1}{2}} \times \left(\sqrt{4p\beta}\tan(\pi(\xi-\frac{1}{2})) + O(\tau) + o(1)\right), \tag{2.244}$$

$$u(x(\xi,t),t) = (T-t)^{-\beta}\beta^{\beta}\left[\cos^{2\beta}(\pi(\xi-\frac{1}{2})) + O(\tau) + o(1)\right]. \tag{2.245}$$

2.8.3 MMPDE5 with variable τ

Instead of being kept constant, there can be advantages to choosing the parameter τ as a solution-dependent function. In particular, it can be chosen such that the MMPDE is dimensionally homogeneous, even when the mesh density function has the general form (2.229). From the dimension equations (2.230) and (2.222), MMPDE5 is dimensionally balanced when τ is chosen as

$$\tau = \kappa u^{\gamma - \frac{1}{\beta}}, \tag{2.246}$$

where $\kappa > 0$ is any dimensionless parameter. With this choice of τ, all of the terms in MMPDE5 are of the same order of magnitude, and the equation has the dominance of equidistribution for κ chosen sufficiently small. In this case, for any choice of $\gamma > 0$, MMPDE5 reduces essentially to the equidistribution relation (cf. (2.232)) , and the corresponding solution is given in (2.240).

2.8.4 Numerical results

We now use MMPDE5 to solve (2.214)–(2.216) with the initial solution

$$u_0(x) = 20\sin(\pi x).$$

The physical PDE is discretized on moving meshes using a cubic spline collocation method, with the coupled system of the physical and mesh PDEs solved simultaneously (see Huang and Russell [187]). A moving mesh with $N = 41$ points is used for all of the computations.

To verify the theoretical results, the scaled solution profile, $u/\|u\|_\infty$, is plotted as a function of ξ for several values of $\|u\|_\infty$. Note that different values of $\|u\|_\infty$ correspond to different instants in time. We also plot $|x_i - x^*|$ against $\|u\|_\infty$ in a logarithmic scale. When MMPDE5 works satisfactorily, the results can be explained from (2.242) and (2.243). Specifically, as $t \to T$,

$$\|u\|_\infty \approx (T-t)^{-\beta}\beta^\beta \quad \text{or} \quad (T-t) \approx \beta \|u\|_\infty^{-\frac{1}{\beta}},$$

$$\frac{u}{\|u\|_\infty} \to \left(1 - 4(\xi - \frac{1}{2})^2\right)^\beta,$$

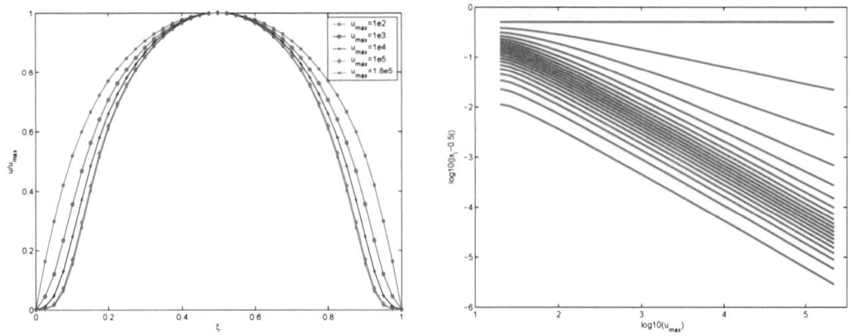

Fig. 2.42 MMPDE5, $M = u^{1.5(p-1)}$, $p = 3$, $\tau = 10^2$, $\beta\gamma = 1.5$. Reprinted from Huang et al. [184], with permission from Elsevier.

$$
\begin{aligned}
\log|x_i - x^*| \to & -\frac{1}{2\beta}\log\|u\|_\infty + \log\frac{4\sqrt{p\beta}(\xi_i - \frac{1}{2})}{\sqrt{1-4(\xi_i-\frac{1}{2})^2}} \\
& + \frac{1}{2}\log\beta + \frac{1}{2}\log\left[\alpha + \frac{1}{\beta}\log\|u\|_\infty - \log\beta\right] \\
\sim & -\frac{1}{2\beta}\log\|u\|_\infty + c_i,
\end{aligned}
$$

where c_i is a constant depending upon ξ_i. Thus, in the limit $t \to T$ the computed solution $u/\|u\|_\infty$ converges to the steady-state profile $(1-4(\xi-\frac{1}{2})^2)^\beta$, while $\log|x_i - x^*|$ becomes linear in $\log\|u\|_\infty$.

The numerical results shown in Figs 2.42–2.44 can be seen to be consistent with the theoretical predictions. Most of mesh points stay in the peak region of blowup while $u/\|u\|_\infty$ converges to a limit profile. The numerical results indicate that MMPDEs in general work satisfactorily when they have the dominance of equidistribution. For completeness, two unsatisfactory situations (not covered by the analysis in this section) are shown in Figs 2.45 and 2.46. For these, fewer and fewer mesh points are concentrated in the peak region of blowup (which is getting narrower as $t \to T$) and $u/\|u\|_\infty$ becomes more like a delta function as $t \to T$. MMPDE5 does not have the dominance of equidistribution for these two cases. Nevertheless, as shown in [184], dominance of equidistribution is a sufficient but generally not a necessary condition for an MMPDE to perform satisfactorily.

2.9 Mesh density functions based on a posteriori error estimates

Mesh density functions can be advantageously defined from a posteriori error estimates using basically the same procedure as that used in §2.4, except that we now

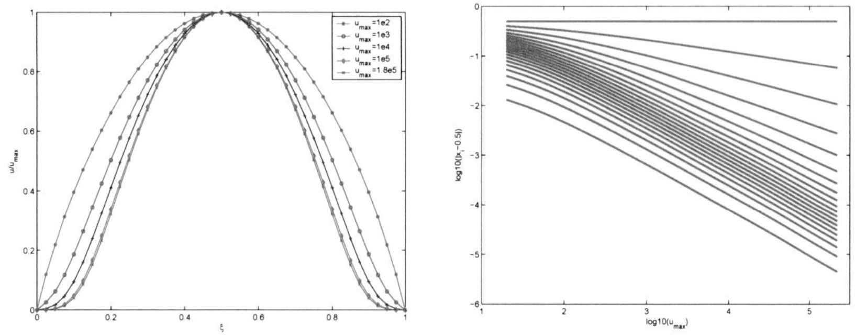

Fig. 2.43 MMPDE5, $M = u^{p-1}$, $p = 3$, $\tau = 10^{-5}$, $\beta\gamma = 1$. Reprinted from Huang et al. [184], with permission from Elsevier.

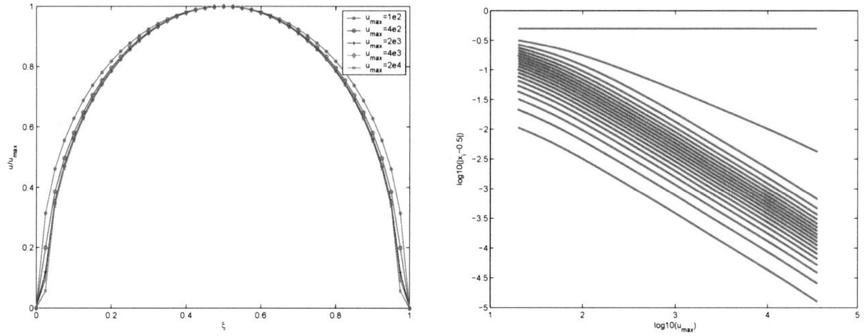

Fig. 2.44 MMPDE5, $M = u^{1.5(p-1)}$, $p = 3$, $\tau = \kappa u^{\gamma - \frac{1}{\beta}}$, $\kappa = 10^{-5}$, $\beta\gamma = 1.5$. Reprinted from Huang et al. [184], with permission from Elsevier.

use a posteriori error estimates instead of estimates for interpolation error. While this procedure is applicable to a wide range of problems, it is illustrated here for a

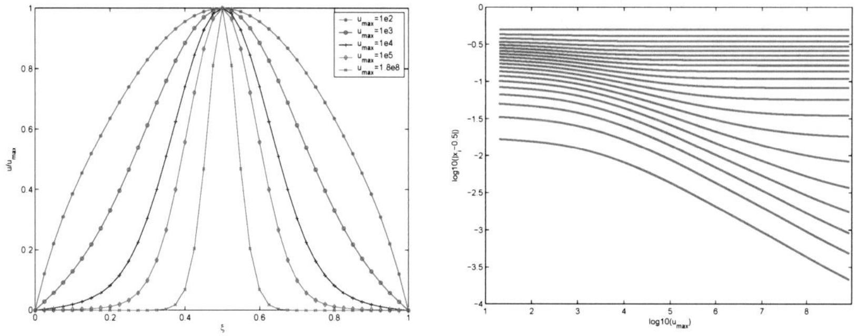

Fig. 2.45 MMPDE5, $M = u^{p-1}$, $p = 2$, $\tau = 10^{2}$, $\beta\gamma = 1$. Reprinted from Huang et al. [184], with permission from Elsevier.

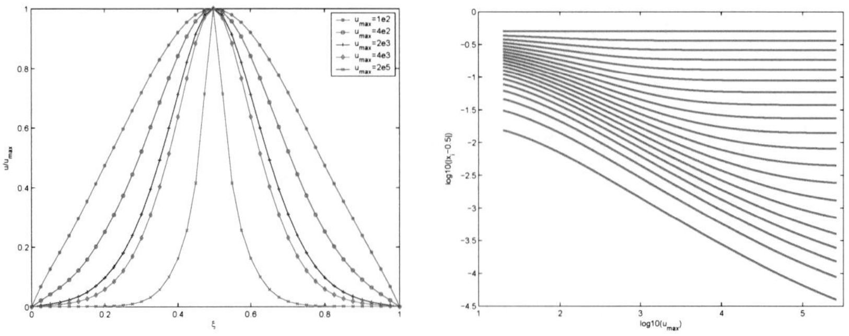

Fig. 2.46 MMPDE5, $M = u^{2(p-1)/3}$, $p = 3$, $\tau = \kappa u^{\gamma - \frac{1}{\beta}}$, $\kappa = 1$, $\beta\gamma = 2/3$. Reprinted from Huang et al. [184], with permission from Elsevier.

very simple linear elliptic differential equation. In this way, the essential features of the procedure are more easily understood in a simple context. The other reason is that theoretical tools for doing a rigorous theoretical analysis of convergence for a more general PDE have yet to be developed.

We consider the linear finite element solution of the differential equation

$$-(au')' + bu' + cu = f, \quad \text{in } \Omega = (0,1) \tag{2.247}$$

subject to the homogeneous Dirichlet boundary conditions

$$u(0) = u(1) = 0. \tag{2.248}$$

Here, we suppose that the coefficients $a(x)$, $b(x)$, $c(x)$, and $f(x)$ satisfy

$$a, b \in W^{1,\infty}(\Omega), \quad c \in L^{\infty}(\Omega), \quad f \in L^2(\Omega) \tag{2.249}$$

and

$$a(x) \geq a_0 > 0, \quad c(x) - \frac{1}{2}b'(x) \geq 0, \quad \forall x \in \Omega \tag{2.250}$$

for some constant a_0, where $W^{1,\infty}(\Omega)$ is the Sobolev space of functions whose derivatives are in $L^{\infty}(\Omega)$.

The variational formulation of BVP (2.247) and (2.248) is to find $u \in V \equiv H_0^1(\Omega)$ such that

$$B(u,v) = (f,v), \quad \forall v \in V \tag{2.251}$$

where

$$B(u,v) = \int_{\Omega} (au'v' + bu'v + cuv)dx, \quad (f,v) = \int_{\Omega} fvdx. \tag{2.252}$$

From Schwarz's inequality and assumptions (2.250), it is not difficult to show that the bilinear form $B(\cdot,\cdot)$ has the properties

$$a_0|v|^2_{H^1(\Omega)} \le B(v,v) \le C|v|^2_{H^1(\Omega)} \quad \forall v \in V \tag{2.253}$$

and

$$B(w,v) \le C|w|_{H^1(\Omega)}|v|_{H^1(\Omega)}, \quad \forall w,v \in V \tag{2.254}$$

where C is a positive constant.

For a given mesh $\mathscr{T}_h$ of N elements, one defines the linear basis functions ϕ_j, $j = 1,...,N$, as in (1.28) and the linear finite element space as $V^h = \text{span}\{\phi_2,...,\phi_{N-1}\}$. The linear finite element solution $u^h \in V^h$ satisfies

$$B(u^h,v^h) = (f,v^h) \qquad \forall v^h \in V^h. \tag{2.255}$$

Let $e^h = u - u^h$, where u is the exact solution of the continuous problem (2.251). Subtracting (2.255) from (2.251), a direct calculation gives the orthogonality property

$$B(e^h,v^h) = 0 \quad \forall v^h \in V^h \tag{2.256}$$

and the error equation

$$B(e^h,v) = (f,v) - B(u^h,v) \quad \forall v \in V. \tag{2.257}$$

2.9.1 An a priori error estimate

For comparison purpose, we derive an a priori error estimate for the finite element solution u^h as well. For this, we use a basic result for piecewise linear interpolation whose proof can be found in most finite element textbooks, e.g., [62, 104].

Lemma 2.9.1 *Denote by Π_h the operator for the piecewise linear interpolation associated with V^h, viz.,*

$$(\Pi_h v)(x) = \sum_{j=1}^{N} v(x_j)\phi_j(x) \quad \forall v \in H^1(\Omega). \tag{2.258}$$

Then, for any j $(2 \le j \le N)$,

$$\|v - \Pi_h v\|_{L^2(I_j)} \le Ch_j|v - \Pi_h v|_{H^1(I_j)}, \quad \forall v \in H^1(I_j) \tag{2.259}$$

$$|v - \Pi_h v|_{H^1(I_j)} \le Ch_j|v|_{H^2(I_j)}, \quad \forall v \in H^2(I_j) \tag{2.260}$$

$$|v - \Pi_h v|_{H^1(I_j)} \le C|v|_{H^1(I_j)}, \quad \forall v \in H^1(I_j). \tag{2.261}$$

Theorem 2.9.1 *If the solution of BVP (2.251) satisfies $u \in H^2(\Omega)$ and the mesh $\mathscr{T}_h$ has the property*

$$h \equiv \max_j h_j \leq \frac{C_1}{N} \tag{2.262}$$

for some positive constant C_1, then the error for the finite element solution u^h is bounded by

$$|u - u^h|_{H^1(\Omega)} \leq \frac{C}{N} |u|_{H^2(\Omega)}, \tag{2.263}$$

where C is a positive constant independent of u and the mesh.

Proof. From (2.253), (2.254), and the orthogonality property (2.256) it follows that

$$\begin{aligned} a_0 |e^h|^2_{H^1(\Omega)} &\leq B(e^h, e^h) \\ &= B(e^h, u - \Pi_h u) \\ &\leq C |e^h|_{H^1(\Omega)} |u - \Pi_h u|_{H^1(\Omega)}. \end{aligned}$$

From this, (2.260), and (2.262) we have

$$\begin{aligned} |e^h|^2_{H^1(\Omega)} &\leq C |u - \Pi_h u|^2_{H^1(\Omega)} \\ &= C \sum_{j=2}^{N} |u - \Pi_h u|^2_{H^1(I_j)} \\ &\leq C \sum_{j=2}^{N} h_j^2 |u|^2_{H^2(I_j)} \\ &\leq \frac{C}{N^2} |u|^2_{H^2(\Omega)}, \end{aligned}$$

which gives (2.263). □

Note that one obvious example of a mesh satisfying (2.262) is a uniform one, so the error bound (2.263) holds in this case.

2.9.2 An a posteriori error estimate

We can now derive a residual-based a posteriori error bound to be used to define the mesh density function. A general procedure for this type of error estimation can be found, for example, in [8, 26, 335].

Lemma 2.9.2 *The error for the finite element solution u^h is bounded by*

$$|u-u^h|^2_{H^1(\Omega)} \le C\eta_h^2 \equiv C\sum_{j=2}^{N} h_j^2 \|r_h\|^2_{L^2(I_j)}, \tag{2.264}$$

where r_h is the residual function, i.e.,

$$r_h = f + a'(u^h)' - b(u^h)' - cu^h. \tag{2.265}$$

Proof. From the orthogonality property (2.256) and the error equation (2.257), for any $v \in V$ we have

$$B(e^h,v) = B(e^h, v-\Pi_h v) = (f, v-\Pi_h v) - B(u^h, v-\Pi_h v).$$

From (2.252) it follows that

$$\begin{aligned} B(e^h,v) &= \int_\Omega (f - b(u^h)' - cu^h)(v-\Pi_h v)dx - \int_\Omega a(u^h)'(v-\Pi_h v)'dx \\ &= \sum_j \int_{I_j} (f - b(u^h)' - cu^h)(v-\Pi_h v)dx - \sum_j \int_{I_j} a(u^h)'(v-\Pi_h v)'dx \\ &= \sum_j \int_{I_j} r_h(v-\Pi_h v)dx, \end{aligned}$$

where the second term has been integrated by parts in the last step. Using Schwarz's inequality and Lemma 2.9.1 we get

$$\begin{aligned} B(e^h,v) &\le \sum_j \|r_h\|_{L^2(I_j)} \|v-\Pi_h v\|_{L^2(I_j)} \\ &\le C\sum_j \|r_h\|_{L^2(I_j)} h_j |v|_{H^1(I_j)} \\ &\le C\left(\sum_j h_j^2 \|r_h\|^2_{L^2(I_j)}\right)^{\frac{1}{2}} |v|_{H^1(\Omega)}. \end{aligned}$$

Then (2.264) follows by taking $v = e^h$ in the above inequality and using (2.253). □

2.9.3 Optimal mesh density function and convergence results

Once the a posteriori error bound (2.264) has been obtained, the procedure used in §2.4.2 can be used for defining the optimal mesh density function and establishing the corresponding convergence results (as in Theorems 2.4.2 and 2.4.4). Specifically, we can regularize the bound η_h as

$$\eta_h^2 \equiv \sum_{j=2}^{N} h_j^2 \|r_h\|_{L^2(I_j)}^2 \le \alpha_h \sum_{j=2}^{N} h_j^3 \left(1 + \frac{1}{\alpha_h} \langle r_h \rangle_{L^2(I_j)}^2 \right),$$

where $\langle r_h \rangle_{L^2(I_j)}$ is the L^2 average of r_h over I_j and $\alpha_h > 0$ is the regularization parameter. The optimal mesh density function and α_h can be found to be

$$\rho_j = \left(1 + \frac{1}{\alpha_h} \langle r_h \rangle_{L^2(I_j)}^2 \right)^{\frac{1}{3}}, \quad j = 2, \ldots, N \tag{2.266}$$

$$\alpha_h = \left(\sum_j h_j \langle r_h \rangle_{L^2(I_j)}^{\frac{2}{3}} \right)^3. \tag{2.267}$$

The following theorem [170] can be proved in a similar fashion as Theorems 2.4.2 and 2.4.4.

Theorem 2.9.2 *Suppose that the solution of BVP (2.251) satisfies $u \in H^2(\Omega)$. Define the mesh density function and α_h as in (2.266) and (2.267), respectively. For any equidistributing mesh satisfying*

$$h_j \rho_j = \frac{\sigma_h}{N-1}, \quad j = 2, \ldots, N \tag{2.268}$$

the error for the linear finite element solution u^h is bounded by

$$|u - u^h|_{H^1(\Omega)} \le \frac{C\sqrt{\alpha_h}}{N}, \tag{2.269}$$

where α_h satisfies

$$\lim_{N\to\infty} \sqrt{\alpha_h} = \|r\|_{L^{\frac{2}{3}}(\Omega)} \tag{2.270}$$

and r is the continuous residual function defined as

$$r = f + a'u' - bu' - cu. \tag{2.271}$$

If further $r \in L^2(\Omega) \cap W^{1,1}(\Omega)$, there exists a constant c such that, for $N > c$,

$$\sqrt{\alpha_h} \le \left(1 - \left(\frac{c}{N}\right)^{\frac{2}{3}}\right)^{-\frac{3}{2}} \left(\|r\|_{L^{\frac{2}{3}}(\Omega)}^{\frac{2}{3}} + \left(\frac{c\|r'\|_{L^1(\Omega)}}{N}\right)^{\frac{2}{3}} \right)^{\frac{3}{2}}. \tag{2.272}$$

It is interesting to point out that the residual function r is proportional to u'', i.e., $r = -au'' \sim u''$. This observation shows that the results in this theorem and those in Theorems 2.4.2 and 2.4.4 (with $k = 1$ and $m = 1$) are in good agreement although they are based on different bounds on interpolation error and solution error.

This could also be explained from the dominance of interpolation error in the finite element solution of elliptic differential equations.

2.9.4 Iterative algorithm for computing equidistributing meshes and numerical examples

The mesh density function and α_h defined in (2.266) and (2.267) depend only upon the (computable) residual, and their computation requires recovery of solution derivatives. However, the finite element solution u^h also depends upon the mesh, so u^h and $\mathscr{T}_h$ cannot be solved separately. Indeed, they are coupled through equations (2.255) and (2.268) (and the boundary conditions $x_1 = 0$ and $x_N = 1$). He and Huang [170] give a proof of the existence of the equidistributing mesh and propose the following iterative scheme to compute it.

Iterative algorithm for computing equidistributing mesh. Given an integer $N > 0$ and an initial mesh $\mathscr{T}_{h^{(0)}}$, for $k = 0, 1, ...,$ do:

(i) Find the finite element solution $u^{h^{(k)}}$ using mesh $\mathscr{T}_{h^{(k)}}$:

$$B(u^{h^{(k)}}, v^h) = (f, v^h) \quad \forall v^h \in V^{h^{(k)}}.$$

(ii) Generate the new mesh $\mathscr{T}_{h^{(k+1)}}$ using the equidistribution relation:

$$\rho_j^{(k)} h_j^{(k+1)} = \frac{\sigma_{h^{(k)}}}{N-1}, \quad j = 2, ..., N$$

where $\rho^{(k)}$ is calculated as in (2.266) based on $u^{h^{(k)}}$ and $\mathscr{T}_{h^{(k)}}$.

The stopping criteria are

$$\max_j |x_j^{(k+1)} - x_j^{(k)}| \leq TOL$$

and

$$\max_j Q_{eq,j}^{(k)} \equiv \max_j \frac{(N-1)\rho_j^{(k)} h_j^{(k)}}{\sigma_{h^{(k)}}} \leq TOL_{eq}$$

for some prescribed tolerances $TOL > 0$ and $TOL_{eq} > 1$. The numerical results presented below are obtained with $TOL = 10^{-5}$ and $TOL_{eq} = 1.01$.

Example 2.9.1 The first example is a reaction-diffusion equation

$$-\varepsilon u'' + u = -2\varepsilon - x(1-x) - 1 \tag{2.273}$$

Table 2.9 Example 2.9.1. *Iter* is the number of iterations required to reach the stopping criterion $\max_i Q_{eq,i} \leq 1.01$ or the maximum allowed number (1000 for these computations). $|u_h - u|_{H^1(\Omega)}$ is the error obtained for the final mesh in each case.

N	21	41	81	161	321	641
Iter	1000	1000	39	4	3	2
$\|u_h - u\|_{H^1(\Omega)}$	3.07	1.41	6.86e-1	3.39e-1	1.69e-1	8.46e-2

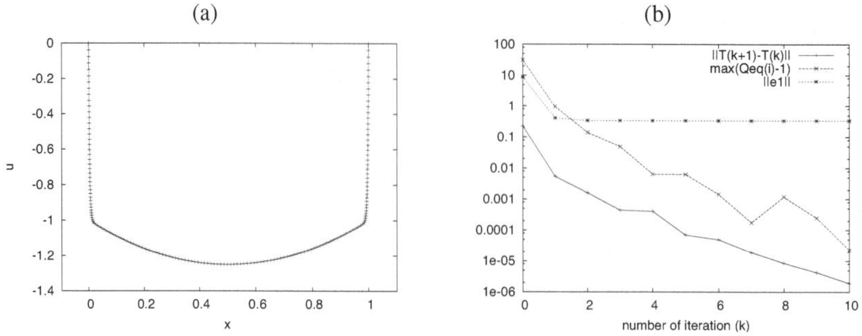

Fig. 2.47 Example 2.9.1. (a) An adaptive mesh of $N = 161$ points is plotted on the graph of the computed solution. (b) The difference between consecutive meshes ($\|\mathcal{T}_h^{(k+1)} - \mathcal{T}_h^{(k)}\|_\infty$), the equidistribution quality measure ($\max_i(Q_{eq,i} - 1)$), and the solution error ($|u_h - u|_{H^1(\Omega)}$) are plotted against the number of iterations k.

subject to the boundary condition (2.248). Using $\varepsilon = 10^{-5}$, the exact solution given by

$$u = \frac{1}{1 - e^{-\frac{2}{\sqrt{\varepsilon}}}} \left(e^{-\frac{1-x}{\sqrt{\varepsilon}}} - e^{-\frac{1+x}{\sqrt{\varepsilon}}} + e^{-\frac{x}{\sqrt{\varepsilon}}} - e^{-\frac{2-x}{\sqrt{\varepsilon}}} \right) - x(1-x) - 1 \tag{2.274}$$

exhibits boundary layers at both ends of interval $[0, 1]$.

Numerical results are shown in Table 2.9 and Figure 2.47. One can see that the iterative algorithm for computing the equidistributing mesh is convergent, at least for relatively large N. Moreover, it converges faster for larger N. This is consistent with algorithms in §2.2 for computing equidistributing meshes for an analytical mesh density function. Furthermore, the numerical results confirm the theoretical prediction that the error in the H^1 semi-norm decreases at the rate of $O(\frac{1}{N})$ as N increases. □

Example 2.9.2 The second example is a convection-dominated differential equation

$$-\varepsilon u'' + (1 - \frac{1}{2}\varepsilon)u' + \frac{1}{4}\left(1 - \frac{1}{4}\varepsilon\right) u = e^{-\frac{x}{4}}, \tag{2.275}$$

Table 2.10 Example 2.9.2. *Iter* is the number of iterations required to reach the stopping criterion $\max_i Q_{eq,i} \le 1.01$ or the maximum allowed number (1000). $|u_h - u|_{H^1(\Omega)}$ is the error obtained for the final mesh in each case.

N	21	41	81	161	321	641
Iter	1000	327	83	9	5	3
$\|u_h - u\|_{H^1(\Omega)}$	1.20	5.07e-1	2.54e-1	1.20e-1	5.95e-2	2.96e-2

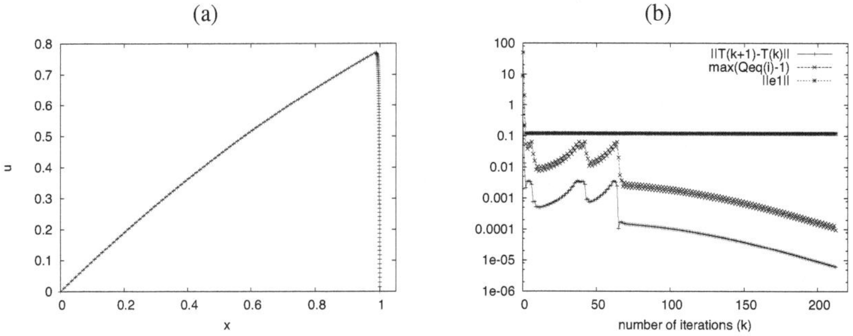

Fig. 2.48 Example 2.9.2. (a) An adaptive mesh of $N = 161$ points is plotted on the graph of the computed solution. (b) The difference between consecutive meshes ($\|\mathcal{T}_h^{(k+1)} - \mathcal{T}_h^{(k)}\|_\infty$), the equidistribution quality measure ($\max_i(Q_{eq,i} - 1)$), and the solution error ($|u_h - u|_{H^1(\Omega)}$) are plotted against the number of iterations k.

where $\varepsilon = 2 \times 10^{-3}$. For boundary conditions (2.248), the exact solution is given by

$$u = e^{-\frac{x}{4}} \left(x - \frac{e^{-\frac{1-x}{\varepsilon}} - e^{-\frac{1}{\varepsilon}}}{1 - e^{-\frac{1}{\varepsilon}}} \right), \tag{2.276}$$

which has the boundary layer at $x = 1$ when ε is small. Numerical results are shown in Table 2.10 and Figure 2.48. □

Example 2.9.3 This example, used by Babuška and Rheinboldt [25], has the form

$$-((x+\alpha)^p u')' + (x+\alpha)^q u = f, \tag{2.277}$$

where f is chosen such that the exact solution of the boundary value problem (with boundary condition (2.248)) is

$$u = (x+\alpha)^r - (\alpha^r(1-x) + (1+\alpha)^r x). \tag{2.278}$$

In our computation, the parameters are taken as $p = 2$, $q = 1$, $r = -1$, and $\alpha = 1/100$. Numerical results are shown in Table 2.11 and Figure 2.49. □

Table 2.11 Example 2.9.3. *Iter* is the number of iterations required to reach the stopping criterion $\max_i Q_{eq,i} \le 1.01$ or the maximum allowed number (1000). $|u_h - u|_{H^1(\Omega)}$ is the error obtained for the final mesh in each case.

N	21	41	81	161	321	641
Iter	4	3	3	2	2	2
$\lvert u_h - u\rvert_{H^1(\Omega)}$	2.15e2	1.13e2	5.73e1	2.88e1	1.44e1	7.20

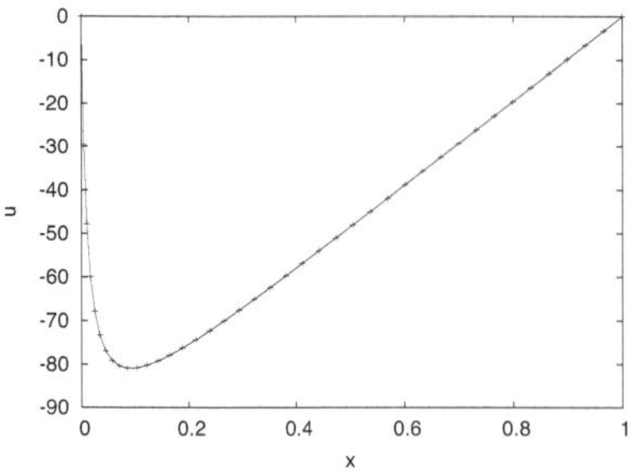

Fig. 2.49 Example 2.9.3. An adaptive mesh of $N = 161$ points is plotted on the graph of the computed solution.

2.10 Biographical notes

The concept of equidistribution is first introduced by Burchard [73] for finding variable nodes for optimal spline approximations. Some early studies of equidistribution, mostly in one spatial dimension, can be found in Rice [284], de Boor [115, 116], Dodson [121], Sacks and Ylvisaker [290, 291, 292], Pereyra and Sewell [273], Babuška and Rheinboldt [24], and Russell and Christiansen [285]. In particular, Babuška and Rheinboldt [24] lay much of the early groundwork for mesh optimality.

Approaches similar to that in §2.1.4 are used by D'Azevedo and Simpson [112], Huang and Sloan [192], and Huang [176] for developing multi-dimensional equidistribution relations.

The simplest algorithm for computing an equidistributing mesh is attributed to de Boor [116]. Various discrete and continuous forms of equidistribution, the associated two-point boundary value problem method, as well as the MMPDE approach can be found in Ren and Russell [282] and Huang, Ren, and Russell [185, 186]. The reader's attention is also brought to the relation between the MMPDE approach and the DAE approach using Baumgarte regularization [39] and the implications of solving DAEs. Furthermore, a number of MMPDEs are derived in [186] under the unifying framework of the MMPDE approach, with some of them being closely related to previous 1D moving mesh methods. For example, MM-

PDE5 (2.52) is used by Anderson [11] while the methods of Adjerid and Flaherty [5, 6], Madsen [248], and Greenberg [158], which are based on attraction and repulsion pseudo-forces, can be recast into the form of MMPDE6 (2.55). The method of Hyman and Larrouturou [197, 198] also simulates attraction and repulsion pseudo-forces but is linked to MMPDE4 (2.54). Other related methods include those of [10, 108, 124, 147, 172, 281, 282].

A group of 1D methods has been motivated by the Lagrange method in fluid dynamics. They are typically formulated by minimizing convection or convection-like terms; e.g., see the review article [169] and Chapter 7 (for discussion of velocity-based methods in multi-dimensions). The method of Petzold [174] defines the mesh velocity by minimizing the time variation of both the unknown variable and the spatial coordinate in computational coordinates.

Error estimates for interpolation have been frequently used for mesh adaptation in the past, e.g., see [9, 162] and the early works [73, 115, 116, 121, 284]. In addition, Dinh and Carey [87] define adaptive coordinate transformations as minimizers of interpolation error bounds, Chen [97] obtains mesh density functions by minimizing some error estimates, and Ji [206] determines optimal mesh density functions to minimize the error in approximating integrals. A systematic study of optimal mesh density functions in multi-dimensions is given by Huang and Sun [193] and Huang [178, 179]. Overall mesh quality measures are introduced in [178].

Beckett and Mackenzie [42] appear to be the first to use a global or integral definition of the adaptation intensity parameter α. It is extensively studied and extended to multi-dimensions in [175, 195].

The issue of the efficiency or the cost-effectiveness of moving mesh methods is addressed in a large number of works; e.g., see Huang [175], Beckett et al. [45], Huang et al. [195], and the more extensive study by Lapenta and Chacón [225]. The efficiency of a moving mesh method can be improved by using a two-level-mesh strategy [175, 195] where the physical PDE is solved on a fine mesh obtained by interpolating a coarser, moving mesh.

A rezoning approach has been developed and used by Tang and his coworkers for discretizing physical PDEs; e.g., see [228, 229, 316, 318].

The first works on the numerical simulation of solution blowup are Nakagawa [266] and Nakagawa and Ushijima [267], where finite difference and finite element schemes on a uniform mesh are employed and analyzed for blowup for PDE (2.214) with $p = 2$. A mesh refinement strategy is proposed by Berger and Kohn [49] and a moving mesh method is presented by Budd et al. [67] for the numerical solution of blowup problems. A key idea in [67] is to define the mesh density function using the scaling invariance argument. The concept is generalized in Huang et al. [184] where dominance of equidistribution is introduced and shown to be sufficient for an MMPDE to work satisfactorily. An early survey on the topic is given by Bandle and Brunner [34]. Other recent works include [3, 59, 65, 66, 71].

In addition to (2.214), a variety of PDEs with blowup solution have been solved using moving mesh methods. They include other reaction-diffusion type PDEs [67], degenerate parabolic problems [67], porous medium equations [64, 67, 151, 152, 352], nonlinear Schrödinger equations [63, 65, 280], chemotaxis equations [66], and Ginzburg-Landau equations [70].

Convergence analysis for moving mesh methods is still in the very early stages, and most results have been limited to the situation where the mesh is known a priori. For example, Qiu and Sloan [278] and Qiu et al. [279] study the convergence of finite difference schemes for singularly perturbed two-point boundary value problems on equidistributing meshes generated using a singular part of the exact solution. The approach is adopted by a number of other researchers; e.g., see [42, 43, 44, 100, 177, 244].

There are a few convergence analyses where a posteriori equidistributing meshes, or equidistributing meshes determined by the computed solution, are considered. Most noticeably, Babuška and Rheinboldt [25] consider the linear finite element solution of a 1D linear elliptic problem and determine the optimal coordinate transformation as a minimizer of a functional derived from a residual-based a posteriori error estimate using asymptotic approximation. Using such a coordinate transformation, they show that a mesh is asymptotically optimal if the residual-based error estimate is evenly distributed among the mesh elements. Kopteva and Stynes [219] study an upwind finite difference discretization of 1D quasi-linear convection-diffusion problems without turning points and develop a convergence analysis for the discretization where the mesh is determined by the computed solution through the equidistribution principle and the arc-length mesh density function. The approach used in §2.9 is adopted from He and Huang [170] where equidistributing meshes are determined and the corresponding convergence is analyzed based on a posteriori error bounds for the linear finite element solution of 1D linear elliptic problems.

A general framework for convergence analysis for parabolic problems on moving meshes can be seen in Dupont [127], Bank and Santos [35], Ferreira [143], and Liu et al. [128, 242], where moving meshes are assumed to satisfy certain smoothness conditions. The stability issue is investigated by Ferreira [143], Formaggia and Nobile [149], and Mackenzie and Mekwi [245].

In addition to finite difference and finite element methods, collocation, finite volume, and spectral methods have also been used for moving mesh methods. Examples include Huang and Russell [187] and Russell et al. [286] (collocation), Farhat et al. [140, 141] and Wang et al. [336] (finite volume), and Mulholland et al. [263], Wang and Shen [338], Feng et al. [142], and Tee and Trefethen [320] (spectral).

2.11 Exercises

1. Find the equidistributing coordinate transformation for the mesh density function

$$\rho(x) = \frac{1}{\pi\sqrt{1-x^2}} \quad \text{on } [-1,1].$$

Show that the maximal and minimal spacings for an equidistributing mesh generated using the coordinate transformation on a uniform mesh of $N+1$ points in ξ are of the orders $O(N^{-1})$ and $O(N^{-2})$. Where do these spacings occur? Verify your answers using the graphs of the coordinate transformation and its inverse.

2. Prove Theorems B.0.11 and B.0.12 for the case where $m=2$, $w_1 = w_2 = 1/2$, $s=1$, and $t=2$.

3. (Ji [206]) Consider a composite k-point Gauss quadrature for approximating $\int_a^b f(x)dx$ by applying the k-point Gauss quadrature to each subinterval of a non-uniform mesh of N points. It can be shown that, assuming that $f(x)$ is sufficiently smooth, the quadrature error is estimated by

$$E_{N,k}(f) \le C_k \sum_{j=2}^{N} h_j^{2k+1} \left| f^{(2k)}(\bar{x}_j) \right|,$$

where C_k is a constant, $h_j = x_j - x_{j-1}$, and $\bar{x}_j \in (x_{j-1}, x_j)$. Show that the error bound can formally be written in an asymptotic form as

$$\frac{C_k}{(N-1)^{2k}} \int_a^b \left(\frac{dx}{d\xi}\right)^{2k} \left| f^{(2k)} \right| dx.$$

If $\left| f^{(2k)} \right|$ does not vanish on (a,b), use Theorem 2.1.1 to find the optimal mesh density function and the corresponding optimal error bound. Discuss what can be done if $\left| f^{(2k)} \right|$ vanishes at some points in (a,b).

4. Solve the BVP (2.31) and (2.32) for $\rho(x) = \frac{4}{\pi(1+x^2)}$ defined on $(0,1)$.

5. For the functional (2.38), i.e.,

$$I[\xi] = \frac{1}{2}\int_a^b \frac{1}{\rho(x,t)} \left(\frac{\partial \xi}{\partial x}\right)^2 dx,$$

show that its functional derivative is

$$\frac{\delta I}{\delta \xi} = -\frac{\partial}{\partial x}\left(\frac{1}{\rho}\frac{\partial \xi}{\partial x}\right).$$

6. Show that (2.31) is the Euler-Lagrange equation for the functional (2.33).

7. Derive (2.49) and (2.50) by differentiating the relation $x = x(\xi(x,t),t)$ with respect to x and t.

8. Transform (2.31) into (2.28) by interchanging the roles of dependent and independent variables.
9. Given a non-uniform mesh $x_1 = a < x_2 < \cdots < x_{N-1} < x_N = b$, consider the piecewise constant interpolation defined by

$$\Pi_1 u(x) = \frac{1}{h_j} \int_{x_{j-1}}^{x_j} u(\tilde{x}) d\tilde{x}, \quad \forall x \in (x_{j-1}, x_j), \quad j = 2, ..., N, \quad \forall u \in H^1(a,b).$$

Show that the L^2 interpolation error is bounded by

$$\|u - \Pi_1 u\|_{L^2(a,b)}^2 \leq C \sum_{j=2}^{N} h_j^3 \langle u \rangle_{H^1(x_{j-1},x_j)}^2 .$$

(Hint: Use Poincaré's inequality (A.10).)

10. Assume that the quantities $\langle u \rangle_{H^1(x_{j-1},x_j)}$, $j = 2,...,N$ in the error bound of Problem 9 do not vanish. Find the optimal mesh density function for the error bound. What is the error bound for the equidistributing mesh corresponding to the optimal mesh density function? Compare it with the error bound on a uniform mesh of the same number of mesh points.
11. Define α_h such that

$$\sigma_h \equiv \sum_j h_j \rho_j = 2(b-a),$$

where ρ_j is defined in (2.71). Prove that the α_h defined this way exists and is unique provided that $\langle u \rangle_{H^{k+1}(I_j)}$ $(j = 2, ..., N)$ are not all zero. Give the tightest possible lower and upper bounds on α_h.

12. Prove (2.106) using Hölder's inequality.
13. For piecewise linear interpolation with uniform and adaptive meshes for function $u(x) = e^{-\frac{x}{\varepsilon}}$ on $[0, 1]$, find the asymptotic error bounds ($m = 0$ and $m = 1$) in terms of small ε. (Hint: See Example 2.5.1.)
14. Derive (2.171) from (2.169).
15. Consider the reaction-diffusion equation

$$u_t = u_{xx} + e^u.$$

Show that the equation can be transformed into

$$v_t = v_{xx} - \frac{1}{v} v_x^2 + v^2,$$

where $v = e^u$. If MMPDE5 (2.52) with mesh density function $\rho = v^\gamma$ and a small constant $\tau > 0$ is used for solving this equation, determine via dimensional analysis for what values of γ MMPDE5 has dominance of equidistribution.

16. Consider the differential equation

$$u_t = u_{xx} + \frac{1}{1-u},$$

which can have a quenching solution – the solution approaches one at a point in a finite time T; e.g., see Pao [270]. Let $v = 1/(1-u)$. Show that the equation can be transformed into

$$v_t = v_{xx} - \frac{2}{v} v_x^2 + v^3.$$

Use dimensional analysis and dominance of equidistribution to explain how to choose the mesh density function of the form $\rho = v^\gamma$ for MMPDE5 applied to the numerical solution of the transformed equation.

Chapter 3
Discretization of PDEs on Time-Varying Meshes

We have seen in the previous chapters for the 1D case that the formulation of mesh movement strategies (such as for MMPDEs) is generally independent of the specific type of physical PDE being solved, and their connection to the physical solution is through the mesh density function. As well, the discretization of the physical PDE and the overall solution procedure can to a large extent be described separately from mesh movement strategies by generally assuming that a moving mesh is available. This separation is also true for multidimensional moving mesh methods. For these reasons, we assume in this chapter that a moving mesh is given and consider the problem of discretizing PDEs and overall solution procedures in the higher dimensional case. The later chapters then return to an examination of the theory and applications of multidimensional moving mesh methods.

Generally speaking, discretization of a PDE on a moving mesh can be carried out using either the quasi-Lagrange or the rezoning approach. For the former, the mesh is considered to move continuously in time, and time derivatives are typically transformed to derivatives along mesh trajectories. For the latter, the mesh is considered to change intermittently at various time levels, and the physical solution must be interpolated from one mesh to the next. These two approaches for discretization shall be described here for finite difference and finite element methods for a general parabolic PDE

$$u_t + \nabla \cdot \boldsymbol{f} = \nabla \cdot (a \nabla u) + s, \qquad \text{in } \Omega \tag{3.1}$$

subject to the Dirichlet boundary condition

$$u = g, \qquad \text{on } \partial\Omega \tag{3.2}$$

where $a = a(\boldsymbol{x},t) \geq \alpha > 0$, $\boldsymbol{f} = \boldsymbol{f}(u,\boldsymbol{x},t)$, $s = s(u,\boldsymbol{x},t)$, and $g = g(\boldsymbol{x},t)$ are given functions. Equation (3.1) involves terms modeling diffusion, convection, and reaction processes.

W. Huang and R.D. Russell, *Adaptive Moving Mesh Methods*, Applied Mathematical Sciences 174, DOI 10.1007/978-1-4419-7196-2_3, © Springer Science+Business Media, LLC 2011

It is important to emphasize that the discretization of the physical PDE and the overall solution procedure depend upon the type of PDE being solved. Throughout this chapter (and the book as a whole), moving mesh methods are normally described for parabolic types of PDEs; nevertheless, it can be relatively straightforward to apply them to other types of PDEs with appropriate discretization schemes.

3.1 Coordinate transformations

3.1.1 Coordinate transformation as a mesh

Although the mesh generation problem is an inherently discrete one, the approach taken throughout this book is generally to consider it in relation to the computation of a continuous coordinate transformation. For the higher dimensional case, suppose that Ω is an open, simply connected physical domain in 3D and Ω_c is a computational domain chosen for the purpose of mesh generation. From the continuous viewpoint, a mesh is generated as the image of a computational (or reference) mesh under an invertible coordinate transformation $\boldsymbol{x} = \boldsymbol{x}(\boldsymbol{\xi}) : \Omega_c \to \Omega$. The computational mesh is typically taken to be uniform or quasi-uniform – see Figure 3.1 for the 2D case.

In the sense that generating the mesh is straightforward after determining the coordinate transformation – one simply evaluates $\boldsymbol{x}(\boldsymbol{\xi})$ at the stationary fixed points in Ω_c – a coordinate transformation is viewed as being equivalent to a mesh. An implication of this interpretation of adaptivity is that it is necessary to incorporate the coordinate transformation into the discretization of the physical PDE. As well, a facility at switching between dependent and independent variables consisting of the physical variables $\boldsymbol{x}$ and the computational variables $\boldsymbol{\xi}$ will be required so that the physical PDE can be written in terms of either set of variables. Preliminary to this, it is useful to first review some basic background material on transformation relations.

3.1.2 Transformation relations

For a coordinate transformation $\boldsymbol{x} = \boldsymbol{x}(\boldsymbol{\xi}) : \ \Omega_c \to \Omega$ and its inverse $\boldsymbol{\xi} = \boldsymbol{\xi}(\boldsymbol{x}) : \Omega \to \Omega_c$, denote the physical and computational coordinates by $\boldsymbol{x} = (x_1, x_2, x_3)^T$ and $\boldsymbol{\xi} = (\xi_1, \xi_2, \xi_3)^T$, respectively. The Jacobian matrix and its determinant (called simply the Jacobian) are denoted by

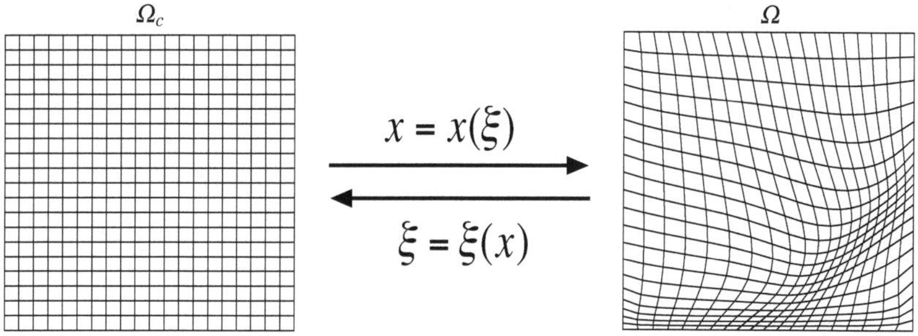

Fig. 3.1 A mesh on Ω is generated as the image of a computational mesh on Ω_c under coordinate transformation $\boldsymbol{x} = \boldsymbol{x}(\boldsymbol{\xi})$.

$$\boldsymbol{J} = \frac{\partial \boldsymbol{x}}{\partial \boldsymbol{\xi}}, \quad J = \det(\boldsymbol{J}).$$

The covariant and contravariant base vectors are defined as

$$\boldsymbol{a}_i = \frac{\partial \boldsymbol{x}}{\partial \xi_i} = \frac{\partial}{\partial \xi_i}\begin{bmatrix} x_1 \\ x_2 \\ x_3 \end{bmatrix}, \qquad \boldsymbol{a}^i = \nabla \xi_i = \begin{bmatrix} \frac{\partial}{\partial x_1} \\ \frac{\partial}{\partial x_2} \\ \frac{\partial}{\partial x_3} \end{bmatrix} \xi_i, \qquad i = 1,2,3 \tag{3.3}$$

where ∇ is the gradient operator with respect to the physical coordinates. To see the relations between these base vectors, we express the Jacobian matrices as

$$\boldsymbol{J} \equiv \frac{\partial \boldsymbol{x}}{\partial \boldsymbol{\xi}} = \frac{\partial(x_1,x_2,x_3)}{\partial(\xi_1,\xi_2,\xi_3)} = \begin{bmatrix} \frac{\partial x_1}{\partial \xi_1} & \frac{\partial x_1}{\partial \xi_2} & \frac{\partial x_1}{\partial \xi_3} \\ \frac{\partial x_2}{\partial \xi_1} & \frac{\partial x_2}{\partial \xi_2} & \frac{\partial x_2}{\partial \xi_3} \\ \frac{\partial x_3}{\partial \xi_1} & \frac{\partial x_3}{\partial \xi_2} & \frac{\partial x_3}{\partial \xi_3} \end{bmatrix} = [\boldsymbol{a}_1, \boldsymbol{a}_2, \boldsymbol{a}_3] \tag{3.4}$$

and

$$\frac{\partial(\xi_1,\xi_2,\xi_3)}{\partial(x_1,x_2,x_3)} = \begin{bmatrix} \frac{\partial \xi_1}{\partial x_1} & \frac{\partial \xi_1}{\partial x_2} & \frac{\partial \xi_1}{\partial x_3} \\ \frac{\partial \xi_2}{\partial x_1} & \frac{\partial \xi_2}{\partial x_2} & \frac{\partial \xi_2}{\partial x_3} \\ \frac{\partial \xi_3}{\partial x_1} & \frac{\partial \xi_3}{\partial x_2} & \frac{\partial \xi_3}{\partial x_3} \end{bmatrix} = \begin{bmatrix} (\boldsymbol{a}^1)^T \\ (\boldsymbol{a}^2)^T \\ (\boldsymbol{a}^3)^T \end{bmatrix}. \tag{3.5}$$

The chain rule implies

$$\frac{\partial(x_1,x_2,x_3)}{\partial(\xi_1,\xi_2,\xi_3)} \cdot \frac{\partial(\xi_1,\xi_2,\xi_3)}{\partial(x_1,x_2,x_3)} = I \quad \text{or} \quad \frac{\partial(\xi_1,\xi_2,\xi_3)}{\partial(x_1,x_2,x_3)} = \boldsymbol{J}^{-1}, \tag{3.6}$$

where I is the identity matrix. Notice that

$$\boldsymbol{J}^{-1} = \frac{\text{Adjoint}(\boldsymbol{J})}{J} = \frac{1}{J}\left[\boldsymbol{a}_2 \times \boldsymbol{a}_3, \boldsymbol{a}_3 \times \boldsymbol{a}_1, \boldsymbol{a}_1 \times \boldsymbol{a}_2\right]^T,$$

and $J = \det(\boldsymbol{J}) = \boldsymbol{a}_1 \cdot (\boldsymbol{a}_2 \times \boldsymbol{a}_3)$. Thus, the base vectors are related by

$$\boldsymbol{a}^i = \frac{1}{J}\boldsymbol{a}_j \times \boldsymbol{a}_k, \quad \boldsymbol{a}_i = J\boldsymbol{a}^j \times \boldsymbol{a}^k, \quad \boldsymbol{a}_i \cdot \boldsymbol{a}^j = \delta_{i,j}, \qquad (i,j,k) \text{ cyclic} \tag{3.7}$$

where $\delta_{i,j}$ is the Kronecker delta ($\delta_{i,j} = 0$ for $i \neq j$ and $\delta_{i,j} = 1$ for $i = j$).

Theorem 3.1.1 (Gradient operator in computational variables) *The gradient operator can be expressed in the computational coordinate system as*

$$\nabla = \sum_i \boldsymbol{a}^i \frac{\partial}{\partial \xi_i} \qquad \textit{(non-conservative form)} \tag{3.8}$$

$$= \frac{1}{J}\sum_i \frac{\partial}{\partial \xi_i} J\boldsymbol{a}^i. \qquad \textit{(conservative form)} \tag{3.9}$$

Consequently, the gradient of any function $u = u(\boldsymbol{x}) = u(\boldsymbol{x}(\boldsymbol{\xi}))$ *is*

$$\nabla u = \sum_i \boldsymbol{a}^i \frac{\partial u}{\partial \xi_i} = \frac{1}{J}\sum_i \frac{\partial}{\partial \xi_i}(Ju\boldsymbol{a}^i)$$

and the divergence of any vector field $\boldsymbol{v} = \boldsymbol{v}(\boldsymbol{x}) = \boldsymbol{v}(\boldsymbol{x}(\boldsymbol{\xi}))$ *is*

$$\nabla \cdot \boldsymbol{v} = \sum_i \boldsymbol{a}^i \cdot \frac{\partial \boldsymbol{v}}{\partial \xi_i} = \frac{1}{J}\sum_i \frac{\partial}{\partial \xi_i}(J\boldsymbol{a}^i \cdot \boldsymbol{v}).$$

Proof. Equation (3.8) follows from using the chain rule to determine each component of ∇u, viz.,

$$\frac{\partial u}{\partial x_j} = \sum_i \frac{\partial u}{\partial \xi_i}\frac{\partial \xi_i}{\partial x_j}$$

and $\partial \xi_i / \partial x_j$ is just the j-th component of $\boldsymbol{a}^i$. Equation (3.9) follows directly from (3.8) and the identity

$$\sum_i \frac{\partial}{\partial \xi_i}\left(J\boldsymbol{a}^i\right) = 0, \tag{3.10}$$

which in turn results from

$$\begin{aligned}
\sum_i \frac{\partial}{\partial \xi_i}\left(J\boldsymbol{a}^i\right) &= \sum_i \frac{\partial}{\partial \xi_i}(\boldsymbol{a}_j \times \boldsymbol{a}_k) \qquad \text{(from } (i,j,k) \text{ cyclic)} \\
&= \sum_i \frac{\partial^2 \boldsymbol{x}}{\partial \xi_i \partial \xi_j} \times \frac{\partial \boldsymbol{x}}{\partial \xi_k} + \sum_i \frac{\partial \boldsymbol{x}}{\partial \xi_j} \times \frac{\partial^2 \boldsymbol{x}}{\partial \xi_i \partial \xi_k} \\
&= \sum_i \frac{\partial^2 \boldsymbol{x}}{\partial \xi_i \partial \xi_j} \times \frac{\partial \boldsymbol{x}}{\partial \xi_k} - \sum_i \frac{\partial^2 \boldsymbol{x}}{\partial \xi_k \partial \xi_i} \times \frac{\partial \boldsymbol{x}}{\partial \xi_j} \qquad \text{(from } (i,j,k) \text{ cyclic)} \\
&= 0,
\end{aligned}$$

where the last step follows after relabeling indices in the second sum. □

It is often desirable for the analysis and/or practical computation to write PDEs in conservative form. For this reason, throughout this chapter we typically have two forms for the differential operators written in terms of the transformed variables, a conservative form as in (3.9) and non-conservative form as in (3.8).

We now consider the time-dependent case, where the coordinate transformation and its inverse depend upon time, i.e.,

$$t = \tau, \quad \boldsymbol{x} = \boldsymbol{x}(\boldsymbol{\xi}, \tau) \tag{3.11}$$

and

$$\tau = t, \quad \boldsymbol{\xi} = \boldsymbol{\xi}(\boldsymbol{x}, t). \tag{3.12}$$

The independent variables are transformed from $(\boldsymbol{x}, t)$ to $(\boldsymbol{\xi}, \tau)$, or vice versa. For any function $u = u(\boldsymbol{x}, t) = u(\boldsymbol{x}(\boldsymbol{\xi}, \tau), \tau)$, denote

$$u_t = \left.\frac{\partial u}{\partial t}\right|_{\boldsymbol{x}}, \qquad \dot{u} = \left.\frac{\partial u}{\partial \tau}\right|_{\boldsymbol{\xi}}. \tag{3.13}$$

That is, u_t is the time derivative when the physical coordinate $\boldsymbol{x}$ is fixed (and is used in the case where u is considered as a function of t and $\boldsymbol{x}$) and $\dot{u}$ is the time derivative of u when the computational coordinate $\boldsymbol{\xi}$ is fixed (and is used in the case when u is considered as a function of τ and $\boldsymbol{\xi}$). The time derivative $\dot{u}$ is often referred to as a time derivative along mesh trajectories.

In the next section, we write the physical PDE in terms of both the physical and computational variables. To prepare for this, the next theorem gives the form of the time derivative of the solution in terms of the computational variables.

Theorem 3.1.2 (Time derivative in computational variables) *In terms of the transformed variables $(\boldsymbol{\xi}, \tau)$, the time derivative u_t is*

$$u_t = \dot{u}\frac{\partial \tau}{\partial t} + \sum_i \frac{\partial u}{\partial \xi_i}\frac{\partial \xi_i}{\partial t} = \dot{u} + \nabla_{\boldsymbol{\xi}} u \cdot \boldsymbol{\xi}_t \quad \textit{(non-conservative form)} \tag{3.14}$$

$$= \frac{1}{J}\dot{\overline{(Ju)}} + \frac{1}{J}\sum_i \frac{\partial}{\partial \xi_i}\left(Ju\frac{\partial \xi_i}{\partial t}\right) = \frac{1}{J}\dot{\overline{(Ju)}} + \frac{1}{J}\nabla_{\boldsymbol{\xi}} \cdot (Ju\boldsymbol{\xi}_t), \quad \textit{(conservative form)} \tag{3.15}$$

where $\dot{\overline{(Ju)}}$ denotes the time derivative of the product term Ju and $\nabla_{\boldsymbol{\xi}}$ is the gradient operator with respect to the computational coordinate $\boldsymbol{\xi}$.

Proof. The proof of (3.14) is obvious.

To prove (3.15), first note that from the chain rule

$$0 = \frac{\partial \boldsymbol{x}}{\partial t} = \dot{\boldsymbol{x}}\frac{\partial \tau}{\partial t} + \frac{\partial \boldsymbol{x}}{\partial \boldsymbol{\xi}}\frac{\partial \boldsymbol{\xi}}{\partial t} = \dot{\boldsymbol{x}} + \boldsymbol{J}\boldsymbol{\xi}_t,$$

so the relation between $\dot{\boldsymbol{x}}$ and $\boldsymbol{\xi}_t$ is

$$\dot{\boldsymbol{x}} = -\boldsymbol{J}\boldsymbol{\xi}_t, \quad \boldsymbol{\xi}_t = -\boldsymbol{J}^{-1}\dot{\boldsymbol{x}}. \tag{3.16}$$

Differentiating $J = \det(\boldsymbol{J}) = \boldsymbol{a}_1 \cdot (\boldsymbol{a}_2 \times \boldsymbol{a}_3)$ with respect to time and using (3.7) gives

$$\begin{aligned} \dot{J} &= \sum_i \dot{\boldsymbol{a}}_i \cdot (\boldsymbol{a}_j \times \boldsymbol{a}_k) \qquad (i,j,k) \text{ cyclic} \\ &= J \sum_i \boldsymbol{a}^i \cdot \dot{\boldsymbol{a}}_i \\ &= J \sum_i \boldsymbol{a}^i \cdot \frac{\partial \dot{\boldsymbol{x}}}{\partial \xi_i}. \end{aligned}$$

Since from (3.8) this implies

$$\dot{J} = J \nabla \cdot \dot{\boldsymbol{x}}, \tag{3.17}$$

from (3.4), (3.7), (3.9), and (3.16) we obtain

$$\begin{aligned} \dot{J} &= J \nabla \cdot \dot{\boldsymbol{x}} \\ &= J \nabla \cdot (-\boldsymbol{J}\boldsymbol{\xi}_t) \\ &= -\sum_i \frac{\partial}{\partial \xi_i} \left(J \boldsymbol{a}^i \cdot \boldsymbol{J}\boldsymbol{\xi}_t \right) \\ &= -\sum_i \frac{\partial}{\partial \xi_i} \left(J [\boldsymbol{a}^i \cdot \boldsymbol{a}_1, \, \boldsymbol{a}^i \cdot \boldsymbol{a}_2, \, \boldsymbol{a}^i \cdot \boldsymbol{a}_3] \boldsymbol{\xi}_t \right) \\ &= -\sum_i \frac{\partial}{\partial \xi_i} \left(J \frac{\partial \xi_i}{\partial t} \right). \end{aligned}$$

Thus,

$$\dot{J} = -\sum_i \frac{\partial}{\partial \xi_i} \left(J \frac{\partial \xi_i}{\partial t} \right) = -\nabla_{\boldsymbol{\xi}} \cdot (J \boldsymbol{\xi}_t), \tag{3.18}$$

which together with (3.14) implies (3.15). □

It is also useful to express u_t in terms of $\dot{\boldsymbol{x}}$. The following corollary can be obtained from the above theorem and (3.16).

Corollary 3.1.1

$$u_t = \dot{u} - \nabla u \cdot \dot{\boldsymbol{x}} \qquad \textit{(non-conservative form)} \tag{3.19}$$

$$= \frac{1}{J}\dot{\overline{(Ju)}} - \nabla \cdot (u \dot{\boldsymbol{x}}). \qquad \textit{(conservative form)} \tag{3.20}$$

The identities (3.17) and (3.18) are frequently used when discretizing physical PDEs. For easy reference we summarize them in the following lemma.

Lemma 3.1.1

$$\dot{J} = J\nabla \cdot \dot{\boldsymbol{x}}, \tag{3.21}$$

$$\dot{J} = -\nabla_{\boldsymbol{\xi}} \cdot (J\boldsymbol{\xi}_t). \tag{3.22}$$

The key identity (3.21) or (3.22) is a mathematical representation of the principle of conservation of space (area in 2D, volume in 3D) under the mapping $\boldsymbol{x}(\boldsymbol{\xi},t)$, so it is often referred to as a Geometric Conservation Law (GCL). To see this, consider a fixed cell A_c in Ω_c and the corresponding moving cell $A(t)$ in Ω under the transformation (3.11). Integrating (3.21) over A_c gives

$$\int_{A_c} \dot{J} d\boldsymbol{\xi} = \int_{A_c} J\nabla \cdot \dot{\boldsymbol{x}} d\boldsymbol{\xi}.$$

Note that

$$\int_{A_c} \dot{J} d\boldsymbol{\xi} = \frac{d}{dt}\int_{A_c} J d\boldsymbol{\xi} = \frac{d}{dt}\int_{A(t)} d\boldsymbol{x}$$

and

$$\int_{A_c} J\nabla \cdot \dot{\boldsymbol{x}} d\boldsymbol{\xi} = \int_{A(t)} \nabla \cdot \dot{\boldsymbol{x}} d\boldsymbol{x} = \int_{\partial A(t)} \dot{\boldsymbol{x}} \cdot \boldsymbol{n} dS,$$

where $\int_{\partial A(t)} dS$ denotes a surface integral and $\boldsymbol{n}$ is the outward normal to the boundary $\partial A(t)$. Hence,

$$\frac{d}{dt}\int_{A(t)} d\boldsymbol{x} = \int_{\partial A(t)} \dot{\boldsymbol{x}} \cdot \boldsymbol{n} dS, \tag{3.23}$$

a mathematical expression of the fact that the increase in the volume of the cell $A(t)$ is equal to the volume gain through the movement of the boundary. This will prove important later, especially when considering the moving mesh methods in Chapter 7.

More generally, multiplying (3.20) by J and integrating the resulting equation over A_c we get

$$\int_{A_c} J u_t d\boldsymbol{\xi} = \int_{A_c} \dot{\overline{(Ju)}} d\boldsymbol{\xi} - \int_{A_c} J\nabla \cdot (u\dot{\boldsymbol{x}}) d\boldsymbol{\xi}.$$

Change of variables gives

$$\int_{A(t)} u_t d\boldsymbol{x} = \frac{d}{dt}\int_{A(t)} u d\boldsymbol{x} - \int_{A(t)} \nabla \cdot (u\dot{\boldsymbol{x}}) d\boldsymbol{x}$$

or

$$\frac{d}{dt}\int_{A(t)} u d\boldsymbol{x} = \int_{A(t)} \left(u_t + \nabla \cdot (u\dot{\boldsymbol{x}})\right) d\boldsymbol{x}. \tag{3.24}$$

Applying Gauss's Theorem to the second term on the right-hand side, we obtain the following lemma.

Lemma 3.1.2 *For any moving cell $A(t)$ in Ω, there holds*

$$\frac{d}{dt}\int_{A(t)} u d\boldsymbol{x} = \int_{A(t)} u_t d\boldsymbol{x} + \int_{\partial A(t)} u\dot{\boldsymbol{x}}\cdot\boldsymbol{n} dS, \tag{3.25}$$

where $\boldsymbol{n}$ denotes the outward normal to the boundary $\partial A(t)$ of $A(t)$.

Equation (3.25) is actually the Leibniz integral rule or Reynolds transport theorem in calculus. It reduces to (3.23) when $u = 1$.

3.1.3 Transformed structure of PDEs

We now write the PDE (3.1) in terms of the computational coordinate system. From (3.8) and (3.9), the diffusion term is transformed into

$$\begin{aligned}\nabla\cdot(a\nabla u) &= \nabla\cdot\left(a\sum_j \boldsymbol{a}^j\frac{\partial u}{\partial \xi_j}\right)\\ &= \frac{1}{J}\sum_i \frac{\partial}{\partial \xi_i}\left(J\boldsymbol{a}^i\cdot a\sum_j \boldsymbol{a}^j\frac{\partial u}{\partial \xi_j}\right)\\ &= \frac{1}{J}\sum_{i,j} \frac{\partial}{\partial \xi_i}\left(aJ\boldsymbol{a}^i\cdot \boldsymbol{a}^j\frac{\partial u}{\partial \xi_j}\right),\end{aligned}$$

and the convection term becomes

$$\nabla\cdot\boldsymbol{f} = \frac{1}{J}\sum_i \frac{\partial}{\partial \xi_i}\left(J\boldsymbol{a}^i\cdot\boldsymbol{f}\right).$$

From (3.15), PDE (3.1) can be written in the conservative form

$$\dot{\overline{(Ju)}} + \sum_i \frac{\partial}{\partial \xi_i}\left(J\boldsymbol{a}^i\cdot\boldsymbol{f} + Ju\frac{\partial \xi_i}{\partial t}\right) = \sum_{i,j} \frac{\partial}{\partial \xi_i}\left(aJ\boldsymbol{a}^i\cdot \boldsymbol{a}^j\frac{\partial u}{\partial \xi_j}\right) + J\,s. \tag{3.26}$$

The GCL is enforced in this conservative form. We show this for the special *uniform flow* case of (3.1) with $\boldsymbol{f} = 0$, $s = 0$ and the exact solution given by $u(\boldsymbol{x},t) = 1$ (also see §3.2 for the concept of uniform flow reproduction). For this situation, (3.26) reduces to (3.22) and thus reproduces the GCL.

In a similar fashion, one can derive the non-conservative form

$$\dot{u}+\sum_i\left(\boldsymbol{a}^i\cdot\frac{\partial \boldsymbol{f}}{\partial \xi_i}+\frac{\partial u}{\partial \xi_i}\frac{\partial \xi_i}{\partial t}\right)$$
$$=\sum_{i,j}(\boldsymbol{a}^i\cdot\boldsymbol{a}^j)\frac{\partial}{\partial \xi_i}\left(a\frac{\partial u}{\partial \xi_j}\right)+a\sum_i\left(\sum_j \boldsymbol{a}^j\cdot\frac{\partial \boldsymbol{a}^i}{\partial \xi_j}\right)\frac{\partial u}{\partial \xi_i}+s. \tag{3.27}$$

This does not reproduce the GCL. It is called the Chain Rule Conservative Law Form (CRCLF) by Hindman [171], who shows that this non-conservative form can capture shock waves even without enforcing the GCL.

3.1.4 Transformation relations in 2D

Since much of our attention shall focus on 2D problems, in this subsection we give the simplified form of the transformation relations for this case. The principle for obtaining the 2D transformation relations from the general 3D relations is fairly easily: one simply sets the third base vector to be the unit vector $\boldsymbol{a}_3=\boldsymbol{a}^3=[0,0,1]^T$ and drops the third component from the final results. For instance, if we denote the 2D physical and computational coordinates by $\boldsymbol{x}=(x,y)^T$ and $\boldsymbol{\xi}=(\xi,\eta)^T$ and the base vectors by

$$\boldsymbol{a}_1=\begin{bmatrix}x_\xi\\ y_\xi\end{bmatrix},\quad \boldsymbol{a}_2=\begin{bmatrix}x_\eta\\ y_\eta\end{bmatrix},\quad \boldsymbol{a}^1=\begin{bmatrix}\xi_x\\ \xi_y\end{bmatrix},\quad \boldsymbol{a}^2=\begin{bmatrix}\eta_x\\ \eta_y\end{bmatrix},$$

we have

$$J=\det\begin{bmatrix}x_\xi & x_\eta & 0\\ y_\xi & y_\eta & 0\\ * & * & 1\end{bmatrix}=x_\xi y_\eta-x_\eta y_\xi.$$

From (3.7),

$$\begin{aligned}\boldsymbol{a}^1&=\frac{1}{J}\boldsymbol{a}_2\times\boldsymbol{a}_3\\ &=\frac{1}{J}(x_\eta\boldsymbol{i}+y_\eta\boldsymbol{j}+(*)\boldsymbol{k})\times\boldsymbol{k}\\ &=\frac{1}{J}(-x_\eta\boldsymbol{j}+y_\eta\boldsymbol{i})\\ &=\frac{1}{J}\begin{bmatrix}y_\eta\\ -x_\eta\end{bmatrix},\end{aligned} \tag{3.28}$$

and similarly,

$$\boldsymbol{a}^2=\frac{1}{J}\begin{bmatrix}-y_\xi\\ x_\xi\end{bmatrix}. \tag{3.29}$$

Thus, we obtain the relation

$$\begin{bmatrix} \xi_x & \xi_y \\ \eta_x & \eta_y \end{bmatrix} = \frac{1}{J} \begin{bmatrix} y_\eta & -x_\eta \\ -y_\xi & x_\xi \end{bmatrix} \tag{3.30}$$

for the Jacobian and inverse Jacobian matrices

$$\boldsymbol{J} = \begin{bmatrix} x_\xi & x_\eta \\ y_\xi & y_\eta \end{bmatrix}, \qquad \boldsymbol{J}^{-1} = \begin{bmatrix} \xi_x & \xi_y \\ \eta_x & \eta_y \end{bmatrix}. \tag{3.31}$$

The other transformation relations follow similarly. The gradient operator is

$$\begin{aligned} \nabla &= \begin{bmatrix} \xi_x \\ \xi_y \end{bmatrix} \frac{\partial}{\partial \xi} + \begin{bmatrix} \eta_x \\ \eta_y \end{bmatrix} \frac{\partial}{\partial \eta} \qquad &\text{(non-conservative form)} \\ &= \frac{1}{J} \frac{\partial}{\partial \xi} J \begin{bmatrix} \xi_x \\ \xi_y \end{bmatrix} + \frac{1}{J} \frac{\partial}{\partial \eta} J \begin{bmatrix} \eta_x \\ \eta_y \end{bmatrix}. \qquad &\text{(conservative form)} \end{aligned} \tag{3.32}$$

For time derivatives, we have the relations

$$\begin{bmatrix} \xi_t \\ \eta_t \end{bmatrix} = - \begin{bmatrix} \xi_x & \xi_y \\ \eta_x & \eta_y \end{bmatrix} \begin{bmatrix} \dot{x} \\ \dot{y} \end{bmatrix} = -\frac{1}{J} \begin{bmatrix} y_\eta & -x_\eta \\ -y_\xi & x_\xi \end{bmatrix} \begin{bmatrix} \dot{x} \\ \dot{y} \end{bmatrix} \tag{3.33}$$

and

$$u_t = \dot{u} + \frac{\partial u}{\partial \xi} \xi_t + \frac{\partial u}{\partial \eta} \eta_t \qquad \text{(non-conservative form)} \tag{3.34}$$

$$= \frac{1}{J} \dot{\overline{(Ju)}} + \frac{1}{J} \frac{\partial}{\partial \xi} (J u \xi_t) + \frac{1}{J} \frac{\partial}{\partial \eta} (J u \eta_t). \qquad \text{(conservative form)} \tag{3.35}$$

Denoting $\boldsymbol{f} = (f_1, f_2)^T$, it follows from (3.26) and (3.27) that PDE (3.1) in conservative form is

$$\begin{aligned} &\dot{\overline{(Ju)}} + \frac{\partial}{\partial \xi} (J\xi_x f_1 + J\xi_y f_2 + Ju\xi_t) + \frac{\partial}{\partial \eta} (J\eta_x f_1 + J\eta_y f_2 + Ju\eta_t) \\ &\quad = \frac{\partial}{\partial \xi} \left(aJ(\xi_x^2 + \xi_y^2) \frac{\partial u}{\partial \xi} \right) + \frac{\partial}{\partial \xi} \left(aJ(\xi_x \eta_x + \xi_y \eta_y) \frac{\partial u}{\partial \eta} \right) \\ &\quad\quad + \frac{\partial}{\partial \eta} \left(aJ(\xi_x \eta_x + \xi_y \eta_y) \frac{\partial u}{\partial \xi} \right) + \frac{\partial}{\partial \eta} \left(aJ(\eta_x^2 + \eta_y^2) \frac{\partial u}{\partial \eta} \right) + J\, s \end{aligned} \tag{3.36}$$

and in non-conservative form is

$$\dot{u}+\left(\xi_x\frac{\partial f_1}{\partial \xi}+\xi_y\frac{\partial f_2}{\partial \xi}+\xi_t\frac{\partial u}{\partial \xi}\right)+\left(\eta_x\frac{\partial f_1}{\partial \eta}+\eta_y\frac{\partial f_2}{\partial \eta}+\eta_t\frac{\partial u}{\partial \eta}\right)$$
$$=(\xi_x^2+\xi_y^2)\frac{\partial}{\partial \xi}\left(a\frac{\partial u}{\partial \xi}\right)+(\xi_x\eta_x+\xi_y\eta_y)\frac{\partial}{\partial \xi}\left(a\frac{\partial u}{\partial \eta}\right)$$
$$+(\xi_x\eta_x+\xi_y\eta_y)\frac{\partial}{\partial \eta}\left(a\frac{\partial u}{\partial \xi}\right)+(\eta_x^2+\eta_y^2)\frac{\partial}{\partial \eta}\left(a\frac{\partial u}{\partial \eta}\right)$$
$$+a\left(\xi_x\frac{\partial \xi_x}{\partial \xi}+\xi_y\frac{\partial \xi_y}{\partial \xi}+\eta_x\frac{\partial \xi_x}{\partial \eta}+\eta_y\frac{\partial \xi_y}{\partial \eta}\right)\frac{\partial u}{\partial \xi}$$
$$+a\left(\xi_x\frac{\partial \eta_x}{\partial \xi}+\xi_y\frac{\partial \eta_y}{\partial \xi}+\eta_x\frac{\partial \eta_x}{\partial \eta}+\eta_y\frac{\partial \eta_y}{\partial \eta}\right)\frac{\partial u}{\partial \eta}+s. \quad (3.37)$$

Remark. In Riemannian geometry, $\boldsymbol{J}^T\boldsymbol{J}$ is often called the *metric tensor*, or simply the *metric*, since it can be used to measure distance in the target space of the coordinate transformation. As a consequence, we refer to terms involving derivatives of the coordinate transformation, such as $J\xi_x$ and $\xi_x\eta_x$, as *transformation metric terms*, or simply *metric terms*.

3.2 Finite difference methods

In this section, we describe in detail how to construct a central finite difference discretization of a physical PDE for a structured moving rectangular mesh. Given a computational domain Ω_c which is a unit square, a rectangular mesh is defined by

$$\mathscr{T}_h^c: \quad \xi_j=(j-1)\Delta\xi,\ \eta_k=(k-1)\Delta\eta, \quad j=1,\ldots,J,\ k=1,\ldots,K$$

where $\Delta\xi=1/(J-1)$, $\Delta\eta=1/(K-1)$ and J and K are given positive integers. The corresponding structured moving mesh is then determined from the given coordinate transformation $x=x(\xi,\eta,t)$, $y=y(\xi,\eta,t): \Omega_c\to\Omega$ by

$$\mathscr{T}_h(t): \quad x_{j,k}(t)=x(\xi_j,\eta_k,t), \quad y_{j,k}(t)=y(\xi_j,\eta_k,t),$$
$$j=1,\ldots,J,\ k=1,\ldots,K. \quad (3.38)$$

Since the mesh $\mathscr{T}_h(t)$ is generally non-rectangular, a finite difference discretization on it becomes quite awkward. As a consequence, the finite difference discretization is typically constructed instead on the computational domain using the simple rectangular mesh $\mathscr{T}_h^c$. Unfortunately, this requires using the more complicated forms of the transformed PDEs (3.36) and (3.37).

3.2.1 The quasi-Lagrange approach

For the quasi-Lagrange approach the physical mesh is viewed as a continuous function of time, and the time derivative term u_t in the PDE is transformed into a term along mesh trajectories, $\dot{u}$. The transformed PDE involves derivatives of the coordinate transformations.

The conservative form. We first consider a central finite difference discretization for the conservative form (3.36). Rewrite (3.36) as

$$\dot{\overline{(Ju)}} + \text{Conv(I)} + \text{Conv(II)} = \text{Diff(I)} + \text{Diff(II)} + \text{Diff(III)} + \text{Diff(IV)} + J\,s, \quad (3.39)$$

where

$$\begin{cases} \text{Conv(I)} &= \frac{\partial}{\partial \xi}\left(J\xi_x f_1 + J\xi_y f_2 + Ju\xi_t\right), \\ \text{Conv(II)} &= \frac{\partial}{\partial \eta}\left(J\eta_x f_1 + J\eta_y f_2 + Ju\eta_t\right), \\ \text{Diff(I)} &= \frac{\partial}{\partial \xi}\left(aJ(\xi_x^2 + \xi_y^2)\frac{\partial u}{\partial \xi}\right), \\ \text{Diff(II)} &= \frac{\partial}{\partial \xi}\left(aJ(\xi_x\eta_x + \xi_y\eta_y)\frac{\partial u}{\partial \eta}\right), \\ \text{Diff(III)} &= \frac{\partial}{\partial \eta}\left(aJ(\xi_x\eta_x + \xi_y\eta_y)\frac{\partial u}{\partial \xi}\right), \\ \text{Diff(IV)} &= \frac{\partial}{\partial \eta}\left(aJ(\eta_x^2 + \eta_y^2)\frac{\partial u}{\partial \eta}\right). \end{cases} \quad (3.40)$$

Denote by $u_{j,k}(t)$ the approximation to the solution at the node $(x_{j,k}(t), y_{j,k}(t))$, i.e., $u_{j,k}(t) \approx u(x_{j,k}(t), y_{j,k}(t), t)$. The convection and diffusion terms are approximated using central finite differences by

$$\begin{aligned} \text{Conv(I)}_{j,k} = \frac{1}{\Delta\xi}\Bigg[& (J\xi_x)_{j+\frac{1}{2},k}\frac{f_{1,j,k}+f_{1,j+1,k}}{2} - (J\xi_x)_{j-\frac{1}{2},k}\frac{f_{1,j,k}+f_{1,j-1,k}}{2} \\ & + (J\xi_y)_{j+\frac{1}{2},k}\frac{f_{2,j,k}+f_{2,j+1,k}}{2} - (J\xi_y)_{j-\frac{1}{2},k}\frac{f_{2,j,k}+f_{2,j-1,k}}{2} \\ & + (J\xi_t)_{j+\frac{1}{2},k}\frac{u_{j,k}+u_{j+1,k}}{2} - (J\xi_t)_{j-\frac{1}{2},k}\frac{u_{j,k}+u_{j-1,k}}{2}\Bigg], \end{aligned} \quad (3.41)$$

$$\begin{aligned} \text{Conv(II)}_{j,k} = \frac{1}{\Delta\eta}\Bigg[& (J\eta_x)_{j,k+\frac{1}{2}}\frac{f_{1,j,k}+f_{1,j,k+1}}{2} - (J\eta_x)_{j,k-\frac{1}{2}}\frac{f_{1,j,k}+f_{1,j,k-1}}{2} \\ & + (J\eta_y)_{j,k+\frac{1}{2}}\frac{f_{2,j,k}+f_{2,j,k+1}}{2} - (J\eta_y)_{j,k-\frac{1}{2}}\frac{f_{2,j,k}+f_{2,j,k-1}}{2} \\ & + (J\eta_t)_{j,k+\frac{1}{2}}\frac{u_{j,k}+u_{j,k+1}}{2} - (J\eta_t)_{j,k-\frac{1}{2}}\frac{u_{j,k}+u_{j,k-1}}{2}\Bigg], \end{aligned} \quad (3.42)$$

$$\text{Diff(I)}_{j,k} = \frac{1}{\Delta\xi^2}\left[\frac{a_{j+\frac{1}{2},k}((J\xi_x)^2+(J\xi_y)^2)_{j+\frac{1}{2},k}}{J_{j+\frac{1}{2},k}}(u_{j+1,k}-u_{j,k})\right.$$
$$\left.-\frac{a_{j-\frac{1}{2},k}((J\xi_x)^2+(J\xi_y)^2)_{j-\frac{1}{2},k}}{J_{j-\frac{1}{2},k}}(u_{j,k}-u_{j-1,k})\right], \qquad (3.43)$$

$$\text{Diff(IV)}_{j,k} = \frac{1}{\Delta\eta^2}\left[\frac{a_{j,k+\frac{1}{2}}((J\eta_x)^2+(J\eta_y)^2)_{j,k+\frac{1}{2}}}{J_{j,k+\frac{1}{2}}}(u_{j,k+1}-u_{j,k})\right.$$
$$\left.-\frac{a_{j,k-\frac{1}{2}}((J\eta_x)^2+(J\eta_y)^2)_{j,k-\frac{1}{2}}}{J_{j,k-\frac{1}{2}}}(u_{j,k}-u_{j,k-1})\right], \qquad (3.44)$$

$$\text{Diff(II)}_{j,k} = \frac{1}{4\Delta\xi\Delta\eta}\left[\frac{a_{j+\frac{1}{2},k}((J\xi_x)(J\eta_x)+(J\xi_y)(J\eta_y))_{j+\frac{1}{2},k}}{J_{j+\frac{1}{2},k}}\right.$$
$$\times(u_{j,k+1}-u_{j,k-1}+u_{j+1,k+1}-u_{j+1,k-1})$$
$$-\frac{a_{j-\frac{1}{2},k}((J\xi_x)(J\eta_x)+(J\xi_y)(J\eta_y))_{j-\frac{1}{2},k}}{J_{j-\frac{1}{2},k}}$$
$$\left.\times(u_{j,k+1}-u_{j,k-1}+u_{j-1,k+1}-u_{j-1,k-1})\right], \qquad (3.45)$$

$$\text{Diff(III)}_{j,k} = \frac{1}{4\Delta\xi\Delta\eta}\left[\frac{a_{j,k+\frac{1}{2}}((J\xi_x)(J\eta_x)+(J\xi_y)(J\eta_y))_{j,k+\frac{1}{2}}}{J_{j,k+\frac{1}{2}}}\right.$$
$$\times(u_{j+1,k}-u_{j-1,k}+u_{j+1,k+1}-u_{j-1,k+1})$$
$$-\frac{a_{j,k-\frac{1}{2}}((J\xi_x)(J\eta_x)+(J\xi_y)(J\eta_y))_{j,k-\frac{1}{2}}}{J_{j,k-\frac{1}{2}}}$$
$$\left.\times(u_{j+1,k}-u_{j-1,k}+u_{j+1,k-1}-u_{j-1,k-1})\right]. \qquad (3.46)$$

Here, outer differentiation operators have been approximated using central finite differences at half points (points midway between mesh points), and function values at half points have been evaluated as an average of the neighboring nodal values. These half-point approximations avoid the use of a stencil wider than $[j-1,j,j+1]\times[k-1,k,k+1]$ in approximating the transformation metric terms (as seen in detail below). Also, note that the diffusion terms incorporate J into the terms like $J\xi_x$ or $J\eta_y$, so their difference approximations are simplified using (3.30). The finite difference semi-discretization of the conservative form (3.36) is thus

$$\overline{(Ju)}^{\cdot}_{j,k} + \text{Conv(I)}_{j,k} + \text{Conv(II)}_{j,k}$$
$$= \text{Diff(I)}_{j,k} + \text{Diff(II)}_{j,k} + \text{Diff(III)}_{j,k} + \text{Diff(IV)}_{j,k} + J_{j,k}\, s_{j,k}. \quad (3.47)$$

It is necessary for the transformation metric terms in (3.41)–(3.46) to be computed in such a way that certain consistency conditions are satisfied. These can be explained using the concept of *uniform flow reproduction* introduced by Hindman [171]. Consider a PDE of the form (3.1) where $s = 0$ and $\boldsymbol{f}$ is any function depending only upon u. If the initial and boundary conditions are of the form $u(\boldsymbol{x}, 0) \equiv U$ and $u|_{\partial\Omega} = U$, where U is a constant, then the PDE has the uniform flow solution

$$u(\boldsymbol{x}, t) \equiv U, \quad \forall \boldsymbol{x} \in \Omega \cup \partial\Omega. \quad (3.48)$$

Consistency of the numerical algorithm requires that a uniform flow reproduction condition holds, i.e, the numerical solution of the PDE reproduces this uniform flow, or

$$u^n_{j,k} = U \quad \forall (j,k) \quad \Longrightarrow \quad u^{n+1}_{j,k} = U \quad \forall (j,k)$$

for each time level $t = t_n$. In the semi-continuous formulation, we thus require

$$u_{j,k}(t) = U \quad \forall (j,k) \quad \Longrightarrow \quad \dot{u}_{j,k} = 0 \quad \forall (j,k). \quad (3.49)$$

Applying this condition to the scheme (3.47), note first that the diffusion terms vanish for the uniform flow solution $u_{j,k}(t) = U$. For the convection terms, the first two terms in (3.41) can be rewritten as

$$\begin{aligned}
&(J\xi_x)_{j+\frac{1}{2},k} \frac{f_{1,j,k} + f_{1,j+1,k}}{2} - (J\xi_x)_{j-\frac{1}{2},k} \frac{f_{1,j,k} + f_{1,j-1,k}}{2} \\
&= \left[(J\xi_x)_{j+\frac{1}{2},k} - (J\xi_x)_{j-\frac{1}{2},k} \right] f_{1,j,k} + \frac{1}{2} (J\xi_x)_{j+\frac{1}{2},k} \left[f_{1,j+1,k} - f_{1,j,k} \right] \\
&\quad + \frac{1}{2} (J\xi_x)_{j-\frac{1}{2},k} \left[f_{1,j,k} - f_{1,j-1,k} \right] \\
&= \left[(J\xi_x)_{j+\frac{1}{2},k} - (J\xi_x)_{j-\frac{1}{2},k} \right] f_{1,j,k},
\end{aligned}$$

where we have used the fact that $f_{1,j+1,k} - f_{1,j,k} = 0$ and $f_{1,j,k} - f_{1,j-1,k} = 0$ for the uniform flow solution. The other terms can be rewritten similarly. Thus, we have

$$\mathrm{Conv(I)}_{j,k} = \frac{(J\xi_x)_{j+\frac{1}{2},k} - (J\xi_x)_{j-\frac{1}{2},k}}{\Delta\xi} f_{1,j,k} + \frac{(J\xi_y)_{j+\frac{1}{2},k} - (J\xi_y)_{j-\frac{1}{2},k}}{\Delta\xi} f_{2,j,k}$$
$$+ \frac{(J\xi_t)_{j+\frac{1}{2},k} - (J\xi_t)_{j-\frac{1}{2},k}}{\Delta\xi} u_{j,k}, \tag{3.50}$$

$$\mathrm{Conv(II)}_{j,k} = \frac{(J\eta_x)_{j,k+\frac{1}{2}} - (J\eta_x)_{j,k-\frac{1}{2}}}{\Delta\eta} f_{1,j,k} + \frac{(J\eta_y)_{j,k+\frac{1}{2}} - (J\eta_y)_{j,k-\frac{1}{2}}}{\Delta\eta} f_{2,j,k}$$
$$+ \frac{(J\eta_t)_{j,k+\frac{1}{2}} - (J\eta_t)_{j,k-\frac{1}{2}}}{\Delta\eta} u_{j,k}. \tag{3.51}$$

Substituting these into (3.47), with $\overline{(Ju)}^{\cdot}_{j,k} = \dot{J}_{j,k}u_{j,k} + J_{j,k}\dot{u}_{j,k}$ we obtain

$$J_{j,k}\dot{u}_{j,k} + \left[\frac{(J\xi_x)_{j+\frac{1}{2},k} - (J\xi_x)_{j-\frac{1}{2},k}}{\Delta\xi} + \frac{(J\eta_x)_{j,k+\frac{1}{2}} - (J\eta_x)_{j,k-\frac{1}{2}}}{\Delta\eta}\right] f_{1,j,k}$$
$$+ \left[\frac{(J\xi_y)_{j+\frac{1}{2},k} - (J\xi_y)_{j-\frac{1}{2},k}}{\Delta\xi} + \frac{(J\eta_y)_{j,k+\frac{1}{2}} - (J\eta_y)_{j,k-\frac{1}{2}}}{\Delta\eta}\right] f_{2,j,k}$$
$$+ \left[\dot{J}_{j,k} + \frac{(J\xi_t)_{j+\frac{1}{2},k} - (J\xi_t)_{j-\frac{1}{2},k}}{\Delta\xi} + \frac{(J\eta_t)_{j,k+\frac{1}{2}} - (J\eta_t)_{j,k-\frac{1}{2}}}{\Delta\eta}\right] u_{j,k}$$
$$= 0. \tag{3.52}$$

Since $f_{1,j,k}$, $f_{2,j,k}$, and $u_{j,k}$ $(\equiv U)$ are arbitrary, we conclude that $\dot{u}_{j,k} = 0$, or the uniform flow condition is satisfied if the transformation metric terms satisfy the conditions

$$\frac{(J\xi_x)_{j+\frac{1}{2},k} - (J\xi_x)_{j-\frac{1}{2},k}}{\Delta\xi} + \frac{(J\eta_x)_{j,k+\frac{1}{2}} - (J\eta_x)_{j,k-\frac{1}{2}}}{\Delta\eta} = 0, \tag{3.53}$$

$$\frac{(J\xi_y)_{j+\frac{1}{2},k} - (J\xi_y)_{j-\frac{1}{2},k}}{\Delta\xi} + \frac{(J\eta_y)_{j,k+\frac{1}{2}} - (J\eta_y)_{j,k-\frac{1}{2}}}{\Delta\eta} = 0, \tag{3.54}$$

$$\dot{J}_{j,k} + \frac{(J\xi_t)_{j+\frac{1}{2},k} - (J\xi_t)_{j-\frac{1}{2},k}}{\Delta\xi} + \frac{(J\eta_t)_{j,k+\frac{1}{2}} - (J\eta_t)_{j,k-\frac{1}{2}}}{\Delta\eta} = 0. \tag{3.55}$$

The first two conditions place restrictions on how the metric terms $J\xi_x$, $J\xi_y$, $J\eta_x$, and $J\eta_y$ must be evaluated. It is easy to verify that they are satisfied by using the central finite difference approximations

$$\begin{cases} (J\xi_x)_{j\pm\frac{1}{2},k} &= +(y_\eta)_{j\pm\frac{1}{2},k} = +\frac{1}{4\Delta\eta}\left(y_{j,k+1} - y_{j,k-1} + y_{j\pm1,k+1} - y_{j\pm1,k-1}\right), \\ (J\xi_y)_{j\pm\frac{1}{2},k} &= -(x_\eta)_{j\pm\frac{1}{2},k} = -\frac{1}{4\Delta\eta}\left(x_{j,k+1} - x_{j,k-1} + x_{j\pm1,k+1} - x_{j\pm1,k-1}\right), \\ (J\eta_x)_{j,k\pm\frac{1}{2}} &= -(y_\xi)_{j,k\pm\frac{1}{2}} = -\frac{1}{4\Delta\xi}(y_{j+1,k} - y_{j-1,k} + y_{j+1,k\pm1} - y_{j-1,k\pm1}), \\ (J\eta_y)_{j,k\pm\frac{1}{2}} &= +(x_\xi)_{j,k\pm\frac{1}{2}} = +\frac{1}{4\Delta\xi}(x_{j+1,k} - x_{j-1,k} + x_{j+1,k\pm1} - x_{j-1,k\pm1}), \end{cases} \tag{3.56}$$

where we have used the transformation relations (3.30).

The third condition, (3.55), is a finite difference approximation to the Geometric Conservation Law (3.22). Thomas and Lombard [322] are apparently the first to recognize the need for a numerical scheme to preserve the GCL when using a conservative form of the transformed PDE. To preserve it for the finite difference semi-discretization (3.47), the Jacobian is treated as an unknown variable determined from solving its evolution equation. As a consequence, to determine $J_{j,k}$ at a new time level, it should not be evaluated by simply using the definition $J = x_\xi y_\eta - x_\eta y_\xi$. Instead, it should be updated by integrating the expression (3.55) using the same integration scheme as that used for (3.47). The related metric terms can be evaluated by

$$\begin{cases} (J\xi_t)_{j\pm\frac{1}{2},k} &= -(J\xi_x)_{j\pm\frac{1}{2},k}\frac{\dot{x}_{j,k}+\dot{x}_{j\pm1,k}}{2} - (J\xi_y)_{j\pm\frac{1}{2},k}\frac{\dot{y}_{j,k}+\dot{y}_{j\pm1,k}}{2}, \\ (J\eta_t)_{j,k\pm\frac{1}{2}} &= -(J\eta_x)_{j,k\pm\frac{1}{2}}\frac{\dot{x}_{j,k}+\dot{x}_{j,k\pm1}}{2} - (J\eta_y)_{j,k\pm\frac{1}{2}}\frac{\dot{y}_{j,k}+\dot{y}_{j,k\pm1}}{2}, \end{cases} \tag{3.57}$$

where we have used the transformation relations (3.33).

The approximations (3.43)–(3.46) to the diffusion terms involve evaluations of the Jacobian at half points. Since these evaluations do not affect the preservation of the GCL, they can be done more directly. For example, using the definition of J, take

$$
\begin{cases}
J_{j+\frac{1}{2},k} &= (x_\xi y_\eta - y_\xi x_\eta)_{j+\frac{1}{2},k} \\
&= \frac{1}{4\Delta\xi\Delta\eta}\big[(x_{j+1,k}-x_{j,k})(y_{j,k+1}-y_{j,k-1}+y_{j+1,k+1}-y_{j+1,k-1}) \\
&\quad -(y_{j+1,k}-y_{j,k})(x_{j,k+1}-x_{j,k-1}+x_{j+1,k+1}-x_{j+1,k-1})\big], \\
J_{j-\frac{1}{2},k} &= (x_\xi y_\eta - y_\xi x_\eta)_{j-\frac{1}{2},k} \\
&= \frac{1}{4\Delta\xi\Delta\eta}\big[(x_{j,k}-x_{j-1,k})(y_{j,k+1}-y_{j,k-1}+y_{j-1,k+1}-y_{j-1,k-1}) \\
&\quad -(y_{j,k}-y_{j-1,k})(x_{j,k+1}-x_{j,k-1}+x_{j-1,k+1}-x_{j-1,k-1})\big], \\
J_{j,k+\frac{1}{2}} &= (x_\xi y_\eta - y_\xi x_\eta)_{j,k+\frac{1}{2}} \\
&= \frac{1}{4\Delta\xi\Delta\eta}\big[(x_{j+1,k}-x_{j-1,k}+x_{j+1,k+1}-x_{j-1,k+1})(y_{j,k+1}-y_{j,k}) \\
&\quad -(y_{j+1,k}-y_{j-1,k}+y_{j+1,k+1}-y_{j-1,k+1})(x_{j,k+1}-x_{j,k})\big], \\
J_{j,k-\frac{1}{2}} &= (x_\xi y_\eta - y_\xi x_\eta)_{j,k-\frac{1}{2}} \\
&= \frac{1}{4\Delta\xi\Delta\eta}\big[(x_{j+1,k}-x_{j-1,k}+x_{j+1,k-1}-x_{j-1,k-1})(y_{j,k}-y_{j,k-1}) \\
&\quad -(y_{j+1,k}-y_{j-1,k}+y_{j+1,k-1}-y_{j-1,k-1})(x_{j,k}-x_{j,k-1})\big].
\end{cases}
\tag{3.58}
$$

For the other factors in the diffusion terms, discretize by

$$
\begin{cases}
(J\xi_x)_{j,k+\frac{1}{2}} = +(y_\eta)_{j,k+\frac{1}{2}} = +\frac{y_{j,k+1}-y_{j,k}}{\Delta\eta}, \\
(J\xi_y)_{j,k+\frac{1}{2}} = -(x_\eta)_{j,k+\frac{1}{2}} = -\frac{x_{j,k+1}-x_{j,k}}{\Delta\eta}, \\
(J\eta_x)_{j+\frac{1}{2},k} = -(y_\xi)_{j+\frac{1}{2},k} = -\frac{y_{j+1,k}-y_{j,k}}{\Delta\xi}, \\
(J\eta_y)_{j+\frac{1}{2},k} = +(x_\xi)_{j+\frac{1}{2},k} = +\frac{x_{j+1,k}-x_{j,k}}{\Delta\xi},
\end{cases}
\tag{3.59}
$$

with similar formulas for $(J\xi_x)_{j,k-\frac{1}{2}}$, $(J\xi_y)_{j,k-\frac{1}{2}}$, $(J\eta_x)_{j-\frac{1}{2},k}$, and $(J\eta_y)_{j-\frac{1}{2},k}$.

The non-conservative form. Now consider a central finite difference discretization for the non-conservative form (3.37). Denoting

$$
\begin{cases}
\alpha &= \xi_x\frac{\partial\xi_x}{\partial\xi} + \xi_y\frac{\partial\xi_y}{\partial\xi} + \eta_x\frac{\partial\xi_x}{\partial\eta} + \eta_y\frac{\partial\xi_y}{\partial\eta}, \\
\beta &= \xi_x\frac{\partial\eta_x}{\partial\xi} + \xi_y\frac{\partial\eta_y}{\partial\xi} + \eta_x\frac{\partial\eta_x}{\partial\eta} + \eta_y\frac{\partial\eta_y}{\partial\eta},
\end{cases}
\tag{3.60}
$$

the standard discretization of (3.37) is

$$\dot{u}_{j,k} + \left[(\xi_x)_{j,k} \frac{f_{1,j+1,k} - f_{1,j-1,k}}{2\Delta\xi} + (\xi_y)_{j,k} \frac{f_{2,j+1,k} - f_{2,j-1,k}}{2\Delta\xi} + (\xi_t)_{j,k} \frac{u_{j+1,k} - u_{j-1,k}}{2\Delta\xi} \right]$$
$$+ \left[(\eta_x)_{j,k} \frac{f_{1,j,k+1} - f_{1,j,k-1}}{2\Delta\eta} + (\eta_y)_{j,k} \frac{f_{2,j,k+1} - f_{2,j,k-1}}{2\Delta\eta} + (\eta_t)_{j,k} \frac{u_{j,k+1} - u_{j,k-1}}{2\Delta\eta} \right]$$
$$= \frac{(\xi_x^2 + \xi_y^2)_{j,k}}{\Delta\xi^2} \left[a_{j+\frac{1}{2},k}(u_{j+1,k} - u_{j,k}) - a_{j-\frac{1}{2},k}(u_{j,k} - u_{j-1,k}) \right]$$
$$+ \frac{(\xi_x\eta_x + \xi_y\eta_y)_{j,k}}{4\Delta\xi\Delta\eta} \left[a_{j+1,k}(u_{j+1,k+1} - u_{j+1,k-1}) - a_{j-1,k}(u_{j-1,k+1} - u_{j-1,k-1}) \right]$$
$$+ \frac{(\xi_x\eta_x + \xi_y\eta_y)_{j,k}}{4\Delta\xi\Delta\eta} \left[a_{j,k+1}(u_{j+1,k+1} - u_{j-1,k+1}) - a_{j,k-1}(u_{j+1,k-1} - u_{j-1,k-1}) \right]$$
$$+ \frac{(\eta_x^2 + \eta_y^2)_{j,k}}{\Delta\eta^2} \left[a_{j,k+\frac{1}{2}}(u_{j,k+1} - u_{j,k}) - a_{j,k-\frac{1}{2}}(u_{j,k} - u_{j,k-1}) \right]$$
$$+ a_{j,k} \frac{\alpha_{j,k}}{2\Delta\xi}(u_{j+1,k} - u_{j-1,k}) + a_{j,k} \frac{\beta_{j,k}}{2\Delta\eta}(u_{j,k+1} - u_{j,k-1}) + s_{j,k}. \tag{3.61}$$

Regarding uniform flow reproduction, one can readily see that a uniform flow solution is always reproduced with the scheme (3.61) regardless of how the transformation metric terms are evaluated, and their evaluations are not subject to any other constraint conditions.

The transformation metric terms can be discretized via central finite differences using the transformation relations (3.30) and (3.33). Specifically,

$$\begin{cases} J_{j,k} &= \frac{1}{4\Delta\xi\Delta\eta} \left[(x_{j+1,k} - x_{j-1,k})(y_{j,k+1} - y_{j,k-1}) \right. \\ & \qquad \left. - (x_{j,k+1} - x_{j,k-1})(y_{j+1,k} - y_{j-1,k}) \right], \\ (\xi_x)_{j,k} &= \frac{1}{2J_{j,k}\Delta\eta}(y_{j,k+1} - y_{j,k-1}), \\ (\xi_y)_{j,k} &= -\frac{1}{2J_{j,k}\Delta\eta}(x_{j,k+1} - x_{j,k-1}), \\ (\eta_x)_{j,k} &= -\frac{1}{2J_{j,k}\Delta\xi}(y_{j+1,k} - y_{j-1,k}), \\ (\eta_y)_{j,k} &= \frac{1}{2J_{j,k}\Delta\xi}(x_{j+1,k} - x_{j-1,k}), \\ (\xi_t)_{j,k} &= -(\xi_x)_{j,k}\dot{x}_{j,k} - (\xi_y)_{j,k}\dot{y}_{j,k}, \\ (\eta_t)_{j,k} &= -(\eta_x)_{j,k}\dot{x}_{j,k} - (\eta_y)_{j,k}\dot{y}_{j,k}. \end{cases} \tag{3.62}$$

Now the quantities α and β can be discretized as

$$\begin{cases} \alpha_{j,k} &= (\xi_x)_{j,k} \frac{(\xi_x)_{j+\frac{1}{2},k} - (\xi_x)_{j-\frac{1}{2},k}}{\Delta\xi} + (\xi_y)_{j,k} \frac{(\xi_y)_{j+\frac{1}{2},k} - (\xi_y)_{j-\frac{1}{2},k}}{\Delta\xi} \\ & + (\eta_x)_{j,k} \frac{(\xi_x)_{j,k+\frac{1}{2}} - (\xi_x)_{j,k-\frac{1}{2}}}{\Delta\eta} + (\eta_y)_{j,k} \frac{(\xi_y)_{j,k+\frac{1}{2}} - (\xi_y)_{j,k-\frac{1}{2}}}{\Delta\eta}, \\ \beta_{j,k} &= (\xi_x)_{j,k} \frac{(\eta_x)_{j+\frac{1}{2},k} - (\eta_x)_{j-\frac{1}{2},k}}{\Delta\xi} + (\xi_y)_{j,k} \frac{(\eta_y)_{j+\frac{1}{2},k} - (\eta_y)_{j-\frac{1}{2},k}}{\Delta\xi} \\ & + (\eta_x)_{j,k} \frac{(\eta_x)_{j,k+\frac{1}{2}} - (\eta_x)_{j,k-\frac{1}{2}}}{\Delta\eta} + (\eta_y)_{j,k} \frac{(\eta_y)_{j,k+\frac{1}{2}} - (\eta_y)_{j,k-\frac{1}{2}}}{\Delta\eta}, \end{cases} \tag{3.63}$$

where ξ_x, ξ_y, η_x, and η_y are evaluated at half points in the same manner as in (3.62) and the specific approximations (3.56), (3.58), and (3.59) are used.

Solution procedure. The boundary condition (3.2) is discretized as

$$u_{j,k} = g(x_{j,k}, y_{j,k}, t), \quad (x_{j,k}, y_{j,k}) \in \partial\Omega. \tag{3.64}$$

Denoting $U = \{u_{j,k}\}$, $X = \{x_{j,k}\}$, and $Y = \{y_{j,k}\}$, the discretization for the PDE (3.47) or (3.61), together with the boundary conditions (3.64), can be written in the abstract form

$$\dot{U} = F(U, X, Y, \dot{X}, \dot{Y}, t). \tag{3.65}$$

This system can be solved either simultaneously or alternately with the system of equations for the mesh generation. As we see in later chapters, a PDE-based mesh generation system often has the form

$$\dot{X} = G_1(U, X, Y, t), \quad \dot{Y} = G_2(U, X, Y, t). \tag{3.66}$$

A one-dimensional example of such a system is given in (1.14) in the continuous form and (1.18) in the discrete form. Equations (3.65) and (3.66) form a coupled system which in principle can be conveniently solved with a simultaneous solution procedure, integrating for both the physical solution U and the mesh (X, Y) using an existing ODE solver or a scheme like a Runge-Kutta method (cf. Figure 1.11). However, as can be seen from the discretization (3.61), for instance, the coupling between U and (X, Y) is highly nonlinear even when the original PDE is linear. This often makes the extended system challenging to solve.

As another implementation option, a PDE-based or an optimization-based mesh generator (cf. Chapter 5) can be used with an alternate solution procedure as illustrated in Figure 1.12. More specifically, assume that the solution U^n and the mesh (X^n, Y^n) are known at current time level $t = t_n$. With the alternate solution procedure, the new mesh is obtained with the mesh movement strategy using the information (U^n, X^n, Y^n) at $t = t_n$, to obtain

$$(X^{n+1}, Y^{n+1}) = G(U^n, X^n, Y^n). \tag{3.67}$$

The mesh can then be calculated via linear interpolation over the time interval $[t_n, t_{n+1}]$,

$$\begin{cases} X(t) = \frac{(t_{n+1} - t)}{(t_{n+1} - t_n)} X^n + \frac{(t - t_n)}{(t_{n+1} - t_n)} X^{n+1}, \\ Y(t) = \frac{(t_{n+1} - t)}{(t_{n+1} - t_n)} Y^n + \frac{(t - t_n)}{(t_{n+1} - t_n)} Y^{n+1}, \end{cases} \tag{3.68}$$

and the physical PDE (3.65) integrated on the moving mesh (3.68) from $t = t_n$ to $t = t_{n+1}$. As for the 1D case in Chapter 2, the key feature of this alternate solution procedure is that the semi-discrete physical PDE (3.65) and the mesh generation equation (3.67) are treated separately. This uncoupling of the semi-discrete physical PDE from the mesh equations so that it can be solved relatively efficiently is particularly beneficial in higher dimensional cases.

3.2.2 The rezoning approach

Recall from Chapter 2 that for the rezoning approach, the mesh is changed intermittently at discrete time levels. The approach is typically associated with an alternate solution procedure and is usually characterized by interpolating the physical solution from the old mesh to the new one, and then integrating the physical PDE on this fixed new mesh.

We illustrate this for the non-conservative form of the transformed PDE (3.37). Assume that the meshes (X^n, Y^n) and (X^{n+1}, Y^{n+1}) and the physical solution U^n are given. The first step in the rezoning approach is to interpolate U^n from the old mesh (X^n, Y^n) to the new one (X^{n+1}, Y^{n+1}), a topic discussed in detail in §3.5. Denoting this interpolated solution by $\tilde{U}^n$, the second step is to discretize the physical PDE on the new mesh (X^{n+1}, Y^{n+1}). Since the mesh will be held fixed in this time step, there is no need to transform the time derivative from u_t to $\dot{u}$. However, since the mesh (X^{n+1}, Y^{n+1}) is generally non-rectangular, we still transform the PDE from the physical domain to the computational domain for the finite difference discretization. For this, (3.37) can be used with $\dot{u}$ replaced by u_t and ξ_t and η_t set to be zero. The resulting central finite difference scheme similar to (3.61) is

$$
\begin{aligned}
\frac{du_{j,k}}{dt} &+ \left[(\xi_x)_{j,k} \frac{f_{1,j+1,k} - f_{1,j-1,k}}{2\Delta\xi} + (\xi_y)_{j,k} \frac{f_{2,j+1,k} - f_{2,j-1,k}}{2\Delta\xi} \right] \\
&+ \left[(\eta_x)_{j,k} \frac{f_{1,j,k+1} - f_{1,j,k-1}}{2\Delta\eta} + (\eta_y)_{j,k} \frac{f_{2,j,k+1} - f_{2,j,k-1}}{2\Delta\eta} \right] \\
&= \frac{(\xi_x^2 + \xi_y^2)_{j,k}}{\Delta\xi^2} \left[a_{j+\frac{1}{2},k}(u_{j+1,k} - u_{j,k}) - a_{j-\frac{1}{2},k}(u_{j,k} - u_{j-1,k}) \right] \\
&+ \frac{(\xi_x \eta_x + \xi_y \eta_y)_{j,k}}{4\Delta\xi\Delta\eta} \left[a_{j+1,k}(u_{j+1,k+1} - u_{j+1,k-1}) - a_{j-1,k}(u_{j-1,k+1} - u_{j-1,k-1}) \right] \\
&+ \frac{(\xi_x \eta_x + \xi_y \eta_y)_{j,k}}{4\Delta\xi\Delta\eta} \left[a_{j,k+1}(u_{j+1,k+1} - u_{j-1,k+1}) - a_{j,k-1}(u_{j+1,k-1} - u_{j-1,k-1}) \right] \\
&+ \frac{(\eta_x^2 + \eta_y^2)_{j,k}}{\Delta\eta^2} \left[a_{j,k+\frac{1}{2}}(u_{j,k+1} - u_{j,k}) - a_{j,k-\frac{1}{2}}(u_{j,k} - u_{j,k-1}) \right] \\
&+ \frac{\alpha_{j,k}}{2\Delta\xi}(u_{j+1,k} - u_{j-1,k}) + \frac{\beta_{j,k}}{2\Delta\eta}(u_{j,k+1} - u_{j,k-1}) + s_{j,k}, \qquad (3.69)
\end{aligned}
$$

where $u_{j,k}(t) \approx u(x_{j,k}^{n+1}, y_{j,k}^{n+1}, t)$, $x_{j,k}(t) = x_{j,k}^{n+1}$, and $y_{j,k}(t) = y_{j,k}^{n+1}$. This system, along with the discrete boundary conditions, can be written in the general form

$$\frac{dU}{dt} = F(U, X^{n+1}, Y^{n+1}, t), \quad t \in (t_n, t_{n+1}]. \tag{3.70}$$

Supplemented with the initial condition $U(t_n) = \tilde{U}^n$, (3.70) can be integrated using any standard integration scheme.

Since the physical PDE is discretized on a fixed mesh for the rezoning approach, mesh movement does not complicate the structure of the physical PDE. Nevertheless, caution should be taken in the interpolation of the physical solution from the old mesh to the new one. Experience shows that for a dissipative PDE, a reasonably accurate interpolation scheme will work. However, for a less dissipative or a conservative PDE, a conservative interpolation scheme which preserves some solution quantities is often necessary. This topic is addressed in §3.5.

3.3 Finite element methods

Finite element methods provide considerable flexibility in that they can easily be used with either structured or unstructured meshes, where each mesh node has basically the same connectivity for the former (and not for the latter). They can also be applied on either the computational domain or directly on the physical domain. Since the physical PDE takes a much simpler form on the physical domain, a moving mesh finite element method is typically employed directly on this physical domain.

Although the finite element method is only illustrated in this section for linear elements on triangular meshes, the approach readily generalizes for any family of affine finite elements on a polygonal or polyhedral domain.

3.3.1 Concepts of unstructured meshes and finite elements

Consider a polyhedral (or polygonal) domain $\Omega \subset \mathbb{R}^d$ ($d \geq 1$). We consider triangulations on Ω defined in the classical sense (e.g., see [104]). Specifically, a triangulation $\mathscr{T}_h$ is formed by subdividing $\overline{\Omega}$, the closure of Ω, into a finite number of open polyhedral (or polygonal) sub-domains K, called mesh elements, in such a way that they satisfy (i) $\overline{\Omega} = \bigcup_{K \in \mathscr{T}_h} \overline{K}$; (ii) every element is non-empty; (iii) any two different elements do not overlap; (iv) any vertex of an element cannot be in the interior of a edge or on a face of another element. The last condition guarantees that there are no hanging vertices, or nodes, as illustrated in Figure 3.2. A triangulation is generally an unstructured mesh since the connectivity properties from one vertex

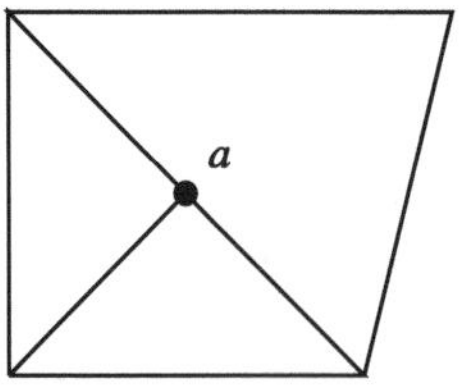

Fig. 3.2 Point $\boldsymbol{a}$ is a hanging node.

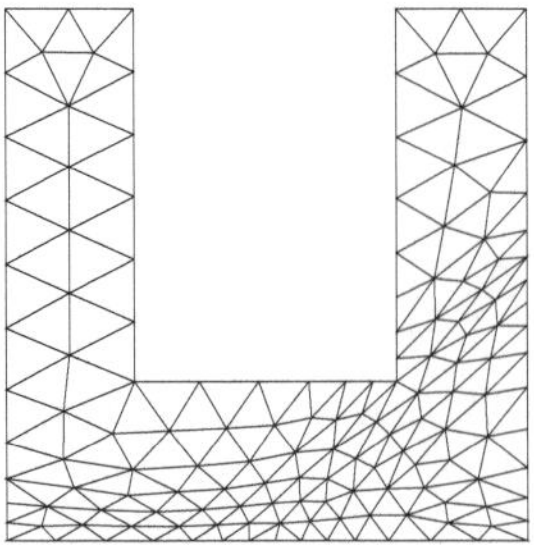

Fig. 3.3 A triangulation of $N = 236$ triangular elements.

to others varies from vertex to vertex. We denote the number of elements of $\mathscr{T}_h$ by N. A typical triangulation with $N = 236$ triangular elements is shown in Figure 3.3. Often we assume that a family of triangulations $\{\mathscr{T}_h\}$ is given on Ω with a parameter $h > 0$ characterizing the family in terms of the size of the elements, as described below.

A *uniform* mesh for Ω is a triangulation whose elements are equilateral and of the same size. Define the diameter of an element K as

$$h_K \equiv \operatorname{diam}(K) = \max_{\boldsymbol{x},\boldsymbol{y} \in K} d(\boldsymbol{x},\boldsymbol{y}), \tag{3.71}$$

where $d(\boldsymbol{x},\boldsymbol{y})$ is the (Euclidean) distance between $\boldsymbol{x}$ and $\boldsymbol{y}$. For a simplex (defined in §3.3.2), h_K is the length of its longest edge(s), and for a uniform mesh

$$h_K \approx \left(\frac{|\Omega|}{N}\right)^{\frac{1}{d}} = O(N^{-\frac{1}{d}}), \qquad \forall K \in \mathscr{T}_h \tag{3.72}$$

where $|\Omega|$ is the area of Ω. For a family of uniform meshes, the parameter h defined by $h = \max_{K \in \mathscr{T}_h} h_K$ satisfies

$$h \to 0 \quad \Longleftrightarrow \quad N \to \infty. \tag{3.73}$$

A generalization of uniform meshes is a *regular* family of triangulations which satisfies the following conditions:

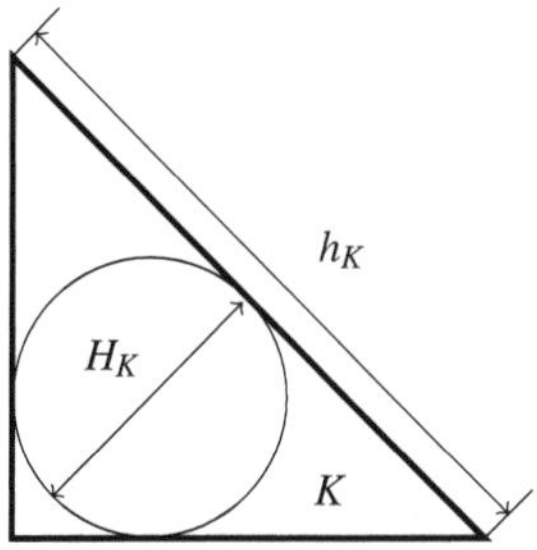

Fig. 3.4 Illustration of the diameter h_K and in-diameter H_K for an element K.

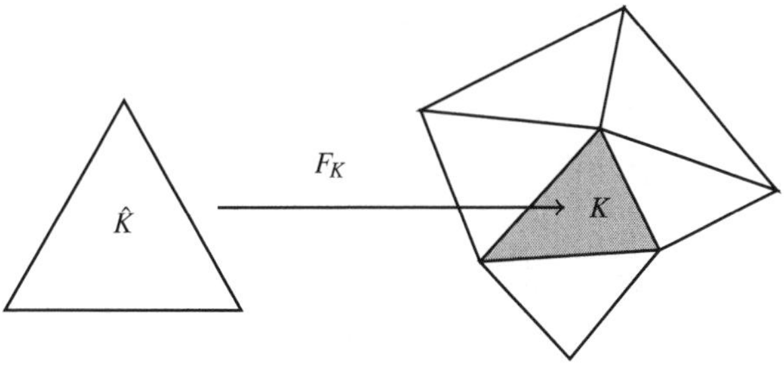

Fig. 3.5 A sketch of affine mapping $F_K : \hat{K} \to K$.

(i) There exists a constant c such that

$$\frac{h_K}{H_K} \le c, \qquad \forall K \in \bigcup_h \mathscr{T}_h$$

where the in-diameter of K is $H_K = \sup\{\text{diam}(B) : B \text{ is a ball contained in } K\}$.

(ii)

$$h = \max_{K \in \mathscr{T}_h} h_K \to 0.$$

An illustration of h_K and H_K for a triangle K is given in Figure 3.4.

A *quasi-uniform* mesh refers to a mesh whose elements are shape regular and of basically the same size. A quasi-uniform mesh is a close generalization of a uniform mesh in the sense that a family of such meshes satisfies relations (3.72) and (3.73).

A family of triangulations $\mathscr{T}_h$ is called *an affine family* if there exists an element $\hat{K}$ such that for any element K, an invertible affine mapping from $\hat{K}$ to K exists. In other words, for each element $K \in \{\mathscr{T}_h\}$, there exists an invertible affine mapping $F_K : \hat{K} \to K$ such that $K = F_K(\hat{K})$ – see Figure 3.5. The element $\hat{K}$ is called *the reference* or *master element* for the family of triangulations. The following result follows from the observation that the relationship among elements defined through invertible affine mappings is transitive.

Proposition 3.3.1 *Any element of a given affine family of triangulations, or its image under any invertible affine transformation, can be chosen as the reference element for the family.*

From this proposition one can see that the reference element need not belong to the family. Moreover, it can be chosen to be as simple or convenient as possible. Indeed, the reference element for an affine family of simplicial triangulations is often chosen as the unit d-simplex defined as $\{(\xi_1,\ldots,\xi_d)^T \mid \sum_i \xi_i \leq 1,\ \xi_i \geq 0 \text{ for all } i\}$ or an equilateral d-simplex having a unitary volume (where a d-simplex is defined in §3.3.2).

Throughout this book, we make the following assumptions:

$$\left\{ \begin{array}{l} \textbf{H1.}\ \{\mathscr{T}_h\} \text{ is an } \textit{affine} \text{ family of triangulations for } \Omega. \\ \textbf{H2.}\ \text{The reference element is chosen to be equilateral with } |\hat{K}| = 1. \end{array} \right. \tag{3.74}$$

We are now prepared to introduce the concept of finite elements. A *finite element* is defined as a triple (K, P_K, Σ_K), where K is a mesh element, P_K is a finite-dimensional linear space of functions defined on K, and Σ_K is a finite set of linearly independent conditions uniquely determining any function in P_K. It is emphasized that a finite element comprises more than simply a mesh element. The latter is used to describe a geometric object only whereas the former contains, in addition to the geometric object, the information about the approximation function space and the conditions determining its members.

Often P_K is chosen to contain a full polynomial space $P_k(K)$ (the space of polynomials of degree at most k) for some integer $k \geq 1$, a property crucial for the convergence of finite element computations. Denote by Ψ_i, $i = 1,\ldots,m$, the linear operators associated with the linear conditions in Σ_K. By definition, for any real values α_i, $i = 1,\ldots,m$, there exists a unique function $p \in P_K$ satisfying

$$\Psi_i(p) = \alpha_i, \quad i = 1,\ldots,m.$$

Consequently, there exist a set of basis functions $\phi_i \in P_K$, $i = 1,\ldots,m$, such that

$$p = \sum_{i=1}^{m} \Psi_i(p)\phi_i, \quad \forall p \in P_K. \tag{3.75}$$

The linear operators Ψ_i, $i = 1,\ldots,m$, are called the degrees of freedom and Σ_K the set of degrees of freedom of the finite element.

Example 3.3.1 An example of linear finite elements in two dimensions is (K, P_K, Σ_K), where K is a triangle with vertices $\boldsymbol{a}_i$, $i = 1,2,3$; P_K consists of all linear functions defined on K, i.e.,

$$P_K = P_1(K) \equiv \{p \mid p = ax + by + c, \quad \forall a,b,c \in \mathbb{R} \quad \forall (x,y) \in K\};$$

and $\Sigma_K = \{\Psi_i(p) \equiv p(\boldsymbol{a}_i),\ i = 1,2,3\}$, i.e., each function in P_K is uniquely determined by its values at the vertices of K. In this case the basis functions of P_K can be chosen as the linear functions ϕ_i, $i = 1,2,3$ satisfying

$$\phi_i(\boldsymbol{a}_j) = \delta_{i,j}, \quad i,j = 1,2,3$$

where δ is the Kronecker delta function. Then (3.75) reduces to

$$p = \sum_{i=1}^{3} \Psi_i(p)\phi_i = \sum_{i=1}^{3} p(\boldsymbol{a}_i)\phi_i, \quad \forall p \in P_K.$$

□

Example 3.3.2 An example of quadratic finite elements in two dimensions is (K, P_K, Σ_K), where K is a triangle with vertices $\boldsymbol{a}_i$, $i = 1,2,3$; P_K consists of all quadratic functions defined on K, i.e., $P_K = P_2(K)$, where

$$P_2(K) \equiv \{p \mid p = ax^2 + bxy + cy^2 + dx + ey + f, \quad \forall a,b,c,d,e,f \in \mathbb{R}, \quad \forall (x,y) \in K\};$$

and $\Sigma = \{\Psi_i(p) \equiv p(\boldsymbol{a}_i),\ i = 1,2,3;\ \Psi_{ij}(p) \equiv p\left((\boldsymbol{a}_i + \boldsymbol{a}_j)/2\right),\ 1 \le i < j \le 3\}$, i.e., each function in $\hat{P}$ is uniquely determined by its values at the vertices of K and the midpoints of its edges. The basis functions of P_K can be chosen as the linear functions ϕ_i, $i = 1,2,3$ and quadratic functions ϕ_{ij}, $1 \le i < j \le 3$ satisfying

$$\phi_i(\boldsymbol{a}_j) = \delta_{i,j}, \quad i,j = 1,2,3$$

$$\begin{cases} \phi_{ij}(\boldsymbol{a}_k) = 0; \quad 1 \le i < j \le 3,\ k = 1,2,3 \\ \phi_{ij}((\boldsymbol{a}_k + \boldsymbol{a}_l)/2) = \delta_{i,k}\delta_{j,l}, \quad 1 \le i < j \le 3,\ 1 \le k < l \le 3. \end{cases}$$

Then (3.75) reads as

$$\begin{aligned} p &= \sum_{i=1}^{3} \Psi_i(p)\phi_i + \sum_{1 \le i < j \le 3} \Psi_{ij}(p)\phi_{ij} \\ &= \sum_{i=1}^{3} p(\boldsymbol{a}_i)\phi_i + \sum_{1 \le i < j \le 3} p((\boldsymbol{a}_i + \boldsymbol{a}_j)/2)\phi_{ij}, \ \forall p \in P_K. \end{aligned}$$

□

Finite element computation can often be significantly simplified by using the concept of affine-equivalence since most of it can be done on a single finite element. Two finite elements are said to be *affine-equivalent* if their mesh elements, finite dimensional function spaces, and sets of degrees of freedom can be mapped to each other through invertible affine mappings. More specifically, consider two affine-equivalent finite elements $(\hat{K},\hat{P},\hat{\Sigma})$ and (K,P_K,Σ_K). Let the invertible affine mapping between them be F_K. Then,

$$K = F_K(\hat{K}). \tag{3.76}$$

Moreover, $\hat{P}$ and P_K are related by

$$P_K = \{p \mid p = \hat{p}\circ F_K^{-1},\ \forall \hat{p} \in \hat{P}\}, \tag{3.77}$$

where F_K^{-1} denotes the inverse mapping of F_K. In particular, their basis functions, upon possible reordering, satisfy

$$\phi_i = \hat{\phi}_i \circ F_K^{-1}, \quad i = 1,...,m. \tag{3.78}$$

Furthermore, the sets of the degrees of freedom of these finite elements, once again upon possible reordering, are related by

$$\Sigma_K = \{\Psi_i(p) = \hat{\Psi}_i(p\circ F_K),\ \forall p \in P_K,\ i = 1,...,m\}. \tag{3.79}$$

This can be seen from (3.77). It can also be derived in a slightly different but more rigorous way. Indeed, recall that functions in $\hat{P}$ can be expressed in terms of basis functions as

$$\hat{p} = \sum_{i=1}^{m} \hat{\Psi}_i(\hat{p})\hat{\phi}_i, \quad \forall \hat{p} \in \hat{P}. \tag{3.80}$$

Composing this with F_K^{-1} from the right, we get

$$\hat{p}\circ F_K^{-1} = \sum_{i=1}^{m} \hat{\Psi}_i(\hat{p})\hat{\phi}_i \circ F_K^{-1}.$$

Using the relation between $\hat{P}$ and P_K and noticing that $\hat{p} = p\circ F_K$, we have

$$p = \sum_{i=1}^{m} \hat{\Psi}_i(p\circ F_K)\phi_i.$$

Relation (3.79) then follows from comparing this with (3.75) and from the uniqueness of the expression.

Example 3.3.3 Suppose that a reference quadratic finite element $(\hat{K},\hat{P},\hat{\Sigma})$ is defined as in Example 3.3.2. Then a finite element (K,P_K,Σ_K) is affine-equivalent to $(\hat{K},\hat{P},\hat{\Sigma})$ if there exists an invertible affine mapping $F_K: \hat{K} \to K$ such that

$$\left\{\begin{array}{l} K = F_K(\hat{K}), \\ P_K = \left\{p \mid p = \hat{p} \circ F_K^{-1},\ \forall \hat{p} \in \hat{P}\right\}, \\ \Sigma_K = \left\{\begin{array}{l} \Psi_i(p) = \hat{\Psi}_i(p \circ F_K) \equiv p(F_K(\hat{\boldsymbol{a}}_i)),\ i = 1,2,3; \\ \Psi_{ij}(p) = \hat{\Psi}_{ij}(p \circ F_K) \equiv p\left(F_K\left((\hat{\boldsymbol{a}}_i + \hat{\boldsymbol{a}}_j)/2\right)\right),\ 1 \le i < j \le 3 \end{array}\right\}. \end{array}\right. \tag{3.81}$$

□

An *affine family of finite elements* is defined as a family of finite elements that are all affine-equivalent to a single finite element, which is called the *reference finite element* and denoted by $(\hat{K}, \hat{P}, \hat{\Sigma})$. It is not difficult to show that affine-equivalence between finite elements is also transitive. Consequently, any finite element in the family, or its image under any invertible affine transformation, can be taken as the reference finite element, and the reference finite element need not belong to the family. Moreover, we have the following theorem.

Theorem 3.3.1 *Suppose that an affine family of triangulations is given. Then any affine family of finite elements associated with this family of triangulations is invariant under any invertible affine transformation of the reference finite element.*

Proof. To be specific, we denote the given affine family of triangulation by $\{\mathscr{T}_h\}$ and the reference finite element of an associated affine family of finite elements by $(\hat{K}, \hat{P}, \hat{\Sigma})$, where $\hat{P} = \{\hat{p}\} = \text{span}\{\hat{\phi}_1, \ldots, \hat{\phi}_m\}$ and $\hat{\Sigma} = \{\hat{\Psi}_1, \ldots, \hat{\Psi}_m\}$. Let the generic finite element be (K, P_K, Σ_K), which is related to $(\hat{K}, \hat{P}, \hat{\Sigma})$ through the invertible affine mapping F_K and (3.76), (3.77), and (3.79).

We now assume that a new reference finite element $(\check{K}, \check{P}, \check{\Sigma})$ is formed through an invertible affine mapping G:

$$\left\{\begin{array}{l} \check{K} = G(\hat{K}); \\ \check{P} = \{\check{p} \mid \check{p} = \hat{p} \circ G^{-1},\ \forall \hat{p} \in \hat{P}\}; \\ \check{\Sigma} = \{\check{\Psi}_i(\check{p}) = \hat{\Psi}_i(\check{p} \circ G),\ \forall \check{p} \in \check{P},\ i = 1, \ldots, m\}. \end{array}\right. \tag{3.82}$$

With this new reference finite element, we denote the generic finite element by $(K, \check{P}_K, \check{\Sigma}_K)$, where

$$\left\{\begin{array}{l} K = \check{F}_K(\check{K}); \\ \check{P}_K = \{q \mid q = \check{p} \circ \check{F}_K^{-1},\ \forall \check{p} \in \check{P}\}; \\ \check{\Sigma}_K = \{\Phi_i(q) = \check{\Psi}_i(q \circ \check{F}_K),\ \forall q \in \check{P}_K,\ i = 1, \ldots, m\}. \end{array}\right. \tag{3.83}$$

(The relations between K, $\hat{K}$, and $\check{K}$ are illustrated in Figure 3.6.) We need to prove that (K, P_K, Σ_K) is equal to $(K, \check{P}_K, \check{\Sigma}_K)$.

First, note from Proposition 3.3.1 that the choice of $\check{K}$ does not affect $\{\mathscr{T}_h\}$.

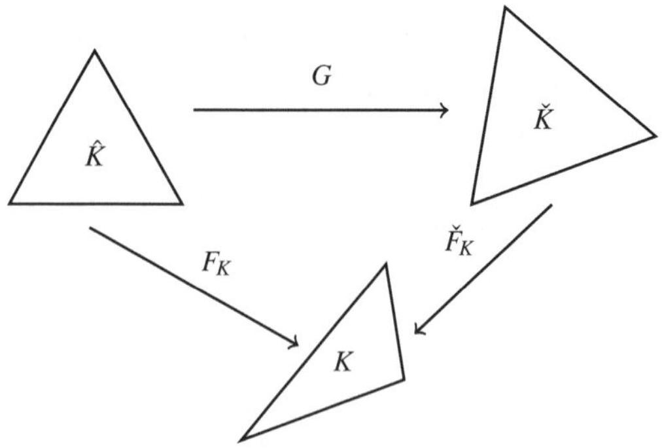

Fig. 3.6 Illustration of the relations between K, $\hat{K}$, and $\check{K}$.

Next, we show that $P_K = \check{P}_K$. Since K is affine-equivalent to $\check{K}$, for any $q \in \check{P}_K$ there exists a function $\check{p} \in \check{P}$ such that

$$q = \check{p} \circ \check{F}_K^{-1}.$$

Similarly, there exist functions $\hat{p} \in \hat{P}$ and $p \in P_K$ satisfying

$$\check{p} = \hat{p} \circ G^{-1}, \quad \hat{p} = p \circ F_K.$$

Combining these and noticing that all mappings involved are linear, we get

$$q = p \circ F_K \circ G^{-1} \circ \check{F}_K^{-1}. \tag{3.84}$$

Note from Figure 3.6 that

$$\check{F}_K^{-1} = G \circ F_K^{-1},$$

so

$$q = p \in P_K, \tag{3.85}$$

implying that $\check{P}_K \subset P_K$. Similarly, we can show $P_K \subset \check{P}_K$ so $P_K = \check{P}_K$.

To show $\check{\Sigma}_K = \Sigma_K$, we notice that for any $\Phi_i \in \check{\Sigma}_K$, there exist $\check{\Psi}_j \in \check{\Sigma}$ and $\hat{\Psi}_k \in \hat{\Sigma}$ such that

$$\Phi_i(q) = \check{\Psi}_j(q \circ \check{F}_K) = \hat{\Psi}_k(q \circ \check{F}_K \circ G).$$

Inserting (3.84) into this equation gives

$$\Phi_i(q) = \hat{\Psi}_k(p \circ F_K).$$

From (3.79), we have

$$\Phi_i(q) = \hat{\Psi}_k(p \circ F_K) = \Psi_k(p).$$

Since $q = p$ (cf. (3.85)) and q is arbitrary, this implies $\Phi_i \in \Sigma_K$ and thus $\check{\Sigma}_K \subset \Sigma_K$. Similarly we can show $\Sigma_K \subset \check{\Sigma}_K$, so $\check{\Sigma}_K = \Sigma_K$.

□

Since (K, P_K, Σ_K) is equal to $(K, \check{P}_K, \check{\Sigma}_K)$, the reference finite element $(\hat{K}, \hat{F}, \hat{\Sigma})$, just like the reference element $\hat{K}$, can be chosen as simple or convenient as possible. Moreover, for a given affine family of triangulations, any finite element computation based on an associated affine family of finite elements will have an invariant outcome under any invertible affine transformation of $(\hat{K}, \hat{P}, \hat{\Sigma})$. In the later chapters we shall find this property extremely useful in designing mesh adaptation algorithms for which we would like the same invariance to hold.

3.3.2 Simplicial elements and d-simplexes

We give a brief description of *simplicial elements*, which are commonly used mesh elements in finite element computation. A simplicial element K in $\mathbb{R}^d$, a *d-simplex*, is defined as the convex hull of any $(d+1)$ points $\boldsymbol{a}_j = [a_{1j}, ..., a_{dj}]^T \in \mathbb{R}^d$, $j = 1, ..., d+1$, such that the matrix

$$A \equiv \begin{bmatrix} a_{11} & a_{12} & \cdots & a_{1,d+1} \\ a_{21} & a_{22} & \cdots & a_{2,d+1} \\ \vdots & \vdots & & \vdots \\ a_{d1} & a_{d2} & \cdots & a_{d,d+1} \\ 1 & 1 & \cdots & 1 \end{bmatrix} \tag{3.86}$$

is non-singular. Mathematically, K can be expressed as

$$K = \left\{ \boldsymbol{x} = \sum_{j=1}^{d+1} \lambda_j \boldsymbol{a}_j : 0 \le \lambda_j \le 1, \ j = 1, ..., d+1, \ \sum_{j=1}^{d+1} \lambda_j = 1 \right\}. \tag{3.87}$$

Notice that a 0-simplex is a point, a 1-simplex is a line segment, a 2-simplex is a triangle, and a 3-simplex is a tetrahedron.

The points $\boldsymbol{a}_j$, $j = 1, ..., d+1$, constitute the vertices of K. For any m with $0 \le m \le d$, an m-face of the d-simplex K is any m-simplex whose $(m+1)$ vertices are also vertices of K. The number of m-faces of K is given by the binomial coefficient

$$C(d+1, m+1) = \binom{d+1}{m+1} = \frac{(d+1)!}{(m+1)!(d-m)!}.$$

Any $(d-1)$-face is simply called a *face* and any 1-face is called an *edge*. Note that there are $(d+1)$ faces and $d(d+1)/2$ edges. Moreover, the straight line segment connecting any two vertices forms an edge of K.

For any given point $\boldsymbol{x} \in \mathbb{R}^d$, the values of $\lambda_j, j = 1,...,d+1$, can be found by solving the algebraic system

$$\sum_{j=1}^{d+1} \lambda_j \boldsymbol{a}_j = \boldsymbol{x}, \qquad \sum_{j=1}^{d+1} \lambda_j = 1. \tag{3.88}$$

Thus, λ_j, $j = 1,...,d+1$, are functions of $\boldsymbol{x}$, i.e., $\lambda_j = \lambda_j(\boldsymbol{x})$. They are called the *barycentric coordinates* of point $\boldsymbol{x} \in \mathbb{R}^d$. On any m-face of K, there are $(d-m)$ vanishing barycentric coordinates.

The volume of a d-simplex (cf. pages 123–126 of [305]) is

$$V = \frac{1}{d!}\det(A) = \frac{1}{d!}\det(\boldsymbol{a}_2 - \boldsymbol{a}_1, ..., \boldsymbol{a}_{d+1} - \boldsymbol{a}_d). \tag{3.89}$$

For a d-simplex with constant edge length a, the volume and the radii of the inscribed and circumscribed spheres [134] are respectively

$$V = \frac{a^d}{d!}\sqrt{\frac{d+1}{2^d}}, \quad H = \frac{2a}{\sqrt{2d(d+1)}}, \quad h = a\sqrt{\frac{2d}{d+1}}. \tag{3.90}$$

From (3.74), we assume that the reference element $\hat{K}$ is taken as an equilateral d-simplex with a unitary volume. For this case, the constant edge length, in-diameter, and diameter of $\hat{K}$ can be found from the above formulas as

$$\begin{cases} \hat{a} = \sqrt{2}\left(\frac{d!}{\sqrt{d+1}}\right)^{\frac{1}{d}}, \quad \hat{H} = \frac{2}{\sqrt{d(d+1)}}\left(\frac{d!}{\sqrt{d+1}}\right)^{\frac{1}{d}}, \\ \hat{h} = 2\sqrt{\frac{d}{d+1}}\left(\frac{d!}{\sqrt{d+1}}\right)^{\frac{1}{d}}. \end{cases} \tag{3.91}$$

3.3.3 The quasi-Lagrange approach

We now construct the basic finite element approximation on a moving triangulation using the quasi-Lagrange approach. To do this, assume that a moving, affine triangulation $\mathscr{T}_h(t) = \{K(t)\}$ is given on Ω, so there exists a reference element $\hat{K}$ (independent of t) such that, for any element $K(t)$, there is an invertible affine mapping $F_{K(t)} : K(t) = F_{K(t)}(\hat{K})$ – see Figure 3.7. The family of meshes $\{\mathscr{T}_h(t), \forall t\}$ maintains its connectivity while allowing the vertices to move continuously, so each element $K(t_0) \in \mathscr{T}_h(t_0)$ for some time t_0 corresponds to a unique element $K(t_1) \in \mathscr{T}_h(t_1)$ at a later time t_1.

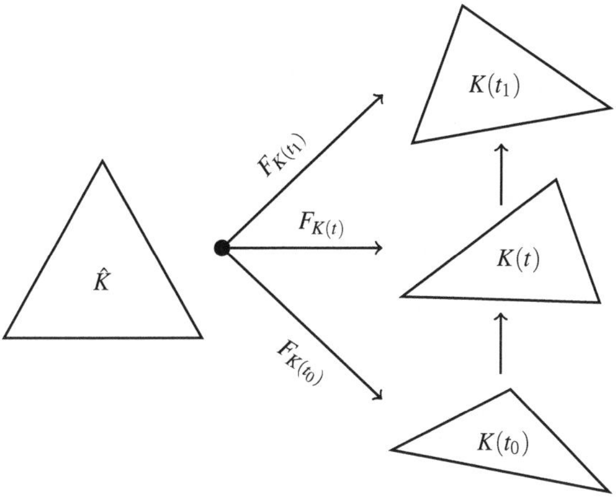

Fig. 3.7 Illustration of $F_{K(t)} : \hat{K} \to K(t)$.

Given a finite-dimensional space of functions $\hat{P}$ on $\hat{K}$ and a basis $\{\hat{\phi}_i(\boldsymbol{\xi})\}$ for it, a time-dependent approximation space $\mathscr{S}^h(t)$ can be defined by

$$\mathscr{S}^h(t) = \left\{ v \in H^1(\Omega) \;\middle|\; v|_{K(t)} \circ F_{K(t)} \in \hat{P}, \quad \forall K(t) \in \mathscr{T}_h(t) \right\}, \tag{3.92}$$

where $v|_{K(t)}$ is the restriction of v on $K(t)$ and $v|_{K(t)} \in P_{K(t)}$. A basis $\{\phi_j(\boldsymbol{x},t)\}$ for $\mathscr{S}^h(t)$ can be chosen in such a way that, for each basis function $\phi_j(\boldsymbol{x},t)$, either $\phi_j\big|_{K(t)}(\boldsymbol{x},t) \equiv 0$ or there exists an index i such that

$$\phi_j(\boldsymbol{x},t) = \hat{\phi}_i\left(F_{K(t)}^{-1}(\boldsymbol{x})\right), \quad \forall \boldsymbol{x} \in K(t). \tag{3.93}$$

(See the relation between affine-equivalent finite elements in (3.81).) Equation (3.93) can be written equivalently as

$$\phi_j\left(F_{K(t)}(\boldsymbol{\xi}),t\right) = \hat{\phi}_i(\boldsymbol{\xi}), \quad \forall \boldsymbol{\xi} \in \hat{K}. \tag{3.94}$$

Differentiating (3.94) with respect to t while keeping $\boldsymbol{\xi}$ fixed, we obtain

$$\frac{\partial \phi_j}{\partial t} + \nabla \phi_j \cdot \dot{F}_{K(t)} = 0 \quad \text{on} \quad K(t)$$

or

$$\frac{\partial \phi_j}{\partial t} = -\nabla \phi_j \cdot \dot{F}_{K(t)} \quad \text{on} \quad K(t). \tag{3.95}$$

Notice that (3.95) is also true when $\phi_j\big|_{K(t)}(\boldsymbol{x},t) \equiv 0$. Thus, the relation holds for any (linear or higher-order) basis function ϕ_j defined on affine-equivalent finite elements. Moreover, in the case of simplicial elements, since it is affine, the mapping $F_{K(t)}$ is actually a linear interpolant of the coordinates of the vertices of the element

$K(t)$, and $\dot{F}_{K(t)}$ is a linear interpolant of the mesh speeds at the vertices. Let $\Pi_1\dot{\boldsymbol{x}}$ be the piecewise linear mesh velocity satisfying

$$(\Pi_1\dot{\boldsymbol{x}})|_{K(t)} = \dot{F}_{K(t)}, \quad \forall K(t) \in \mathscr{T}_h(t). \tag{3.96}$$

Since $K(t)$ is arbitrary, from (3.95) we obtain the following lemma:

Lemma 3.3.1 *For any (linear or higher-order) basis function ϕ_j defined on a simplicial mesh for Ω, it holds that*

$$\frac{\partial \phi_j}{\partial t} = -\nabla \phi_j \cdot \Pi_1\dot{\boldsymbol{x}}, \quad \forall \boldsymbol{x} \in \Omega \tag{3.97}$$

where $\Pi_1\dot{\boldsymbol{x}}$ is the piecewise linear interpolant for velocities of mesh vertices.

From the way the basis is constructed, any function $u^h = u^h(\boldsymbol{x},t) \in \mathscr{S}^h(t)$ has the representation

$$u^h(\boldsymbol{x},t) = \sum_j u_j(t)\phi_j(\boldsymbol{x},t),$$

and it follows from (3.97) that for any $\boldsymbol{x} \in \Omega$,

$$\begin{aligned}\frac{\partial u^h}{\partial t}(\boldsymbol{x},t) &= \sum_j \frac{du_j}{dt}(t)\phi_j(\boldsymbol{x},t) + \sum_j u_j(t)\frac{\partial \phi_j}{\partial t}(\boldsymbol{x},t) \\ &= \sum_j \frac{du_j}{dt}(t)\phi_j(\boldsymbol{x},t) - \nabla u^h(\boldsymbol{x},t)\cdot \Pi_1\dot{\boldsymbol{x}}. \end{aligned} \tag{3.98}$$

Note that viewing $\sum_j \frac{du_j}{dt}(t)\phi_j(\boldsymbol{x},t)$ as $\dot{u}^h$, (3.98) reduces to the continuous relation between u_t and $\dot{u}$ in (3.19).

We now consider the finite element semi-discretization of (3.1) and (3.2). Assume that the boundary data g can be extended to a function $g(\boldsymbol{x},t) \in \mathscr{S}^h(t)$ defined inside Ω. Let $\mathscr{S}_0^h(t) = \{v \mid v \in \mathscr{S}^h(t),\ v|_{\partial\Omega} = 0\}$. Then the finite element approximation problem is to find $u^h(\boldsymbol{x},t) - g(\boldsymbol{x},t) \in \mathscr{S}_0^h(t)$ such that

$$\begin{aligned}&\int_\Omega \left[\sum_j \frac{du_j}{dt}(t)\phi_j(\boldsymbol{x},t) - \nabla u^h(\boldsymbol{x},t)\cdot \Pi_1\dot{\boldsymbol{x}} + \nabla\cdot \boldsymbol{f}(u^h,\boldsymbol{x},t)\right] v d\boldsymbol{x} \\ &\quad = \int_\Omega \left[-a\nabla u^h \cdot \nabla v + s(u^h,\boldsymbol{x},t)v\right] d\boldsymbol{x}, \qquad \forall v \in \mathscr{S}_0^h(t). \end{aligned} \tag{3.99}$$

A system of ODEs for the vector $U(t)$ of unknown functions $\{u_j(t)\}$ can be obtained by taking v to be the basis functions of $\mathscr{S}_0^h(t)$. Denoting by $\boldsymbol{X}$ and $\dot{\boldsymbol{X}}$ the node location and the nodal mesh velocity, respectively, we write this ODE system in the abstract form

$$B(\boldsymbol{X},t)\dot{U} = F(U,\boldsymbol{X},\dot{\boldsymbol{X}},t), \tag{3.100}$$

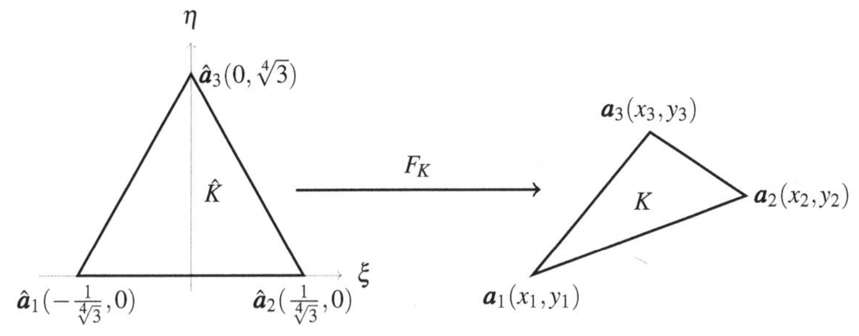

Fig. 3.8 The reference element $\hat{K}$ for a triangular mesh and the affine mapping F_K.

where $B(\boldsymbol{X},t)$ is essentially the mass matrix, modified to incorporate the boundary conditions.

As for the finite difference case, (3.100) can be solved simultaneously or alternately with a mesh equation, assumed here to be of the form

$$\dot{\boldsymbol{X}} = G(U, \boldsymbol{X}, t). \tag{3.101}$$

As before, for a simultaneous solution procedure, the extended system (3.100) and (3.101) is integrated simultaneously for the physical solution U and the mesh $\boldsymbol{X}$ using an ODE solver or a suitable marching scheme such as a Runge-Kutta method.

If an alternate solution procedure is used instead, (3.100) and (3.101) are solved alternately for U and $\boldsymbol{X}$. Taking the procedure shown in Figure 1.14 for illustration, assume that approximations for the physical solution U^n and mesh $\boldsymbol{X}^n$ are given at time $t = t_n$. A new mesh $\boldsymbol{X}^{n+1}$ is first generated from $(U^n, \boldsymbol{X}^n)$. This can be done by solving a mesh system, either PDE or optimization-based, to give $\boldsymbol{X}^{n+1}$ at time level t_{n+1}. Then the physical PDE (3.100) is integrated from $t = t_n$ to $t = t_{n+1}$ on the moving mesh

$$\boldsymbol{X}(t) = \frac{(t_{n+1} - t)}{(t_{n+1} - t_n)} \boldsymbol{X}^n + \frac{(t - t_n)}{(t_{n+1} - t_n)} \boldsymbol{X}^{n+1}, \quad \forall t \in [t_n, t_{n+1}].$$

Note that if the family of meshes $\{\mathscr{T}_h(t), \forall t\}$ is given, Theorem 3.3.1 implies that the solution $u^h(\boldsymbol{x},t)$ of (3.99) or (3.100) is invariant under any invertible affine transformation of $(\hat{K}, \hat{P}, \hat{\Sigma})$. Moreover, when the mesh is generated through a mesh adaptation algorithm such as (3.101), the invariance will still hold provided that the mesh adaptation is also invariant under any invertible affine transformation of $\hat{K}$.

Linear finite elements on triangular meshes. For illustrative purposes, we consider the special case of linear finite elements on a triangular mesh $\mathscr{T}_h$ and give the key equations needed for the finite element implementation. We choose $\hat{K}$ to be equilateral, as shown in Figure 3.8, together with the affine mapping F_K for an element K. Note that the current choice of $\hat{K}$ is different from the traditional unit triangle with vertices (0,0), (0,1), and (1,0), which is not equilateral. But as previously discussed, this should not affect the outcome of a finite element computation since the two choices of $\hat{K}$ can be mapped to each other through an invertible affine mapping. For the current case, $\hat{P}$ is the set of linear functions, i.e.,

$$\hat{P} = \{v \mid v = a\xi + b\eta + c, \forall a,b,c \in \mathbb{R}\}.$$

The mapping F_K can be expressed using the barycentric coordinates on $\hat{K}$:

$$\begin{cases} \hat{\phi}_1(\xi,\eta) &= \frac{1}{\hat{D}} \begin{vmatrix} 1 & \xi & \eta \\ 1 & \frac{1}{\sqrt[4]{3}} & 0 \\ 1 & 0 & \sqrt[4]{3} \end{vmatrix} = \frac{1}{2} - \frac{\sqrt[4]{3}}{2}\xi - \frac{1}{2\sqrt[4]{3}}\eta, \\ \hat{\phi}_2(\xi,\eta) &= \frac{1}{\hat{D}} \begin{vmatrix} 1 & -\frac{1}{\sqrt[4]{3}} & 0 \\ 1 & \xi & \eta \\ 1 & 0 & \sqrt[4]{3} \end{vmatrix} = \frac{1}{2} + \frac{\sqrt[4]{3}}{2}\xi - \frac{1}{2\sqrt[4]{3}}\eta, \\ \hat{\phi}_3(\xi,\eta) &= \frac{1}{\hat{D}} \begin{vmatrix} 1 & -\frac{1}{\sqrt[4]{3}} & 0 \\ 1 & \frac{1}{\sqrt[4]{3}} & 0 \\ 1 & \xi & \eta \end{vmatrix} = \frac{1}{\sqrt[4]{3}}\eta, \end{cases} \tag{3.102}$$

where

$$\hat{D} = \begin{vmatrix} 1 & -\frac{1}{\sqrt[4]{3}} & 0 \\ 1 & \frac{1}{\sqrt[4]{3}} & 0 \\ 1 & 0 & \sqrt[4]{3} \end{vmatrix} = 2.$$

These coordinates satisfy

$$\hat{\phi}_i(\hat{\boldsymbol{a}}_j) = \delta_{ij}, \quad i,j = 1,2,3 \tag{3.103}$$

$$0 \le \hat{\phi}_i(\xi,\eta) \le 1, \quad \forall(\xi,\eta) \in \hat{K} \tag{3.104}$$

$$\hat{\phi}_1(\xi,\eta) + \hat{\phi}_2(\xi,\eta) + \hat{\phi}_3(\xi,\eta) = 1, \quad \forall(\xi,\eta) \in \hat{K}. \tag{3.105}$$

The affine mapping F_K shown in Figure 3.8 can be expressed as

$$\begin{bmatrix} x \\ y \end{bmatrix} = F_K(\xi,\eta) = \sum_{i=1}^{3} \hat{\phi}_i(\xi,\eta) \begin{bmatrix} x_i \\ y_i \end{bmatrix}. \tag{3.106}$$

Differentiating F_K with respect to ξ and η, we obtain its Jacobian matrix

$$F_K' = \begin{bmatrix} \frac{\sqrt[4]{3}}{2}(x_2 - x_1) & \frac{1}{2\sqrt[4]{3}}(2x_3 - x_1 - x_2) \\ \frac{\sqrt[4]{3}}{2}(y_2 - y_1) & \frac{1}{2\sqrt[4]{3}}(2y_3 - y_1 - y_2) \end{bmatrix}. \tag{3.107}$$

The inverse mapping of F_K, F_K^{-1}, can be obtained by solving (3.106) for ξ and η, but a more compact derivation uses the barycentric coordinates on K to compute ξ and η directly. Letting

$$D = \begin{vmatrix} 1 & x_1 & y_1 \\ 1 & x_2 & y_2 \\ 1 & x_3 & y_3 \end{vmatrix} = (x_2 - x_1)(y_3 - y_2) - (x_3 - x_2)(y_2 - y_1),$$

the barycentric coordinates on K are given by

$$\begin{cases} \lambda_1(x,y) &= \frac{1}{D}\begin{vmatrix} 1 & x & y \\ 1 & x_2 & y_2 \\ 1 & x_3 & y_3 \end{vmatrix} = \frac{1}{D}\left[(x_2 - x)(y_3 - y_2) - (x_3 - x_2)(y_2 - y)\right], \\ \lambda_2(x,y) &= \frac{1}{D}\begin{vmatrix} 1 & x_1 & y_1 \\ 1 & x & y \\ 1 & x_3 & y_3 \end{vmatrix} = \frac{1}{D}\left[(x_3 - x)(y_1 - y_3) - (x_1 - x_3)(y_3 - y)\right], \\ \lambda_3(x,y) &= \frac{1}{D}\begin{vmatrix} 1 & x_1 & y_1 \\ 1 & x_2 & y_2 \\ 1 & x & y \end{vmatrix} = \frac{1}{D}\left[(x_1 - x)(y_2 - y_1) - (x_2 - x_1)(y_1 - y)\right]. \end{cases} \tag{3.108}$$

Then F_K^{-1} has the form

$$\begin{bmatrix} \xi \\ \eta \end{bmatrix} = F_K^{-1}(x,y) = \begin{bmatrix} -\frac{1}{\sqrt[4]{3}} \\ 0 \end{bmatrix} \lambda_1(x,y) + \begin{bmatrix} \frac{1}{\sqrt[4]{3}} \\ 0 \end{bmatrix} \lambda_2(x,y) + \begin{bmatrix} 0 \\ \sqrt[4]{3} \end{bmatrix} \lambda_3(x,y). \tag{3.109}$$

A direct calculation gives

$$\left(F_K'\right)^{-1} = \frac{1}{D}\begin{bmatrix} \frac{1}{\sqrt[4]{3}}(2y_3 - y_1 - y_2) & -\frac{1}{\sqrt[4]{3}}(2x_3 - x_1 - x_2) \\ -\sqrt[4]{3}(y_2 - y_1) & \sqrt[4]{3}(x_2 - x_1) \end{bmatrix}. \tag{3.110}$$

We can now construct the basis functions for the finite element space of piecewise linear functions $\mathscr{S}^h$ in (3.92). Denote the vertices of $\mathscr{T}_h$ by $\{\boldsymbol{a}_1, \ldots, \boldsymbol{a}_{N_v}\}$ and let ω_j be the patch of elements which contain $\boldsymbol{a}_j$ as a vertex. For each $K \in \omega_j$, let $\hat{a}_{i_K}$ be the vertex of $\hat{K}$ corresponding to $\boldsymbol{a}_j$ under the mapping F_K, i.e., $\boldsymbol{a}_j = F_K(\hat{\boldsymbol{a}}_{i_K})$. Then the basis function ϕ_j associated with vertex $\boldsymbol{a}_j$ can be expressed as

$$\phi_j(x,y) = \begin{cases} 0, & \forall (x,y) \notin \omega_j \\ \hat{\phi}_{i_K}(F_K^{-1}(x,y)) = \lambda_{i_K}(x,y), & \forall (x,y) \in K, \quad \forall K \in \omega_j. \end{cases} \tag{3.111}$$

Table 3.1 Parameters for numerical integration schemes on a triangle.

m	$(\hat{\phi}_1^{(j)}, \hat{\phi}_2^{(j)}, \hat{\phi}_3^{(j)})$	$w^{(j)}$	Degree of Precision
1	$(\frac{1}{3}, \frac{1}{3}, \frac{1}{3})$	1	1
3	$(0, \frac{1}{2}, \frac{1}{2})$, $(\frac{1}{2}, 0, \frac{1}{2})$, $(\frac{1}{2}, \frac{1}{2}, 0)$	$\frac{1}{3}$	2
7	$(1,0,0)$, $(0,1,0)$, $(0,0,1)$ $(0, \frac{1}{2}, \frac{1}{2})$, $(\frac{1}{2}, 0, \frac{1}{2})$, $(\frac{1}{2}, \frac{1}{2}, 0)$ $(\frac{1}{3}, \frac{1}{3}, \frac{1}{3})$	$\frac{1}{20}$ $\frac{2}{15}$ $\frac{9}{20}$	3

The integrals in (3.99) must generally be computed approximately using numerical quadrature schemes, which on a triangle take the form

$$\int_K f(x,y)dxdy \approx |K| \sum_{j=1}^{m} w^{(j)} f\left(\sum_{i=1}^{3} \hat{\phi}_i^{(j)} \begin{bmatrix} x_i \\ y_i \end{bmatrix} \right), \tag{3.112}$$

where $w^{(j)}$ are the weights and $(\hat{\phi}_1^{(j)}, \hat{\phi}_2^{(j)}, \hat{\phi}_3^{(j)})$ are the barycentric coordinates on $\hat{K}$. These parameters are given in Table 3.1 for three schemes. The choice of m in practice depends upon the user's desired level of accuracy and code complexity. A computationally efficient choice is often $m = 3$.

3.3.4 The rezoning approach

As for finite differences, using the rezoning approach the physical solution U^n must be interpolated from the old mesh $\mathscr{T}_h(t_n)$ to the new one $\mathscr{T}_h(t_{n+1})$; see §3.5. The physical solution is then discretized on $\mathscr{T}_h(t_{n+1})$, which is held fixed for the integration step over $(t_n, t_{n+1}]$. The discrete equation is similar to (3.99), except that now the approximation space is defined on the fixed mesh $\mathscr{T}_h(t_{n+1})$, so the basis functions are time independent and the mesh speed term $\Pi_1 \dot{\boldsymbol{x}}$ is set to zero.

3.4 Two-mesh strategy for mesh movement

Generally speaking, we do not have to calculate the mesh points to an accuracy as high as for the physical PDE. For this reason, we can perform mesh movement on a relatively coarse mesh and solve the physical PDE on a fine mesh obtained from the coarse mesh via refinement. It has indeed been shown in Huang [175] (also see Example 6.6.1 of §6.6) that this two-mesh strategy can dramatically improve the efficiency of moving mesh methods. A similar idea has been used by Fiedler and Trapp [144] for the dynamic generation of adaptive meshes using an elliptic differential equation system and by Mulholland et al. [263] where an adaptive finite difference mesh is used for the pseudo-spectral solution of near-singular problems.

The strategy can be used with structured and unstructured meshes. Obviously, the simplest way to obtain a fine mesh is to uniformly refine all of the elements of the coarse mesh. For example, a fine triangular mesh can be obtained by dividing each triangle of a coarse mesh into four triangles by connecting the edge midpoints. The situation for a 2D logically rectangular mesh is slightly more complicated. Denote the coarse mesh by $\{(x^c_{jk}, y^c_{jk}),\ j=1,...,J^c,\ k=1,...,K^c\}$. Given two positive integers JM and KM, the fine mesh $\{(x_{jk}, y_{jk}),\ j=1,...,J,\ k=1,...,K\}$ can be obtained using uniform refinement as follows. First of all, let

$$J-1=(J^c-1)\cdot JM, \quad K-1=(K^c-1)\cdot KM,$$

or

$$J=1+(J^c-1)\cdot JM, \quad K=1+(K^c-1)\cdot KM.$$

Then, set

$$\begin{cases} x_{1+(j-1)\cdot JM, 1+(k-1)\cdot KM} = x^c_{j,k}, \\ y_{1+(j-1)\cdot JM, 1+(k-1)\cdot KM} = y^c_{j,k}, \end{cases} \qquad j=1,...,J^c,\ k=1,...,K^c.$$

Note that each element of the coarse mesh is a macro-element of the fine mesh. We next compute the equidistant points ($JM+1$ points in j direction and $KM+1$ points in k direction) on the boundary segments of each macro-element. By connecting the points on the opposite boundary segments of the macro-elements, we find the position of all mesh points.

To capture the fine structures of the physical solution, the monitor function should first be computed on the fine mesh and then projected to the coarse mesh. Moreover, the coarse mesh cannot be chosen too coarse to catch the fine structures of the physical solution. On the other hand, it cannot be chosen too fine to reduce the necessary overhead of mesh movement. The particular choice of the coarse mesh depends of course upon the specific applications.

3.5 Interpolation on moving meshes

As we have seen in previous sections, with a rezoning approach the physical solution needs to be interpolated between moving mesh points. Interpolation for moving meshes is easier than for general meshes because the number of elements and the mesh connectivity are kept the same at any two time levels. In this section we limit discussion to this interpolation problem itself, describing two types of interpolation scheme – a traditional one (a linear interpolation scheme) and a PDE-based one – in two dimensions. The discussion is similar to that in §2.6.3 for one dimension except that multi-dimensional interpolation is generally more complicated and less efficient.

Denote the old and new meshes by $\mathscr{T}_h^0 = \{(x_j^0, y_j^0)\}$ and $\mathscr{T}_h^1 = \{(x_j^1, y_j^1)\}$, respectively. Assume that the physical solution $u = u(x,y)$ is available at the vertices of $\mathscr{T}_h^0$, i.e., $u_j^0 = u(x_j^0, y_j^0)$ are given. The task is to find approximations $u_j^1 \approx u(x_j^1, y_j^1)$ using interpolation of these known values.

3.5.1 Linear interpolation

The major difficulty for interpolation generally lies in point location, i.e., finding the elements that contain the new mesh points. For the general situation where the meshes may have different topologies, this point location problem has been a topic of active research in its own right, and many open questions remain in three and higher dimensions; e.g., see [114] for general discussion on this topic and [283] for a fast nearest-point searching method for scattered data.

Fortunately, in our situation the points can be reasonably efficiently located. Since the new and old meshes have the same topology, the new mesh can be viewed as a deformation of the old one. Thus, the search for the location of a mesh point (x_j^1, y_j^1) on the old mesh can be started with the neighboring elements of the corresponding mesh point (x_j^0, y_j^0). If the desired element is not found there, we expand the search to the neighboring elements of these neighboring elements, continuing the search until the element that contains (x_j^1, y_j^1) is found. In the context of moving meshes, the new node (x_j^1, y_j^1) is typically not very far from the old one (x_j^0, y_j^0). Thus, the point location process typically terminates after a few tests. Note that a point is in a triangle if and only if its barycentric coordinates on the triangle are non-negative. For example, consider the triangle element K shown in Figure 3.8. A point (x,y) is in K only if the quantities $\lambda_1(x,y)$, $\lambda_2(x,y)$, and $\lambda_3(x,y)$ defined in (3.108) are all non-negative. After the triangle K containing the point (x,y) is determined, the linear interpolation value can simply be calculated using

$$u(x,y) \approx \lambda_1(x,y)u(x_1,y_1) + \lambda_2(x,y)u(x_2,y_2) + \lambda_3(x,y)u(x_3,y_3). \qquad (3.113)$$

3.5.2 PDE-based interpolation

Like the one-dimensional case (cf. §2.6.3), the current interpolation problem on a moving mesh can be formulated as an equivalent problem of solving a differential equation. Define a time continuation from the old mesh to the new one by

$$\mathscr{T}_h(t): \quad x_j(t) = (1-t)x_j^0 + tx_j^1, \quad y_j(t) = (1-t)y_j^0 + ty_j^1, \qquad t \in [0,1]. \quad (3.114)$$

Then the interpolation of u from $\mathscr{T}_h^0$ to $\mathscr{T}_h^1$ is equivalent to finding on the moving mesh $\mathscr{T}_h(t)$ the solution $v = v(x,y,t)$ of the differential equation

$$\frac{\partial v}{\partial t} = 0, \qquad t \in (0,1] \qquad (3.115)$$

subject to the initial condition $v(x_j(0), y_j(0), 0) = u_j^0$. The sought values u_j^1 are related to v by $u_j^1 = v_j^1 \approx v(x_j(1), y_j(1), 1)$. From (3.34) and (3.35), (3.115) can be transformed into the computational domain as

$$\dot{v} + \frac{\partial v}{\partial \xi}\xi_t + \frac{\partial v}{\partial \eta}\eta_t = 0, \qquad \text{(non-conservative form)} \qquad (3.116)$$

$$\dot{\overline{(Jv)}} + \frac{\partial}{\partial \xi}(Jv\xi_t) + \frac{\partial}{\partial \eta}(Jv\eta_t) = 0. \qquad \text{(conservative form)} \qquad (3.117)$$

Note that these are linear convection equations. Moreover, when a finite element method is used, (3.115) can be discretized directly on the moving mesh (cf. §3.3.2) and the resulting equation, a simple variation of (3.99), is essentially equivalent to a convection equation.

Once again, finite differences, finite elements, finite volumes, and other methods can be used for solving PDE (3.115), (3.116), or (3.117), but caution should be taken. While the effect of the convection terms caused by the mesh movement is not well understood in moving mesh methods, a use of some upwinding is recommended to avoid a possible stability problem. Moreover, a conservative interpolation scheme is often necessary when the physical problem exhibits a strong hyperbolic feature. Consequently, a conservative scheme should generally be used for solving the interpolation PDEs. To this end, use of the conservation form (3.117) or the Leibniz integral rule (3.25) is helpful; see §2.6.3 and Tang [318].

3.6 Biographical notes

A more complete list of transformation relations can be seen in Thompson et al. [325].

As discussed in the biographical notes of Chapter 2, in addition to finite difference and finite element methods, collocation, finite volume, and spectral methods have also been used for moving mesh methods. Examples include [187, 286] for collocation, [140, 141, 316] for finite volume, and [263, 338, 142, 320] for spectral methods. The stability issue is investigated by Ferreira [143], Formaggia and Nobile [149], and Mackenzie and Mekwi [245].

In this chapter only parabolic-type differential equations have been discussed. However, for hyperbolic problems or problems exhibiting strong hyperbolic features (such as convection-dominated PDEs), a method with some numerical dissipation such as the ENO (essentially non-oscillatory), the WENO (weighted ENO), or a discontinuous Galerkin (DG) scheme (e.g, see Shu [301, 302]) should be used. The reader is referred to the review article by Tang [318] for the numerical solution of hyperbolic differential equations on moving meshes.

3.7 Exercises

1. Verify the 2D version of (3.10) using (3.28) and (3.29).
2. Derive (3.19) from (3.20).
3. Derive (3.21) and (3.20) from (3.23) and (3.25), respectively.
4. Show that (3.26) and (3.27) are mathematically equivalent.
5. Derive in detail the 2D versions (3.36) and (3.37) of the conservative and non-conservative forms of the physical PDE in equations (3.26) and (3.27).
6. Give the coordinates of the vertices of the unit 2-simplex (triangle) or an equilateral 2-simplex (triangle) having a unitary volume.
7. Give the coordinates of the vertices of the unit 3-simplex (tetrahedron) or an equilateral 3-simplex (tetrahedron) having a unitary volume.
8. Describe two different constant-polynomial finite elements associated with a triangle K with vertices $\boldsymbol{a}_i$, $i = 1,2,3$.
9. Verify by direct calculation that (3.97) holds for 1D piecewise linear basis functions defined in (1.33).
10. Prove properties (3.103)–(3.105).
11. Prove that the degree of precision of quadrature (3.112) with $m = 3$ (cf. Table 3.1) is 2.
12. Verify the result $\hat{\phi}_{i_K}(F_K^{-1}(x,y)) \equiv \lambda_{i_K}(x,y)$ given in (3.111).

Chapter 4
Basic Principles of Multidimensional Mesh Adaptation

We have seen in Chapter 2 the crucial role that the equidistribution principle plays in designing adaptive mesh algorithms, where in fact a mesh can be fully determined from it in 1D. The situation becomes much more complicated in multidimensions. The equidistribution principle, specifying only the volume of mesh elements, is no longer sufficient for determining a multidimensional mesh. An additional condition is needed for specifying the shape and orientation of mesh elements.

The major objective of this chapter is to study the basic principles of multidimensional mesh adaptivity, including the needed alignment condition. While controlled by the mesh density function in 1D, the mesh adaptivity is now driven by a solution-dependent, matrix-valued monitor function M, which defines a metric on the physical domain and specifies the size, shape, and orientation of mesh elements throughout. Three fundamental interpretations of the multidimensional mesh adaptivity are presented in this chapter. The first one views mesh adaptation as a technique to generate an M-uniform mesh (a uniform mesh in the metric space) while the second and third consider mesh adaptation from the perspectives of mesh control and function approximation, respectively. The interpretations lead naturally to the equidistribution and alignment conditions for meshes, with the role of the latter condition being to ensure that the mesh is properly aligned with behavior of the physical solution. These conditions are analyzed and related to the mesh quality in a mathematically precise way. Interpretations are also given from both a discrete (mesh) and continuous (coordinate transformation) perspective.

We assume in this chapter that a monitor function has been chosen. However, like the mesh density function in the 1D case, the proper choice of a monitor function is instrumental in the successful use of mesh adaptation principles. This topic is considered at length in Chapter 5. In Chapters 6 and 7 we consider the various types of methods for generating adaptive methods. The basic theory covered in this chapter will be important there in providing many of the needed tools for analyzing

W. Huang and R.D. Russell, *Adaptive Moving Mesh Methods*, Applied Mathematical Sciences 174, DOI 10.1007/978-1-4419-7196-2_4, © Springer Science+Business Media, LLC 2011

the basic theoretical properties (like solution existence) and practical features of these various methods.

It bears emphasizing that while the mesh adaptation principles are developed in this chapter for the purpose of adaptive mesh movement, they are applicable for the general areas of mesh generation and adaptation as well. For example, they can be the basis for determining a metric tensor for unstructured mesh adaptation [179].

4.1 Mesh adaptation from perspective of uniform meshes in a metric space

We first study multidimensional mesh adaptation from the perspective of constructing uniform meshes in a metric space in $\Omega \subset \mathbb{R}^d$ $(d \geq 1)$, where an adaptive mesh is generated as a uniform one in some metric space in Ω with a matrix-valued function $M = M(\boldsymbol{x})$ defined on it. (See §2.1.3 for the 1D case.) Such a mesh is referred to as an *M-uniform mesh*. In the definition, M can be understood for standing for the matrix-valued function M and/or the metric. The function $M = M(\boldsymbol{x})$ is called the *monitor function*, and $\rho = \sqrt{\det(M(\boldsymbol{x}))}$ is referred to as the corresponding *mesh density function*. In the literature M is occasionally called the *metric tensor* since it plays the same role as a metric tensor for a Riemannian manifold. Like the mesh density function in the 1D case, the monitor function is always chosen to be symmetric and positive definite, and it is normally a measure of the difficulty in approximating the solution. In this section we assume that the monitor function is given and defer the discussion of how it is chosen to the next chapter.

An M-uniform mesh is generally non-uniform in the Euclidean space. For illustrative purpose, we show in Figure 4.1(a) a mesh which is visually non-uniform but uniform in the metric specified by

$$M = \begin{bmatrix} 4 & 0 \\ 0 & 1 \end{bmatrix} \tag{4.1}$$

and in Figure 4.1(b) a mesh which is visually uniform but non-uniform in the metric M.

We shall see that viewing an adaptive mesh as a uniform one in some metric space has the advantage that such a mesh can be described both geometrically and analytically in a relatively simple manner. Such a description in turn facilitates (a) a precise characterization of mesh alignment with the monitor function (and thus with the solution when M is defined as a function of u – cf. Theorem 4.2.4 in §4.2), (b) a proper choice of the monitor function based on error estimates (cf. §5.2), and (c) a consequent development of reliable algorithms for adaptive mesh generation (cf. Chapter 6).

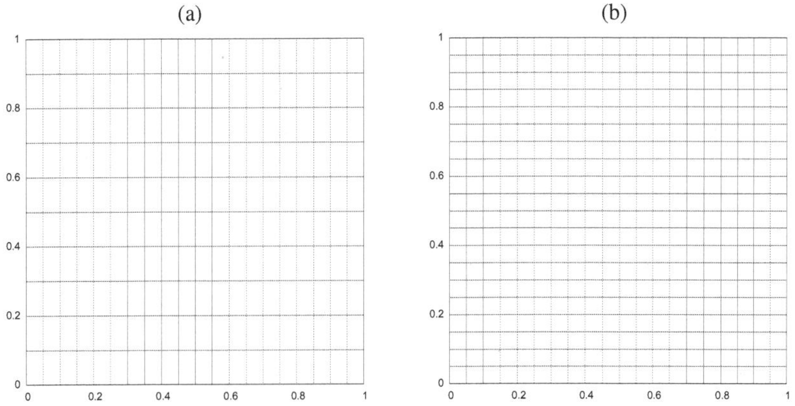

Fig. 4.1 (a) A uniform mesh and (b) a non-uniform mesh in the metric specified by M defined in (4.1).

4.1.1 Mathematical description of M-uniform meshes

To give the mathematical description for an M-uniform mesh, we assume that the physical domain $\Omega \subset \mathbb{R}^d$ is polyhedral. Moreover, we consider an affine family of simplicial meshes $\{\mathscr{T}_h\}$ for Ω, with the reference element $\hat{K}$ being chosen as an equilateral d-simplex with unit volume. (The edge length, diameter, and in-diameter of $\hat{K}$ are given in (3.91).) Recall that, by definition, for any element K in $\{\mathscr{T}_h\}$ there exists a unique invertible affine mapping $F_K : \hat{K} \to K$ such that $K = F_K(\hat{K})$.

By definition, an M-uniform mesh is uniform in the metric M. In other words, a mesh $\mathscr{T}_h$ is M-uniform if and only if all its elements have the same volume and are equilateral in the metric. Mathematically, these two conditions can be described as follows.

(i) All elements have a constant volume in the metric M:

$$\int_K \rho(\boldsymbol{x}) d\boldsymbol{x} = \frac{\sigma}{N}, \qquad \forall K \in \mathscr{T}_h \tag{4.2}$$

where N is the number of the elements of $\mathscr{T}_h$ and $\sigma = \int_\Omega \rho(\boldsymbol{x}) d\boldsymbol{x}$.

(ii) All elements are equilateral in the metric M: Let $\gamma_1, \ldots, \gamma_{d(d+1)/2}$ be the edges of an element K. (For simplicity, the dependence of the edges on K is omitted.) This condition requires

$$|\gamma_1|_M = \cdots = |\gamma_{d(d+1)/2}|_M, \quad \forall K \in \mathscr{T}_h \tag{4.3}$$

where $|\gamma_i|_M$ denotes the length of edge γ_i in the metric M. If $\boldsymbol{x} = \boldsymbol{\phi}_i(s)$, $s \in [s_0^i, s_1^i]$, is a parameterized equation for γ_i, then this length can be expressed by

$$|\gamma_i|_M = \int_{s_0^i}^{s_1^i} \sqrt{\left(\frac{d\boldsymbol{\phi}_i}{ds}\right)^T M(\boldsymbol{\phi}_i(s)) \frac{d\boldsymbol{\phi}_i}{ds}} ds. \tag{4.4}$$

For a given continuous monitor function $M = M(\boldsymbol{x})$, (4.2) and (4.3) give a highly nonlinear system of equations for the coordinates of the vertices of $\mathscr{T}_h$. The nonlinearity makes the system difficult (if not impossible) to solve, which in turn makes it inefficient to generate an M-uniform mesh in practice. More importantly, it turns out that the conditions (4.2) and (4.3) are not convenient to use in actual error analysis. In the next subsection, we consider a simplification of the conditions which can avoid these difficulties.

4.1.2 Equidistribution and alignment conditions

We consider here a type of approximate M-uniform mesh which has a simpler definition and is more convenient to use in error analysis (cf. §5.2). The simplification comes from the replacement of $M(\boldsymbol{x})$ in (4.2) and (4.3) by its average on K, i.e.,

$$M_K = \frac{1}{|K|} \int_K M(\boldsymbol{x}) d\boldsymbol{x}. \tag{4.5}$$

The conditions become

$$\int_K \rho_K d\boldsymbol{x} = \frac{\sigma_h}{N}, \quad \forall K \in \mathscr{T}_h \tag{4.6}$$

$$|\gamma_1|_{M_K} = \cdots = |\gamma_{d(d+1)/2}|_{M_K}, \quad \forall K \in \mathscr{T}_h \tag{4.7}$$

where

$$\rho_K = \sqrt{\det(M_K)}, \quad \sigma_h = \sum_{K \in \mathscr{T}_h} \int_K \rho_K d\boldsymbol{x}. \tag{4.8}$$

These conditions can be simplified. Notice that

$$\int_K \rho_K d\boldsymbol{x} = \rho_K \, |K|,$$

where $|K|$ is the volume of K (in the Euclidean sense). Moreover, since K $(= F_K(\hat{K}))$ is a d-simplex, its edges are straight line segments and $\frac{d}{ds}\boldsymbol{\phi}_i$ is a constant vector. As a result,

$$|\gamma_i|_{M_K} = \sqrt{\left(\frac{d\boldsymbol{\phi}_i}{ds}\right)^T M_K \frac{d\boldsymbol{\phi}_i}{ds}} \, (s_1^i - s_0^i) = \sqrt{\boldsymbol{\gamma}_i^T M_K \boldsymbol{\gamma}_i},$$

where $\boldsymbol{\gamma}_i$ denotes a vector corresponding to the edge γ_i. Conditions (4.6) and (4.7) can then be expressed as

$$\rho_K |K| = \frac{\sigma_h}{N}, \tag{4.9}$$

$$\boldsymbol{\gamma}_1^T M_K \boldsymbol{\gamma}_1 = \cdots = \boldsymbol{\gamma}_{d(d+1)/d}^T M_K \boldsymbol{\gamma}_{d(d+1)/2}. \tag{4.10}$$

Denote the edges of the reference element $\hat{K}$ by $\hat{\gamma}_1, ..., \hat{\gamma}_{d(d+1)/2}$. We assume that they have been reordered such that they in turn correspond to the edges of K, γ_1, ..., $\gamma_{d(d+1)/2}$. It is easy to see that the edges of K and $\hat{K}$ are related by

$$\boldsymbol{\gamma}_i = F_K' \hat{\boldsymbol{\gamma}}_i, \quad i = 1, ..., \frac{d(d+1)}{2}$$

where F_K' denotes the Jacobian matrix of mapping F_K. Inserting this into (4.10) yields

$$\hat{\boldsymbol{\gamma}}_1^T (F_K')^T M_K F_K' \hat{\boldsymbol{\gamma}}_1 = \cdots = \hat{\boldsymbol{\gamma}}_{d(d+1)/2}^T (F_K')^T M_K F_K' \hat{\boldsymbol{\gamma}}_{d(d+1)/2}. \tag{4.11}$$

We need the following lemma to further simplify (4.11).

Lemma 4.1.1 *Suppose that S is a $d \times d$ symmetric and positive definite constant matrix and θ is a positive number. There exists a d-simplex K_d with constant edge length l_d such that*

$$\boldsymbol{\gamma}^T S \boldsymbol{\gamma} = \theta \quad \textit{for any edge vector } \boldsymbol{\gamma} \textit{ of } K_d \tag{4.12}$$

if and only if

$$S = \frac{\theta}{l_d^2} I_d, \tag{4.13}$$

where I_d is the $d \times d$ identity matrix.

Proof. It is easy to see that (4.13) is sufficient for (4.12).

We prove by induction that (4.13) is also necessary for (4.12). Obviously, (4.12) implies (4.13) for $d = 1$. For any positive integer $n \geq 1$, assume that (4.12) implies (4.13) for $d = n - 1$. We want to prove that this also holds for $d = n$.

Note that the conclusion remains unchanged if K_n is rotated by any angle and in any direction. Thus we take K_n as an n-simplex formed by adding a new vertex $\boldsymbol{a}_{n+1} = (a_{1,n+1}, ..., a_{n-1,n+1}, a_{n,n+1})^T$ to an $(n-1)$-simplex K_{n-1} which lies on the coordinate super-plane formed by the first $(n-1)$ coordinate axes. Without loss of generality we assume $a_{n,n+1} > 0$. The vertices of K_{n-1} have the form

$$\boldsymbol{a}_j = (a_{1j}, ..., a_{n-1,j}, 0)^T, \quad j = 1, ..., n. \tag{4.14}$$

Since K_n has a constant edge length l_n, the length of the edge connecting $\boldsymbol{a}_j$ ($j = 1, ..., n$) and $\boldsymbol{a}_{n+1}$ is equal to l_n, i.e.,

$$\sum_{i=1}^{n-1} (a_{i,n+1} - a_{ij})^2 + a_{n,n+1}^2 = l_n^2, \quad j = 1, ..., n. \tag{4.15}$$

Moreover, the projection of vertex $\boldsymbol{a}_{n+1}$ onto K_{n-1} should by symmetry coincide with this center, viz.,

$$a_{i,n+1} = \frac{1}{n} \sum_{j=1}^{n} a_{ij}, \quad i = 1, ..., n-1. \tag{4.16}$$

Condition (4.12) can be written as

$$(\boldsymbol{a}_j - \boldsymbol{a}_k)^T S (\boldsymbol{a}_j - \boldsymbol{a}_k) = \theta, \quad j,k = 1,...,n, \quad j \neq k \tag{4.17}$$

$$(\boldsymbol{a}_j - \boldsymbol{a}_{n+1})^T S (\boldsymbol{a}_j - \boldsymbol{a}_{n+1}) = \theta, \quad j = 1,...,n. \tag{4.18}$$

Let

$$\begin{cases} \tilde{\boldsymbol{a}}_j = \begin{bmatrix} a_{1j} \\ \vdots \\ a_{n-1,j} \end{bmatrix} \quad (j = 1,...,n+1), \\ \tilde{\boldsymbol{s}} = \begin{bmatrix} s_{1n} \\ \vdots \\ s_{n-1,n} \end{bmatrix}, \quad \tilde{S} = \begin{bmatrix} s_{11} & \cdots & s_{1,n-1} \\ \vdots & & \vdots \\ s_{n-1,1} & \cdots & s_{n-1,n-1} \end{bmatrix}. \end{cases} \tag{4.19}$$

From (4.14), (4.17) reduces to

$$(\tilde{\boldsymbol{a}}_j - \tilde{\boldsymbol{a}}_k)^T \tilde{S} (\tilde{\boldsymbol{a}}_j - \tilde{\boldsymbol{a}}_k) = \theta, \quad j,k = 1,...,n, \quad j \neq k$$

which represents the same condition as (4.12) but for $(n-1)$-simplex K_{n-1} with constant edge length l_n. By the induction assumption, this implies

$$\tilde{S} = \frac{\theta}{l_n^2} I_{n-1}. \tag{4.20}$$

Using this, we can rewrite (4.18) as

$$\sum_{i=1}^{n-1} (a_{ij} - a_{i,n+1})^2 \frac{\theta}{l_n^2} - 2 \sum_{i=1}^{n-1} (a_{ij} - a_{i,n+1}) a_{n,n+1} s_{in} + a_{n,n+1}^2 s_{nn} = \theta, \quad j = 1,...,n$$

which together with (4.15) implies

$$-2 \sum_{i=1}^{n-1} (a_{ij} - a_{i,n+1}) a_{n,n+1} s_{in} + a_{n,n+1}^2 s_{nn} = a_{n,n+1}^2 \frac{\theta}{l_n^2}, \quad j = 1,...,n. \tag{4.21}$$

Summing from $j = 1$ to n and using (4.16) and the assumption $a_{n,n+1} > 0$, we have

$$s_{nn} = \frac{\theta}{l_n^2}. \tag{4.22}$$

Inserting this back into (4.21) yields

$$\sum_{i=1}^{n-1}(a_{ij}-a_{i,n+1})s_{in}=0, \quad j=1,...,n$$

or

$$(\tilde{\boldsymbol{a}}_j-\tilde{\boldsymbol{a}}_{n+1})^T\tilde{\boldsymbol{s}}=0, \quad j=1,...,n. \tag{4.23}$$

By subtracting one equation from another in (4.23) we get

$$(\tilde{\boldsymbol{a}}_j-\tilde{\boldsymbol{a}}_{j+1})^T\tilde{\boldsymbol{s}}=0, \quad j=1,...,n-1. \tag{4.24}$$

Recall that $\tilde{\boldsymbol{a}}_j$, $j=1,...,n$ are the vertices of $(n-1)$-simplex $\hat{K}_{n-1}$. From the definition of $(n-1)$-simplexes, the vectors $(\tilde{\boldsymbol{a}}_j-\tilde{\boldsymbol{a}}_{j+1})$, $j=1,...,n-1$ are linearly independent. Consequently, the linear system (4.24) has the unique zero solution $\tilde{\boldsymbol{s}}=0$. Combining this with (4.20) and (4.22) we obtain

$$S=\frac{\theta}{l_n^2}I_n,$$

and the proof is complete. □

Note now that (4.11) implies (4.12) with $\theta := \hat{\boldsymbol{\gamma}}_1^T(F_K')^T M_K F_K' \hat{\boldsymbol{\gamma}}_1$. Let $\theta_K = \theta/\hat{a}^2$, where $\hat{a}$ is the constant edge length of $\hat{K}$ (cf. (3.91)). Lemma 4.1.1 then implies that (4.11) is equivalent to

$$(F_K')^T M_K F_K' = \theta_K I. \tag{4.25}$$

As a consequence, the matrix $(F_K')^T M_K F_K'$ has equal eigenvalues. From the arithmetic-mean geometric-mean inequality (cf. Theorem B.0.11), it is easy to prove that (4.25), and therefore (4.11), is equivalent to

$$\frac{1}{d}\operatorname{tr}((F_K')^T M_K F_K') = \det((F_K')^T M_K F_K')^{\frac{1}{d}}, \quad \forall K \in \mathscr{T}_h \tag{4.26}$$

where $\operatorname{tr}(\cdot)$ and $\det(\cdot)$ denote the trace and determinant of a matrix, respectively. These quantities are in turn equal to the sum and product of the eigenvalues.

When $(F_K')^T M_K F_K'$ has equal eigenvalues, so does its inverse matrix. Therefore, another condition equivalent to (4.11) is

$$\frac{1}{d}\operatorname{tr}((F_K')^{-1} M_K^{-1} (F_K')^{-T}) = \det((F_K')^{-1} M_K^{-1} (F_K')^{-T})^{\frac{1}{d}}, \quad \forall K \in \mathscr{T}_h. \tag{4.27}$$

Thus far, we have shown that a mesh satisfying (4.6) and (4.7) also satisfies (4.9) and (4.26), i.e.,

$$\rho_K\,|K| = \frac{\sigma_h}{N}, \quad \forall K \in \mathscr{T}_h \tag{4.28}$$

$$\frac{1}{d}\mathrm{tr}((F_K')^T M_K F_K') = \det((F_K')^T M_K F_K')^{\frac{1}{d}}, \quad \forall K \in \mathscr{T}_h \tag{4.29}$$

where σ_h is defined in (4.8). The condition (4.28) states that the volume of K is proportional to the reciprocal of ρ_K. It obviously controls the size of mesh elements through the mesh density function ρ_K. (The larger ρ_K, the smaller $|K|$; and vice versa.) Moreover, (4.28) is a natural multidimensional generalization of (2.5), a variant of the 1D equidistribution principle. For these reasons, (4.28) is referred to as the (multidimensional) *equidistribution condition*. On the other hand, the condition (4.29) is an approximation to (4.3), requiring that element K be equilateral in the metric M_K. It is shown in §4.2 that any element K satisfying (4.29) is aligned with M_K and its shape and orientation are controlled to some extent by M_K. The condition (4.29) (or its equivalent (4.27)) is thus called the *alignment condition*.

Note that a mesh satisfying (4.28) and (4.29) is only approximately M-uniform since these conditions are just approximations to (4.2) and (4.3). Moreover, it is unclear if such a mesh exists in multidimensions for a given monitor function and a given number of elements.[1] Nevertheless, there are advantages to using the equidistribution and alignment conditions (4.28) and (4.29). These conditions are convenient to use in error analysis; e.g., see §5.2. Also, as shown in §5.2, interpolation error bounds obtained for meshes satisfying (4.28) and (4.29) are stable with respect to the mesh in the sense that error bounds with the same optimal convergence order and the same optimal solution dependent factor can be obtained for meshes satisfying only the approximate conditions

$$\frac{N\,|K|\rho_K}{\sigma_h} \le \kappa_{eq}, \quad \forall K \in \mathscr{T}_h \tag{4.30}$$

$$\frac{\mathrm{tr}((F_K')^T M_K F_K')}{d\,\det((F_K')^T M_K F_K')^{\frac{1}{d}}} \le \kappa_{ali}, \quad \forall K \in \mathscr{T}_h \tag{4.31}$$

for some constants $\kappa_{ali} > 1$ and $\kappa_{eq} > 1$ of small or moderate size (cf. (2.88)). Satisfaction of (4.28) and (4.29) can serve in practice as a goal (rather than mandatory conditions) for the mesh generation or as a tool for use in understanding existing adaptive mesh methods (see Chapters 6 and 7). In addition, these conditions can be used directly to construct objective functions optimized locally or globally to improve the adaptivity and alignment of the current mesh. For example, the sum of the left-hand-side term of (4.31) over the elements in the element patch ω_i associated with an interior vertex $\boldsymbol{a}_i$,

$$\sum_{K\in\omega_i} \frac{\mathrm{tr}((F_K')^T M_K F_K')}{d\,\det((F_K')^T M_K F_K')^{\frac{1}{d}}}, \tag{4.32}$$

[1] In 1D, (4.29) is always true and (4.28) reduces to (2.5). The existence of an equidistributing mesh satisfying (2.5) is proven in Proposition 2.1.1.

can be minimized over the coordinates of the vertex to improve the alignment of elements in ω_i.

For these reasons, we shall hereafter consider only meshes satisfying the equidistribution and alignment conditions instead of the exact M-uniform conditions. For notational simplicity, approximate M-uniform meshes satisfying (4.28) and (4.29) will still be referred to simply as *M-uniform* meshes, and meshes satisfying (4.30) and (4.31) are called *quasi M-uniform* meshes. (See (2.88) for the definition of a quasi-equidistributing mesh.)

To conclude this section, we summarize below several useful equivalents to (4.28) and (4.29), one of their invariance properties, and briefly discuss related linearly varying error measures. More geometric interpretations of the conditions shall be given in the next section.

Theorem 4.1.1 *Conditions (4.28) and (4.29) are mathematically equivalent to each of the following conditions:*

$$(F_K')^T M_K F_K' = \left(\frac{\sigma_h}{N}\right)^{\frac{2}{d}} I, \quad \forall K \in \mathscr{T}_h \tag{4.33}$$

$$F_K'(F_K')^T = \left(\frac{\sigma_h}{N}\right)^{\frac{2}{d}} M_K^{-1}, \quad \forall K \in \mathscr{T}_h \tag{4.34}$$

$$(F_K')^{-T}(F_K')^{-1} = \left(\frac{\sigma_h}{N}\right)^{-\frac{2}{d}} M_K, \quad \forall K \in \mathscr{T}_h \tag{4.35}$$

$$(F_K')^{-1} M_K^{-1} (F_K')^{-T} = \left(\frac{\sigma_h}{N}\right)^{-\frac{2}{d}} I, \quad \forall K \in \mathscr{T}_h. \tag{4.36}$$

Proof. The equivalence of (4.33) to any of conditions (4.34)–(4.36) can be easily obtained by direct algebraic calculations.

To show that (4.28) and (4.29) imply (4.33), we recall that (4.25) is equivalent to (4.29), and it suffices to show how the constant θ_K is obtained from compatibility and (4.28). Notice that

$$|K| = \int_K d\boldsymbol{x} = \int_{\hat{K}} \det(F_K') d\boldsymbol{\xi} = \det(F_K')\, |\hat{K}| = \det(F_K'). \tag{4.37}$$

Taking the determinant in (4.25), we get

$$\rho_K^2 \det(F_K')^2 = \theta_K^d,$$

or from (4.37),

$$\theta_K^{\frac{d}{2}} = \rho_K\, |K|.$$

Summing this over all elements and using (4.8) gives

$$\theta_K = \left(\frac{\sigma_h}{N}\right)^{\frac{2}{d}}. \tag{4.38}$$

Thus, (4.33) follows from (4.25).

For the converse, (4.28) can be obtained by taking determinants of both sides of (4.33), and (4.29) results from the arithmetic-mean geometric-mean inequality (cf. Theorem B.0.11) and the observation that (4.33) implies that the eigenvalues of $(F_K')^T M_K F_K'$ are equal to each other. Thus, (4.28) and (4.29) are also necessary conditions for (4.33). □

Theorem 4.1.2 *Conditions (4.28) and (4.29) are invariant under a scaling transformation of M: $M \to cM$ for any positive constant c.*

Proof. This can be verified directly. □

Note that (4.29) has a stronger invariance, i.e., it is invariant under $M_K \to \theta(\boldsymbol{x}) M_K$ for any strictly positive function $\theta = \theta(\boldsymbol{x})$.

The simple invariance in this last theorem is a useful property when determining a suitable monitor function, as we see in §5.2. In particular, it means that one can multiply M by an arbitrary positive constant without changing the mesh adaptivity.

Linearly varying error measures. As in 1D, the interpolation error bounds obtained in §5.2 for M-uniform meshes are proportional to a power of σ_h (or $\alpha_h \sigma_h$ when the monitor function is scaled by a regularization parameter $\alpha_h > 0$). Thus, from the equidistribution condition (4.28), we may view σ_h (or $\alpha_h \sigma_h$) as the total "error" over the physical domain and the mesh density function ρ_K (or $\alpha_K \rho_K$) as an "error" density. Moreover, $\rho_K |K|$ (or $\alpha_K \rho_K |K|$) defines a linearly varying (in volume) error measure; cf. §2.1.2 and the references [16] (Chapter 9) and [227, 273] for more discussion on the topic.

4.2 Mesh control perspective

From the perspective of mesh control, the basic task of mesh adaptation is to generate a mesh such that the size, shape, and orientation of its elements are specified from a user-prescribed monitor function $M = M(\boldsymbol{x})$. To investigate such a specification, we first show how the mesh elements are determined from F_K', the Jacobian matrix of F_K (for any $K \in \mathscr{T}_h$), for an affine family of simplicial meshes.

4.2.1 Jacobian matrix and size, shape, and orientation of mesh elements

Denote the edges of K and $\hat{K}$ by $\gamma_1, ..., \gamma_{d(d+1)/2}$ and $\hat{\gamma}_1, ..., \hat{\gamma}_{d(d+1)/2}$, respectively, where the edges are ordered such that

$$\boldsymbol{\gamma}_i = F_K' \hat{\boldsymbol{\gamma}}_i, \quad i = 1, ..., \frac{d(d+1)}{2}. \tag{4.39}$$

Let the singular value decomposition (SVD) of F_K' be

$$F_K' = U \Sigma V^T, \tag{4.40}$$

where $U = [\boldsymbol{u}_1, ..., \boldsymbol{u}_d]$ (the left singular vectors) and $V = [\boldsymbol{v}_1, ..., \boldsymbol{v}_d]$ (the right singular vectors) are orthogonal matrices, $\Sigma = \text{diag}(\sigma_1, ..., \sigma_d)$, and σ_i's are the (positive) singular values of the matrix. Note that the dependence on K is suppressed in (4.40) for simplicity.

Size. From (4.37) we have

$$|K| = \det(F_K') = \sigma_1 \cdots \sigma_d, \tag{4.41}$$

so the size of K is determined by the determinant of F_K'.

Shape. The shape of K is completely determined by the lengths of its edges. From (4.39) and (4.40), these edge lengths are given by

$$|\gamma_i|^2 = \boldsymbol{\gamma}_i^T \boldsymbol{\gamma}_i = \hat{\boldsymbol{\gamma}}_i^T (F_K')^T F_K' \hat{\boldsymbol{\gamma}}_i = \hat{\boldsymbol{\gamma}}_i^T V \Sigma^2 V^T \hat{\boldsymbol{\gamma}}_i, \quad i = 1, ..., \frac{d(d+1)}{2}. \tag{4.42}$$

Notice that $\hat{\boldsymbol{\gamma}}_i$, $i = 1, ..., d(d+1)/2$, are constant vectors since the chosen reference element $\hat{K}$ is fixed. Then (4.42) indicates how the edge lengths, and thus the shape of K, are determined by Σ (the singular values) and V (the right singular vectors) of F_K'.

Orientation. The orientation of K can be defined as the direction of the longest edge or a combination of the edge directions. In this sense, the orientation of an element is fully determined by the directions of the edges. From (4.39),

$$\boldsymbol{\gamma}_i = F_K' \hat{\boldsymbol{\gamma}}_i = U \Sigma V^T \hat{\boldsymbol{\gamma}}_i, \quad i = 1, ..., \frac{d(d+1)}{2}. \tag{4.43}$$

Thus, all three components of the SVD of F_K', U, Σ, and V, are needed to completely determine the orientation of K.

Circumscribed ellipsoids. It is often useful to consider mesh alignment in terms of the elements' circumscribed ellipsoids since the size, shape, and orientation of the latter can be uniquely defined through the directions and lengths of their principal axes (cf. §4.3).

Theorem 4.2.1 *The equation for the circumscribed ellipsoid of K is given by*

$$\mathscr{E}: \quad (\boldsymbol{x}-\boldsymbol{x}_K)^T (F_K')^{-T} (F_K')^{-1} (\boldsymbol{x}-\boldsymbol{x}_K)^T = \frac{\hat{h}^2}{4}, \tag{4.44}$$

where $\hat{h}$ is the diameter of $\hat{K}$ (cf. (3.91)). The principal axes of the circumscribed ellipsoid are formed by the left singular vectors $\boldsymbol{u}_i$, $i = 1,...,d$ of F_K' while their semi-lengths are determined in terms of the singular values by

$$a_i = \frac{\hat{h}\sigma_i}{2} \quad i = 1,...,d \tag{4.45}$$

where σ_i's are the singular values of F_K'.

Proof. Denote by $\mathscr{E}$ and $\hat{\mathscr{E}}$ the circumscribed ellipsoids of K and $\hat{K}$, respectively (cf. Figure 4.5). Let the centers of $\hat{K}$ and K be $\boldsymbol{\xi}_{\hat{K}}$ and $\boldsymbol{x}_K$. Since $\hat{K}$ is an equilateral d-simplex, its circumscribed ellipsoid $\hat{\mathscr{E}}$ is a sphere with radius $\hat{h}/2$, i.e.,

$$\hat{\mathscr{E}}: \quad |\boldsymbol{\xi} - \boldsymbol{\xi}_{\hat{K}}|^2 = \frac{\hat{h}^2}{4}. \tag{4.46}$$

We have $\mathscr{E} = F_K(\hat{\mathscr{E}})$ because $K = F_K(\hat{K})$. The equation for $\mathscr{E}$, (4.44), is then obtained from (4.46) and the fact that

$$\boldsymbol{x} - \boldsymbol{x}_K = F_K(\boldsymbol{\xi}) - F_K(\boldsymbol{\xi}_{\hat{K}}) = F_K'(\boldsymbol{\xi} - \boldsymbol{\xi}_{\hat{K}}).$$

Using (4.40), (4.44) can be rewritten as

$$\mathscr{E}: \quad (\boldsymbol{x}-\boldsymbol{x}_K)^T U\Sigma^{-2}U^T (\boldsymbol{x}-\boldsymbol{x}_K)^T = \frac{\hat{h}^2}{4}.$$

Letting $\boldsymbol{y} = U^T(\boldsymbol{x} - \boldsymbol{x}_K)$, we have

$$\mathscr{E}: \quad \sum_{i=1}^{d} \frac{y_i^2}{\left(\frac{\hat{h}\sigma_i}{2}\right)^2} = 1.$$

The remaining conclusions of the theorem can be drawn from this equation and the fact that the axes ($\boldsymbol{e}_i$, $i = 1,...,d$) of the $\boldsymbol{y}$-coordinate system correspond to the left singular vectors $\boldsymbol{u}_i = U\boldsymbol{e}_i$, $i = 1,...,d$ of F_K' in the $\boldsymbol{x}$-coordinate system. □

The theorem is illustrated in Figure 4.2. From the definition of the diameter h_K of K, it is obvious that

$$h_K \leq 2a_{max} = \hat{h}\sigma_{max}, \tag{4.47}$$

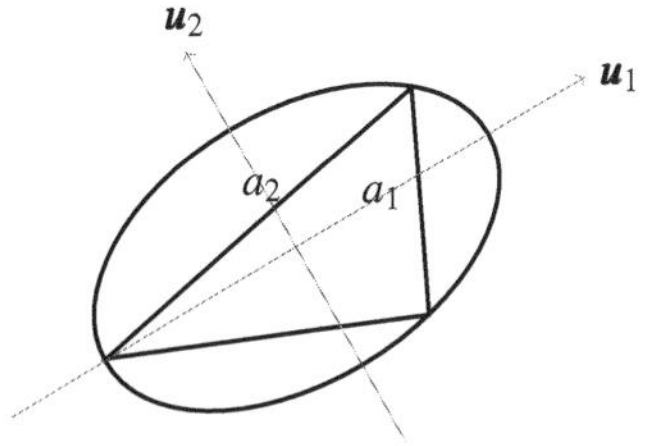

Fig. 4.2 The circumscribed ellipsoid $\mathscr{E}$ of K is determined by F_K'. Here, $\boldsymbol{u}_1$ and $\boldsymbol{u}_2$ are the left singular vectors of F_K', and the semi-lengths a_1 and a_2 of the principal axes are given in (4.45).

where $a_{max} = \max_i a_i$ and $\sigma_{max} = \max_i \sigma_i$ Moreover, from (4.41) we have

$$\sigma_{min}^d \leq |K| = \sigma_1 \cdots \sigma_d \leq \sigma_{max}^d, \tag{4.48}$$

where $\sigma_{min} = \min_i \sigma_i$.

From the above analysis we can conclude that a complete determination of the size, shape, and orientation of element K requires a full specification of the Jacobian matrix F_K'. Such a specification is generally overkill from the standpoint of mesh adaptation. To explain why, we first notice that F_K' depends upon $\hat{K}$ so a different F_K' has to be chosen if $\hat{K}$ is transformed by an invertible affine mapping (e.g., a rotation or dilation transformation). However, $\hat{K}$ is only an auxiliary tool used in finite element computation, and more importantly, for a given affine family of triangulations, the outcome of a finite element computation using an associated affine family of finite elements is unaffected by any invertible affine transformation of the reference finite element (cf. Theorem 3.3.1 and §3.3.3). So it is natural to require that mesh adaptation be invariant under any invertible affine transformation of $\hat{K}$. This is clearly impossible when mesh adaptation is realized by specifying the Jacobian matrix of the coordinate transformation. Another indication of overkill is that the Jacobian specification can cause inconsistency between the corresponding coordinate transformation or mesh with the domain boundary (cf. §6.5.4 and D'Azevedo and Simpson [112]). Nevertheless, some researchers consider the specification of the Jacobian matrix as a most direct control method for mesh adaptation. Methods along this line typically enforce the Jacobian specification in some weak (such as least squares) sense – see the discussion of the Jacobian-weighted and reference Jacobian methods in §6.5.4.

4.2.2 Mesh adaptation via metric specification

We consider here a weaker mesh control (than the Jacobian specification) for which the resulting mesh adaptation is invariant under affine transformation of the reference element $\hat{K}$. It is motivated by the concept of the *metric of a coordinate transformation* in the context of differential geometry. The idea is to instead of using F_K' directly, specify $(F_K')^{-T}(F_K')^{-1}$, the metric of the inverse coordinate transformation of F_K, through the monitor function $M = M(\boldsymbol{x})$. Together with compatibility, this gives

$$(F_K')^{-T}(F_K')^{-1} = \left(\frac{\sigma_h}{N}\right)^{-\frac{2}{d}} M_K, \quad \forall K \in \mathscr{T}_h. \tag{4.49}$$

Notice that (4.49) is the same as (4.35), so by Theorem 4.1.1 this metric specification results in an M-uniform mesh.

To see more clearly how the mesh elements are controlled via (4.49), using (4.40) we get

$$U \left(\frac{\sigma_h}{N}\right)^{\frac{2}{d}} \Sigma^{-2} U^T = M_K. \tag{4.50}$$

It implies that the eigenvectors of M_K determine U and the eigenvalues determine $(\sigma_h/N)^{\frac{2}{d}}\,\Sigma^{-2}$. Denote the eigen-decomposition of M_K by

$$M_K = QDQ^T, \tag{4.51}$$

where Q is an orthogonal matrix of eigenvectors, $D = \text{diag}(\lambda_1, ..., \lambda_d)$, and λ_i's are the eigenvalues of M_K. Upon a possible reordering of the eigenvalues and eigenvectors, from (4.50) we have

$$U = Q, \quad \Sigma = \left(\frac{\sigma_h}{N}\right)^{\frac{1}{d}} D^{-\frac{1}{2}}. \tag{4.52}$$

Inserting this into (4.40) yields

$$F_K' = \left(\frac{\sigma_h}{N}\right)^{\frac{1}{d}} Q D^{-\frac{1}{2}} V^T. \tag{4.53}$$

This reveals how the elements of an M-uniform mesh are controlled by the monitor function M: all the components of the SVD of F_K' except V are determined by M_K. The right-singular vector matrix V, which represents a rotation acting on the reference element $\hat{K}$, is left undetermined from the specification condition (4.49). Its determination is left up to the specific algorithm used in mesh generation, complying with necessary mesh topology and consistency between the mesh and the domain boundary.

The indeterminacy of V from (4.49) implies that the corresponding mesh adaptation is invariant under rotation of $\hat{K}$. In fact, this invariance holds for any invertible affine transformation of $\hat{K}$. To show this, we first recall from Theorem 4.1.1 that (4.49) is equivalent to (4.28) and (4.29). The latter two conditions have been obtained simply under the assumptions that $\hat{K}$ is equilateral and has unit volume. The unit volume assumption can easily be removed by modifying (4.49) to be

$$(F_K')^{-T}(F_K')^{-1} = \left(\frac{\sigma_h}{N\,|\hat{K}|}\right)^{-\frac{2}{d}} M_K, \quad \forall K \in \mathscr{T}_h. \tag{4.54}$$

(Interestingly, the conditions (4.28) and (4.29) remain the same for this situation.) But the condition (4.49) takes a more complicated form when $\hat{K}$ is not equilateral. To see this, define the edge matrix of $\hat{K}$ as

$$\hat{E} = (\hat{\boldsymbol{a}}_2 - \hat{\boldsymbol{a}}_1, ..., \hat{\boldsymbol{a}}_{d+1} - \hat{\boldsymbol{a}}_d), \tag{4.55}$$

where $\hat{\boldsymbol{a}}_1, ..., \hat{\boldsymbol{a}}_{d+1}$ are the vertices of $\hat{K}$. Consider a d-simplex $\check{K}$ with a constant unitary edge length and denote its vertices and edge matrix by $\check{\boldsymbol{a}}_i$, $i = 1, ..., d+1$ and $\check{E}$, respectively. Let the mapping from $\hat{K}$ to $\check{K}$ be G. Upon possible reordering of the vertices, we have

$$\check{\boldsymbol{a}}_{i+1} - \check{\boldsymbol{a}}_i = G'(\hat{\boldsymbol{a}}_{i+1} - \hat{\boldsymbol{a}}_i), \quad i = 1, ..., d$$

and therefore

$$\check{E} = G'\hat{E}. \tag{4.56}$$

Then the mapping $\check{F}_K : \check{K} \to K$ (cf. Figure 3.6) satisfies (4.54), viz.,

$$(\check{F}_K')^{-T}(\check{F}_K')^{-1} = \left(\frac{\sigma_h}{N\,|\check{K}|}\right)^{-\frac{2}{d}} M_K, \quad \forall K \in \mathscr{T}_h. \tag{4.57}$$

From Figure 3.6 and (4.56) it follows that

$$\check{F}_K = F_K \circ G^{-1}, \quad \check{F}_K' = F_K'(G')^{-1} = F_K'\hat{E}(\check{E})^{-1}.$$

Combining this with (4.57) gives the modification of (4.49) for the case with a non-equilateral reference element, i.e., for any $K \in \mathscr{T}_h$,

$$(F_K')^{-T}\hat{E}^{-T}\check{E}^T\check{E}\hat{E}^{-1}(F_K')^{-1} = \left(\frac{\sigma_h}{N\,|\hat{K}|\cdot\det(\check{E})\det(\hat{E})^{-1}}\right)^{-\frac{2}{d}} M_K. \tag{4.58}$$

Moreover, noticing that $E_K = \check{F}_K'\check{E}$, where $E_K = (\boldsymbol{a}_2 - \boldsymbol{a}_1, ..., \boldsymbol{a}_{d+1} - \boldsymbol{a}_d)$ is the edge matrix of K, from (4.57) we have

$$E_K^{-T}\check{E}^T\check{E}E_K^{-1} = \left(\frac{\sigma_h}{N\,|\check{K}|}\right)^{-\frac{2}{d}} M_K, \quad \forall K \in \mathscr{T}_h \tag{4.59}$$

which is independent of $\hat{K}$, implying that (4.58) is invariant under any invertible affine transformation of $\hat{K}$.

On the other hand, one may not want to deal with this complication since an equilateral reference element can easily be chosen in practice. For this reason, we shall stick with the form (4.49), and we show below that it is invariant under invertible affine transformations preserving the shape of $\hat{K}$. Such transformations include rotation, translation, dilation transformations, and any combination thereof (but exclude linear shear transformations).

Theorem 4.2.2 *Conditions (4.28) and (4.29) are invariant under rotation, translation, and dilation transformations of* $\hat{K}$.

Proof. The proof can be done for the equivalent condition (4.49) using edge matrices as in (4.59), but here we take a different approach and deal directly with (4.28) and (4.29). Let $\check{K}$ be a new reference element formed by an invertible affine mapping $G: \hat{K} \to \check{K}$ which preserves the shape of $\hat{K}$ (see Figure 3.6). It is not difficult to see that G can be expressed as

$$G(\boldsymbol{\xi}) = G'\boldsymbol{\xi} + \boldsymbol{\xi}_0, \tag{4.60}$$

where $\boldsymbol{\xi}_0$ is a constant vector and G' is a scalar orthogonal matrix satisfying

$$(G')^{-T}(G')^{-1} = cI \tag{4.61}$$

for some positive constant c. Denote the affine mapping from the new reference element $\check{K}$ to element K by $\check{F}_K$, as shown in Figure 3.6. We need to show that conditions (4.28) and (4.29) have the same form for $\check{F}_K$. To begin with, note that (4.28) involves no factors related to the reference element and hence holds for any reference element. Next, note that F_K and $\check{F}_K$ are related by

$$F_K = \check{F}_K \circ G, \quad F_K' = \check{F}_K'\, G'.$$

Inserting this into (4.29), we have

$$\frac{1}{d}\mathrm{tr}((G')^T(\check{F}_K')^T M_K \check{F}_K' G') = \det((G')^T(\check{F}_K')^T M_K \check{F}_K' G')^{\frac{1}{d}},$$

implying that $(G')^T(\check{F}_K')^T M_K \check{F}_K' G'$ has equal eigenvalues, viz.,

$$(G')^T(\check{F}_K')^T M_K \check{F}_K' G' = \theta I$$

for some constant $\theta > 0$. It follows that

$$(\check{F}_K')^T M_K \check{F}_K' = \theta (G')^{-T}(G')^{-1}.$$

Since $(G')^{-T}(G')^{-1}$ is a scalar matrix, $(\check{F}_K')^T M_K \check{F}_K'$ is as well, so

$$\frac{1}{d}\text{tr}((\check{F}_K')^T M_K \check{F}_K') = \det((\check{F}_K')^T M_K \check{F}_K')^{\frac{1}{d}}.$$

This has the same form as (4.29), except that F_K has been replaced by $\check{F}_K$. Hence, (4.28) and (4.29) maintain the same form for the new reference element and the new mapping $\check{F}_K$, and the proof is complete. □

4.2.3 Geometric interpretations of mesh equidistribution and alignment

Geometrically, (4.53) together with (4.41) imply that for a given monitor function, the size of K for the corresponding M-uniform mesh is completely determined from M_K. This is not true for the shape and orientation since they are measured in terms of the edge lengths and directions which are fully determined only by the Jacobian matrix F_K'. We now determine the extent to which the shape and orientation of the elements of an M-uniform mesh are controlled by the monitor function. Recall that (4.49) implies (4.52), i.e.,

$$\boldsymbol{u}_i = \boldsymbol{q}_i, \quad \sigma_i = \left(\frac{\sigma_h}{N}\right)^{\frac{1}{d}} \frac{1}{\sqrt{\lambda_i}}, \quad i = 1, ..., d \tag{4.62}$$

where $(\lambda_i, \boldsymbol{q}_i)$'s are eigenpairs of M_K and σ_i's and $\boldsymbol{u}_i$'s are singular values and left singular vectors of F_K', respectively. Combining this with Theorem 4.2.1 and (4.47), we have the following theorem.

Theorem 4.2.3 *The size and circumscribed ellipsoids of the elements of an M-uniform mesh $\mathscr{T}_h$ satisfying (4.28) and (4.29) are completely determined from the monitor function M. More specifically, the size of mesh elements is inversely proportional to $\rho_K = \sqrt{det(M_K)}$. Also, the principal axes of the circumscribed ellipsoid of any element $K \in \mathscr{T}_h$ are formed by the eigenvectors of M_K, and their semi-lengths are determined by*

$$a_i = \frac{\hat{h}}{2}\left(\frac{\sigma_h}{N}\right)^{\frac{1}{d}} \frac{1}{\sqrt{\lambda_i}}, \quad i = 1, ..., d. \tag{4.63}$$

Furthermore,

$$h_K \le \hat{h}\left(\frac{\sigma_h}{N}\right)^{\frac{1}{d}} \frac{1}{\sqrt{\lambda_{min}}}, \quad \forall K \in \mathscr{T}_h \tag{4.64}$$

where $\hat{h}$ is given in (3.91) and λ_{min} is the minimum eigenvalue of M_K.

A 2D illustration of the result is given in Figure 4.3.

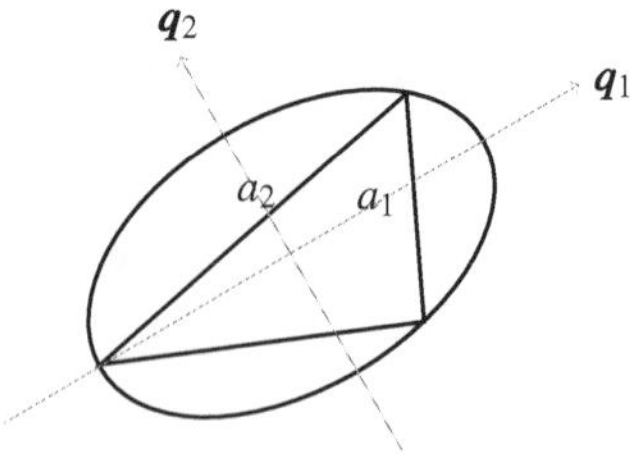

Fig. 4.3 The circumscribed ellipsoid $\mathscr{E}$ of K is determined by M_K. Here, $\boldsymbol{q}_1$ and $\boldsymbol{q}_2$ are the eigenvectors of M_K and the semi-lengths a_1 and a_2 of the principal axes are given in (4.63).

It is emphasized that Theorem 4.2.3 provides a precise description on how an element K of an M-uniform mesh is aligned with the matrix-valued monitor function M_K in terms of its circumscribed ellipsoid. This description is useful in practical computation since a mesh can be aligned with the geometry of the solution by properly choosing a monitor function depending upon the solution. In fact, such a description can be obtained for an element satisfying only the alignment condition (4.29). To see this, note that (4.29) implies that all the eigenvalues of matrix $(F_K')^T M_K F_K'$ are equal, with their size being undetermined. This gives

$$(F_K')^{-T}(F_K')^{-1} = \theta_K M_K, \tag{4.65}$$

or from the SVDs (4.40) and (4.51),

$$\boldsymbol{u}_i = \boldsymbol{q}_i, \quad \sigma_i = \frac{1}{\sqrt{\theta_K \lambda_i}}, \quad i = 1, ..., d \tag{4.66}$$

where $\theta_K > 0$ is an arbitrary constant (depending upon K). For easy reference, this description of mesh alignment is given in the following theorem.

Theorem 4.2.4 *The shape and orientation of any simplicial element K satisfying the alignment condition (4.29) is determined by M_K in the following sense: The principal axes of its circumscribed ellipsoid are formed by the eigenvectors of M_K, and their semi-lengths are determined by*

$$a_i = \frac{\hat{h}}{2} \frac{1}{\sqrt{\theta_K \lambda_i}}, \quad i = 1, ..., d \tag{4.67}$$

where $\theta_K > 0$ is an arbitrary constant.

As long as $\sigma_h \le C_1$ and $\lambda_{min} \ge C_2 > 0$ for some constants C_1 and C_2, (4.64) implies that

$$h = \max_{K \in \mathscr{T}_h} h_K \to 0 \quad \text{as} \quad N \to \infty, \tag{4.68}$$

a property often needed to ensure that an error bound on a family of triangulations $\mathscr{T}_h$ converges.

Furthermore, we can see from Theorem 4.2.3 that the metric specification (4.49) offers considerable freedom in choosing elements in the mesh generation process. Indeed, elements having the same volume and the same circumscribed ellipsoid determined by M are all possible candidates. This property allows the elements the flexibility to align themselves according to some prescribed mesh topology and with the boundary of the spatial domain. On the flip side, not all such elements are optimal, and neither is the resulting mesh necessarily optimal. Fortunately, as we see in Chapter 5 (in §5.2), an M-uniform mesh associated with a proper choice of M is asymptotically optimal in the sense that it leads to an interpolation error bound with an optimal convergence order and an optimal solution-dependent factor for a sufficiently large number of elements.

Example 4.2.1 To illustrate the above, we show in Figure 4.4(b) three triangular elements having the same area and the same circumscribed ellipse. They are generated by taking the reference element $\hat{K}$ (shown in Figure 4.4(a)) to have a constant unit edge length, $M_K = \begin{bmatrix} 4 & 0 \\ 0 & 1 \end{bmatrix}$, and $(\sigma_h/N)^{-\frac{2}{d}} = 1$ in (4.53) and (4.49). From (4.44) and (4.49), the equation of the circumscribed ellipse for this case is

$$4x^2 + y^2 = \frac{1}{3}.$$

The triangles in Figure 4.4(b) are generated for three choices of rotation matrix V using the mapping $K = F_K(\hat{K}) = (M_K^{-\frac{1}{2}} V^T)\hat{K}$, which is derived from (4.53) with $D = M_K$ and $Q_K = I$. Consequently, while their shape and orientation are obviously different, these elements all satisfy the specification condition (4.49) and have the same circumscribed ellipse (determined completely by M) and the same area. They would all be possible candidates for use in actual mesh generation with M. □

4.2.4 *Special case: scalar monitor functions*

An interesting special case is a monitor function in the form of a scalar matrix function

$$M_K = w_K I, \tag{4.69}$$

where $w = w(\boldsymbol{x})$ is a positive scalar function and $w_K = (1/|K|) \int_K w(\boldsymbol{x}) d\boldsymbol{x}$. Conditions (4.29) and (4.35) simplify to

$$\frac{1}{d}\mathrm{tr}((F_K')^T F_K') = \det((F_K')^T F_K')^{\frac{1}{d}}, \tag{4.70}$$

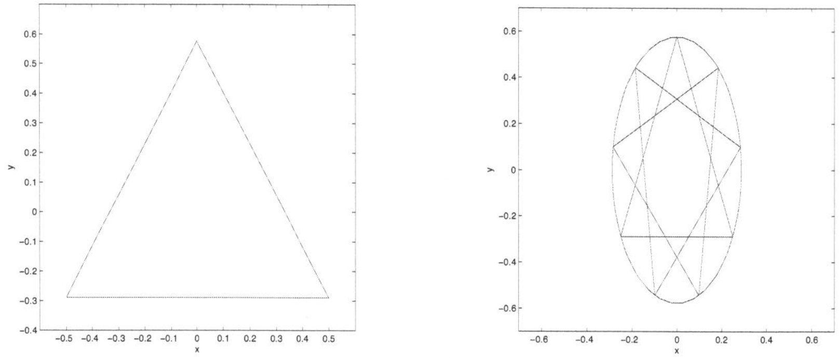

(a) Reference element (b) Elements with same circumscribed ellipse

Fig. 4.4 Reference and physical elements.

$$(F_K')^{-T}(F_K')^{-1} = \left(\frac{\sigma_h}{N}\right)^{-\frac{2}{d}} w_K I. \tag{4.71}$$

In this case, $(F_K')^{-T}(F_K')^{-1}$ is a scalar matrix, meaning that the corresponding circumscribed ellipsoids are actually spherical. In other words, the mesh is isotropic. The form (4.69) for the monitor function allows for generation of a non-uniform mesh, but it is too restrictive to permit mesh alignment since the eigenvalues of M_K are all equal. In §5.2 we see how such a monitor function arises naturally when optimizing an isotropic error bound.

4.3 Continuous perspective

From a continuous point of view, mesh generation is viewed as being mathematically equivalent to the determination of a coordinate transformation, where the mesh is generated as the image of a reference mesh under the transformation. An advantage of putting mesh generation in the continuous context is that mesh elements can be viewed simply as ellipsoids (see Figure 4.5) for which the size, shape, and orientation can be uniquely defined as follows: the size is quantified by the volume, the shape is determined by the ratios between the lengths of the principal axes, and the orientation is controlled by the principal directions. Another advantage is that whereas the equidistribution and alignment conditions (4.28) and (4.29) are valid only for simplicial meshes, their continuous analogs developed below will hold for any type of meshes at least in the asymptotic sense. This is especially convenient for quadrilateral meshes, which we shall see frequently in Chapters 6 and 7.

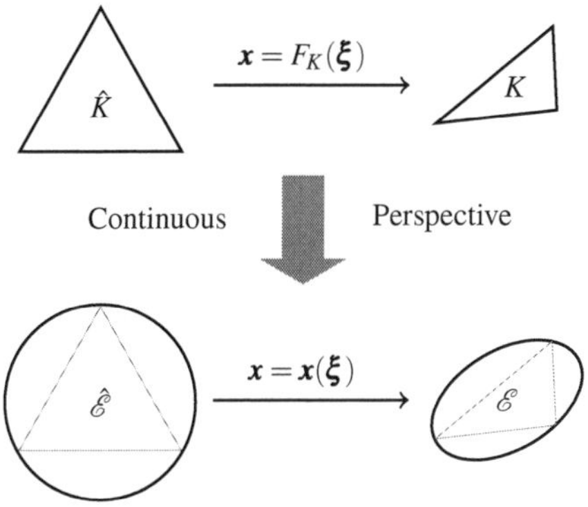

Fig. 4.5 A continuous view of mesh elements: ellipsoids (or spheres).

From the continuous viewpoint, the size, shape, and orientation of mesh elements (i.e., ellipsoids) will be seen to be completely determined by $\boldsymbol{J}^{-T}\boldsymbol{J}^{-1}$, the metric of the inverse coordinate transformation, where $\boldsymbol{J}$ is the Jacobian matrix of the coordinate transformation $\boldsymbol{x} = \boldsymbol{x}(\boldsymbol{\xi}) : \Omega_c \to \Omega$ for some computational domain Ω_c artificially chosen for the purpose of mesh generation. This is distinct from the discrete situation where, as shown in §4.2.1, the shape and orientation of mesh elements are only partially determined by $(F_K')^{-T}(F_K')^{-1}$. Still, the analysis for the current situation is very similar to that in the previous section. We take an arbitrary element $\hat{\mathscr{E}}$ with center $\boldsymbol{\xi}_0 \in \Omega_c$. As commonly done in practical computation, we assume that the fixed mesh on the reference domain Ω_c is uniform. Under this assumption, element $\hat{\mathscr{E}}$ can be viewed as a sphere defined by the equation

$$\hat{\mathscr{E}}: \quad (\boldsymbol{\xi} - \boldsymbol{\xi}_0)^T (\boldsymbol{\xi} - \boldsymbol{\xi}_0) = \hat{r}^2,$$

where the radius $\hat{r}$ is a constant related to the number of mesh elements N by $\hat{r}^d N \propto |\Omega_c|$ or $\hat{r} \propto (|\Omega_c|/N)^{1/d}$. Linearizing the coordinate transformation about $\boldsymbol{\xi}_0$, we obtain

$$x(\boldsymbol{\xi}) = x(\boldsymbol{\xi}_0) + \boldsymbol{J}(\boldsymbol{\xi}_0)(\boldsymbol{\xi} - \boldsymbol{\xi}_0) + O(|\boldsymbol{\xi} - \boldsymbol{\xi}_0|^2).$$

Thus, the equation for the corresponding element $\mathscr{E} = \boldsymbol{x}(\hat{\mathscr{E}})$ in Ω satisfies

$$\mathscr{E}: \quad (\boldsymbol{x} - \boldsymbol{x}_0)^T \boldsymbol{J}^{-T}(\boldsymbol{\xi}_0)\boldsymbol{J}^{-1}(\boldsymbol{\xi}_0)(\boldsymbol{x} - \boldsymbol{x}_0) \approx \hat{r}^2, \tag{4.72}$$

where $\boldsymbol{x}_0 = x(\boldsymbol{\xi}_0)$. As for (4.44) and Theorem 4.2.1, one can show that the principal axes of element $\mathscr{E}$ are formed by the left singular vectors of $\boldsymbol{J}(\boldsymbol{\xi}_0)$, and their semi-lengths are given by the product of $\hat{r}$ times the singular values. Since the left singular vectors and reciprocals of the singular values of $\boldsymbol{J}(\boldsymbol{\xi}_0)$ squared are respectively the eigenvectors and eigenvalues of $\boldsymbol{J}^{-T}\boldsymbol{J}^{-1}(\boldsymbol{\xi}_0)$, we conclude that the mesh

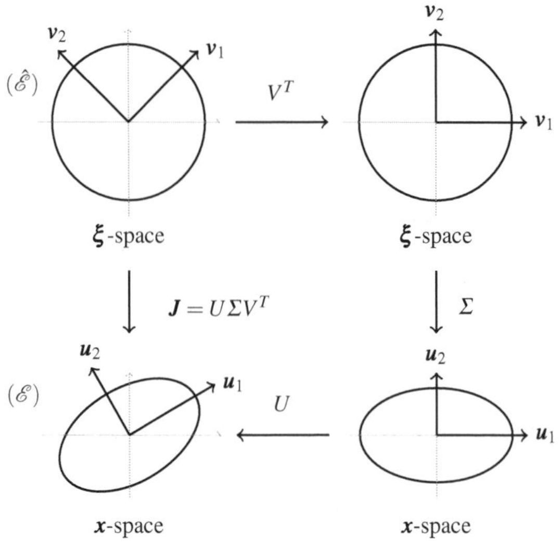

Fig. 4.6 Illustration of the 2D mapping of a computational element ($\hat{\mathscr{E}}$, a sphere) to a physical mesh element ($\mathscr{E}$, an ellipsoid) under $\boldsymbol{x} = \boldsymbol{x}(\boldsymbol{\xi}_0) + \boldsymbol{J}(\boldsymbol{\xi}_0)(\boldsymbol{\xi} - \boldsymbol{\xi}_0) : \Omega_c \to \Omega$, where both $\boldsymbol{x}(\boldsymbol{\xi}_0)$ and $\boldsymbol{\xi}_0$ are taken here to be zero. Reprinted from Huang [180], with permission from Global Science Press Ltd.

element $\mathscr{E}$, or more precisely, its size, shape, and orientation, are determined from the eigenvalues and eigenvectors of the metric $\boldsymbol{J}^{-T}\boldsymbol{J}^{-1}(\boldsymbol{\xi}_0)$.

Stating things slightly differently, we take the SVD for $\boldsymbol{J}$,

$$\boldsymbol{J} = U\Sigma V^T,$$

and examine its role in the mapping of $\hat{\mathscr{E}}$ to $\mathscr{E}$. As illustrated in Figure 4.6 for $d = 2$, the orientation of $\mathscr{E}$ is determined by the left singular vectors $U = [\boldsymbol{u}_1, ..., \boldsymbol{u}_d]$, the size and shape are controlled by the singular values $\Sigma = \text{diag}(\sigma_1, ..., \sigma_d)$, and the right singular vectors $V = [\boldsymbol{v}_1, ..., \boldsymbol{v}_d]$ play no role in describing the geometry in Ω. Thus, we see how the eigenvalues σ_i^{-2} and corresponding eigenvectors $\boldsymbol{u}_i$ of $\boldsymbol{J}^{-T}\boldsymbol{J}^{-1}(\boldsymbol{\xi}_0)$ specify the size, shape, and orientation of $\mathscr{E}$ and thereby determine the metric.

In the above analysis, the element $\hat{\mathscr{E}}$ and its center $\boldsymbol{\xi}_0$ are arbitrary, so the argument is valid for all mesh elements. Consequently, in what follows we replace $\boldsymbol{\xi}_0$ by a general point $\boldsymbol{\xi}$.

Since the size, shape, and orientation of mesh elements can be completely determined by the metric $\boldsymbol{J}^{-T}\boldsymbol{J}^{-1}$ in the continuous sense, mesh adaptation or mesh control can in principle be realized by specifying the metric from the monitor function. This leads to a continuous analog of (4.49) or (4.35),

$$J^{-T}J^{-1} = \left(\frac{\sigma}{|\Omega_c|}\right)^{-\frac{2}{d}} M(\boldsymbol{x}), \tag{4.73}$$

where $\sigma = \int_\Omega \rho(\boldsymbol{x})d\boldsymbol{x}$ and $\rho(\boldsymbol{x}) = \sqrt{\det(M(\boldsymbol{x}))}$. Similarly, we have the following continuous version of Theorem 4.1.1.

Theorem 4.3.1 *Conditions*

$$J\rho = \frac{\sigma}{|\Omega_c|}, \quad \forall \boldsymbol{x} \in \Omega \tag{4.74}$$

$$\frac{1}{d} tr(\boldsymbol{J}^T M \boldsymbol{J}) = det(\boldsymbol{J}^T M \boldsymbol{J})^{\frac{1}{d}}, \quad \forall \boldsymbol{x} \in \Omega \tag{4.75}$$

are mathematically equivalent to each of the following conditions:

$$\boldsymbol{J}^T M \boldsymbol{J} = \left(\frac{\sigma}{|\Omega_c|}\right)^{\frac{2}{d}} I, \quad \forall \boldsymbol{x} \in \Omega \tag{4.76}$$

$$\boldsymbol{J}\boldsymbol{J}^T = \left(\frac{\sigma}{|\Omega_c|}\right)^{\frac{2}{d}} M^{-1}, \quad \forall \boldsymbol{x} \in \Omega \tag{4.77}$$

$$\boldsymbol{J}^{-T}\boldsymbol{J}^{-1} = \left(\frac{\sigma}{|\Omega_c|}\right)^{-\frac{2}{d}} M, \quad \forall \boldsymbol{x} \in \Omega \tag{4.78}$$

$$\boldsymbol{J}^{-1}M^{-1}\boldsymbol{J}^{-T} = \left(\frac{\sigma}{|\Omega_c|}\right)^{-\frac{2}{d}} I, \quad \forall \boldsymbol{x} \in \Omega. \tag{4.79}$$

Like the discrete case, (4.74) and (4.75) are called the *equidistribution* and *alignment conditions*, respectively, and a coordinate transformation satisfying these conditions will be referred to as an *M-uniform coordinate transformation.*

It is easy to see that (4.75) is equivalent to

$$\frac{1}{d}\text{tr}(\boldsymbol{J}^{-1}M^{-1}\boldsymbol{J}^{-T}) = \det(\boldsymbol{J}^{-1}M^{-1}\boldsymbol{J}^{-T})^{\frac{1}{d}}, \quad \forall \boldsymbol{x} \in \Omega. \tag{4.80}$$

Finally, we give a continuous analog of Theorem 4.2.2.

Theorem 4.3.2 *Conditions (4.74) and (4.75) are invariant under rotation, translation, and dilation transformations of the computational coordinate* $\boldsymbol{\xi}$.

Proof. If $\boldsymbol{\xi}$ is transformed into $\check{\boldsymbol{\xi}}$ through rotation, translation, and/or dilation, then $\check{\boldsymbol{\xi}}$ is related to $\boldsymbol{\xi}$ by

$$\check{\boldsymbol{\xi}} = \theta Q \boldsymbol{\xi} + \boldsymbol{c},$$

where θ is a non-zero constant, $\boldsymbol{c}$ a constant vector, and Q a constant orthogonal matrix. Under this mapping, Ω_c is mapped into $\check{\Omega}_c$, with the volume $|\check{\Omega}_c| =$

$|\theta \det(Q)||\Omega_c|$. Moreover, by the chain rule we have

$$\boldsymbol{J} \equiv \frac{\partial \boldsymbol{x}}{\partial \boldsymbol{\xi}} = \frac{\partial \boldsymbol{x}}{\partial \check{\boldsymbol{\xi}}} \frac{\partial \check{\boldsymbol{\xi}}}{\partial \boldsymbol{\xi}} = \frac{\partial \boldsymbol{x}}{\partial \check{\boldsymbol{\xi}}} \theta Q \equiv \theta \check{\boldsymbol{J}} Q.$$

From these, conditions (4.74) and (4.75) become

$$\check{J} \rho = \frac{\sigma}{|\check{\Omega}_c|}, \tag{4.81}$$

$$\frac{1}{d} \text{tr}(Q^T \check{\boldsymbol{J}}^T M \check{\boldsymbol{J}} Q) = \det(Q^T \check{\boldsymbol{J}}^T M \check{\boldsymbol{J}} Q)^{\frac{1}{d}}. \tag{4.82}$$

Since $Q^T \check{\boldsymbol{J}}^T M \check{\boldsymbol{J}} Q$ has the same eigenvalues as $\check{\boldsymbol{J}}^T M \check{\boldsymbol{J}}$ does, it follows that

$$\text{tr}(Q^T \check{\boldsymbol{J}}^T M \check{\boldsymbol{J}} Q) = \text{tr}(\check{\boldsymbol{J}}^T M \check{\boldsymbol{J}}), \quad \det(Q^T \check{\boldsymbol{J}}^T M \check{\boldsymbol{J}} Q) = \det(\check{\boldsymbol{J}}^T M \check{\boldsymbol{J}}),$$

and (4.82) reduces to

$$\frac{1}{d} \text{tr}(\check{\boldsymbol{J}}^T M \check{\boldsymbol{J}}) = \det(\check{\boldsymbol{J}}^T M \check{\boldsymbol{J}})^{\frac{1}{d}}. \tag{4.83}$$

Note that (4.81) and (4.83) have the same form as (4.74) and (4.75). Hence, (4.74) and (4.75) are invariant under rotation, translation, and dilation transformations of the computational coordinate $\boldsymbol{\xi}$. □

4.4 Function approximation perspective

Mesh adaptation can also be studied from the perspective of function approximation. The basic idea is the same as that used for the 1D case in §2.1.4: Given a function $u = u(\boldsymbol{x})$ defined on $\Omega \subset \mathbb{R}^d$ $(d \geq 1)$, we seek a coordinate transformation $\boldsymbol{x} = \boldsymbol{x}(\boldsymbol{\xi}) : \Omega_c \to \Omega$ such that $u(\boldsymbol{x}(\boldsymbol{\xi}))$ can be efficiently approximated on a uniform mesh in the new coordinate system. For simplicity, we explain this using approximation with piecewise constant interpolation. Let $\mathscr{T}_{c,h}$ be a uniform mesh on Ω_c. Define a piecewise constant approximation to $u(\boldsymbol{x}(\boldsymbol{\xi}))$ by

$$u(\boldsymbol{x}(\boldsymbol{\xi})) \approx u(\boldsymbol{x}(\boldsymbol{\xi}_{K_c})), \quad \forall \boldsymbol{\xi} \in K_c, \forall K_c \in \mathscr{T}_{c,h}$$

where $\boldsymbol{\xi}_{K_c}$ is the center of K_c. From Taylor's theorem, the error for this approximation is

$$u(\boldsymbol{x}(\boldsymbol{\xi})) - u(\boldsymbol{x}(\boldsymbol{\xi}_{K_c})) = (\boldsymbol{\xi} - \boldsymbol{\xi}_{K_c})^T \nabla_{\boldsymbol{\xi}} u(\boldsymbol{x}(\boldsymbol{\xi}_{K_c})) + O(|\boldsymbol{\xi} - \boldsymbol{\xi}_{K_c}|^2), \quad \forall \boldsymbol{\xi} \in K_c.$$

It follows that

$$\max_{\boldsymbol{\xi}\in K_c} |u(\boldsymbol{x}(\boldsymbol{\xi})) - u(\boldsymbol{x}(\boldsymbol{\xi}_{K_c}))| = \max_{\boldsymbol{\xi}\in K_c} \left|(\boldsymbol{\xi} - \boldsymbol{\xi}_{K_c})^T \nabla_{\boldsymbol{\xi}} u(\boldsymbol{x}(\boldsymbol{\xi}_{K_c}))\right| + O(h_c^2),$$

where h_c is the element diameter of the uniform mesh $\mathscr{T}_{c,h}$ (cf. Figure 3.4). If a coordinate transformation can be chosen such that

$$\max_{\boldsymbol{\xi}\in K_c} \left|(\boldsymbol{\xi} - \boldsymbol{\xi}_{K_c})^T \nabla_{\boldsymbol{\xi}} u(\boldsymbol{x}(\boldsymbol{\xi}_{K_c}))\right| = c h_c, \tag{4.84}$$

for a suitable positive constant c independent of K_c and h_c, then

$$\max_{\boldsymbol{\xi}\in K_c} |u(\boldsymbol{x}(\boldsymbol{\xi})) - u(\boldsymbol{x}(\boldsymbol{\xi}_{K_c}))| = c h_c + O(h_c^2). \tag{4.85}$$

Since the right-hand-side terms are element independent, the approximation has the same level of error on all of the elements. In this sense, a uniform mesh is efficient in resolving $u(\boldsymbol{x}(\boldsymbol{\xi}))$ via piecewise constant interpolation.

Noticing that $h_c = \max_{\boldsymbol{\xi}\in K_c} |\boldsymbol{\xi} - \boldsymbol{\xi}_{K_c}|$, we can rewrite (4.84) as

$$\max_{\boldsymbol{\xi}\in K_c} \left|(\boldsymbol{\xi} - \boldsymbol{\xi}_{K_c})^T \nabla_{\boldsymbol{\xi}} u(\boldsymbol{x}(\boldsymbol{\xi}_{K_c}))\right| = c \max_{\boldsymbol{\xi}\in K_c} |\boldsymbol{\xi} - \boldsymbol{\xi}_{K_c}|.$$

A stronger version of this condition is

$$\left|(\boldsymbol{\xi} - \boldsymbol{\xi}_{K_c})^T \nabla_{\boldsymbol{\xi}} u(\boldsymbol{x}(\boldsymbol{\xi}_{K_c}))\right| = c\,|\boldsymbol{\xi} - \boldsymbol{\xi}_{K_c}|, \quad \forall \boldsymbol{\xi} \in K_c. \tag{4.86}$$

Generally speaking, the condition (4.86) is too strong to be satisfied by a coordinate transformation since it requires that $u(\boldsymbol{x}(\boldsymbol{\xi}))$ have a constant rate of change in all directions at $\boldsymbol{\xi}_{K_c}$. Nevertheless, it is very useful in practice, serving as a goal for the mesh generation rather than a mandatory condition.

A continuous form of (4.86) is

$$\left|d\boldsymbol{\xi}^T \nabla_{\boldsymbol{\xi}} u(\boldsymbol{x}(\boldsymbol{\xi}))\right| = c\,|d\boldsymbol{\xi}|,$$

where $\boldsymbol{\xi}_{K_c}$ has been replaced by an arbitrary point $\boldsymbol{\xi}$, and $d\boldsymbol{\xi}$ is a differential element. From the transformation relation $\nabla_{\boldsymbol{\xi}} = \boldsymbol{J}^T \nabla$, we have

$$\left|d\boldsymbol{\xi}^T \boldsymbol{J}^T \nabla u\right| = c\,|d\boldsymbol{\xi}|,$$

or

$$d\boldsymbol{\xi}^T \boldsymbol{J}^T \nabla u \nabla u^T \boldsymbol{J} d\boldsymbol{\xi} = c^2 d\boldsymbol{\xi}^T d\boldsymbol{\xi}.$$

Since $d\boldsymbol{\xi}$ is arbitrary, it follows that

$$\boldsymbol{J}^T \nabla u \nabla u^T \boldsymbol{J} = c^2 I.$$

As discussed in more detail later, the equation is in practice replaced by the regularized form

$$\boldsymbol{J}^T \left[\nabla u \nabla u^T + \alpha^2 I\right] \boldsymbol{J} = c^2 I,$$

where $\alpha > 0$ is a parameter, or

$$\boldsymbol{J}^T \left[I + \frac{1}{\alpha^2} \nabla u \nabla u^T\right] \boldsymbol{J} = \frac{c^2}{\alpha^2} I. \tag{4.87}$$

From the compatibility condition, the constant c satisfies

$$\frac{c^2}{\alpha^2} = \left(\frac{\sigma}{|\Omega_c|}\right)^{\frac{2}{d}},$$

where $\sigma = \int_\Omega \rho(\boldsymbol{x}) d\boldsymbol{x}$, $\rho(\boldsymbol{x}) = \sqrt{\det(M(\boldsymbol{x}))}$, and

$$M(\boldsymbol{x}) = I + \frac{1}{\alpha^2} \nabla u \nabla u^T. \tag{4.88}$$

Inserting this into (4.87), we have

$$\boldsymbol{J}^T M(\boldsymbol{x}) \boldsymbol{J} = \left(\frac{\sigma}{|\Omega_c|}\right)^{\frac{2}{d}} I. \tag{4.89}$$

Note that (4.89) is equivalent to (4.73), the adaptation equation derived from the perspective of mesh control.

4.5 Mesh quality measures

Standard ways to assess a given mesh are given by its regularity and its level of adaptivity. Mesh regularity is a geometric property indicating how close mesh elements are to being equilateral (cf. §3.5). It can be measured in a number of ways. Commonly used measures include the minimum angle [356], the maximum angle [22], and the aspect ratio. The first two are used primarily for triangular elements, although they can be extended to tetrahedral elements [220]. The aspect ratio of an element is defined as the ratio of the radii of its circumscribed and inscribed circles (or spheres). Several other regularity measures and their relationships are discussed by Liu and Joe [240] for tetrahedral elements (also see (4.97) below).

Mesh adaptivity characterizes how well a mesh adapts to the solution and is thus a property related to the particular physical problem being solved. The analysis in the previous subsection suggests that mesh adaptivity be quantified via equidistribution and alignment, i.e., by measuring how closely conditions (4.28) and (4.29) are satisfied by a given mesh for a given monitor function $M = M(\boldsymbol{x})$. Thus, *the equidistribution (quality) measure* is defined from (4.28) as

$$Q_{eq}(K) = \frac{N\rho_K \, |K|}{\sigma_h} \quad \forall K \in \mathscr{T}_h. \tag{4.90}$$

It follows that

$$Q_{eq}(K) > 0 \quad \forall K \in \mathscr{T}_h \tag{4.91}$$

and

$$\frac{1}{N} \sum_K Q_{eq}(K) = 1. \tag{4.92}$$

Consequently, $\max_K Q_{eq}(K) = 1$ if and only if the mesh satisfies the equidistribution condition (4.28) exactly. Moreover, the larger the value of $\max_K Q_{eq}(K)$, the more the quantity $|K| \, N\rho_K$ changes, and in this sense, the farther the mesh is from satisfying (4.28).

The alignment (quality) measure is defined from (4.29) as

$$Q_{ali}(K) = \left[\frac{\operatorname{tr}\left(F_K'^T M_K F_K'\right)}{d \det\left(F_K'^T M_K F_K'\right)^{\frac{1}{d}}} \right]^{\frac{d}{2(d-1)}}, \quad \forall K \in \mathscr{T}_h \tag{4.93}$$

where the reason for the exponent $\frac{d}{2(d-1)}$ will be apparent from Theorem 4.5.1 (below) when relating quality measures to the aspect ratio of mesh elements. The arithmetic-geometric mean inequality (Appendix B) implies

$$Q_{ali}(K) \geq 1, \quad \forall K \in \mathscr{T}_h$$

with $Q_{ali}(K) = 1$ for all $K \in \mathscr{T}_h$ if and only if (4.29) is satisfied exactly. The larger $Q_{ali}(K)$, the more the eigenvalues of $F_K'^T M_K F_K'$ differ from each other and the farther the alignment condition (4.29) is from being satisfied.

The alignment condition (4.93), defined using the Jacobian matrix F_K', can instead be defined in terms of its inverse. From (4.27), the alignment measure based on the inverse Jacobian matrix can be defined as

$$\hat{Q}_{ali}(K) = \left[\frac{\operatorname{tr}\left(F_K'^{-1} M_K^{-1} F_K'^{-T}\right)}{d \det\left(F_K'^{-1} M_K^{-1} F_K'^{-T}\right)^{\frac{1}{d}}} \right]^{\frac{d}{2(d-1)}} \quad \forall K \in \mathscr{T}_h. \tag{4.94}$$

As before, we have

$$\hat{Q}_{ali}(K) \geq 1, \quad \forall K \in \mathscr{T}_h,$$

and $\hat{Q}_{ali}(K) = 1$ for all $K \in \mathscr{T}_h$ if and only if (4.29), or equivalently (4.27), is satisfied exactly.

Note that from Theorems 4.1.2 and 4.2.2, Q_{eq}, Q_{ali}, and $\hat{Q}_{ali}$ are invariant under a scaling transformation of M and under rotation, translation, and dilation transformations of $\hat{K}$.

Since Q_{ali} and $\hat{Q}_{ali}$ measure the shape regularity and alignment of mesh elements in the metric $M(\boldsymbol{x})$, it is not surprising that for $M = I$ there are corresponding mesh geometric (quality) measures. These *geometric (quality) measures* are respectively

$$Q_{geo}(K) = \left[\frac{\operatorname{tr}\left(F_K'^T F_K' \right)}{d \det\left(F_K'^T F_K' \right)^{\frac{1}{d}}} \right]^{\frac{d}{2(d-1)}}, \quad \forall K \in \mathscr{T}_h \tag{4.95}$$

$$\hat{Q}_{geo}(K) = \left[\frac{\operatorname{tr}\left(F_K'^{-1} F_K'^{-T} \right)}{d \det\left(F_K'^{-1} F_K'^{-T} \right)^{\frac{1}{d}}} \right]^{\frac{d}{2(d-1)}}, \quad \forall K \in \mathscr{T}_h. \tag{4.96}$$

Like their alignment measure counterparts, Q_{geo} and $\hat{Q}_{geo}$ have the properties

$$Q_{geo}(K) \ge 1, \quad \hat{Q}_{geo}(K) \ge 1, \quad \forall K \in \mathscr{T}_h$$

and $Q_{geo}(K) = 1$ (or $\hat{Q}_{geo}(K) = 1$) for all $K \in \mathscr{T}_h$ if and only if mesh elements are equilateral.

Liu and Joe [239, 240] have studied several shape measures for tetrahedral elements. One of them, defined in terms of edge matrices (cf. (4.55)), is

$$Q_{geo,LJ}(K) = \frac{3\det((E\hat{E}^{-1})^T (E\hat{E}^{-1}))^{\frac{1}{3}}}{\operatorname{tr}((E\hat{E}^{-1})^T (E\hat{E}^{-1}))}, \tag{4.97}$$

where E and $\hat{E}$ are the edge matrices of K and $\hat{K}$, respectively, and the reference element $\hat{K}$ is an equilateral tetrahedron having the same volume as K. From the relation $E = F_K' \hat{E}$ (cf. (4.56)) one can see that for $d = 3$ the geometric measure $Q_{geo}(K)$ defined in (4.95) is related to this shape measure by

$$Q_{geo}(K) = \left[\frac{1}{Q_{geo,LJ}(K)} \right]^{\frac{3}{4}}.$$

The quality measures can be defined similarly in the continuous form. For easy reference they are given below:

$$Q_{eq}(\boldsymbol{x}) = \frac{\rho J |\Omega_c|}{\sigma}, \tag{4.98}$$

$$Q_{ali}(\boldsymbol{x}) = \left[\frac{\mathrm{tr}\left(\boldsymbol{J}^T M \boldsymbol{J}\right)}{d \det\left(\boldsymbol{J}^T M \boldsymbol{J}\right)^{\frac{1}{d}}} \right]^{\frac{d}{2(d-1)}}, \tag{4.99}$$

$$\hat{Q}_{ali}(\boldsymbol{x}) = \left[\frac{\mathrm{tr}\left(\boldsymbol{J}^{-1} M^{-1} \boldsymbol{J}^{-T}\right)}{d \det\left(\boldsymbol{J}^{-1} M^{-1} \boldsymbol{J}^{-T}\right)^{\frac{1}{d}}} \right]^{\frac{d}{2(d-1)}}, \tag{4.100}$$

$$Q_{geo}(\boldsymbol{x}) = \left[\frac{\mathrm{tr}\left(\boldsymbol{J}^T \boldsymbol{J}\right)}{d \det\left(\boldsymbol{J}^T \boldsymbol{J}\right)^{\frac{1}{d}}} \right]^{\frac{d}{2(d-1)}}, \tag{4.101}$$

$$\hat{Q}_{geo}(\boldsymbol{x}) = \left[\frac{\mathrm{tr}\left(\boldsymbol{J}^{-1} \boldsymbol{J}^{-T}\right)}{d \det\left(\boldsymbol{J}^{-1} \boldsymbol{J}^{-T}\right)^{\frac{1}{d}}} \right]^{\frac{d}{2(d-1)}}. \tag{4.102}$$

We end this subsection by deriving several key relationships between the mesh aspect ratio and the geometric and alignment measures. For brevity, results are given only for the discrete formulas for the quality measure, but the analogous ones are valid for the continuous forms.

Theorem 4.5.1 *For any $K \in \mathscr{T}_h$,*

$$1 \le Q_{ali}(K) \le \frac{\mu_{max}}{\mu_{min}} \le \left[\sqrt{d(d-1)\left(Q_{ali}^{\frac{2(d-1)}{d}}(K) - 1\right)} + 1 \right]^d, \tag{4.103}$$

$$1 \le \hat{Q}_{ali}(K) \le \frac{\mu_{max}}{\mu_{min}} \le \left[\sqrt{d(d-1)\left(\hat{Q}_{ali}^{\frac{2(d-1)}{d}}(K) - 1\right)} + 1 \right]^d, \tag{4.104}$$

where μ_{max} and μ_{min} are the maximum and minimum singular values of $M_K^{\frac{1}{2}} F_K'$.

Proof. Denote the singular values of $M_K^{\frac{1}{2}} F_K'$ by μ_i, $i = 1, ..., d$. Then the eigenvalues of $(F_K')^T M_K F_K'$ are μ_i^2, $i = 1, ..., d$, and $Q_{ali}(K)$ can be expressed as

$$Q_{ali}(K) = \left[\frac{\sum_i \mu_i^2}{d \left(\prod_i \mu_i^2\right)^{\frac{1}{d}}} \right]^{\frac{d}{2(d-1)}}. \tag{4.105}$$

It follows that

$$Q_{ali}(K) \le \left[\frac{d\, \mu_{\max}^2}{d\left(\mu_{\max}^2 \mu_{\min}^{2(d-1)}\right)^{\frac{1}{d}}} \right]^{\frac{d}{2(d-1)}} = \frac{\mu_{\max}}{\mu_{\min}}.$$

Using a refined version of the arithmetic-geometric mean inequality, (B.1), we also have

$$
\begin{aligned}
Q_{ali}^{\frac{2(d-1)}{d}}(K) - 1 &\geq \frac{1}{d(d-1)} \frac{\sum_{i<j}(\mu_i - \mu_j)^2}{(\prod_i \mu_i^2)^{\frac{1}{d}}} \\
&\geq \frac{1}{d(d-1)} \frac{(\mu_{\max} - \mu_{\min})^2}{\mu_{\min}^{\frac{2}{d}} \mu_{\max}^{\frac{2(d-1)}{d}}} \\
&= \frac{1}{d(d-1)} \left[\left(\frac{\mu_{\max}}{\mu_{\min}} \right)^{\frac{1}{d}} - \left(\frac{\mu_{\min}}{\mu_{\max}} \right)^{\frac{d-1}{d}} \right]^2 \\
&\geq \frac{1}{d(d-1)} \left[\left(\frac{\mu_{\max}}{\mu_{\min}} \right)^{\frac{1}{d}} - 1 \right]^2,
\end{aligned}
$$

and (4.103) follows easily.

Inequality (4.104) can be obtained similarly. □

Notice that when $M = I$, Q_{ali} reduces to Q_{geo}, $M_K^{\frac{1}{2}} F_K'$ equals F_K', and $\mu_i = \sigma_i$, $i = 1, \ldots, d$. (Recall that σ_i's are the singular values of F_K'.) Moreover, from Theorem 4.2.1 we know that $\sigma_{max}/\sigma_{min}$ is the aspect ratio of the circumscribed ellipsoid of K. Thus, the above theorem implies that $Q_{geo}(K)$ and $\hat{Q}_{geo}(K)$ are equivalent to the aspect ratio of the circumscribed ellipsoid of K. Similarly, we can conclude that $Q_{ali}(K)$ and $\hat{Q}_{ali}(K)$ are equivalent to the aspect ratio of the circumscribed ellipsoid in the metric specified by M_K.

The precise mathematical relationships between the alignment and geometric measures Q_{ali} and Q_{geo} developed below will prove useful in defining the optimal monitor function and in developing mesh adaptation algorithms. To derive these relationships, we first establish some basic properties of matrix traces.

Lemma 4.5.1 *For any matrix $A \in \mathbb{R}^{d \times d}$,*

$$
tr(A^T A) = tr(A A^T) = \|A\|_F^2, \tag{4.106}
$$

where $\|\cdot\|_F$ denotes the Frobenius matrix norm.

Proof. Equality (4.106) follows directly from the definition of the Frobenius norm. □

Lemma 4.5.2 *If S is a $d \times d$ symmetric matrix, then for any matrix $A \in \mathbb{R}^{d \times d}$,*

$$
|tr(A^T S A)| \leq tr(A^T A)\, \|S\|, \tag{4.107}
$$

where $\|S\|$ denotes the L_2 matrix norm of S. If S is also positive definite, then

$$\|S\|^{-1}\, tr(A^TSA) \le tr(A^TA) \le tr(A^TSA)\, \|S^{-1}\|. \tag{4.108}$$

Proof. Denote the eigen-decomposition of S by

$$S = Q\Sigma Q^T,$$

where Q is an orthogonal matrix, $\Sigma = \text{diag}(\lambda_1, ..., \lambda_d)$, and λ_i, $i = 1, ..., d$, are the eigenvalues of S. Writing

$$A^TQ = [\boldsymbol{q}_1, ..., \boldsymbol{q}_d],$$

then

$$A^TSA = (A^TQ)\Sigma(Q^TA) = [\boldsymbol{q}_1, ..., \boldsymbol{q}_d]\Sigma[\boldsymbol{q}_1, ..., \boldsymbol{q}_d]^T = \sum_i \lambda_i \boldsymbol{q}_i \boldsymbol{q}_i^T.$$

It follows that

$$\begin{aligned} |\text{tr}(A^TSA)| &= |\sum_i \lambda_i \text{tr}(\boldsymbol{q}_i\boldsymbol{q}_i^T)| \\ &= |\sum_i \lambda_i \|\boldsymbol{q}_i\|^2| \\ &\le \sum_i \|\boldsymbol{q}_i\|^2 \cdot |\lambda|_{max} \\ &= \text{tr}(A^TA)\, \|S\|, \end{aligned}$$

which gives (4.107). Now (4.108) follows since

$$\text{tr}(A^TA) = \text{tr}(A^TS^{\frac{1}{2}}S^{-1}S^{\frac{1}{2}}A) \le \text{tr}(A^TSA)\, \|S^{-1}\|. \tag{4.109}$$

□

Corollary 4.5.1 *The geometric and alignment measures are related by*

$$\|M_K\|^{-1}\, \rho_K^{\frac{2}{d}}\, Q_{ali}^{\frac{2(d-1)}{d}}(K) \le Q_{geo}^{\frac{2(d-1)}{d}}(K) \le \|M_K^{-1}\|\, \rho_K^{\frac{2}{d}}\, Q_{ali}^{\frac{2(d-1)}{d}}(K), \tag{4.110}$$

$$\|M_K^{-1}\|^{-1}\, \rho_K^{-\frac{2}{d}}\, \hat{Q}_{ali}^{\frac{2(d-1)}{d}}(K) \le \hat{Q}_{geo}^{\frac{2(d-1)}{d}}(K) \le \|M_K\|\, \rho_K^{-\frac{2}{d}}\, \hat{Q}_{ali}^{\frac{2(d-1)}{d}}(K), \tag{4.111}$$

where $\rho_K = \sqrt{det(M_K)}$.

Proof. The proof follows directly from (4.108) and the definitions of the geometric and alignment measures. □

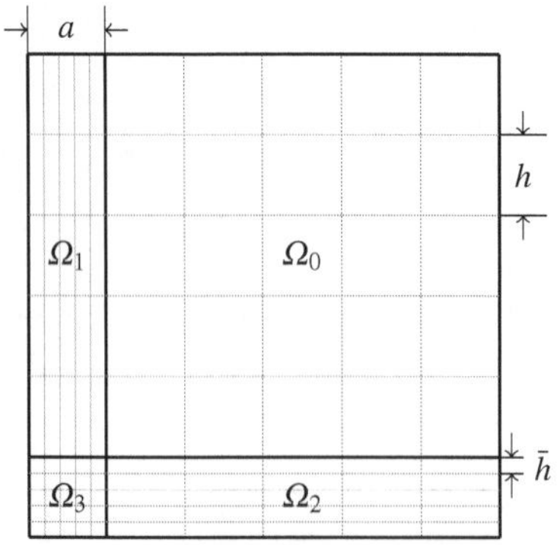

Fig. 4.7 A Shishkin-type mesh. The mesh parameters are given by $a = a_0\varepsilon|\ln\varepsilon|$ for a constant $a_0 \geq 2$, $h = (1-a)/L$, and $\bar{h} = a/L$, where L is a positive integer. The total number of mesh points in each direction is $(2L+1)$.

4.6 Analytical and numerical examples

In this section we present an analytical and two numerical examples to illustrate the equidistribution and alignment conditions (4.28) and (4.29) (or (4.74) and (4.75) in the continuous form) and the quality measures defined in the previous section.

Example 4.6.1 We first consider a two-dimensional example consisting of the singularly perturbed partial differential equation $-\varepsilon^2\Delta u + u = 0$ with $0 < \varepsilon \ll 1$ and boundary conditions such that the solution is

$$u(x,y) = e^{-\frac{x}{\varepsilon}} + e^{-\frac{y}{\varepsilon}}, \quad \text{in } \Omega = [0,1]\times[0,1]. \tag{4.112}$$

This example is selected because the solution, exhibiting boundary layers near $x = 0$ and near $y = 0$, is easy to deal with analytically, and because it is known that a simple piecewise uniform mesh of Shishkin-type [300] resolves the boundary layers. Since we are mainly concerned with the asymptotic behavior of the quality measures, we use their continuous forms for this example.

We choose the monitor function

$$M = \begin{bmatrix} 1+|u_{xx}| & 0 \\ 0 & 1+|u_{yy}| \end{bmatrix} = \begin{bmatrix} 1+\frac{1}{\varepsilon^2}e^{-\frac{x}{\varepsilon}} & 0 \\ 0 & 1+\frac{1}{\varepsilon^2}e^{-\frac{y}{\varepsilon}} \end{bmatrix}, \tag{4.113}$$

which will be seen in §5.2 to be optimal if we use linear interpolation to approximate $u(x,y)$ and measure the error in the H^1 semi-norm. The quality measures are studied for two types of meshes, uniform and Shishkin-type. A Shishkin-type mesh and its parameters are shown in Figure 4.7. For simplicity, the coordinate transformation associated with the Shishkin-type mesh is considered to be a piecewise linear mapping from a uniform rectangular mesh on the unit square to this adaptive mesh. All calculations are done for the asymptotic limit as $\varepsilon \to 0$.

Since the mesh elements are rectangular and the coordinate transformation is assumed to be piecewise linear, it is not difficult to see that for an arbitrary element K in Ω, the Jacobian matrix is

$$\boldsymbol{J} = (2L)\begin{bmatrix} h_{K,x} & 0 \\ 0 & h_{K,y} \end{bmatrix}, \quad \forall (x,y) \in K \tag{4.114}$$

where $h_{K,x}$ and $h_{K,y}$ denote the lengths of K in the x and y directions, respectively, and the spacing of the uniform mesh is $1/(2L)$. For the Shishkin-type mesh in Figure 4.7,

$$J = (2L)^2 h_{K,x} h_{K,y} = \begin{cases} h^2(2L)^2 = O(1), & \text{in } \Omega_0 \\ h\bar{h}(2L)^2 = O(\varepsilon|\ln\varepsilon|), & \text{in } \Omega_1, \Omega_2 \\ \bar{h}^2(2L)^2 = O(\varepsilon^2|\ln\varepsilon|^2), & \text{in } \Omega_3. \end{cases} \tag{4.115}$$

The mesh density function $\rho = \sqrt{\det(M)}$ can be estimated by

$$\begin{aligned} \rho(x,y) &= \sqrt{(1+\frac{1}{\varepsilon^2}e^{-\frac{x}{\varepsilon}})(1+\frac{1}{\varepsilon^2}e^{-\frac{y}{\varepsilon}})} \\ &= \begin{cases} O(1), & \text{in } \Omega_0 \\ O(\frac{1}{\varepsilon}e^{-\frac{x}{2\varepsilon}}) = O(\varepsilon^{-1}), & \text{in } \Omega_1 \\ O(\frac{1}{\varepsilon}e^{-\frac{y}{2\varepsilon}}) = O(\varepsilon^{-1}), & \text{in } \Omega_2 \\ O(\frac{1}{\varepsilon^2}e^{-\frac{x}{2\varepsilon}}e^{-\frac{y}{2\varepsilon}}) = O(\varepsilon^{-2}), & \text{in } \Omega_3 \end{cases} \end{aligned}$$

from which it follows that

$$\sigma = \int_\Omega \rho(x,y)dxdy = O(1). \tag{4.116}$$

Thus, the equidistribution measure for the Shishkin-type mesh is

$$Q_{eq} = \frac{\rho J |\Omega_c|}{\sigma} = \begin{cases} O(1), & \text{in } \Omega_0 \\ O(|\ln\varepsilon|), & \text{in } \Omega_1, \Omega_2 \\ O(|\ln\varepsilon|^2), & \text{in } \Omega_3. \end{cases} \tag{4.117}$$

From (4.101) and (4.114), the corresponding geometric measure is

$$Q_{geo} = \frac{h_{K,x}^2 + h_{K,y}^2}{2h_{K,x}h_{K,y}}. \tag{4.118}$$

From (4.99) and (4.113), the corresponding alignment measure is

Table 4.1 Quality measures for Shishkin-type and uniform meshes for the monitor function given in (4.113).

	Uniform Mesh			Shishkin-type Mesh		
	Ω_0	Ω_1, Ω_2	Ω_3	Ω_0	Ω_1, Ω_2	Ω_3
$Q_{geo}(x,y)$	O(1)	O(1)	O(1)	O(1)	$O(\varepsilon^{-1}\|\ln\varepsilon\|^{-1})$	O(1)
$Q_{ali}(x,y)$	O(1)	$O(\varepsilon^{-1})$	O(1)	O(1)	$O(\|\ln\varepsilon\|)$	O(1)
$Q_{eq}(x,y)$	O(1)	$O(\varepsilon^{-1})$	$O(\varepsilon^{-2})$	O(1)	$O(\|\ln\varepsilon\|)$	$O(\|\ln\varepsilon\|^2)$

$$Q_{ali} = \frac{h_{K,x}^2(1+\frac{1}{\varepsilon^2}e^{-\frac{x}{\varepsilon}}) + h_{K,y}^2(1+\frac{1}{\varepsilon^2}e^{-\frac{y}{\varepsilon}})}{2\rho h_{K,x}h_{K,y}}. \tag{4.119}$$

These can be estimated separately in the regions Ω_i, $i = 0,1,2,3$. The results are summarized in Table 4.1. For comparison purposes, the results for a uniform mesh are also included in the table. The values of Q_{geo} indicate that the elements of the uniform mesh have a perfect regularity, whereas some elements of the Shishkin-type mesh have a large aspect ratio. On the other hand, the results for Q_{eq} and Q_{ali} show that for the given monitor function the Shishkin-type mesh has a much better level of equidistribution and alignment than the uniform one. In the terminology of §4.1, the Shishkin-type mesh will be much closer to being M-uniform than the uniform one when ε is small. We return to this example in §5.2, where a more precise comparison based on a general analysis of the errors for adaptive versus nonadaptive meshes is made. □

Example 4.6.2 In this example we consider rectangular meshes shown in Figure 4.1 for the constant monitor function defined in (4.1). For the 41×21 mesh in Figure 4.1(a), it is easy to find analytically that $\|Q_{geo}\|_\infty = 1.25$, $\|Q_{ali}\|_\infty = 1.0$, and $\|Q_{eq}\|_\infty = 1.0$. This indicates that the mesh satisfies exactly both the equidistribution and alignment conditions and thus an M-uniform mesh for M defined in (4.1). On the other hand, for the 21×21 mesh in Figure 4.1(b), $\|Q_{geo}\|_\infty = 1.0$, $\|Q_{ali}\|_\infty = 1.25$, $\|Q_{eq}\|_\infty = 1.0$. Thus, it does not satisfy the alignment condition and is not M-uniform for M defined (4.1). Note that these results will remain the same when these meshes are refined by simultaneously doubling the numbers of subintervals in both x and y directions.

□

Example 4.6.3 In this example M-uniform unstructured meshes of Delaunay-type (e.g., see [150]) are generated for $\Omega = (0,1) \times (0,1)$ using the following monitor functions:

Table 4.2 Example 4.6.3: The quality measures for unstructured meshes obtained with various monitor functions.

M_1	N	226	1078	2322	4452	9180	18292
	$\|\|Q_{geo}\|\|_\infty$	1.3	1.4	1.4	1.4	1.4	1.4
	$\|\|Q_{ali}\|\|_\infty$	1.3	1.4	1.4	1.4	1.4	1.4
	$\|\|Q_{eq}\|\|_\infty$	1.4	1.4	1.7	1.5	1.5	1.6
M_2	N	238	1084	2314	4566	9092	17960
	$\|\|Q_{geo}\|\|_\infty$	1.8	2.4	2.1	2.3	2.5	2.4
	$\|\|Q_{ali}\|\|_\infty$	1.4	1.5	1.4	1.4	1.4	1.4
	$\|\|Q_{eq}\|\|_\infty$	1.4	1.6	1.7	1.7	1.6	1.5
M_3	N	230	1136	2274	4508	8282	17330
	$\|\|Q_{geo}\|\|_\infty$	10	10	9.8	9.9	11	12
	$\|\|Q_{ali}\|\|_\infty$	1.3	1.3	1.4	1.4	1.4	1.5
	$\|\|Q_{eq}\|\|_\infty$	1.6	1.7	1.6	1.6	1.6	1.7
M_4	N	342	1366	2607	4944	9685	18788
	$\|\|Q_{geo}\|\|_\infty$	7.9	13	15	17	16	18
	$\|\|Q_{ali}\|\|_\infty$	1.4	1.5	1.3	1.4	1.5	1.4
	$\|\|Q_{eq}\|\|_\infty$	1.6	1.7	1.7	1.7	1.8	1.7

$$\begin{cases} M_1 = \begin{bmatrix} 1 & 0 \\ 0 & 1 \end{bmatrix}, \quad M_2 = \begin{bmatrix} 4 & 0 \\ 0 & 1 \end{bmatrix}, \\ M_3 = \begin{bmatrix} 100 & 0 \\ 0 & 1 \end{bmatrix}, \quad M_4 = \begin{bmatrix} 1 & 0 \\ 0 & 1 \end{bmatrix} + \nabla v \nabla v^T, \end{cases} \tag{4.120}$$

where $v = \tanh(10(y - 0.5sin(\pi x) - 0.25))$. Monitor function M_1 is included here to check how far a quasi-uniform mesh generated by the computer code is from being uniform. The results are shown in Table 4.2 and Figure 4.8. It can be seen that meshes obtained with M_2, M_3, and M_4 have the same values of $||Q_{ali}||_\infty$ and $||Q_{eq}||_\infty$ as those generated with M_1. Thus, we can conclude that the same code produces quasi M-uniform meshes satisfying the equidistribution and alignment conditions for monitor functions M_2, M_3, and M_4 as closely as the generated quasi-uniform meshes satisfy the uniform conditions. □

4.7 Biographical notes

The alignment condition (4.29) is first introduced in a continuous form by Huang [176], who considers isotropy or conformity conditions in terms of eigenvalues.

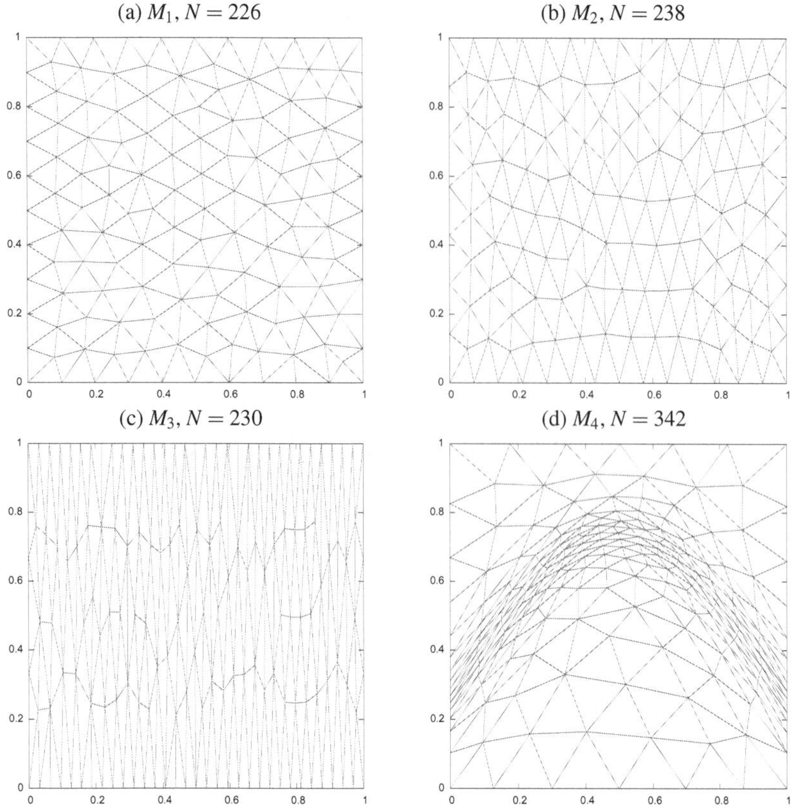

Fig. 4.8 Example 4.6.3. Quasi M-uniform unstructured meshes generated with the various monitor functions defined in (4.120).

The relations between the equidistribution and alignment conditions and other conditions have been studied in [180].

Mesh assessment has been extensively studied in the context of finite elements. For example, the minimum angle [356], the maximum angle [22, 205, 220, 310], and the aspect ratio have all been widely used to characterize the shape of elements in the traditional (isotropic) error analysis. A review of mesh quality measures is given by Apel et al. [15]. Shape measures and their relations for tetrahedral elements are summarized and studied by Liu and Joe [239, 240], Parthasarathy et al. [271], and Dompierre et al. [123]. An algebraic framework for mesh quality measures is developed by Knupp [216] based on the Jacobian matrix.

A number of mesh quality measures taking into account both the mesh quality and solution behavior have been developed. An example is the measure proposed by Berzins [50] for triangular and tetrahedral meshes. A so-called matching function is used by Kunert [221] to measure the correspondence of a mesh to the anisotropic features of the solution. The mesh quality measures in §4.5, which are also solution-dependent (through the monitor function), are developed by Huang in [178].

4.8 Exercises

1. Verify Lemma 4.1.1 directly for $d = 2$ and for $d = 3$.
2. Prove that (4.25) and (4.26) are mathematically equivalent.
3. Prove that (4.26) and (4.27) are mathematically equivalent.
4. Give a proof of Theorem 4.1.2.
5. Derive in detail inequalities (4.104).
6. Verify equation (4.92).
7. Show that if $\max_K Q_{eq}(K) = 1$, then $Q_{eq}(K) = 1$ for all $K \in \mathscr{T}_h$, i.e., the mesh satisfies the equidistribution condition exactly.
8. Complete the details for the proof of Lemma 4.5.1.
9. Complete the details for the first example in §4.6, including finding the explicit bound in (4.115) for the Shishkin mesh.

Chapter 5
Monitor Functions

From the previous chapter we have seen that an adaptive mesh can conveniently be viewed as an M-uniform mesh, or a uniform one in a metric space equipped with metric tensor or monitor function $M = M(\boldsymbol{x})$. A key to the success to this approach of mesh adaptation is in the selection of a proper M. In this chapter, we shall study how to define the monitor function based on estimates for interpolation error, (semi-) a posteriori error bounds for the solution, or some other geometric and physical considerations. For a given monitor function M, the central issue for mesh adaptation then becomes generating an M-uniform mesh. This issue is addressed in Chapters 6 and 7.

The monitor function can generally be chosen based on error estimates or special problem considerations. In this chapter the objective is to explore how the monitor function can be defined based on interpolation error. The error is first considered for Taylor polynomial approximation in §5.1.1. This motivates the interpolation error form for the general d-dimensional problem in Sobolev spaces, which involves a fairly complicated derivation that is given in §§5.1.2-5.2. The reader not interested in such details may simply skip the Sobolev analysis and find the summary of the interpolation error bounds and corresponding formulas for optimal monitor functions in §5.2.5.

We shall take two distinct approaches in developing adaptive mesh generation strategies in multidimensions. The first is the *isotropic* approach, a prerequisite for it being that the shape of the mesh elements is regular (or close to being equilateral), and the size is determined by equidistribution of an error estimate or indicator. When isotropic meshes are generated this way, all of the elements have a small aspect ratio (defined as the ratio of the radii of their circumscribed and inscribed spheres). The other is the *anisotropic* approach. It takes full advantage of mesh adaptation by allowing the size, shape, and orientation of mesh elements to adapt to the solution behavior. For most of the problems of interest to us here, the adaptation generally produces an anisotropic mesh, where some elements have a very large aspect ratio

W. Huang and R.D. Russell, *Adaptive Moving Mesh Methods*, Applied Mathematical Sciences 174, DOI 10.1007/978-1-4419-7196-2_5,

because the solution varies markedly over the region. In this case, since the equidistribution principle alone is insufficient for determining the mesh, one of necessity requires additional conditions for the adaptive mesh.

5.1 Interpolation theory in Sobolev spaces

The primary goal below is to obtain interpolation error estimates which can be used in determining the optimal monitor function, based on the mesh adaptation principles discussed in the previous chapter. The development is similar to the one in §2.4 for the 1D case, where Taylor series approximation provides a convenient way to obtain basic error bounds and motivates the case of general interpolation. However, the formulas and analysis for the multi-dimensional situation become much more complicated. Interpolation theory in Sobolev spaces is studied for a general setting of finite elements in this section.

5.1.1 Error estimates for linear Lagrange interpolation at vertices

We first consider a formal (non-rigorous) analysis of the simple case of linear Lagrange interpolation at the vertices of an arbitrary simplicial element K with vertices $\boldsymbol{a}_i$, $i = 1, \dots, d+1$. For a function $u = u(\boldsymbol{x})$, the linear Lagrange interpolant is

$$\Pi_1 u(\boldsymbol{x}) = \sum_{i=1}^{d+1} u(\boldsymbol{a}_i)\phi_i(\boldsymbol{x}), \tag{5.1}$$

where the basis functions $\{\phi_i\}$ are linear Lagrange polynomials satisfying

$$0 \le \phi_i(\boldsymbol{x}) \le 1, \quad \phi_i(\boldsymbol{a}_j) = \delta_{i,j}, \quad \sum_i \phi_i(\boldsymbol{x}) = 1, \quad \sum_i \phi_i(\boldsymbol{x})\boldsymbol{a}_i = \boldsymbol{x}. \tag{5.2}$$

Denote the center of K by $\boldsymbol{x}_K$. Assuming that $u(\boldsymbol{x})$ is sufficiently smooth on $K \cup \partial K$, by Taylor's theorem we can expand $u(\boldsymbol{x})$ about $\boldsymbol{x}_K$ as

$$u(\boldsymbol{x}) = u(\boldsymbol{x}_K) + \nabla u^T(\boldsymbol{x}_K)(\boldsymbol{x}-\boldsymbol{x}_K) + \frac{1}{2}(\boldsymbol{x}-\boldsymbol{x}_K)^T H(u,\boldsymbol{x}_K)(\boldsymbol{x}-\boldsymbol{x}_K) + h.o.t., \tag{5.3}$$

where $H(u,\boldsymbol{x}_K)$ denotes the Hessian of u evaluated at $\boldsymbol{x}_K$, and $h.o.t.$ stands for higher order terms. Taking $\boldsymbol{x} = \boldsymbol{a}_i$,

$$u(\boldsymbol{a}_i) = u(\boldsymbol{x}_K) + \nabla u^T(\boldsymbol{x}_K)(\boldsymbol{a}_i-\boldsymbol{x}_K) + \frac{1}{2}(\boldsymbol{a}_i-\boldsymbol{x}_K)^T H(u,\boldsymbol{x}_K)(\boldsymbol{a}_i-\boldsymbol{x}_K) + h.o.t. \tag{5.4}$$

Inserting this into (5.1), subtracting from (5.3), and using (5.2), we obtain

$$u(\boldsymbol{x}) - \Pi_1 u(\boldsymbol{x}) = \frac{1}{2}\sum_i \phi_i(\boldsymbol{x})(\boldsymbol{x}-\boldsymbol{a}_i)^T H(u,\boldsymbol{x}_K)(\boldsymbol{a}_i - \boldsymbol{x}_K) + h.o.t. \tag{5.5}$$

The lemma below is needed to estimate the right-hand-side term. For any symmetric matrix S, we define

$$|S| = \sqrt{S^2}. \tag{5.6}$$

If the eigen-decomposition of S is $S = Q\Sigma Q^T$, where Q is an orthogonal matrix and $\Sigma = \mathrm{diag}(\lambda_1, ..., \lambda_d)$, then it is easy to show that $|S|$ can be expressed as

$$|S| = Q\mathrm{diag}(|\lambda_1|, ..., |\lambda_d|)Q^T. \tag{5.7}$$

Lemma 5.1.1 *Suppose that S is a $d \times d$ symmetric matrix. Then*

$$|\boldsymbol{u}^T S \boldsymbol{v}| \le \frac{1}{2}(\boldsymbol{u}^T |S| \boldsymbol{u} + \boldsymbol{v}^T |S| \boldsymbol{v}), \quad \forall \boldsymbol{u}, \boldsymbol{v} \in \mathbb{R}^d. \tag{5.8}$$

Proof. Decompose the matrix Σ into

$$\Sigma = \Sigma_+ - \Sigma_-,$$

where

$$\Sigma_+ = \mathrm{diag}(\max\{0, \lambda_1\}, ..., \max\{0, \lambda_d\}),$$
$$\Sigma_- = \mathrm{diag}(\max\{0, -\lambda_1\}, ..., \max\{0, -\lambda_d\}),$$

so that $|S| = Q(\Sigma_+ + \Sigma_-)Q^T$. The result follows since

$$\begin{aligned}
|\boldsymbol{u}^T S \boldsymbol{v}| &= |(Q^T\boldsymbol{u})^T \Sigma (Q^T\boldsymbol{v})| \\
&\le |(Q^T\boldsymbol{u})^T \Sigma_+ (Q^T\boldsymbol{v})| + |(Q^T\boldsymbol{u})^T \Sigma_- (Q^T\boldsymbol{v})| \\
&= |(\Sigma_+^{\frac{1}{2}} Q^T\boldsymbol{u})^T (\Sigma_+^{\frac{1}{2}} Q^T\boldsymbol{v})| + |(\Sigma_-^{\frac{1}{2}} Q^T\boldsymbol{u})^T (\Sigma_-^{\frac{1}{2}} Q^T\boldsymbol{v})| \\
&\le \frac{1}{2}\left(\|\Sigma_+^{\frac{1}{2}} Q^T\boldsymbol{u}\|^2 + \|\Sigma_+^{\frac{1}{2}} Q^T\boldsymbol{v}\|^2 + \|\Sigma_-^{\frac{1}{2}} Q^T\boldsymbol{u}\|^2 + \|\Sigma_-^{\frac{1}{2}} Q^T\boldsymbol{v}\|^2 \right) \\
&= \frac{1}{2}\left(\boldsymbol{u}^T Q\Sigma_+ Q^T \boldsymbol{u} + \boldsymbol{v}^T Q\Sigma_+ Q^T \boldsymbol{v} + \boldsymbol{u}^T Q\Sigma_- Q^T \boldsymbol{u} + \boldsymbol{v}^T Q\Sigma_- Q^T \boldsymbol{v}\right) \\
&= \frac{1}{2}\left(\boldsymbol{u}^T |S| \boldsymbol{u} + \boldsymbol{v}^T |S| \boldsymbol{v}\right).
\end{aligned}$$

□

If F_K is the mapping from the reference element $\hat{K}$ to K and $\hat{\boldsymbol{a}}_i$, $i = 1, ..., (d+1)$ and $\boldsymbol{\xi}_c$ are the vertices and center of $\hat{K}$, respectively, then from the lemma we have

$$
\begin{aligned}
&|(\boldsymbol{x}-\boldsymbol{a}_i)^T H(u,\boldsymbol{x}_K)(\boldsymbol{a}_i-\boldsymbol{x}_K)| \\
&\le \frac{1}{2}(\boldsymbol{x}-\boldsymbol{a}_i)^T |H(u,\boldsymbol{x}_K)|(\boldsymbol{x}-\boldsymbol{a}_i) + \frac{1}{2}(\boldsymbol{a}_i-\boldsymbol{x}_K)^T |H(u,\boldsymbol{x}_K)|(\boldsymbol{a}_i-\boldsymbol{x}_K) \\
&= \frac{1}{2}(\boldsymbol{\xi}-\hat{\boldsymbol{a}}_i)^T (F_K')^T |H(u,\boldsymbol{x}_K)| F_K'(\boldsymbol{\xi}-\hat{\boldsymbol{a}}_i) \\
&\qquad + \frac{1}{2}(\hat{\boldsymbol{a}}_i-\boldsymbol{\xi}_c)^T (F_K')^T |H(u,\boldsymbol{x}_K)| F_K'(\hat{\boldsymbol{a}}_i-\boldsymbol{\xi}_c) \\
&\le \frac{1}{2}\|(F_K')^T |H(u,\boldsymbol{x}_K)| F_K'\| \cdot \|\boldsymbol{\xi}-\hat{\boldsymbol{a}}_i\|^2 + \frac{1}{2}\|(F_K')^T |H(u,\boldsymbol{x}_K)| F_K'\| \cdot \|\hat{\boldsymbol{a}}_i-\boldsymbol{\xi}_c\|^2 \\
&\le \frac{1}{2}\mathrm{tr}\left((F_K')^T |H(u,\boldsymbol{x}_K)| F_K'\right) \cdot \|\boldsymbol{\xi}-\hat{\boldsymbol{a}}_i\|^2 \\
&\qquad + \frac{1}{2}\mathrm{tr}\left((F_K')^T |H(u,\boldsymbol{x}_K)| F_K'\right) \cdot \|\hat{\boldsymbol{a}}_i-\boldsymbol{\xi}_c\|^2.
\end{aligned} \tag{5.9}
$$

Since $\hat{K}$ is equilateral and has a unitary volume by assumption, it follows from (5.9) that

$$
|(\boldsymbol{x}-\boldsymbol{a}_i)^T H(u,\boldsymbol{x}_K)(\boldsymbol{a}_i-\boldsymbol{x}_K)| \le C\,\mathrm{tr}\left((F_K')^T |H(u,\boldsymbol{x}_K)| F_K'\right), \tag{5.10}
$$

where the constant $C = O(1)$. Using (5.2) and (5.5) leads to

$$
|u(\boldsymbol{x}) - \Pi_1 u(\boldsymbol{x})| \le C\,\mathrm{tr}\left((F_K')^T |H(u,\boldsymbol{x}_K)| F_K'\right) + h.o.t.,
$$

and more generally, for any $q \in [1,\infty]$,

$$
\|u - \Pi_1 u\|_{L^q(K)} \le C|K|^{\frac{1}{q}} \cdot \mathrm{tr}\left((F_K')^T |H(u,\boldsymbol{x}_K)| F_K'\right) + h.o.t. \tag{5.11}
$$

A bound on the gradient of the error can be obtained as follows: Differentiating (5.5) gives

$$
\begin{aligned}
&\nabla(u-\Pi_1 u) \\
&= \frac{1}{2}\sum_i (\boldsymbol{x}-\boldsymbol{a}_i)^T H(u,\boldsymbol{x}_K)(\boldsymbol{a}_i-\boldsymbol{x}_K)\nabla\phi_i(\boldsymbol{x}) + \frac{1}{2}\sum_i \phi_i(\boldsymbol{x}) H(u,\boldsymbol{x}_K)(\boldsymbol{a}_i-\boldsymbol{x}_K) + h.o.t. \\
&= \frac{1}{2}\sum_i (\boldsymbol{x}-\boldsymbol{a}_i)^T H(u,\boldsymbol{x}_K)(\boldsymbol{a}_i-\boldsymbol{x}_K)\nabla\phi_i(\boldsymbol{x}) + \frac{1}{2}H(u,\boldsymbol{x}_K)(\boldsymbol{x}-\boldsymbol{x}_K) + h.o.t.
\end{aligned} \tag{5.12}
$$

Recall that $\phi_i(F_K(\boldsymbol{\xi}))$ corresponds to a basis function on the reference element $\hat{K}$ (cf. (3.93)), so

$$
\nabla_{\boldsymbol{\xi}}\phi_i(F_K(\boldsymbol{\xi})) = O(1),
$$

where $\nabla_{\boldsymbol{\xi}}$ is the gradient operator with respect to $\boldsymbol{\xi}$. From the chain rule,

$$
\nabla_{\boldsymbol{\xi}}\phi_i(F_K(\boldsymbol{\xi})) = (F_K')^T \nabla\phi_i(\boldsymbol{x}),
$$

so

$$\|\nabla \phi_i(\boldsymbol{x})\| \le C\|(F_K^{'})^{-1}\| \quad \forall \boldsymbol{x} \in K. \tag{5.13}$$

Next, we have

$$\begin{aligned}
\|H(u,\boldsymbol{x}_K)(\boldsymbol{x}-\boldsymbol{x}_K)\| &= \left((\boldsymbol{x}-\boldsymbol{x}_K)^T H(u,\boldsymbol{x}_K)H(u,\boldsymbol{x}_K)(\boldsymbol{x}-\boldsymbol{x}_K)\right)^{\frac{1}{2}} \\
&= \left((\boldsymbol{\xi}-\boldsymbol{\xi}_c)^T (F_K^{'})^T H(u,\boldsymbol{x}_K)H(u,\boldsymbol{x}_K)F_K^{'}(\boldsymbol{\xi}-\boldsymbol{\xi}_c)\right)^{\frac{1}{2}} \\
&= \left((\boldsymbol{\xi}-\boldsymbol{\xi}_c)^T (F_K^{'})^T |H(u,\boldsymbol{x}_K)|\cdot|H(u,\boldsymbol{x}_K)|\,F_K^{'}(\boldsymbol{\xi}-\boldsymbol{\xi}_c)\right)^{\frac{1}{2}} \\
&\le C\left\|(F_K^{'})^T |H(u,\boldsymbol{x}_K)|\cdot|H(u,\boldsymbol{x}_K)|\,F_K^{'}\right\|^{\frac{1}{2}} \\
&\le C\left\|(F_K^{'})^T |H(u,\boldsymbol{x}_K)|F_K^{'}\cdot(F_K^{'})^{-1}(F_K^{'})^{-T}\cdot(F_K^{'})^T |H(u,\boldsymbol{x}_K)|F_K^{'}\right\|^{\frac{1}{2}} \\
&\le C\|(F_K^{'})^T |H(u,\boldsymbol{x}_K)|F_K^{'}\|\cdot\|(F_K^{'})^{-1}\| \\
&\le C\,\mathrm{tr}\left((F_K^{'})^T |H(u,\boldsymbol{x}_K)|F_K^{'}\right)\cdot\|(F_K^{'})^{-1}\|.
\end{aligned}$$

Combining this, (5.13), (5.10), and (5.12), we obtain

$$\|\nabla(u-\Pi_1 u)\| \le C\|(F_K^{'})^{-1}\|\cdot\mathrm{tr}\left((F_K^{'})^T |H(u,\boldsymbol{x}_K)|F_K^{'}\right) + h.o.t.$$

and

$$\|\nabla(u-\Pi_1 u)\|_{L^q(K)} \le C|K|^{\frac{1}{q}}\cdot\|(F_K^{'})^{-1}\|\cdot\mathrm{tr}\left((F_K^{'})^T |H(u,\boldsymbol{x}_K)|F_K^{'}\right) + h.o.t. \tag{5.14}$$

The bounds (5.11) and (5.14), obtained here formally, are essentially the same as those later derived rigorously (see (5.45)). The bound (5.11) is insensitive to the shape of element K, being consistent with the well-known fact that the L^q norm of the finite element error is insensitive to the element shape. On the other hand, the presence of the factor $\|(F_K^{'})^{-1}\|$ in (5.14) indicates that this bound is subject to the minimum angle condition: it grows as the minimum angle of K goes to zero. It is very instructive to observe that the term $\mathrm{tr}\left((F_K^{'})^T |H(u,\boldsymbol{x}_K)|F_K^{'}\right)$ in these bounds couples the main geometric features of the element (size, shape, and orientation) with the Hessian of $u = u(\boldsymbol{x})$. In this sense, the bounds in (5.11) and (5.14) are anisotropic in character, as discussed further in §5.1.4, §5.1.5, and §5.1.6. This is in contrast to what is called an isotropic bound, where only the size of an element is directly coupled with the solution information.

5.1.2 A classical result

This subsection and the next three are devoted to the study of interpolation theory in Sobolev spaces in a general setting of finite elements. The primary goal is to obtain interpolation error estimates which can be used in determining how to define the optimal monitor function for mesh adaptation. The error bounds have a similar form to those in the previous section, except that extra terms arise from having to consider general interpolation operators on finite elements (instead of simply Taylor polynomial approximations).

A classical finite element interpolation result is given in this section, followed by derivations of isotropic and anisotropic error bounds and bounds on element faces. A summary of these error bounds and formulas for optimal monitor functions are then given in §5.2.

We begin by introducing Sobolev space notation. (See Appendix A for a brief review of the standard definitions and properties of Sobolev spaces.) For a given bounded domain $D \subset \mathbb{R}^d$ $(d \geq 1)$, the norm and semi-norm of the Sobolev space $W^{m,p}(D)$ are denoted by $\|\cdot\|_{W^{m,p}(D)}$ and $|\cdot|_{W^{m,p}(D)}$, respectively. The *scaled* semi-norm is defined as $\langle\cdot\rangle_{W^{m,p}(D)} \equiv (1/|D|)^{1/p} |\cdot|_{W^{m,p}(D)}$.

The following two theorems are classical results from the theory of interpolation on Sobolev spaces. The reader is referred to [104] for their proofs.

Theorem 5.1.1 *Let $\hat{K}$ be an open set of $\mathbb{R}^d$. For some integers $k \geq 0$, $0 \leq l \leq k+1$, and $m \geq 0$ and some real numbers $p, q \in [1,\infty]$, let $W^{l,p}(\hat{K})$ and $W^{m,q}(\hat{K})$ be Sobolev spaces satisfying the inclusion*

$$W^{l,p}(\hat{K}) \hookrightarrow W^{m,q}(\hat{K}), \tag{5.15}$$

and let $\hat{\Pi}_k \in \mathscr{L}(W^{l,p}(\hat{K}); W^{m,q}(\hat{K}))$ be a mapping such that

$$\hat{\Pi}_k \hat{p} = \hat{p}, \quad \forall \hat{p} \in P_k(\hat{K}). \tag{5.16}$$

Then

$$|\hat{u} - \hat{\Pi}_k \hat{u}|_{W^{m,q}(\hat{K})} \leq C |\hat{u}|_{W^{l,p}(\hat{K})}, \quad \forall \hat{u} \in W^{l,p}(\hat{K}) \tag{5.17}$$

where $C = C(\hat{\Pi}_k, \hat{K})$ is a constant depending only upon $\hat{\Pi}_k$ and $\hat{K}$.

This theorem on interpolation error bound, given for the reference element $\hat{K}$, is valid for any open set of $\mathbb{R}^d$. The mapping $\hat{\Pi}_k$ is said to be polynomial preserving since it preserves polynomials in $P_k(\hat{K})$ (cf. (5.16)). The inclusion relation (5.15) is discussed after the next theorem.

Theorem 5.1.2 *Let $(\hat{K}, \hat{P}, \hat{\Sigma})$ be a finite element associated with the simplicial reference element $\hat{K}$. Let s be the greatest order of partial derivatives occurring in*

Table 5.1 The parameters contained in Theorems 5.1.1 and 5.1.2.

Parameter	Range	Physical Meaning
k	Integer, $k \geq 0$	Degree of interpolating polynomial, $P_k(K) \subset P_K$
l	Integer, $0 \leq l \leq k+1$	Regularity of interpolated functions, $u \in W^{l,p}(K)$
m	Integer, $0 \leq m \leq l$	Order of derivatives in error norm, $e \in W^{m,q}(K)$
p	Real, $1 \leq p \leq \infty$	Regularity of interpolated functions, $u \in W^{l,p}(K)$
q	Real, $1 \leq q \leq \infty$	Used in the error norm, $e \in W^{m,q}(K)$

$\hat{\Sigma}$. *For some integers m, k, and l:* $0 \leq m \leq l \leq k+1$, *and some numbers* $p, q \in [1, \infty]$, *if*

$$W^{l,p}(\hat{K}) \hookrightarrow C^s(\hat{K}), \tag{5.18}$$

$$W^{l,p}(\hat{K}) \hookrightarrow W^{m,q}(\hat{K}), \tag{5.19}$$

$$P_k(\hat{K}) \subset \hat{P} \subset W^{m,q}(\hat{K}), \tag{5.20}$$

where $P_k(\hat{K})$ *is the space of polynomials of degree* $\leq k$, *then there exists a constant* $C = C(\hat{K}, \hat{P}, \hat{\Sigma})$ *such that, for all affine-equivalent finite elements* (K, P_K, Σ_K),

$$|u - \Pi_{k,K} u|_{W^{m,q}(K)} \leq C \|(F_K^{'})^{-1}\|^m \cdot |det(F_K^{'})|^{\frac{1}{q}} \cdot |\hat{u}|_{W^{l,p}(\hat{K})}, \quad \forall u \in W^{l,p}(K) \tag{5.21}$$

where $\Pi_{k,K} : W^{l,p}(K) \to P_K$ *denotes the* P_K*-interpolation operator on* K, $\hat{u} = u \circ F_K$ *is the composite function defined on* $\hat{K}$, *and* $\|\cdot\|$ *denotes the* l_2 *matrix norm.*

The theorem can roughly be seen as a consequence of Theorem 5.1.1 and Lemma 5.1.3 (see §5.1.3 below). It applies for simplicial finite elements, i.e., where the mesh elements are d-simplices. Recall from §3.3 that two finite elements are affine-equivalent if their mesh elements, finite dimensional function spaces, and sets of degrees of freedom can be mapped to each other through affine mappings.

Note that u and $\hat{u}$ are the same function in (5.21), but where there is no ambiguity we use u to denote the function considered as a function on K and $\hat{u}$ when considered as a function on $\hat{K}$. The error bound in (5.21) is given in terms of derivatives on $\hat{K}$. This is crucial for the study of anisotropic meshes since as we shall see it allows one to develop error bounds directly coupling mesh properties with solution derivatives on K. It should be pointed out that while (5.21) is not optimal when $m \geq 1$, it greatly simplifies the discussion since there is no need to introduce conditions like the maximum angle condition (e.g., see Babuška and Aziz [22]).

Theorems 5.1.1 and 5.1.2 involve the five parameters m, k, l, p, and q. Their physical meanings are summarized in Table 5.1. *Hereafter, we assume that these parameters have been chosen such that Theorem 5.1.2 holds.* For completeness sake, it is worthwhile spelling out precisely when the conditions (5.18)–(5.20) (and (5.15)) hold. For $\hat{K} \subset \mathbb{R}^d$, from the Sobolev embedding theorem [4]

$$\begin{cases} l > \frac{d}{p} + s & \text{for } p > 1 \\ l \geq d + s & \text{for } p = 1 \end{cases} \quad \Longrightarrow \quad W^{l,p}(\hat{K}) \hookrightarrow C^s(\hat{K})$$

$$\begin{cases} l \geq m & \text{for } p \geq q \\ l < \frac{d}{p} + m & \text{for } \frac{1}{q} = \frac{1}{p} - \frac{l-m}{d} \\ l = \frac{d}{p} + m & \text{for } 1 \leq q < \infty \end{cases} \quad \Longrightarrow \quad W^{l,p}(\hat{K}) \hookrightarrow W^{m,q}(\hat{K}). \tag{5.22}$$

Regarding condition (5.20), we simply note that $\hat{P}$ is often chosen as $P_k(\hat{K})$, and in such case it places no further constraints on the parameters m, k, l, p, and q.

Example 5.1.1 Consider the widely used choice of Lagrange interpolation ($s = 0$ since the interpolation function uses only zero order derivatives) with $p = q = 2$. Condition (5.22) becomes $0 \leq m \leq l \leq k+1$ and $l > d/2$. Thus, condition (5.21) holds for functions in $H^1(\hat{K}) \equiv W^{1,2}(\hat{K})$ in one dimension and $H^2(\hat{K}) \equiv W^{2,2}(\hat{K})$ in two and three dimensions. □

5.1.3 Relations between norms on affine-equivalent elements

The task of the next three subsections is to develop from (5.21) bounds for the error over the entire domain Ω. To this end, we derive here estimates of the norm of F_K', the Jacobian matrix of the affine mapping between two simplicial elements $\hat{K}$ and K, and relations of function norms on them. The basic tools are the coordinate transformation and the chain rule.

Denote the physical coordinates on K and the computational coordinates on $\hat{K}$ by $\boldsymbol{x} = (x_1, ..., x_d)^T$ and $\boldsymbol{\xi} = (\xi_1, ..., \xi_d)^T$, respectively. Note that $\partial \boldsymbol{x} / \partial \boldsymbol{\xi} = F_K'$ on K and is piecewise constant on the whole domain Ω. Moreover, for any $p \geq 1$ we define

$$\|F_K'\|_p = \left(\sum_{i,j} \left| \frac{\partial x_i}{\partial \xi_j} \right|^p \right)^{\frac{1}{p}}.$$

Note that $\|\cdot\|_p$ is a matrix norm and thus is equivalent to the l_2 matrix norm, $\|\cdot\|$.

Lemma 5.1.2 *The Jacobian matrix, F_K', of the affine mapping F_K between two simplicial elements $\hat{K}$ and K has the properties*

$$\left| det(F_K') \right| = \frac{|K|}{|\hat{K}|}, \quad \|F_K'\| \leq \frac{h_K}{H_{\hat{K}}}, \quad \|(F_K')^{-1}\| \leq \frac{h_{\hat{K}}}{H_K}, \tag{5.23}$$

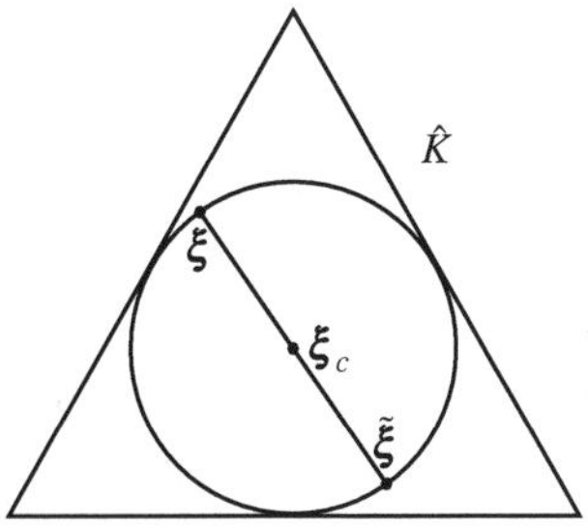

Fig. 5.1 Element $\hat{K}$ and its inscribed circle.

where h_K and H_K are respectively the diameter and in-diameter of K, and $h_{\hat{K}}$ and $H_{\hat{K}}$ are the corresponding quantities for $\hat{K}$.

Proof. The equality in (5.23) follows from

$$|K| = \int_K d\mathbf{x} = \int_{\hat{K}} \left|\det(F_K')\right| d\boldsymbol{\xi} = \left|\det(F_K')\right| \, |\hat{K}|.$$

Let the boundary of the largest inscribed ball of $\hat{K}$ be denoted by the sphere $S(\boldsymbol{\xi}_c, r)$, with center $\boldsymbol{\xi}_c$ and radius r – see Figure 5.1. By definition, the diameter of $S(\boldsymbol{\xi}_c, r)$ is equal to the in-diameter of $\hat{K}$, i.e., $H_{\hat{K}} = 2r$. For any given point $\boldsymbol{\xi}$ on $S(\boldsymbol{\xi}_c, r)$, the conjugate point $\tilde{\boldsymbol{\xi}}$, defined as the intersection of the sphere with the straight line passing through $\boldsymbol{\xi}_c$ and $\boldsymbol{\xi}$, satisfies $\|\boldsymbol{\xi} - \tilde{\boldsymbol{\xi}}\| = H_{\hat{K}}$. It follows that

$$\begin{aligned} \|F_K'\| &= \sup_{\boldsymbol{\xi} \neq 0} \frac{\|F_K' \boldsymbol{\xi}\|}{\|\boldsymbol{\xi}\|} \\ &= \sup_{\boldsymbol{\xi} \in S(\boldsymbol{\xi}_c, r)} \frac{\|F_K'(\boldsymbol{\xi} - \tilde{\boldsymbol{\xi}})\|}{\|\boldsymbol{\xi} - \tilde{\boldsymbol{\xi}}\|} \\ &= \frac{1}{H_{\hat{K}}} \sup_{\boldsymbol{\xi} \in S(\boldsymbol{\xi}_c, r)} \|F_K' \boldsymbol{\xi} - F_K' \tilde{\boldsymbol{\xi}}\|. \end{aligned} \tag{5.24}$$

Since both $F_K' \boldsymbol{\xi}$ and $F_K' \tilde{\boldsymbol{\xi}}$ are in K, we have $\|F_K' \boldsymbol{\xi} - F_K' \tilde{\boldsymbol{\xi}}\| \leq h_K$, and the first inequality in (5.23) follows directly.

The second inequality is obtained by simply interchanging the roles of $\hat{K}$ and K.

□

Lemma 5.1.3 *Assume that $\hat{K}$ and K are affine-equivalent simplicial elements. For any $l \geq 0$ and $1 \leq p \leq \infty$,*

$$|\hat{v}|_{W^{l,p}(\hat{K})} \le C\|F_K'\|^l det\left(F_K'\right)^{-\frac{1}{p}} |v|_{W^{l,p}(K)}, \quad \forall v \in W^{l,p}(K) \tag{5.25}$$

$$|v|_{W^{l,p}(K)} \le C\|(F_K')^{-1}\|^l det\left(F_K'\right)^{\frac{1}{p}} |\hat{v}|_{W^{l,p}(\hat{K})}, \quad \forall v \in W^{l,p}(K) \tag{5.26}$$

where $\hat{v} = v \circ F_K$ and C is a constant depending only upon l and p.

Proof. We prove the lemma only for $p < \infty$. The situation for $p = \infty$ can be proved similarly with corresponding modification of notation.

By changing the variables of integration and using the chain rule we have

$$\begin{aligned}
\sum_{i_1,\dots,i_l} \left| \frac{\partial^l \hat{u}}{\partial \xi_{i_1} \cdots \partial \xi_{i_l}} \right|^p &= \sum_{i_2,\dots,i_l} \sum_{i_1} \left| \sum_{j_1} \frac{\partial x_{j_1}}{\partial \xi_{i_1}} \frac{\partial^l \hat{u}}{\partial x_{j_1} \partial \xi_{i_2} \cdots \partial \xi_{i_l}} \right|^p \\
&\le \sum_{i_2,\dots,i_l} \sum_{i_1} \left(\left| \sum_{j_1} \frac{\partial x_{j_1}}{\partial \xi_{i_1}} \right|^p \right) \left(\sum_{j_1} \left| \frac{\partial^l \hat{u}}{\partial x_{j_1} \partial \xi_{i_2} \cdots \partial \xi_{i_l}} \right|^p \right) \\
&= \|F_K'\|_p^p \sum_{j_1} \sum_{i_2,\dots,i_l} \left| \frac{\partial^l \hat{u}}{\partial x_{j_1} \partial \xi_{i_2} \cdots \partial \xi_{i_l}} \right|^p .
\end{aligned} \tag{5.27}$$

Repeating the above process l times and using the equivalence of matrix norms, we obtain

$$\begin{aligned}
\sum_{i_1,\dots,i_l} \left| \frac{\partial^l \hat{u}}{\partial \xi_{i_1} \cdots \partial \xi_{i_l}} \right|^p &\le \|F_K'\|_p^{lp} \sum_{j_1,\dots,j_l} \left| \frac{\partial^l u}{\partial x_{j_1} \cdots \partial x_{j_l}} \right|^p \\
&\le C\|F_K'\|^{lp} \sum_{j_1,\dots,j_l} \left| \frac{\partial^l u}{\partial x_{j_1} \cdots \partial x_{j_l}} \right|^p ,
\end{aligned}$$

which leads to (5.25).

Inequality (5.26) can be obtained similarly by interchanging the roles of $\hat{K}$ and K. □

Inequality (5.25) can be improved by coupling F_K' with derivatives, which leads to anisotropic error bounds (see §5.1.5 and §5.1.6). We consider below such improvements for separate cases with $l = 1$ and $l \ge 2$.

Lemma 5.1.4 *Assume that $\hat{K}$ and K are affine-equivalent simplicial elements with $|\hat{K}| = O(1)$. For any $1 \le p \le \infty$ and $v \in W^{1,p}(K)$,*

$$|\hat{v}|_{W^{1,p}(\hat{K})} \le C \left[\frac{1}{|K|} \int_K \left[tr((F_K')^T \nabla u(\boldsymbol{x}) \nabla u^T(\boldsymbol{x}) F_K') \right]^{\frac{p}{2}} d\boldsymbol{x} \right]^{\frac{1}{p}}, \tag{5.28}$$

$$\begin{aligned}
|\hat{v}|_{W^{1,p}(\hat{K})} &\le C \left[tr\left((F_K')^T \nabla u_K \nabla u_K^T F_K' \right) \right]^{\frac{1}{2}} \\
&\quad + C \left[tr((F_K')^T F_K') \right]^{\frac{1}{2}} \cdot \langle \nabla u - \nabla u_K \rangle_{L^p(K)},
\end{aligned} \tag{5.29}$$

where $\hat{v} = v \circ F_K$, *C is a constant depending only upon p, and*

$$\nabla u_K = \frac{1}{|K|} \int_K \nabla u(\boldsymbol{x}) d\boldsymbol{x}. \tag{5.30}$$

Proof. Denote the i-th unit vector in $\mathbb{R}^d$ by $\boldsymbol{e}_i$. Then,

$$\begin{aligned}
\sum_i \left| \frac{\partial \hat{u}}{\partial \xi_i} \right|^p &= \sum_i \left| \sum_j \frac{\partial x_j}{\partial \xi_i} \frac{\partial u}{\partial x_j} \right|^p \\
&= \sum_i \left| (F_K' \boldsymbol{e}_i)^T \nabla u(\boldsymbol{x}) \right|^p \\
&\le C \left[\sum_i \left| (F_K' \boldsymbol{e}_i)^T \nabla u(\boldsymbol{x}) \right|^2 \right]^{\frac{p}{2}} \\
&= C \left[\left| (F_K')^T \nabla u(\boldsymbol{x}) \right|^2 \right]^{\frac{p}{2}} \\
&= C \left[\mathrm{tr}((F_K')^T \nabla u(\boldsymbol{x}) \nabla u^T(\boldsymbol{x}) F_K') \right]^{\frac{p}{2}},
\end{aligned}$$

which leads to (5.28).

To show the second bound, we have

$$\begin{aligned}
\sum_i \left| \frac{\partial \hat{u}}{\partial \xi_i} \right|^p &= \sum_i \left\| (F_K' \boldsymbol{e}_i)^T (\nabla u_K + \nabla u(\boldsymbol{x}) - \nabla u_K) \right\|^p \\
&\le C \sum_i \left\| (F_K' \boldsymbol{e}_i)^T \nabla u_K \right\|^p + C \sum_i \left\| (F_K' \boldsymbol{e}_i)^T (\nabla u(\boldsymbol{x}) - \nabla u_K) \right\|^p \\
&\le C \left(\sum_i \left\| (F_K' \boldsymbol{e}_i)^T \nabla u_K \right\|^2 \right)^{\frac{p}{2}} + C \left(\sum_i \left\| F_K' \boldsymbol{e}_i \right\|^2 \right)^{\frac{p}{2}} \|\nabla u(\boldsymbol{x}) - \nabla u_K\|^p \\
&= C \left[\mathrm{tr} \left((F_K')^T \nabla u_K \nabla u_K^T F_K' \right) \right]^{\frac{p}{2}} + C \left[\mathrm{tr} \left((F_K')^T F_K' \right) \right]^{\frac{p}{2}} \|\nabla u(\boldsymbol{x}) - \nabla u_K\|^p.
\end{aligned}$$

Consequently,

$$\begin{aligned}
&|\hat{u}|_{W^{1,p}(\hat{K})} \\
&\le C \left\{ \left[\mathrm{tr} \left((F_K')^T \nabla u_K \nabla u_K^T F_K' \right) \right]^{\frac{p}{2}} + [\mathrm{tr}((F_K')^T F_K')]^{\frac{p}{2}} \frac{1}{|K|} \int_K \|\nabla u(\boldsymbol{x}) - \nabla u_K\|^p d\boldsymbol{x} \right\}^{\frac{1}{p}} \\
&\le C \left[\mathrm{tr} \left((F_K')^T \nabla u_K \nabla u_K^T F_K' \right) \right]^{\frac{1}{2}} + C [\mathrm{tr}((F_K')^T F_K')]^{\frac{1}{2}} \cdot \left[\frac{1}{|K|} \int_K \|\nabla u(\boldsymbol{x}) - \nabla u_K\|^p d\boldsymbol{x} \right]^{\frac{1}{p}},
\end{aligned}$$

which gives (5.29). □

Lemma 5.1.5 *Assume that $\hat{K}$ and K are affine-equivalent simplicial elements with $|\hat{K}| = O(1)$. For any $l \geq 2$, $1 \leq p \leq \infty$, and $v \in W^{l,p}(K)$,*

$$|\hat{u}|_{W^{l,p}(\hat{K})} \leq C\|F_K^{'}\|^{l-2}\left[\frac{1}{|K|}\int_K \left[tr\left((F_K^{'})^T |H(D^{l-2}u)|(\boldsymbol{x})F_K^{'}\right)\right]^p d\boldsymbol{x}\right]^{\frac{1}{p}}, \quad (5.31)$$

$$\begin{aligned} |\hat{u}|_{W^{l,p}(\hat{K})} &\leq C\|F_K^{'}\|^{l-2} tr\left((F_K^{'})^T |H_K(D^{l-2}u)| F_K^{'}\right) \\ &+ C\|F_K^{'}\|^{l-2} tr\left((F_K^{'})^T F_K^{'}\right)\left\langle H(D^{l-2}u) - H_K(D^{l-2}u)\right\rangle_{L^p(K)}, \end{aligned} \quad (5.32)$$

where $\hat{v} = v \circ F_K$, C is a constant depending only upon l and p, and

$$D^{(i_1,\ldots,i_{l-2})}u = \frac{\partial^{l-2}u}{\partial x_{i_1}\ldots\partial x_{i_{l-2}}}, \quad (5.33)$$

$$|H(D^{l-2}u)(\boldsymbol{x})| = \sum_{i_1,\ldots,i_{l-2}} |H(D^{(i_1,\ldots,i_{l-2})}u)(\boldsymbol{x})|, \quad (5.34)$$

$$|H_K| \equiv |H_K(D^{l-2}u)| = \sum_{i_1,\ldots,i_{l-2}} |H_K(D^{(i_1,\ldots,i_{l-2})}u)|, \quad (5.35)$$

$$H_K(D^{(j_1,\ldots,j_{l-2})}u) = \frac{1}{|K|}\int_K H(D^{(j_1,\ldots,j_{l-2})}u)(\boldsymbol{x})d\boldsymbol{x}. \quad (5.36)$$

Proof. Repeating the process in (5.27) l times, we obtain

$$\begin{aligned} &\sum_{i_1,\ldots,i_l}\left|\frac{\partial^l \hat{u}}{\partial \xi_{i_1}\ldots\partial \xi_{i_l}}\right|^p \\ &\leq C\|F_K^{'}\|^{p(l-2)} \sum_{j_1,\ldots,j_{l-2}}\sum_{i_{l-1},i_l}\left|\sum_{j_{l-1},j_l}\frac{\partial x_{j_{l-1}}}{\partial \xi_{i_{l-1}}}\frac{\partial x_{j_l}}{\partial \xi_{i_l}}\frac{\partial^2\left(D^{(j_1,\ldots,j_{l-2})}u\right)}{\partial x_{j_{l-1}}\partial x_{j_l}}\right|^p \\ &= C\|F_K^{'}\|^{p(l-2)} \sum_{j_1,\ldots,j_{l-2}}\sum_{i_{l-1},i_l}\left|(F_K^{'}\boldsymbol{e}_{i_{l-1}})^T H(D^{(j_1,\ldots,j_{l-2})}u)(\boldsymbol{x})(F_K^{'}\boldsymbol{e}_{i_l})\right|^p, \end{aligned} \quad (5.37)$$

where $H(D^{(j_1,\ldots,j_{l-2})}u)$ denotes the Hessian of $D^{(j_1,\ldots,j_{l-2})}u$. Using Lemma 5.1.1 and the equivalence of vector and matrix norms, we have

$$\sum_{j_1,\ldots,j_{l-2}} \sum_{i_{l-1},i_l} \left|(F_K'\boldsymbol{e}_{i_{l-1}})^T H(D^{(j_1,\ldots,j_{l-2})}u)(\boldsymbol{x})(F_K'\boldsymbol{e}_{i_l})\right|^p$$

$$\leq \left[\sum_{j_1,\ldots,j_{l-2}} \sum_{i_{l-1},i_l} \left|(F_K'\boldsymbol{e}_{i_{l-1}})^T H(D^{(j_1,\ldots,j_{l-2})}u)(\boldsymbol{x})(F_K'\boldsymbol{e}_{i_l})\right|\right]^p$$

$$\leq C\Big[\sum_{j_1,\ldots,j_{l-2}} \sum_{i_{l-1},i_l} \Big\{(F_K'\boldsymbol{e}_{i_{l-1}})^T |H(D^{(j_1,\ldots,j_{l-2})}u)|(\boldsymbol{x})(F_K'\boldsymbol{e}_{i_{l-1}})$$

$$+ (F_K'\boldsymbol{e}_{i_l})^T |H(D^{(j_1,\ldots,j_{l-2})}u)|(\boldsymbol{x})(F_K'\boldsymbol{e}_{i_l})\Big\} \Big]^p$$

$$\leq C\left[\sum_{j_1,\ldots,j_{l-2}} \mathrm{tr}\left((F_K')^T |H(D^{(j_1,\ldots,j_{l-2})}u)|(\boldsymbol{x})F_K'\right)\right]^p$$

$$= C\left[\mathrm{tr}\left((F_K')^T |H(D^{l-2}u)|(\boldsymbol{x})F_K'\right)\right]^p.$$

Combining the above results, we can easily obtain (5.31).

Moreover, from Lemmas 4.5.2 and 5.1.1, it follows that

$$\left|(F_K'\boldsymbol{e}_{i_{l-1}})^T H(D^{(j_1,\ldots,j_{l-2})}u)(\boldsymbol{x})(F_K'\boldsymbol{e}_{i_l})\right|$$

$$\leq \left|(F_K'\boldsymbol{e}_{i_{l-1}})^T H_K(D^{(j_1,\ldots,j_{l-2})}u)(F_K'\boldsymbol{e}_{i_l})\right|$$

$$+ \left|(F_K'\boldsymbol{e}_{i_{l-1}})^T \left(H(D^{(j_1,\ldots,j_{l-2})}u)(\boldsymbol{x}) - H_K(D^{(j_1,\ldots,j_{l-2})}u)\right)(F_K'\boldsymbol{e}_{i_l})\right|$$

$$\leq C\Big[(F_K'\boldsymbol{e}_{i_{l-1}})^T |H_K(D^{(j_1,\ldots,j_{l-2})}u)|(F_K'\boldsymbol{e}_{i_{l-1}})$$

$$+ (F_K'\boldsymbol{e}_{i_l})^T |H_K(D^{(j_1,\ldots,j_{l-2})}u)|(F_K'\boldsymbol{e}_{i_l}) \Big]$$

$$+C\Big[(F_K'\boldsymbol{e}_{i_{l-1}})^T \left|H(D^{(j_1,\ldots,j_{l-2})}u)(\boldsymbol{x}) - H_K(D^{(j_1,\ldots,j_{l-2})}u)\right| (F_K'\boldsymbol{e}_{i_{l-1}})$$

$$+ (F_K'\boldsymbol{e}_{i_l})^T \left|H(D^{(j_1,\ldots,j_{l-2})}u)(\boldsymbol{x}) - H_K(D^{(j_1,\ldots,j_{l-2})}u)\right| (F_K'\boldsymbol{e}_{i_l}) \Big]$$

$$\leq C\Big[(F_K'\boldsymbol{e}_{i_{l-1}})^T |H_K(D^{(j_1,\ldots,j_{l-2})}u)|(F_K'\boldsymbol{e}_{i_{l-1}})$$

$$+ (F_K'\boldsymbol{e}_{i_l})^T |H_K(D^{(j_1,\ldots,j_{l-2})}u)|(F_K'\boldsymbol{e}_{i_l}) \Big]$$

$$+C\left[(F_K'\boldsymbol{e}_{i_{l-1}})^T (F_K'\boldsymbol{e}_{i_{l-1}}) + (F_K'\boldsymbol{e}_{i_l})^T (F_K'\boldsymbol{e}_{i_l})\right]$$

$$\times \left\|\left|H(D^{(j_1,\ldots,j_{l-2})}u)(\boldsymbol{x}) - H_K(D^{(j_1,\ldots,j_{l-2})}u)\right|\right\|$$

$$= C\Big[(F_K'\boldsymbol{e}_{i_{l-1}})^T |H_K(D^{(j_1,\ldots,j_{l-2})}u)|(F_K'\boldsymbol{e}_{i_{l-1}})$$

$$+ (F_K'\boldsymbol{e}_{i_l})^T |H_K(D^{(j_1,\ldots,j_{l-2})}u)|(F_K'\boldsymbol{e}_{i_l}) \Big]$$

$$+C\left[(F_K'\boldsymbol{e}_{i_{l-1}})^T (F_K'\boldsymbol{e}_{i_{l-1}}) + (F_K'\boldsymbol{e}_{i_l})^T (F_K'\boldsymbol{e}_{i_l})\right]$$

$$\times \left\|H(D^{(j_1,\ldots,j_{l-2})}u)(\boldsymbol{x}) - H_K(D^{(j_1,\ldots,j_{l-2})}u)\right\|.$$

Combining this with (5.37), we have

$$
\begin{aligned}
&\sum_{i_1,\ldots,i_l} \left| \frac{\partial^l \hat{u}}{\partial \xi_{i_1} \ldots \partial \xi_{i_l}} \right|^p \\
&\le C \|F_K'\|^{p(l-2)} \left[\sum_{j_1,\ldots,j_{l-2}} \sum_{i_{l-1},i_l} (F_K' \boldsymbol{e}_{i_{l-1}})^T |H_K(D^{(j_1,\ldots,j_{l-2})}u)| (F_K' \boldsymbol{e}_{i_{l-1}}) \right]^p \\
&+ C \|F_K'\|^{p(l-2)} \left[\sum_{j_1,\ldots,j_{l-2}} \sum_{i_{l-1},i_l} (F_K' \boldsymbol{e}_{i_l})^T |H_K(D^{(j_1,\ldots,j_{l-2})}u)| (F_K' \boldsymbol{e}_{i_l}) \right]^p \\
&+ C \|F_K'\|^{p(l-2)} \Big[\sum_{j_1,\ldots,j_{l-2}} \sum_{i_{l-1},i_l} \left\{ (F_K' \boldsymbol{e}_{i_{l-1}})^T (F_K' \boldsymbol{e}_{i_{l-1}}) + (F_K' \boldsymbol{e}_{i_l})^T (F_K' \boldsymbol{e}_{i_l}) \right\} \\
&\qquad \times \left\| H(D^{(j_1,\ldots,j_{l-2})}u)(\boldsymbol{x}) - H_K(D^{(j_1,\ldots,j_{l-2})}u) \right\| \Big]^p \\
&\le C \|F_K'\|^{p(l-2)} \left[\mathrm{tr} \left((F_K')^T |H_K(D^{l-2}u)| F_K' \right) \right]^p \\
&+ C \|F_K'\|^{p(l-2)} \left[\mathrm{tr} \left((F_K')^T F_K' \right) \right]^p \|H(D^{l-2}u)(\boldsymbol{x}) - H_K(D^{l-2}u)\|^p,
\end{aligned}
$$

which leads to (5.32). □

5.1.4 Isotropic error bounds

We now develop element-wise bounds on the interpolation error using Theorem 5.1.2. The strategy is to estimate $|\hat{u}|_{W^{l,p}(\hat{K})}$ in (5.21) in terms of the physical derivatives of u on the element K. This is done here with the *isotropic* approach, where terms involving the shape of elements are decoupled from those involving the physical derivatives of u in the error bound. The anisotropic approach, for which the shape and orientation of elements are explicitly coupled with the physical derivatives of u, is then studied in the next two subsections.

Theorem 5.1.3 *Suppose that the assumptions in Theorem 5.1.2 hold. Then, for any $u \in W^{l,p}(K)$,*

$$
|u - \Pi_{k,K} u|_{W^{m,q}(K)} \le C \|(F_K')^{-1}\|^m \cdot \|F_K'\|^l \cdot |K|^{\frac{1}{q} - \frac{1}{p}} \cdot |u|_{W^{l,p}(K)}. \tag{5.38}
$$

Proof. It follows from Theorem 5.1.2 and Lemmas 5.1.2 and 5.1.3. □

The bound (5.38) is isotropic in the sense that the physical derivative term $|u|_{W^{l,p}(K)}$ in (5.38) is not directly coupled with the Jacobian matrix F_K', which characterizes the shape and orientation of K.

It is instructive to see what the bound (5.38) looks like for regular triangulations, particularly uniform ones. To do this, we first derive a basic estimate for the norm of $F_K^{'}$.

From (3.91) we have $h_{\hat{K}} = \hat{h} = O(1)$ and $H_{\hat{K}} = \hat{H} = O(1)$ for the reference element $\hat{K}$. Using Lemma 5.1.2 and the facts that $H_K = O(h_K)$ and $|K| = O(h_K^d)$ for any regular element, it follows from (5.38) that for any element K in a regular, affine triangulation,

$$|u - \Pi_{k,K} u|_{W^{m,q}(K)} \le C h_K^{l-m+\frac{d}{q}-\frac{d}{p}} \cdot |u|_{W^{l,p}(K)}, \qquad \forall u \in W^{l,p}(K). \tag{5.39}$$

If we further assume that $\mathscr{T}_h$ is uniform or quasi-uniform, then $h \approx h_K$ for all $K \in \mathscr{T}_h$ and

$$|u - \Pi_k u|_{W^{m,q}(\Omega)} \equiv \left(\sum_{K \in \mathscr{T}_h} |u - \Pi_{k,K} u|^q_{W^{m,q}(K)} \right)^{\frac{1}{q}} \le C h^{l-m+\frac{d}{q}-\frac{d}{p}} \left(\sum_{K \in \mathscr{T}_h} |u|^q_{W^{l,p}(K)} \right)^{\frac{1}{q}}.$$

From the arithmetic-mean and geometric-mean inequality (cf. Theorem B.0.11), for $p \ge q$ we obtain

$$\left(\sum_{K \in \mathscr{T}_h} |u|^q_{W^{l,p}(K)} \right)^{\frac{1}{q}} \le N^{\frac{1}{q}-\frac{1}{p}} \left(\sum_{K \in \mathscr{T}_h} |u|^p_{W^{l,p}(K)} \right)^{\frac{1}{p}} \le C h^{\frac{d}{p}-\frac{d}{q}} \left(\sum_{K \in \mathscr{T}_h} |u|^p_{W^{l,p}(K)} \right)^{\frac{1}{p}},$$

where N is the number of elements of the triangulation $\mathscr{T}_h$. On the other hand, when $p \le q$, Jensen's inequality (cf. Theorem B.0.12) leads to

$$\left(\sum_{K \in \mathscr{T}_h} |u|^q_{W^{l,p}(K)} \right)^{\frac{1}{q}} \le \left(\sum_{K \in \mathscr{T}_h} |u|^p_{W^{l,p}(K)} \right)^{\frac{1}{p}}.$$

Combining these results, we obtain

$$|u - \Pi_k u|_{W^{m,q}(\Omega)} \le C h^{l-m-\max\{0,\frac{d}{p}-\frac{d}{q}\}} |u|_{W^{l,p}(\Omega)}, \qquad \forall u \in W^{l,p}(\Omega) \tag{5.40}$$

and in particular, when $p \ge q$,

$$|u - \Pi_k u|_{W^{m,q}(\Omega)} \le C h^{l-m} |u|_{W^{l,p}(\Omega)}, \qquad \forall u \in W^{l,p}(\Omega). \tag{5.41}$$

While this type of error bound, found in the traditional finite element literature [104], can be a reasonable one when u is smooth, lack of information about coupling between the solution and the mesh in situations requiring adaptivity limits its applicability. The error bounds developed in later sections for adaptive meshes do not have this drawback.

5.1.5 Anisotropic error bounds: Case $l = 1$

The goal now is to obtain error bounds for which the physical derivative terms are directly coupled to the information about the size, shape, and orientation of mesh elements. Generally speaking, such *anisotropic* error estimates are more difficult to obtain and their bounds are more complicated than the corresponding isotropic ones since they have to incorporate directional changes in the solution. However, they can provide sharper bounds, especially when the physical solution exhibits anisotropic features such as faster change in one direction than the others. As a result, anisotropic error bounds provide the very information needed in the design and analysis of many adaptive algorithms, as we see in subsequent chapters.

We first consider the case $l = 1$. This case occurs for piecewise constant interpolation ($k = 0$) or general interpolation functions which preserve k-th degree polynomials (which are functions in $W^{1,p}(\Omega)$ with $l = 1 < k+1$). The conditions (5.22) can be seen to require that $s = 0$ (where s is the maximal order of derivatives appearing in $\hat{\Sigma}$), $0 \le m \le 1$, $q \le p$, and $p \ge 1$ for $d = 1$ and $p > d$ for $d \ge 2$.

Theorem 5.1.4 *Suppose that the assumptions in Theorem 5.1.2 hold with $l = 1$. Then for any $u \in W^{1,p}(K)$,*

$$|u - \Pi_{k,K}u|_{W^{m,q}(K)}$$
$$\le C|K|^{\frac{1}{q}-\frac{1}{p}} \cdot \|(F_K')^{-1}\|^m \left[\int_K \left[tr\left((F_K')^T \nabla u(\boldsymbol{x}) \nabla u^T(\boldsymbol{x}) F_K' \right) \right]^{\frac{p}{2}} d\boldsymbol{x} \right]^{\frac{1}{p}}, \tag{5.42}$$

or in a different form,

$$|u - \Pi_{k,K}u|_{W^{m,q}(K)} \le C|K|^{\frac{1}{q}} \cdot \|(F_K')^{-1}\|^m \cdot \left[tr\left((F_K')^T \nabla u_K \nabla u_K^T F_K' \right) \right]^{\frac{1}{2}}$$
$$+ C|K|^{\frac{1}{q}} \cdot \|(F_K')^{-1}\|^m \cdot \left[tr\left((F_K')^T F_K' \right) \right]^{\frac{1}{2}} \cdot \langle \nabla u - \nabla u_K \rangle_{L^p(K)}, \tag{5.43}$$

where ∇u_K is defined in (5.30).

Proof. It follows from Theorem 5.1.2 and Lemma 5.1.4. □

Note that (5.43) is also valid if ∇u_K is replaced by $\nabla u(\boldsymbol{x}_K)$. The second term in the bound can be considered as a higher order term (cf. §5.2), which vanishes when u is a quadratic function on K. Also note that the norm, trace, and determinant terms in both (5.42) and (5.43) are all independent of the coordinate system, so the error bounds are as well. Finally, the Jacobian matrix F_K' is directly coupled to the gradient of u, so these bounds can be used in an effective mesh generation process (cf. §5.2) by basing the choice of the shape and orientation of K on the gradient of u.

5.1.6 Anisotropic error bounds: Case $l \geq 2$

Theorem 5.1.5 *Suppose that the assumptions in Theorem 5.1.2 hold. Then, for any $u \in W^{l,p}(K)$,*

$$|u - \Pi_{k,K} u|_{W^{m,q}(K)} \leq C |K|^{\frac{1}{q} - \frac{1}{p}} \cdot \|(F_K^{\prime})^{-1}\|^m \cdot \|F_K^{\prime}\|^{l-2} \times \left[\int_K \left[tr\left((F_K^{\prime})^T |H(D^{l-2}u)|(\boldsymbol{x}) F_K^{\prime} \right) \right]^p d\boldsymbol{x} \right]^{\frac{1}{p}}, \qquad (5.44)$$

or in a different form,

$$\begin{aligned} &|u - \Pi_{k,K} u|_{W^{m,q}(K)} \\ &\leq C |K|^{\frac{1}{q}} \cdot \|(F_K^{\prime})^{-1}\|^m \cdot \|F_K^{\prime}\|^{l-2} \cdot tr\left((F_K^{\prime})^T |H_K(D^{l-2}u)| F_K^{\prime} \right) \\ &+ C |K|^{\frac{1}{q}} \cdot \|(F_K^{\prime})^{-1}\|^m \cdot \|F_K^{\prime}\|^{l-2} \cdot tr\left((F_K^{\prime})^T F_K^{\prime} \right) \\ &\times \left\langle H(D^{l-2}u) - H_K(D^{l-2}u) \right\rangle_{L^p(K)}, \end{aligned} \qquad (5.45)$$

where $|H(D^{l-2}u)(\boldsymbol{x})|$, $|H_K(D^{l-2}u)|$, and $H_K(D^{(j_1,\ldots,j_{l-2})}u)$ are defined in (5.34), (5.35), and (5.36), respectively.

Proof. The theorem is a consequence of Theorem 5.1.2 and Lemma 5.1.5. □

Once again, the Jacobian matrix $F_K^{\prime}$ is directly coupled with the Hessian of the function u in the bounds, and the second term in (5.45) can be considered as a higher order term. As a consequence, when designing an adaptive algorithm based on these bounds, the shape and orientation of mesh elements will be determined using the Hessian of u (see §5.2 for specifics).

The error bounds (5.44) and (5.45) developed for $l = 2$ are independent of the coordinate system. Moreover, they appear to give a lowest bound in its simple form not involving maximum angle type conditions (cf. Babuška and Aziz [22]). However, when $l \geq 3$, the bounds corresponding to (5.44) and (5.45) depend upon the choice of a specific coordinate system. They also may easily be over estimates and thus lose some anisotropic information, as discussed by Cao [80], who gives a more sophisticated estimate for high degree polynomial interpolation in two dimensions.

5.1.7 Interpolation error on element faces

From time to time we need to estimate interpolation error on element faces. Such estimation relies upon use of the trace operator and its properties (see Appendix A).

Theorem 5.1.6 *Let $(\hat{K},\hat{P},\hat{\Sigma})$ be a finite element associated with the simplicial reference element $\hat{K}$ and let s be the greatest order of partial derivatives occurring in $\hat{\Sigma}$. For some integer $k \ge 0$ and some real numbers $p, q \in (1,\infty)$, if*

$$W^{2,p}(\hat{K}) \hookrightarrow C^s(\hat{K}), \tag{5.46}$$

$$W^{2,p}(\hat{K}) \hookrightarrow W^{1,q}(\hat{K}), \tag{5.47}$$

$$P_k(\hat{K}) \subset \hat{P} \subset W^{1,q}(\hat{K}), \tag{5.48}$$

where $P_k(\hat{K})$ is the space of polynomials of degree no more than k, then there exists a constant $C = C(\hat{K},\hat{P},\hat{\Sigma})$ such that, for all affine-equivalent finite elements (K,P_K,Σ_K) and for $u \in W^{2,p}(K)$,

$$\left[\sum_{\gamma\in\partial K} \frac{1}{|\gamma|} \|u - \Pi_{k,K}u\|^q_{L^q(\gamma)}\right]^{\frac{1}{q}} \le C|K|^{-\frac{1}{p}} \left[\int_K \left[tr\left((F_K')^T |H(u)|(\boldsymbol{x}) F_K'\right)\right]^p d\boldsymbol{x}\right]^{\frac{1}{p}}, \tag{5.49}$$

or in a different form,

$$\left[\sum_{\gamma\in\partial K} \frac{1}{|\gamma|} \|u - \Pi_{k,K}u\|^q_{L^q(\gamma)}\right]^{\frac{1}{q}} \le C\, tr\left((F_K')^T |H_K(u)| F_K'\right) + C\, tr\left((F_K')^T F_K'\right) \cdot \langle H(u) - H_K(u)\rangle_{L^p(K)}, \tag{5.50}$$

where $\Pi_{k,K}$ denotes the P_K-interpolation operator on K.

Proof. For a face γ of K and the corresponding face $\hat{\gamma}$ of $\hat{K}$, $\gamma = F_K(\hat{\gamma})$. Since $F_K : \hat{K} \to K$ is affine, its restriction on $\hat{\gamma}$, $F_K|_{\hat{\gamma}} : \hat{\gamma} \to \gamma$, is also affine. Denote the Jacobian of this restricted mapping by $J(\hat{\gamma},\gamma)$. Note that $J(\hat{\gamma},\gamma)$ is a constant, and moreover,

$$|\gamma| = \int_\gamma dS = \int_{\hat{\gamma}} J(\hat{\gamma},\gamma) d\hat{S} = J(\hat{\gamma},\gamma) \int_{\hat{\gamma}} d\hat{S} = J(\hat{\gamma},\gamma)|\hat{\gamma}|.$$

Thus,

$$J(\hat{\gamma},\gamma) = \frac{|\gamma|}{|\hat{\gamma}|}.$$

From this, we have

$$
\begin{aligned}
&\sum_{\gamma\in\partial K}\frac{1}{|\gamma|}\|u-\Pi_{k,K}u\|^q_{L^q(\gamma)}\\
&=\sum_{\gamma\in\partial K}\frac{1}{|\gamma|}\int_\gamma |u-\Pi_{k,K}u|^q dS\\
&=\sum_{\gamma\in\partial K}\frac{1}{|\hat{\gamma}|}\int_{\hat{\gamma}} |\hat{u}-\hat{\Pi}_{k,K}\hat{u}|^q d\hat{S}\\
&\le C\sum_{\gamma\in\partial K}\frac{1}{|\hat{\gamma}|}\left[\int_{\hat{K}}|\hat{u}-\hat{\Pi}_{k,K}\hat{u}|^q d\boldsymbol{\xi}+\int_{\hat{K}}|\nabla_{\boldsymbol{\xi}}(\hat{u}-\hat{\Pi}_{k,K}\hat{u})|^q d\boldsymbol{\xi}\right],
\end{aligned}
$$

where $\hat{\Pi}_{k,K}\hat{u}(\boldsymbol{\xi})=\Pi_{k,K}u(F_K(\boldsymbol{\xi}))$ and Theorem A.0.6 has been used in the last step. Since the reference element $\hat{K}$ is equilateral and of unit volume, $|\hat{\gamma}|=O(1)$. From Theorem 5.1.1 we obtain

$$
\sum_{\gamma\in\partial K}\frac{1}{|\gamma|}\|u-\Pi_{k,K}u\|^q_{L^q(\gamma)}\le C|\hat{u}|^q_{W^{2,p}(\hat{K})}.
$$

Combining this with Lemma 5.1.5 gives (5.49) and (5.50), respectively. □

For any face γ of element K,

$$
|K|\ge|\gamma|\sigma_{\min}, \tag{5.51}
$$

where σ_{min} is the minimum singular value of F_K', so we have

$$
\max_{\gamma\in\partial K}|\gamma|\le\frac{|K|}{\sigma_{\min}}=|K|\cdot\|(F_K')^{-1}\|.
$$

Using this, one obtains slightly different bounds from (5.49) and (5.50); viz., for $u\in W^{2,p}(K)$,

$$
\begin{aligned}
&\|u-\Pi_{k,K}u\|_{L^q(\partial K)}\\
&\quad\le C|K|^{\frac{1}{q}-\frac{1}{p}}\cdot\|(F_K')^{-1}\|^{\frac{1}{q}}\left[\int_K\left[\mathrm{tr}\left((F_K')^T|H(u)|(\boldsymbol{x})F_K'\right)\right]^p d\boldsymbol{x}\right]^{\frac{1}{p}},
\end{aligned}
\tag{5.52}
$$

$$
\begin{aligned}
\|u-\Pi_{k,K}u\|_{L^q(\partial K)}&\le C|K|^{\frac{1}{q}}\cdot\|(F_K')^{-1}\|^{\frac{1}{q}}\mathrm{tr}\left((F_K')^T|H_K(u)|F_K'\right)\\
&\quad+C|K|^{\frac{1}{q}}\cdot\|(F_K')^{-1}\|^{\frac{1}{q}}\mathrm{tr}\left((F_K')^T F_K'\right)\cdot\langle H(u)-H_K(u)\rangle_{L^p(K)}.
\end{aligned}
\tag{5.53}
$$

5.2 Monitor functions based on interpolation error

In this section, the motivation for defining the monitor function is to minimize the error bounds obtained in §5.1.4, §5.1.5 and §5.1.6. The procedure is similar to that used in §2.4 for the 1D case but considerably more complicated.

5.2.1 *Monitor function based on isotropic error estimates*

In this subsection the monitor function is defined based on minimizing the isotropic error bound (5.38). In addition, a global error bound is obtained for a corresponding M-uniform or quasi M-uniform mesh, and its convergence is investigated.

Taking the q-th power on both sides of (5.38) and summing over all of the elements, we obtain, for $u \in W^{l,p}(\Omega)$,

$$|u-\Pi_k u|^q_{W^{m,q}(\Omega)} \leq C\sum_K |K|\cdot \|(F_K^{'})^{-1}\|^{mq}\cdot \|F_K^{'}\|^{lq}\cdot \langle u\rangle^q_{W^{l,p}(K)}, \tag{5.54}$$

where

$$\langle u\rangle_{W^{l,p}(K)} = \left(\frac{1}{|K|}\int_K \sum_{i_1,\ldots,i_l}\left|D^{(i_1,\ldots,i_l)}u\right|^p d\boldsymbol{x}\right)^{\frac{1}{p}} = \left(\frac{1}{|K|}\int_K \|D^l u\|^p_{l_p} d\boldsymbol{x}\right)^{\frac{1}{p}}.$$

Since

$$\|F_K^{'}\| \leq \|F_K^{'}\|_F, \qquad \|(F_K^{'})^{-1}\| \leq \|(F_K^{'})^{-1}\|_F,$$

from Lemma 4.5.1 we have

$$\|F_K^{'}\|^2 \leq \mathrm{tr}\left((F_K^{'})^T(F_K^{'})\right), \quad \|(F_K^{'})^{-1}\|^2 \leq \mathrm{tr}\left((F_K^{'})^{-1}(F_K^{'})^{-T}\right). \tag{5.55}$$

Thus (5.54) can be rewritten as

$$|u-\Pi_k u|^q_{W^{m,q}(\Omega)} \leq C\sum_K |K|\cdot \left[\frac{1}{d}\mathrm{tr}\left((F_K^{'})^{-1}(F_K^{'})^{-T}\right)\right]^{\frac{mq}{2}}$$

$$\times \left[\frac{1}{d}\mathrm{tr}\left((F_K^{'})^T(F_K^{'})\right)\right]^{\frac{lq}{2}}\cdot \langle u\rangle^q_{W^{l,p}(K)}. \tag{5.56}$$

Denote

$$E(\mathscr{T}_h) = N^{\frac{(l-m)q}{d}} \sum_K |K| \cdot \left[\frac{1}{d}\text{tr}\left((F_K')^{-1}(F_K')^{-T}\right)\right]^{\frac{mq}{2}} \times \left[\frac{1}{d}\text{tr}\left((F_K')^T(F_K')\right)\right]^{\frac{lq}{2}} \cdot \langle u\rangle^q_{W^{l,p}(K)}. \quad (5.57)$$

The monitor function can now be defined in such a way that this error bound attains its minimum among all possible M-uniform meshes.

Assume that, for a given monitor function and a given number of elements, an M-uniform mesh can be generated for the physical domain Ω using a meshing strategy for which an M-uniform mesh is viewed as a function of the monitor function, $\mathscr{T}_h = \mathscr{T}_h(M)$. Then the task for finding the optimal monitor function becomes

$$\min_{\text{admissible } M} E(\mathscr{T}_h(M)). \quad (5.58)$$

As for the 1D case (§2.4.2), direct solution of this optimization problem is generally impractical. Here we use the indirect approach given in the proof of Theorem 2.1.2 of first finding a lower bound on $E(\mathscr{T}_h)$ and then showing that it can be attained with an M-uniform mesh for an appropriately chosen (optimal) monitor function.

From the arithmetic-mean geometric-mean inequality (cf. Theorem B.0.11) and the fact that $\det(F_K') = |K|$, we have

$$\begin{aligned} E(\mathscr{T}_h(M)) &= N^{\frac{(l-m)q}{d}} \sum_K |K| \cdot \left[\frac{1}{d}\text{tr}\left((F_K')^{-1}(F_K')^{-T}\right)\right]^{\frac{mq}{2}} \\ &\qquad \times \left[\frac{1}{d}\text{tr}\left((F_K')^T(F_K')\right)\right]^{\frac{lq}{2}} \cdot \langle u\rangle^q_{W^{l,p}(K)} \\ &\geq N^{\frac{(l-m)q}{d}} \sum_K |K| \cdot \left[\det\left((F_K')^{-1}(F_K')^{-T}\right)\right]^{\frac{mq}{2d}} \\ &\qquad \times \left[\det\left((F_K')^T(F_K')\right)\right]^{\frac{lq}{2d}} \cdot \langle u\rangle^q_{W^{l,p}(K)} \quad (5.59) \\ &= N^{\frac{(l-m)q}{d}} \sum_K |K|^{1+\frac{(l-m)q}{d}} \cdot \langle u\rangle^q_{W^{l,p}(K)} \\ &= N^{\frac{d+q(l-m)}{d}} \cdot \frac{1}{N}\sum_K \left[|K| \cdot \langle u\rangle^{\frac{dq}{d+q(l-m)}}_{W^{l,p}(K)}\right]^{\frac{d+q(l-m)}{d}} \\ &\geq N^{\frac{d+q(l-m)}{d}} \cdot \left[\frac{1}{N}\sum_K |K| \cdot \langle u\rangle^{\frac{dq}{d+q(l-m)}}_{W^{l,p}(K)}\right]^{\frac{d+q(l-m)}{d}} \quad (5.60) \\ &= \left[\sum_K |K| \cdot \langle u\rangle^{\frac{dq}{d+q(l-m)}}_{W^{l,p}(K)}\right]^{\frac{d+q(l-m)}{d}}. \quad (5.61) \end{aligned}$$

Note that the lower bound (5.61) is mesh-dependent. Thus, while an M-uniform mesh attaining this bound is not necessarily optimal among all partitions of N elements, it is asymptotically optimal in the sense that the bound converges to a constant lower bound, i.e.,

$$\left[\sum_K |K|\cdot \langle u\rangle_{W^{l,p}(K)}^{\frac{dq}{d+q(l-m)}}\right]^{\frac{d+q(l-m)}{d}} \to \left[\int_\Omega \|D^l u\|_{l^p}^{\frac{dq}{d+q(l-m)}} d\boldsymbol{x}\right]^{\frac{d+q(l-m)}{d}}, \tag{5.62}$$

as long as

$$h = \max_K h_K \to 0 \quad \text{as} \quad N \to \infty. \tag{5.63}$$

As we have seen in the 1D case and shall see later in this section, this last property of the mesh is crucial to ensure that the error bound converges.

We now show how to choose M such that the lower bound (5.61) is attained with a corresponding M-uniform mesh. We first notice that equality in (5.59) holds when the mesh satisfies

$$\frac{1}{d}\text{tr}\left((F_K')^{-1}(F_K')^{-T}\right) = \det\left((F_K')^{-1}(F_K')^{-T}\right) \tag{5.64}$$

and

$$\frac{1}{d}\text{tr}\left((F_K')^{T}(F_K')\right) = \det\left((F_K')^{T}(F_K')\right). \tag{5.65}$$

Comparing these to (4.29) and (4.27), the alignment conditions satisfied by an M-uniform mesh, suggests that M be chosen in the form

$$M_K = \theta_K I \tag{5.66}$$

for some scalar function $\theta = \theta_K$. Next, (5.60) is an equality when the mesh satisfies

$$|K|\cdot \langle u\rangle_{W^{l,p}(K)}^{\frac{dq}{d+q(l-m)}} = C \qquad \forall K \in \mathscr{T}_h$$

for some constant C. Comparing this with the equidistribution condition (4.28) for an M-uniform mesh implies that ρ be chosen as

$$\rho_K = \langle u\rangle_{W^{l,p}(K)}^{\frac{dq}{d+q(l-m)}}. \tag{5.67}$$

From (5.66) and (5.67) and the relation $\rho_K = \sqrt{\det(M_K)}$, the optimal monitor function is

$$M_K = \langle u\rangle_{W^{l,p}(K)}^{\frac{2q}{d+q(l-m)}} I, \qquad \forall K \in \mathscr{T}_h. \tag{5.68}$$

By construction, $E(\mathscr{T}_h)$ attains its lower bound (5.61) with an M-uniform mesh for the monitor function (5.68). The interpolation error for such a mesh is bounded by

$$|u-\Pi_k u|_{W^{m,q}(\Omega)} \le C N^{-\frac{(l-m)}{d}} \left[\sum_K |K| \cdot \langle u \rangle_{W^{l,p}(K)}^{\frac{dq}{d+q(l-m)}} \right]^{\frac{d+q(l-m)}{dq}}. \tag{5.69}$$

Generally speaking, the term $\langle u \rangle_{W^{l,p}(K)}$ is not necessarily positive and can vanish locally, so to ensure that M in (5.68) is positive definite, we regularize using a positive parameter α_h and rewrite (5.56) as

$$\begin{aligned}
|u-\Pi_k u|^q_{W^{m,q}(\Omega)} &\le C \sum_K |K| \cdot \left[\frac{1}{d}\text{tr}\left((F_K')^{-1}(F_K')^{-T}\right)\right]^{\frac{mq}{2}} \\
&\quad \times \left[\frac{1}{d}\text{tr}\left((F_K')^{T}(F_K')\right)\right]^{\frac{lq}{2}} \cdot \left(\alpha_h + \langle u \rangle_{W^{l,p}(K)}\right)^q \\
&\le C\alpha_h^q \sum_K |K| \cdot \left[\frac{1}{d}\text{tr}\left((F_K')^{-1}(F_K')^{-T}\right)\right]^{\frac{mq}{2}} \\
&\quad \times \left[\frac{1}{d}\text{tr}\left((F_K')^{T}(F_K')\right)\right]^{\frac{lq}{2}} \cdot \left(1 + \alpha_h^{-1}\langle u \rangle_{W^{l,p}(K)}\right)^q.
\end{aligned} \tag{5.70}$$

By the same argument as above, minimizing this bound we obtain the monitor function, mesh density function, and error bound as

$$M_K = M_K^{iso} \equiv \left(1+\alpha_h^{-1}\langle u \rangle_{W^{l,p}(K)}\right)^{\frac{2q}{d+q(l-m)}} I, \qquad \forall K \in \mathcal{T}_h \tag{5.71}$$

$$\rho_K = \rho_K^{iso} \equiv \left(1+\alpha_h^{-1}\langle u \rangle_{W^{l,p}(K)}\right)^{\frac{dq}{d+q(l-m)}}, \qquad \forall K \in \mathcal{T}_h \tag{5.72}$$

$$|u-\Pi_k u|_{W^{m,q}(\Omega)} \le C\alpha_h N^{-\frac{(l-m)}{d}} \sigma_h^{\frac{d+q(l-m)}{dq}}, \tag{5.73}$$

where $\sigma_h = \sum_K |K| \rho_K^{iso}$.

The regularization parameter α_h is often referred to as the *intensity parameter* since it controls the level of intensity of mesh concentration.[1] To examine how to choose α_h, we first prove two lemmas which are multidimensional extensions of Lemmas 2.4.1 and 2.4.2.

Lemma 5.2.1 *For any real numbers $\gamma \in (0,1]$ and $p \in [1,\infty)$ and any mesh $\mathcal{T}_h$ for Ω,*

$$\|v\|^{\gamma p}_{L^{\gamma p}(\Omega)} \le \sum_K |K|^{1-\gamma} \|v\|^{\gamma p}_{L^p(K)} \le |\Omega|^{1-\gamma} \|v\|^{\gamma p}_{L^p(\Omega)}, \quad \forall v \in L^p(\Omega). \tag{5.74}$$

Proof. See the proof of Lemma 2.4.1. □

---

[1] As α_h increases, the mesh becomes more uniform.

Lemma 5.2.2 *For any real numbers $\gamma \in (0,1]$ and $p \in [1,\infty)$ and any mesh $\mathscr{T}_h$ for Ω,*

$$\begin{aligned}\|v\|^{\gamma p}_{L^{\gamma p}(\Omega)} &\le \sum_K |K|^{1-\gamma} \|v\|^{\gamma p}_{L^p(K)} \\ &\le c^2_{\gamma p} \|v\|^{\gamma p}_{L^{\gamma p}(\Omega)} + c_{\gamma p}(1+c_{\gamma p}) c^{\gamma p}_{p,p} h^{\gamma p} |\Omega|^{1-\gamma} \|\nabla v\|^{\gamma p}_{L^p(\Omega)} \end{aligned} \quad (5.75)$$

for any $v \in W^{1,p}(\Omega)$, where $h = \max_{K\in\mathscr{T}_h} h_K$ and the constants $c_{\gamma p}$ and $c_{p,p}$ are defined in (B.4) and Theorem A.0.9, respectively.

Proof. The proof is similar to that of Lemma 2.4.2. The left inequality of (5.75) is a consequence of Lemma 5.2.1. To prove the right inequality, let

$$v_K = \frac{1}{|K|}\int_K v d\mathbf{x}.$$

From Corollary B.0.1, Theorem A.0.2 (Hölder's inequality), and Theorem A.0.9 (Poincaré's inequality) we have

$$\begin{aligned}
&\sum_K |K|^{1-\gamma}\|v\|^{\gamma p}_{L^p(K)} - c^2_{\gamma p}\|v\|^{\gamma p}_{L^{\gamma p}(\Omega)} \\
&= \sum_K |K|^{1-\gamma}\|v - v_K + v_K\|^{\gamma p}_{L^p(K)} - c^2_{\gamma p}\sum_K \|v\|^{\gamma p}_{L^{\gamma p}(K)} \\
&\le c_{\gamma p}\sum_K |K|^{1-\gamma}\|v - v_K\|^{\gamma p}_{L^p(K)} + c_{\gamma p}\sum_K |K|^{1-\gamma}\|v_K\|^{\gamma p}_{L^p(K)} - c^2_{\gamma p}\sum_K \|v\|^{\gamma p}_{L^{\gamma p}(K)} \\
&= c_{\gamma p}\sum_K |K|^{1-\gamma}\|v - v_K\|^{\gamma p}_{L^p(K)} + c_{\gamma p}\sum_K |K|\cdot|v_K|^{\gamma p} - c^2_{\gamma p}\sum_K\int_K |v|^{\gamma p} d\mathbf{x} \\
&= c_{\gamma p}\sum_K |K|^{1-\gamma}\|v - v_K\|^{\gamma p}_{L^p(K)} + c_{\gamma p}\sum_K\int_K \left(|v_K|^{\gamma p} - c_{\gamma p}|v|^{\gamma p}\right) d\mathbf{x} \\
&\le c_{\gamma p}\sum_K |K|^{1-\gamma}\|v - v_K\|^{\gamma p}_{L^p(K)} + c^2_{\gamma p}\sum_K\int_K |v_K - v|^{\gamma p} d\mathbf{x} \\
&\le c_{\gamma p}(1+c_{\gamma p})\sum_K |K|^{1-\gamma}\|v - v_K\|^{\gamma p}_{L^p(K)} \\
&\le c_{\gamma p}(1+c_{\gamma p})c^{\gamma p}_{p,p}\sum_K |K|^{1-\gamma}h_K^{\gamma p}\|\nabla v\|^{\gamma p}_{L^p(K)} \\
&\le c_{\gamma p}(1+c_{\gamma p})c^{\gamma p}_{p,p}h^{\gamma p}\sum_K |K|^{1-\gamma}\|\nabla v\|^{\gamma p}_{L^p(K)} \\
&\le c_{\gamma p}(1+c_{\gamma p})c^{\gamma p}_{p,p}h^{\gamma p}|\Omega|^{1-\gamma}\|\nabla v\|^{\gamma p}_{L^p(\Omega)}.
\end{aligned}$$

□

Three different choices of α_h are examined. The first choice is $\alpha_h \to 0$, corresponding in the limit to the non-regularized situation. Taking $\alpha_h \to 0$ can provide a simplified derivation and a guideline for dealing with other situations. For the second choice, α_h is taken as a fixed number depending upon the solution in a way that M_K^{iso} is invariant under a scaling transformation of u. For the third, α_h is taken such that σ_h is bounded by a constant while keeping M_K^{iso} invariant under a scaling transformation of u.

Case 1. We consider

$$\alpha_h \to 0. \tag{5.76}$$

Since by Theorem 4.1.2 the equidistribution and alignment conditions are invariant under a scaling transformation of M_K, choosing M_K to be an arbitrary (positive) constant multiple of M_K^{iso} does not change the underlying M-uniform mesh. In particular, let

$$M_K = \alpha_h^{\frac{2q}{d+q(l-m)}} M_K^{iso} = \left(\alpha_h + \langle u \rangle_{W^{l,p}(K)}\right)^{\frac{2q}{d+q(l-m)}} I, \quad \forall K \in \mathscr{T}_h,$$

so the corresponding mesh density function is

$$\rho_K = \alpha_h^{\frac{dq}{d+q(l-m)}} \rho_K^{iso} = \left(\alpha_h + \langle u \rangle_{W^{l,p}(K)}\right)^{\frac{dq}{d+q(l-m)}}, \quad \forall K \in \mathscr{T}_h.$$

Taking the limit $\alpha_h \to 0$ we obtain the monitor and mesh density functions for the non-regularized situation in (5.67) and (5.68), and the error bound (5.73) reduces to (5.69), i.e.,

$$|u - \Pi_k u|_{W^{m,q}(\Omega)} \leq C N^{-\frac{(l-m)}{d}} \left(\sum_K |K| \, \langle u \rangle_{W^{l,p}(K)}^{\frac{dq}{d+q(l-m)}} \right)^{\frac{d+q(l-m)}{dq}}. \tag{5.77}$$

To ensure that M_K is positive definite, we assume here that there exists a constant β such that

$$\|D^l u(\boldsymbol{x})\|_{l^p} \geq \beta > 0, \quad \text{a.e. in } \Omega \tag{5.78}$$

where "a.e." stands for almost everywhere, or every point except a zero-measure set. Moreover, we assume $q \leq p$ for the analysis which follows.[2]

The key to the convergence analysis is to show that an M-uniform mesh for monitor function (5.68) satisfies (5.63). This follows from Theorem 4.2.3 and Lemma 5.2.1 since

---

[2] Both Lemmas 5.2.1 and 5.2.2 require $\gamma \leq 1$. In the current context, $\gamma = \frac{dq}{p(d+q(l-m))}$, and the assumption $q \leq p$ ensures this condition be satisfied. This remark also applies to the situations discussed in the next two subsections.

$$
\begin{aligned}
h_K &\le C N^{-\frac{1}{d}} \lambda_{min}(M_K)^{-\frac{1}{2}} \sigma_h^{\frac{1}{d}} \\
&\le C N^{-\frac{1}{d}} \beta^{-\frac{1}{2}} \left(\sum_K |K| \, \langle u \rangle_{W^{l,p}(K)}^{\frac{dq}{d+q(l-m)}} \right)^{\frac{1}{d}} \\
&\le C N^{-\frac{1}{d}} \beta^{-\frac{1}{2}} \, |u|_{W^{l,p}(\Omega)}^{\frac{q}{d+q(l-m)}} .
\end{aligned}
\tag{5.79}
$$

From (5.63), it is obvious that

$$
\lim_{N \to \infty} \left(\sum_K |K| \, \langle u \rangle_{W^{l,p}(K)}^{\frac{dq}{d+q(l-m)}} \right)^{\frac{d+q(l-m)}{dq}} \le C |u|_{W^{l,\frac{dq}{d+q(l-m)}}(\Omega)} . \tag{5.80}
$$

For $u \in W^{l+1,p}(\Omega)$, this limit can be refined from Lemma 5.2.2 and (5.79) to

$$
\begin{aligned}
&\left(\sum_K |K| \, \langle u \rangle_{W^{l,p}(K)}^{\frac{dq}{d+q(l-m)}} \right)^{\frac{d+q(l-m)}{dq}} \\
&= \left[\sum_K |K|^{1-\frac{dq}{p(d+q(l-m))}} \, |u|_{W^{l,p}(K)}^{\frac{dq}{d+q(l-m)}} \right]^{\frac{d+q(l-m)}{dq}} \\
&\le C \left[|u|_{W^{l,\frac{dq}{d+q(l-m)}}(\Omega)}^{\frac{dq}{d+q(l-m)}} + h^{\frac{dq}{d+q(l-m)}} \, |u|_{W^{l+1,p}(\Omega)}^{\frac{dq}{d+q(l-m)}} \right]^{\frac{d+q(l-m)}{dq}} \\
&\le C \left[|u|_{W^{l,\frac{dq}{d+q(l-m)}}(\Omega)} + N^{-\frac{1}{d}} \, |u|_{W^{l,p}(\Omega)}^{\frac{q}{d+q(l-m)}} \, |u|_{W^{l+1,p}(\Omega)} \right] .
\end{aligned}
$$

Inserting this into (5.77) gives

$$
\begin{aligned}
&|u - \Pi_k u|_{W^{m,q}(\Omega)} \\
&\quad \le C N^{-\frac{(l-m)}{d}} \left(|u|_{W^{l,\frac{dq}{d+q(l-m)}}(\Omega)} + N^{-\frac{1}{d}} \, |u|_{W^{l,p}(\Omega)}^{\frac{q}{d+q(l-m)}} \, |u|_{W^{l+1,p}(\Omega)} \right) .
\end{aligned}
\tag{5.81}
$$

Case 2. We next consider

$$
\alpha_h = \hat{\alpha} \, \langle u \rangle_{L^p(\Omega)} , \tag{5.82}
$$

where $\hat{\alpha} = O(1)$ is a given positive constant.[3] Notice that for this choice the monitor function is invariant under a scaling transformation of u, i.e., $u \to cu$ for any positive

[3] The dimensionless parameter $\hat{\alpha}$ can often be chosen as $\hat{\alpha} = 1$. Without loss of generality, it is assumed that $\langle u \rangle_{L^p(\Omega)} > 0$ in (5.82).

constant c. Moreover,

$$\begin{aligned}
\sigma_h &= \sum_K |K| \left[1 + \alpha_h^{-1} \langle u \rangle_{W^{l,p}(K)}\right]^{\frac{dq}{d+q(l-m)}} \\
&\le c_{\frac{dq}{d+q(l-m)}} \sum_K |K| \left[1 + \alpha_h^{-\frac{dq}{d+q(l-m)}} \langle u \rangle_{W^{l,p}(K)}^{\frac{dq}{d+q(l-m)}}\right] \\
&= c_{\frac{dq}{d+q(l-m)}} \left[|\Omega| + \alpha_h^{-\frac{dq}{d+q(l-m)}} \sum_K |K| \langle u \rangle_{W^{l,p}(K)}^{\frac{dq}{d+q(l-m)}}\right],
\end{aligned} \tag{5.83}$$

where constant $c_{\frac{dq}{d+q(l-m)}}$ is defined in (B.4). Combining this with (5.73) we obtain

$$\begin{aligned}
&|u - \Pi_k u|_{W^{m,q}(\Omega)} \\
&\quad \le CN^{-\frac{(l-m)}{d}} \left[\hat{\alpha} \|u\|_{L^p(\Omega)} + \left(\sum_K |K| \langle u \rangle_{W^{l,p}(K)}^{\frac{dq}{d+q(l-m)}}\right)^{\frac{d+q(l-m)}{dq}}\right].
\end{aligned} \tag{5.84}$$

To show convergence from this error bound, applying Lemma 5.2.1 to (5.83) yields

$$\sigma_h \le C \left[1 + \left(\frac{|u|_{W^{l,p}(\Omega)}}{\hat{\alpha} \|u\|_{L^p(\Omega)}}\right)^{\frac{dq}{d+q(l-m)}}\right].$$

From Theorem 4.2.3 and the fact that $\lambda_{min}(M_K^{iso}) \ge 1$, we get

$$h_K \le \frac{\hat{h} N^{-\frac{1}{d}} \sigma_h^{\frac{1}{d}}}{\lambda_{min}(M_K^{iso})^{\frac{1}{2}}} \le CN^{-\frac{1}{d}} \left(1 + \left(\frac{|u|_{W^{l,p}(\Omega)}}{\hat{\alpha} \|u\|_{L^p(\Omega)}}\right)^{\frac{q}{d+q(l-m)}}\right). \tag{5.85}$$

Thus, the mesh satisfies the property (5.63), and the sum in the bound (5.84) satisfies (5.80). Moreover, from Lemma 5.2.2 and inequalities (5.84) and (5.85) it follows that, for $u \in W^{l+1,p}(\Omega)$,

$$\begin{aligned}
|u - \Pi_k u|_{W^{m,q}(\Omega)} \le CN^{-\frac{(l-m)}{d}} \Bigg[&\hat{\alpha} \|u\|_{L^p(\Omega)} + |u|_{W^{l,\frac{dq}{d+q(l-m)}}(\Omega)} \\
&+ N^{-\frac{1}{d}} \left(1 + \left(\frac{|u|_{W^{l,p}(\Omega)}}{\hat{\alpha} \|u\|_{L^p(\Omega)}}\right)^{\frac{q}{d+q(l-m)}}\right) |u|_{W^{l+1,p}(\Omega)} \Bigg].
\end{aligned} \tag{5.86}$$

When the l-th order derivatives are large and the related terms in (5.86) dominate,

$$|u - \Pi_k u|_{W^{m,q}(\Omega)} \lessapprox CN^{-\frac{(l-m)}{d}} \left(|u|_{W^{l,\frac{dq}{d+q(l-m)}}(\Omega)} + N^{-\frac{1}{d}} |u|_{W^{l,p}(\Omega)}^{\frac{q}{d+q(l-m)}} |u|_{W^{l+1,p}(\Omega)}\right),$$

which is essentially (5.81).

Case 3. It is recommended in [175] (also cf. §2.4.2) that α_h be chosen such that (a) the monitor function M be invariant under a scaling transformation of u and (b) $\sigma_h \le C$ for some constant C. In the current context, (5.83) implies that by choosing

$$\alpha_h = \alpha_h^{iso} \equiv \left[\frac{1}{|\Omega|} \sum_K |K| \, \langle u \rangle_{W^{l,p}(K)}^{\frac{dq}{d+q(l-m)}} \right]^{\frac{d+q(l-m)}{dq}}, \tag{5.87}$$

we have

$$\sigma_h \le 2\, |\Omega| \, c_{\frac{dq}{d+q(l-m)}} . \tag{5.88}$$

It is easy to verify that M_K^{iso} for this choice of α_h is invariant under a scaling transformation of u.

Since $\lambda_{min}(M_K^{iso}) \ge 1$, an immediate consequence of (5.88) is that any M-uniform mesh satisfies the property (5.63). From (5.73), (5.87), and (5.88) we get

$$|u - \Pi_k u|_{W^{m,q}(\Omega)} \le C N^{-\frac{(l-m)}{d}} \left[\sum_K |K| \, \langle u \rangle_{W^{l,p}(K)}^{\frac{dq}{d+q(l-m)}} \right]^{\frac{d+q(l-m)}{dq}}, \tag{5.89}$$

with (5.80) holding. Moreover, from (5.87) it follows that

$$|\Omega| (\alpha_h^{iso})^{\frac{dq}{d+q(l-m)}} = \sum_K |K| \, \langle u \rangle_{W^{l,p}(K)}^{\frac{dq}{d+q(l-m)}} = \sum_K |K|^{1-\frac{dq}{p(d+q(l-m))}} |u|_{W^{l,p}(K)}^{\frac{dq}{d+q(l-m)}} . \tag{5.90}$$

Assuming that $q \le p$ and $u \in W^{l+1,p}(\Omega)$, Lemma 5.2.2 and the above equation give

$$\begin{aligned} |u|_{W^{l,\frac{dq}{d+q(l-m)}}(\Omega)}^{\frac{dq}{d+q(l-m)}} &\le |\Omega| (\alpha_{iso}^h)^{\frac{dq}{d+q(l-m)}} \\ &\le C \left(|u|_{W^{l,\frac{dq}{d+q(l-m)}}(\Omega)}^{\frac{dq}{d+q(l-m)}} + N^{-\frac{q}{d+q(l-m)}} |u|_{W^{l+1,p}(\Omega)}^{\frac{dq}{d+q(l-m)}} \right). \end{aligned}$$

Combining these with (5.73) we obtain

$$|u - \Pi_k u|_{W^{m,q}(\Omega)} \le C N^{-\frac{(l-m)}{d}} \left(|u|_{W^{l,\frac{dq}{d+q(l-m)}}(\Omega)} + N^{-\frac{1}{d}} |u|_{W^{l+1,p}(\Omega)} \right). \tag{5.91}$$

The above results are summarized in the following theorem.

Theorem 5.2.1 *Suppose that the condition (5.22) is satisfied and that $q \le p$. Suppose also that the mesh density function and the monitor function are chosen as in (5.72) and (5.71), respectively, and consider the corresponding family of M-uniform meshes.*

(a) Choosing α_h such that $\alpha_h \to 0$, if (5.78) holds, then for $u \in W^{l,p}(\Omega)$

$$|u - \Pi_k u|_{W^{m,q}(\Omega)} \le C N^{-\frac{(l-m)}{d}} \left(\sum_K |K| \langle u \rangle_{W^{l,p}(K)}^{\frac{dq}{d+q(l-m)}} \right)^{\frac{d+q(l-m)}{dq}}, \tag{5.92}$$

with

$$\lim_{N \to \infty} \left(\sum_K |K| \langle u \rangle_{W^{l,p}(K)}^{\frac{dq}{d+q(l-m)}} \right)^{\frac{d+q(l-m)}{dq}} \le C |u|_{W^{l,\frac{dq}{d+q(l-m)}}(\Omega)}. \tag{5.93}$$

If further $u \in W^{l+1,p}(\Omega)$, then

$$|u - \Pi_k u|_{W^{m,q}(\Omega)} \le C N^{-\frac{(l-m)}{d}} \left(|u|_{W^{l,\frac{dq}{d+q(l-m)}}(\Omega)} + N^{-\frac{1}{d}} |u|_{W^{l,p}(\Omega)}^{\frac{q}{d+q(l-m)}} |u|_{W^{l+1,p}(\Omega)} \right). \tag{5.94}$$

(b) Choosing $\alpha_h = \hat{\alpha} \langle u \rangle_{L^p(\Omega)}$ for some positive constant $\hat{\alpha}$, then for $u \in W^{l,p}(\Omega)$,

$$|u - \Pi_k u|_{W^{m,q}(\Omega)} \le C N^{-\frac{(l-m)}{d}} \left[\hat{\alpha} \|u\|_{L^p(\Omega)} + \left(\sum_K |K| \langle u \rangle_{W^{l,p}(K)}^{\frac{dq}{d+q(l-m)}} \right)^{\frac{d+q(l-m)}{dq}} \right], \tag{5.95}$$

with (5.93) holding. If further $u \in W^{l+1,p}(\Omega)$, then

$$|u - \Pi_k u|_{W^{m,q}(\Omega)} \le C N^{-\frac{(l-m)}{d}} \Bigg[\hat{\alpha} \|u\|_{L^p(\Omega)} + |u|_{W^{l,\frac{dq}{d+q(l-m)}}(\Omega)} + N^{-\frac{1}{d}} \left(1 + \left(\frac{|u|_{W^{l,p}(\Omega)}}{\hat{\alpha} \|u\|_{L^p(\Omega)}} \right)^{\frac{q}{d+q(l-m)}} \right) |u|_{W^{l+1,p}(\Omega)} \Bigg]. \tag{5.96}$$

(c) Choosing $\alpha_h = \alpha_h^{iso}$, where α_h^{iso} is defined in (5.87), then for $u \in W^{l,p}(\Omega)$,

$$|u - \Pi_k u|_{W^{m,q}(\Omega)} \le C N^{-\frac{(l-m)}{d}} \left(\sum_K |K| \langle u \rangle_{W^{l,p}(K)}^{\frac{dq}{d+q(l-m)}} \right)^{\frac{d+q(l-m)}{dq}}, \tag{5.97}$$

with (5.93) holding. If further $u \in W^{l+1,p}(\Omega)$, then

$$|u - \Pi_k u|_{W^{m,q}(\Omega)} \le C N^{-\frac{(l-m)}{d}} \left(|u|_{W^{l,\frac{dq}{d+q(l-m)}}(\Omega)} + N^{-\frac{1}{d}} |u|_{W^{l+1,p}(\Omega)} \right). \tag{5.98}$$

Note that the error bounds (5.92) and (5.97), obtained for the asymptotically non-regularized ($\alpha_h \to 0$) and regularized ($\alpha_h = \alpha_h^{iso}$) cases, respectively, are the same to within a multiplicative constant. This indicates that α_h^{iso} is of the right magnitude since the regularization does not enlarge the error bound significantly. In fact, it leads to a bound with the best asymptotic behavior among the three cases since (5.98) does not involve the factor $|u|_{W^{l,p}(\Omega)}^{\frac{q}{d+q(l-m)}}$ that appears in the higher order terms of error bounds (5.94) and (5.96) for the other two cases. The above analysis also shows that the optimal monitor function (5.71) resulting from an isotropic error bound is a scalar matrix function (cf. (4.69)). As discussed in §4.1.2, such a monitor function results in an isotropic M-uniform mesh.

Theorem 5.2.1 is only applicable for M-uniform meshes exactly satisfying the equidistribution and alignment conditions. Fortunately, the results can be readily extended to quasi M-uniform meshes satisfying (4.30) and (4.31). For simplicity, we just consider the case with $\alpha_h = \alpha_h^{iso}$ below, although the other cases can be treated similarly.

We first notice that from (4.31) and Theorem 4.5.1 (using the (4.103) bound in (4.104)), a quasi M-uniform mesh also satisfies for any $K \in \mathscr{T}_h$,

$$\frac{\mathrm{tr}\left(F_K'^{-1} M_K^{-1} F_K'^{-T}\right)}{d \det\left(F_K'^{-1} M_K^{-1} F_K'^{-T}\right)^{\frac{1}{d}}} \le \left(\sqrt{d(d-1)(\kappa_{ali}-1)}+1\right)^{2(d-1)} \equiv \hat{\kappa}_{ali} \tag{5.99}$$

for any $K \in \mathscr{T}_h$. Moreover, for the monitor function defined in (5.71), inequalities (4.31) and (5.99) reduce to

$$\frac{\mathrm{tr}((F_K')^T F_K')}{d \det((F_K')^T F_K')^{\frac{1}{d}}} \le \kappa_{ali}, \quad \forall K \in \mathscr{T}_h \tag{5.100}$$

$$\frac{\mathrm{tr}\left(F_K'^{-1} F_K'^{-T}\right)}{d \det\left(F_K'^{-1} F_K'^{-T}\right)^{\frac{1}{d}}} \le \hat{\kappa}_{ali}, \quad \forall K \in \mathscr{T}_h. \tag{5.101}$$

From Theorem 4.5.1 we have

$$\sigma_{max} \le \sigma_{min} \left[\sqrt{d(d-1)(\kappa_{ali}-1)}+1\right]^d \approx \sigma_{min} \kappa_{ali}^{\frac{d}{2}},$$

where σ_{min} and $\sigma_{\max}$ are the minimum and maximum singular values of F_K'. Note that (4.30), together with (5.88) and the fact that $\rho_K \ge 1$, imply that

$$|K| \leq \frac{C\kappa_{eq}}{N}.$$

Thus, from (4.47) and (4.48) we have

$$h_K \leq \hat{h}\sigma_{max} \leq C\kappa_{ali}^{\frac{d}{2}}\sigma_{min} \leq C\kappa_{ali}^{\frac{d}{2}}|K|^{\frac{1}{d}} \leq C\kappa_{ali}^{\frac{d}{2}}\kappa_{eq}^{\frac{1}{d}}N^{-\frac{1}{d}} \to 0 \quad \text{as} \quad N \to \infty. \tag{5.102}$$

Theorem 5.2.2 *Suppose that the condition (5.22) is satisfied and that $q \leq p$. Suppose also that the mesh density function, the monitor function, and the regularization parameter are chosen as in (5.72), (5.71), and (5.87), respectively. For a quasi M-uniform mesh satisfying (4.30) and (4.31) for some constants κ_{eq} and κ_{ali}, the interpolation error for $u \in W^{l,p}(\Omega)$ is bounded by*

$$|u - \Pi_k u|_{W^{m,q}(\Omega)} \leq CN^{-\frac{(l-m)}{d}} \hat{\kappa}_{ali}^{\frac{m}{2}} \kappa_{ali}^{\frac{l}{2}} \kappa_{eq}^{\frac{d+q(l-m)}{dq}} \left[\sum_K |K| \langle u \rangle_{W^{l,p}(K)}^{\frac{dq}{d+q(l-m)}} \right]^{\frac{d+q(l-m)}{dq}}, \tag{5.103}$$

with

$$\lim_{N\to\infty} \left[\sum_K |K| \langle u \rangle_{W^{l,p}(K)}^{\frac{dq}{d+q(l-m)}} \right]^{\frac{d+q(l-m)}{dq}} \leq C|u|_{W^{l,\frac{dq}{d+q(l-m)}}(\Omega)}. \tag{5.104}$$

If further $u \in W^{l+1,p}(\Omega)$, then

$$|u - \Pi_k u|_{W^{m,q}(\Omega)} \leq C\hat{\kappa}_{ali}^{\frac{m}{2}} \kappa_{ali}^{\frac{l}{2}} \kappa_{eq}^{\frac{d+q(l-m)}{dq}} N^{-\frac{(l-m)}{d}} \left(|u|_{W^{l,\frac{dq}{d+q(l-m)}}(\Omega)} + \kappa_{ali}^{\frac{d}{2}} \kappa_{eq}^{\frac{1}{d}} N^{-\frac{1}{d}} |u|_{W^{l+1,p}(\Omega)} \right). \tag{5.105}$$

Proof. From (5.70), (4.30), (5.100), and (5.101), we have

$$
\begin{aligned}
&|u-\Pi_k u|^q_{W^{m,q}(\Omega)}\\
&\le C\alpha_h^q \sum_K |K|\cdot\left[\frac{1}{d}\mathrm{tr}\left((F_K')^{-1}(F_K')^{-T}\right)\right]^{\frac{mq}{2}}\\
&\quad\times\left[\frac{1}{d}\mathrm{tr}\left((F_K')^{T}(F_K')\right)\right]^{\frac{lq}{2}}\cdot\left(1+\alpha_h^{-1}\langle u\rangle_{W^{l,p}(K)}\right)^q\\
&\le C\alpha_h^q \sum_K |K|\cdot\left[\hat{\kappa}_{ali}\det\left((F_K')^{-1}(F_K')^{-T}\right)^{\frac{1}{d}}\right]^{\frac{mq}{2}}\\
&\quad\times\left[\kappa_{ali}\det\left((F_K')^{T}(F_K')\right)^{\frac{1}{d}}\right]^{\frac{lq}{2}}\cdot\left(1+\alpha_h^{-1}\langle u\rangle_{W^{l,p}(K)}\right)^q\\
&= C\alpha_h^q\hat{\kappa}_{ali}^{\frac{mq}{2}}\kappa_{ali}^{\frac{lq}{2}}\sum_K |K|^{1+\frac{(l-m)q}{d}}\cdot\left(1+\alpha_h^{-1}\langle u\rangle_{W^{l,p}(K)}\right)^q\\
&= C\alpha_h^q\hat{\kappa}_{ali}^{\frac{mq}{2}}\kappa_{ali}^{\frac{lq}{2}}\sum_K |K|^{\frac{d+(l-m)q}{d}}\cdot\rho_K^{\frac{d+q(l-m)}{d}}\\
&\le C\alpha_h^q\hat{\kappa}_{ali}^{\frac{mq}{2}}\kappa_{ali}^{\frac{lq}{2}}\sum_K\left(\kappa_{eq}\frac{\sigma_h}{N}\right)^{\frac{d+q(l-m)}{d}}\\
&= C\alpha_h^q\hat{\kappa}_{ali}^{\frac{mq}{2}}\kappa_{ali}^{\frac{lq}{2}}\kappa_{eq}^{\frac{d+q(l-m)}{d}}\sigma_h^{\frac{d+q(l-m)}{d}}N^{-\frac{(l-m)q}{d}}\\
&\le CN^{-\frac{(l-m)q}{d}}\hat{\kappa}_{ali}^{\frac{mq}{2}}\kappa_{ali}^{\frac{lq}{2}}\kappa_{eq}^{\frac{d+q(l-m)}{d}}\left[\sum_K |K|\,\langle u\rangle_{W^{l,p}(K)}^{\frac{dq}{d+q(l-m)}}\right]^{\frac{d+q(l-m)}{d}},
\end{aligned}
$$

which gives (5.103).

Inequalities (5.104) and (5.105) can be proven in a similar manner to that used for Theorem 5.2.1. □

Observe that Theorem 5.2.2 reduces to Theorem 5.2.1 (for the case $\alpha_h = \alpha_h^{iso}$) for an M-uniform mesh for which $\kappa_{eq} = 1$, $\kappa_{ali} = 1$, and $\hat{\kappa}_{ali} = 1$.

5.2.2 Monitor function based on anisotropic error estimates: $l = 1$

The monitor function is now defined based on the anisotropic error bound given in (5.43). Recall that the condition (5.22) requires that $s = 0$, $0 \le m \le 1$, and $p > d$ for $d > 1$ and $p = d$ for $d = 1$. We restrict consideration to $m = 0$, the most common case in practice.

Taking the q-th power on both sides of (5.43) and summing over all the elements, since $m = 0$ we have

$$\|u-\Pi_k u\|^q_{L^q(\Omega)} \le C\sum_K |K| \left[\frac{1}{d}\mathrm{tr}\left((F_K')^T \nabla u_K \nabla u_K^T F_K'\right)\right]^{\frac{q}{2}}$$
$$+ C\sum_K |K| \left[\frac{1}{d}\mathrm{tr}\left((F_K')^T F_K'\right)\right]^{\frac{q}{2}} \langle \nabla u - \nabla u_K\rangle^q_{L^p(K)}$$
$$\equiv E(\mathscr{T}_h) + E_{h.o.t.}(\mathscr{T}_h). \tag{5.106}$$

We define the monitor function based on $E(\mathscr{T}_h)$, the first term in the bound, and later show that the second term $E_{h.o.t.}(\mathscr{T}_h)$ is a higher order term under suitable conditions. Regularizing as before with a constant $\alpha_h > 0$,

$$E(\mathscr{T}_h) \le C\alpha_h^{\frac{q}{2}} \sum_K |K| \left[\frac{1}{d}\mathrm{tr}\left((F_K')^T \left[I+\alpha_h^{-1}\nabla u_K \nabla u_K^T\right] F_K'\right)\right]^{\frac{q}{2}}. \tag{5.107}$$

Comparison of this with the alignment condition (4.29) suggests choosing the monitor function

$$M_K = \theta_K \left[I + \alpha_h^{-1}\nabla u_K \nabla u_K^T\right], \tag{5.108}$$

where θ_K is yet to be determined. Inserting this into (4.29) yields

$$\frac{1}{d}\mathrm{tr}\left((F_K')^T \left[I+\alpha_h^{-1}\nabla u_K \nabla u_K^T\right] F_K'\right) = \det\left((F_K')^T \left[I+\alpha_h^{-1}\nabla u_K \nabla u_K^T\right] F_K'\right)^{\frac{1}{d}}.$$

Using this and

$$\det\left(I+\alpha_h^{-1}\nabla u_K \nabla u_K^T\right) = 1+\alpha_h^{-1}\|\nabla u_K\|^2, \tag{5.109}$$

we get

$$E(\mathscr{T}_h) \le C\alpha_h^{\frac{q}{2}} \sum_K |K| \left[\det\left((F_K')^T \left[I+\alpha_h^{-1}\nabla u_K \nabla u_K^T\right] F_K'\right)^{\frac{1}{d}}\right]^{\frac{q}{2}}$$
$$= C\alpha_h^{\frac{q}{2}} \sum_K |K| \left[|K|^2 (1+\alpha_h^{-1}\|\nabla u_K\|^2)\right]^{\frac{q}{2d}}$$
$$= C\alpha_h^{\frac{q}{2}} \sum_K \left[|K| (1+\alpha_h^{-1}\|\nabla u_K\|^2)^{\frac{q}{2(d+q)}}\right]^{\frac{d+q}{d}}. \tag{5.110}$$

Choosing

$$\rho_K = \rho_K^{ani,1} \equiv (1+\alpha_h^{-1}\|\nabla u_K\|^2)^{\frac{q}{2(d+q)}} \tag{5.111}$$

and using the equidistribution condition (4.28), from (5.110) we have

$$E(\mathscr{T}_h) \le C\alpha_h^{\frac{q}{2}} \sum_K \left(|K|\rho_K^{ani,1}\right)^{\frac{d+q}{d}} = C\alpha_h^{\frac{q}{2}} N^{-\frac{q}{d}} \sigma_h^{\frac{d+q}{d}}. \tag{5.112}$$

It follows easily that since $\rho_K = \sqrt{\det(M_K)}$,

$$M_K = M_K^{ani,1} \equiv (1+\alpha_h^{-1}\|\nabla u_K\|^2)^{-\frac{1}{d+q}}\left[I+\alpha_h^{-1}\nabla u_K \nabla u_K^T\right]. \tag{5.113}$$

Similarly, $E_{h.o.t.}(\mathscr{T}_h)$ can be bounded by

$$\begin{aligned} E_{h.o.t.}(\mathscr{T}_h) &= C\sum_K |K|\left[\frac{1}{d}\mathrm{tr}\left((F_K')^T F_K'\right)\right]^{\frac{q}{2}} \langle \nabla u - \nabla u_K\rangle^q_{L^p(K)} \\ &\le CN^{-\frac{q}{d}}\sigma_h^{\frac{d+q}{d}}\cdot\frac{1}{\sigma_h}\sum_K |K|\rho_K^{ani,1}\langle \nabla u - \nabla u_K\rangle^q_{L^p(K)}. \end{aligned} \tag{5.114}$$

If $u \in W^{2,\infty}(\Omega)$, then from Poincaré's inequality (cf. Theorem A.0.9) we have

$$\langle \nabla u - \nabla u_K\rangle_{L^p(K)} \le Ch_K\left\langle \nabla^2 u\right\rangle_{L^p(K)} \le Ch|u|_{W^{2,\infty}(\Omega)},$$

where $h = \max_K h_K$. Inserting this into (5.114) gives

$$E_{h.o.t.}(\mathscr{T}_h) \le C\,h^q\,N^{-\frac{q}{d}}\sigma_h^{\frac{d+q}{d}}|u|^q_{W^{2,\infty}(\Omega)}. \tag{5.115}$$

A direct comparison of (5.115) with (5.112) shows that $E_{h.o.t.}(\mathscr{T}_h)$ is a higher order term compared to $E(\mathscr{T}_h)$ as long as the mesh satisfies the property

$$h = \max_K h_K \to 0 \quad \text{as} \quad N \to \infty. \tag{5.116}$$

The property (5.116) is crucial to obtain convergence from the error bounds. In the following we shall show that this property can be satisfied by an M-uniform mesh when $u \in W^{1,\infty}(\Omega)$ and α_h is properly chosen. For this, we first need two lemmas similar to Lemmas 5.2.1 and 5.2.2.

Lemma 5.2.3 *For any real numbers $\gamma \in (0,1]$ and $p \in [1,\infty)$ and any mesh $\mathscr{T}_h$ for Ω,*

$$\sum_K |K|\cdot|v_K|^{\gamma p} \le |\Omega|^{1-\gamma}\|v\|^{\gamma p}_{L^p(\Omega)}, \quad \forall v \in L^p(\Omega) \tag{5.117}$$

where $v_K = (1/|K|)\int_K v(\boldsymbol{x})d\boldsymbol{x}$.

Proof.

$$\begin{aligned} \sum_K |K|\cdot|v_K|^{\gamma p} &\le \sum_K |K|\left[\frac{1}{|K|}\int_K |v|d\boldsymbol{x}\right]^{\gamma p} \\ &\le \sum_K |K|\left[\frac{1}{|K|}\int_K |v|^p d\boldsymbol{x}\right]^{\gamma} \\ &= \sum_K |K|^{1-\gamma}\|v\|^{\gamma p}_{L^p(K)}, \end{aligned}$$

and (5.117) follows from Lemma 5.2.1. □

Lemma 5.2.4 *For any real numbers $\gamma \in (0,1]$ and $p \in [1,\infty)$ and any mesh $\mathscr{T}_h$ for Ω,*

$$
\begin{aligned}
\frac{1}{c_{\gamma p}}\|v\|^{\gamma p}_{L^{\gamma p}(\Omega)} - c^{\gamma p}_{p,p} h^{\gamma p}|\Omega|^{1-\gamma}\|\nabla v\|^{\gamma p}_{L^p(\Omega)} &\le \sum_K |K| \cdot |v_K|^{\gamma p} \\
&\le c_{\gamma p}\|v\|^{\gamma p}_{L^{\gamma p}(\Omega)} + c_{\gamma p} c^{\gamma p}_{p,p} h^{\gamma p}|\Omega|^{1-\gamma}\|\nabla v\|^{\gamma p}_{L^p(\Omega)} \qquad (5.118)
\end{aligned}
$$

for any $v \in W^{1,p}(\Omega)$, where $v_K = (1/|K|)\int_K v(\boldsymbol{x})d\boldsymbol{x}$, $h = \max_{K\in\mathscr{T}_h} h_K$, and the constants $c_{\gamma p}$ and $c_{p,p}$ are defined in (B.4) and Theorem A.0.9, respectively.

Proof. Arguing as in the proof of Lemma 5.2.2 we see that

$$
\sum_K |K| \cdot |v_K|^{\gamma p} - c_{\gamma p}\|v\|^{\gamma p}_{L^{\gamma p}(\Omega)} \le c_{\gamma p} \sum_K |K|^{1-\gamma}\|v - v_K\|^{\gamma p}_{L^p(K)}.
$$

Moreover, from (B.3) we have

$$
\begin{aligned}
&\sum_K |K| \cdot |v_K|^{\gamma p} - \frac{1}{c_{\gamma p}}\|v\|^{\gamma p}_{L^{\gamma p}(\Omega)} \\
&= \frac{1}{c_{\gamma p}} \sum_K \int_K \left(c_{\gamma p}|v_K|^{\gamma p} - |v(\boldsymbol{x})|^{\gamma p}\right) d\boldsymbol{x} \\
&\ge -\sum_K \int_K |v_K - v(\boldsymbol{x})|^{\gamma p} d\boldsymbol{x} \\
&\ge -\sum_K |K|^{1-\gamma}\|v - v_K\|^{\gamma p}_{L^p(K)}.
\end{aligned}
$$

Combining these results with Poincaré's inequality (Theorem A.0.9) gives (5.118). □

We now analyze convergence from the error bounds for two choices of α_h. Note that we cannot take $\alpha_h \to 0$ for the current situation since $\nabla u_K \nabla u_K^T$ is always singular and thus M_K cannot be ensured to be positive definite.

Case 1. Take

$$
\alpha_h = \hat{\alpha}^2 \langle u \rangle^2_{L^p(\Omega)} \qquad (5.119)
$$

for some fixed positive constant $\hat{\alpha}$. For this choice, $M_K^{ani,1}$ defined in (5.113) is invariant under a scaling transformation of u.

Notice that

$$\begin{aligned}
\sigma_h &= \sum_K |K| \rho_K^{ani,1} \\
&= \sum_K |K| (1 + \alpha_h^{-1} \|\nabla u_K\|^2)^{\frac{q}{2(d+q)}} \\
&\le \sum_K |K| \left(1 + \alpha_h^{-\frac{q}{2(d+q)}} \|\nabla u_K\|^{\frac{q}{d+q}} \right) \\
&\le |\Omega| + \alpha_h^{-\frac{q}{2(d+q)}} \sum_K |K| \cdot \|\nabla u_K\|^{\frac{q}{d+q}} \qquad (5.120) \\
&= |\Omega| + \left(\frac{\left(\sum_K |K| \cdot \|\nabla u_K\|^{\frac{q}{d+q}} \right)^{\frac{d+q}{q}}}{\hat{\alpha} \langle u \rangle_{L^p(\Omega)}} \right)^{\frac{q}{d+q}} . \qquad (5.121)
\end{aligned}$$

Assuming $u \in W^{1,\infty}(\Omega)$, we have from (5.113) that

$$\begin{aligned}
\lambda_{min}(M_K) &= \left(1 + \alpha_h^{-1} \|\nabla u_K\|^2\right)^{-\frac{1}{d+q}} \\
&\ge \left(1 + \left(\frac{|u|_{W^{1,\infty}(\Omega)}}{\hat{\alpha} \langle u \rangle_{L^p(\Omega)}} \right)^2 \right)^{-\frac{1}{d+q}} .
\end{aligned}$$

This, together with (5.121), Lemma 5.2.3, and Theorem 4.2.3, implies

$$\begin{aligned}
h_K &\le C N^{-\frac{1}{d}} \lambda_{min}(M_K)^{-\frac{1}{2}} \sigma_h^{\frac{1}{d}} \\
&\le C N^{-\frac{1}{d}} \left(1 + \left(\frac{|u|_{W^{1,\infty}(\Omega)}}{\hat{\alpha} \|u\|_{L^p(\Omega)}} \right)^{\frac{1}{d+q}} \right) \left(1 + \left(\frac{\left(\sum_K |K| \cdot \|\nabla u_K\|^{\frac{q}{d+q}} \right)^{\frac{d+q}{q}}}{\hat{\alpha} \|u\|_{L^p(\Omega)}} \right)^{\frac{q}{d(d+q)}} \right) \\
&\le C N^{-\frac{1}{d}} \left(1 + \left(\frac{|u|_{W^{1,\infty}(\Omega)}}{\hat{\alpha} \|u\|_{L^p(\Omega)}} \right)^{\frac{1}{d+q}} \right) \left(1 + \left(\frac{|u|_{W^{1,p}(\Omega)}}{\hat{\alpha} \|u\|_{L^p(\Omega)}} \right)^{\frac{q}{d(d+q)}} \right) , \qquad (5.122)
\end{aligned}$$

so the mesh satisfies property (5.116).

From Lemma 5.2.4, it follows that

$$\left(\sum_K |K| \cdot \|\nabla u_K\|^{\frac{q}{d+q}} \right)^{\frac{d+q}{q}} \le C \left(|u|_{W^{1,\frac{q}{d+q}}(\Omega)} + h |u|_{W^{2,p}(\Omega)} \right) .$$

Combining this with (5.112), (5.115), and (5.121), we have, for $u \in W^{2,\infty}(\Omega)$,

$$\|u-\Pi_k u\|_{L^q(\Omega)} \le CN^{-\frac{1}{d}}\hat{\alpha}\|u\|_{L^p(\Omega)}\left(1+h\frac{|u|_{W^{2,\infty}(\Omega)}}{\hat{\alpha}\|u\|_{L^p(\Omega)}}\right)$$
$$\times\left(1+\frac{|u|_{W^{1,\frac{q}{d+q}}(\Omega)}}{\hat{\alpha}\|u\|_{L^p(\Omega)}}+h\frac{|u|_{W^{2,p}(\Omega)}}{\hat{\alpha}\|u\|_{L^p(\Omega)}}\right)^{\frac{1}{d}}. \quad (5.123)$$

Case 2. For this case, α_h is chosen such that $M_K^{ani,1}$ is invariant under a scaling transformation of u and

$$\sigma_h \le C \quad (5.124)$$

for some constant C. From (5.120), this specifies that

$$\alpha_h = \alpha_h^{ani,1} \equiv \left[\frac{1}{|\Omega|}\sum_K |K|\cdot\|\nabla u_K\|^{\frac{q}{d+q}}\right]^{\frac{2(d+q)}{q}}. \quad (5.125)$$

Assume that $u \in W^{2,\infty}(\Omega)$. Then Lemma 5.2.4 implies

$$\alpha_h \le C\left(|u|_{W^{1,\frac{q}{d+q}}(\Omega)}+h\,|u|_{W^{2,p}(\Omega)}\right)^2. \quad (5.126)$$

Combining this and (5.124) with (5.106), (5.112), and (5.115), we obtain

$$\begin{aligned}\|u-\Pi_k u\|_{L^q(\Omega)} &\le CN^{-\frac{1}{d}}\left(|u|_{W^{1,\frac{q}{d+q}}(\Omega)}+h\,|u|_{W^{2,p}(\Omega)}+h\,|u|_{W^{2,\infty}(\Omega)}\right)\\ &\le CN^{-\frac{1}{d}}\left(|u|_{W^{1,\frac{q}{d+q}}(\Omega)}+h\,|u|_{W^{2,\infty}(\Omega)}\right).\end{aligned} \quad (5.127)$$

To see if the mesh satisfies (5.116), using (5.113) and (5.124) we obtain

$$h_K \le CN^{-\frac{1}{d}}\lambda_{min}(M_K)^{-\frac{1}{2}}\sigma_h^{\frac{1}{d}} \le CN^{-\frac{1}{d}}\left(1+\frac{|u|_{W^{1,\infty}(\Omega)}}{\sqrt{\alpha_h^{ani,1}}}\right)^{\frac{1}{d+q}},$$

or

$$h \le CN^{-\frac{1}{d}}\left(1+\frac{|u|_{W^{1,\infty}(\Omega)}}{\sqrt{\alpha_h^{ani,1}}}\right)^{\frac{1}{d+q}}. \quad (5.128)$$

Applying Lemma 5.2.4 to (5.125) gives

$$\alpha_h \ge C\left(|u|_{W^{1,\frac{q}{d+q}}(\Omega)}-h\,|u|_{W^{2,p}(\Omega)}\right)^2.$$

This unfortunately does not guarantee a positive lower bound for α_h, so we cannot derive the property (5.116) from (5.128).

To avoid this difficulty, we combine choices (5.119) and (5.125), i.e.,

$$\alpha_h = \max\{\alpha_h^{ani,1}, \hat{\alpha}^2 \langle u \rangle_{L^p(\Omega)}^2\}, \tag{5.129}$$

where $\hat{\alpha}$ is a fixed positive constant and $\alpha_h^{ani,1}$ is defined in (5.125). When $\alpha_h^{ani,1} \geq \hat{\alpha}^2 \langle u \rangle_{L^p(\Omega)}^2$, from (5.128) we have

$$h \leq C N^{-\frac{1}{d}} \left(1 + \frac{|u|_{W^{1,\infty}(\Omega)}}{\hat{\alpha}|u|_{L^p(\Omega)}}\right)^{\frac{1}{d+q}} \tag{5.130}$$

and the error is bounded by (5.127). On the other hand, when $\alpha_h^{ani,1} < \hat{\alpha}^2 \langle u \rangle_{L^p(\Omega)}^2$, then as in Case 1 above, the error is bounded by (5.123), where h satisfies (5.122).

The above results are summarized in the following theorem.

Theorem 5.2.3 *Suppose that $q \leq p$, $u \in W^{2,\infty}(\Omega)$, and the condition (5.22) is satisfied with $l = 1$ and $m = 0$. Suppose also that the mesh density function and the monitor function are respectively chosen as in (5.111) and (5.113). Consider a family of corresponding M-uniform meshes.*

(a) If the regularization parameter is taken as $\alpha_h = \hat{\alpha}^2 \langle u \rangle_{L^p(\Omega)}^2$ for some positive constant $\hat{\alpha}$, then

$$\|u - \Pi_k u\|_{L^q(\Omega)} \leq C N^{-\frac{1}{d}} \hat{\alpha} \|u\|_{L^p(\Omega)} \left(1 + h \frac{|u|_{W^{2,\infty}(\Omega)}}{\hat{\alpha}\|u\|_{L^p(\Omega)}}\right) \times \left(1 + \frac{|u|_{W^{1,\frac{q}{d+q}}(\Omega)}}{\hat{\alpha}\|u\|_{L^p(\Omega)}} + h \frac{|u|_{W^{2,p}(\Omega)}}{\hat{\alpha}\|u\|_{L^p(\Omega)}}\right)^{\frac{1}{d}}, \tag{5.131}$$

where

$$h \leq C N^{-\frac{1}{d}} \left(1 + \left(\frac{|u|_{W^{1,\infty}(\Omega)}}{\hat{\alpha}\|u\|_{L^p(\Omega)}}\right)^{\frac{1}{d+q}}\right) \left(1 + \left(\frac{|u|_{W^{1,p}(\Omega)}}{\hat{\alpha}\|u\|_{L^p(\Omega)}}\right)^{\frac{q}{d(d+q)}}\right). \tag{5.132}$$

(b) If instead α_h is taken as

$$\alpha_h = \max\{\alpha_h^{ani,1}, \hat{\alpha}^2 \langle u \rangle_{L^p(\Omega)}^2\}, \tag{5.133}$$

where $\alpha_h^{ani,1}$ is defined in (5.125) and $\hat{\alpha}_h$ is a positive constant, then when $\alpha_h^{ani,1} \geq \hat{\alpha}^2 \langle u \rangle_{L^p(\Omega)}^2$ the interpolation error is bounded by

$$\|u-\Pi_k u\|_{L^q(\Omega)} \le CN^{-\frac{1}{d}}\left(|u|_{W^{1,\frac{q}{d+q}}(\Omega)} + h\,|u|_{W^{2,\infty}(\Omega)}\right), \tag{5.134}$$

where

$$h \le CN^{-\frac{1}{d}}\left(1+\frac{|u|_{W^{1,\infty}(\Omega)}}{\hat{\alpha}|u|_{L^p(\Omega)}}\right)^{\frac{1}{d+q}} \tag{5.135}$$

(and when $\alpha_h^{ani,1} < \hat{\alpha}^2 \langle u\rangle^2_{L^p(\Omega)}$, (5.131) and (5.132) hold).

From the theorem one can see that for both choices of α_h the interpolation error has the asymptotic bound (for large N)

$$\|u-\Pi_k u\|_{L^q(\Omega)} \le CN^{-\frac{1}{d}}\left(\hat{\alpha}\|u\|_{L^p(\Omega)} + |u|_{W^{1,\frac{q}{d+q}}(\Omega)}\right) + h.o.t. \tag{5.136}$$

It is worth pointing out that estimates for interpolation error can also be obtained for a quasi M-uniform mesh satisfying (4.30) and (4.31) using the same procedure as for Theorem 5.2.2. For brevity, such derivations and results are not given for this case (nor for the case $l=2$ discussed in the next subsection).

5.2.3 Monitor function based on anisotropic error estimates: $l=2$

While the analysis can be straightforwardly extended to the general case $l \ge 2$, for simplicity we only consider here the most widely used case $l=2$ with $m=0$ or $m=1$. Our analysis is now based on the bound (5.45).

Taking the q-th power on both sides of (5.45) and summing over all the elements, the same argument as in the previous subsection gives

$$\begin{aligned}
&|u-\Pi_k u|^q_{W^{m,q}(\Omega)} \\
&\le C\sum_K |K|\cdot\left[\frac{1}{d}\mathrm{tr}\left((F_K')^{-1}(F_K')^{-T}\right)\right]^{\frac{mq}{2}}\left[\frac{1}{d}\mathrm{tr}\left((F_K')^T|H_K(u)|F_K'\right)\right]^q \\
&+ C\sum_K |K|\cdot\left[\frac{1}{d}\mathrm{tr}\left((F_K')^{-1}(F_K')^{-T}\right)\right]^{\frac{mq}{2}}\left[\frac{1}{d}\mathrm{tr}\left((F_K')^T F_K'\right)\right]^q \\
&\qquad \times \langle H(u)-H_K(u)\rangle^q_{L^p(K)} \\
&\equiv E(\mathscr{T}_h) + E_{h.o.t.}(\mathscr{T}_h),
\end{aligned} \tag{5.137}$$

where we have used

$$\|(F_K')^{-1}\| \le \|(F_K')^{-1}\|_F = \left[\mathrm{tr}\left((F_K')^{-1}(F_K')^{-T}\right)\right]^{\frac{1}{2}}.$$

The monitor function is defined based on $E(\mathscr{T}_h)$ where once again, we show that $E_{h.o.t.}(\mathscr{T}_h)$ is a higher order term. Regularizing $E(\mathscr{T}_h)$ with a positive constant α_h and using Lemma 4.5.2, we have

$$E(\mathscr{T}_h) \le C\alpha_h^q \sum_K |K| \cdot \left[\frac{1}{d}\mathrm{tr}\left((F_K')^{-1} H_{K,\alpha}^{-1} (F_K')^{-T}\right)\right]^{\frac{mq}{2}} \times \left[\frac{1}{d}\mathrm{tr}\left((F_K')^T H_{K,\alpha} F_K'\right)\right]^{q} \|H_{K,\alpha}\|^{\frac{mq}{2}}, \tag{5.138}$$

where

$$H_{K,\alpha} \equiv I + \alpha_h^{-1} |H_K(u)|. \tag{5.139}$$

A comparison of this with alignment conditions (4.29) and (4.27) suggests that the monitor function minimizing this bound be of the form

$$M_K = \theta_K H_{K,\alpha} \tag{5.140}$$

for a suitably chosen scalar function $\theta = \theta_K$. Inserting this into (4.29) and (4.27) leads to

$$\frac{1}{d}\mathrm{tr}\left((F_K')^T H_{K,\alpha} F_K'\right) = \det\left((F_K')^T H_{K,\alpha} F_K'\right)^{\frac{1}{d}} = |K|^{\frac{2}{d}} \det(H_{K,\alpha})^{\frac{1}{d}}$$

and

$$\frac{1}{d}\mathrm{tr}\left((F_K')^{-1} H_{K,\alpha}^{-1} (F_K')^{-T}\right) = \det\left((F_K')^{-1} H_{K,\alpha}^{-1} (F_K')^{-T}\right)^{\frac{1}{d}} = |K|^{-\frac{2}{d}} \det(H_{K,\alpha})^{-\frac{1}{d}}.$$

Combining these results with (5.138) we obtain

$$\begin{aligned} E(\mathscr{T}_h) &\le C\alpha_h^q \sum_K |K| \cdot \left(|K|^2 \det(H_{K,\alpha})\right)^{\frac{q(2-m)}{2d}} \|H_{K,\alpha}\|^{\frac{mq}{2}} \\ &= C\alpha_h^q \sum_K \left(|K| \det(H_{K,\alpha})^{\frac{q(2-m)}{2(d+q(2-m))}} \|H_{K,\alpha}\|^{\frac{mqd}{2(d+q(2-m))}} \right)^{\frac{d+q(2-m)}{d}}. \end{aligned} \tag{5.141}$$

A direct comparison with equidistribution condition (4.28) suggests that ρ_K be chosen as

$$\rho_K = \rho_K^{ani,2} \equiv \det(H_{K,\alpha})^{\frac{q(2-m)}{2(d+q(2-m))}} \|H_{K,\alpha}\|^{\frac{mqd}{2(d+q(2-m))}}. \tag{5.142}$$

Using the relation $\rho_K = \det(M_K)^{\frac{1}{2}}$ and (5.140) we get

$$M_K = M_K^{ani,2} \equiv \det(H_{K,\alpha})^{-\frac{1}{d+q(2-m)}} \|H_{K,\alpha}\|^{\frac{mq}{d+q(2-m)}} H_{K,\alpha}. \tag{5.143}$$

It is not difficult to show that for an M-uniform mesh associated with $M_K^{ani,2}$, the bounds for $E(\mathscr{T}_h)$ and $E_{h.o.t.}(\mathscr{T}_h)$ are

$$E(\mathscr{T}_h) \le C N^{-\frac{q(2-m)}{d}} \alpha_h^q \sigma_h^{\frac{d+q(2-m)}{d}}, \tag{5.144}$$

$$E_{h.o.t.}(\mathscr{T}_h) \le C N^{-\frac{q(2-m)}{d}} \sigma_h^{\frac{d+q(2-m)}{d}} \cdot \frac{1}{\sigma_h} \sum_K |K| \, \rho_K^{ani,2} \langle H(u) - H_K(u) \rangle_{L^p(K)}^q. \tag{5.145}$$

If $u \in W^{3,\infty}(\Omega)$, from Poincaré's inequality (cf. Theorem A.0.9) we get

$$\langle H(u) - H_K(u) \rangle_{L^p(K)} \le C h_K \left\langle \nabla^3 u \right\rangle_{L^p(K)} \le C h |u|_{W^{3,\infty}(\Omega)},$$

where ∇^3 is the tensor of third derivatives. Combining this with (5.145),

$$E_{h.o.t.}(\mathscr{T}_h) \le C h^q N^{-\frac{q(2-m)}{d}} \sigma_h^{\frac{d+q(2-m)}{d}} |u|_{W^{3,\infty}(\Omega)}^q. \tag{5.146}$$

As before, this implies that $E_{h.o.t.}(\mathscr{T}_h)$ is a higher order term compared to $E(\mathscr{T}_h)$ provided that the mesh satisfies the property

$$h = \max_K h_K \to 0 \quad \text{as} \quad N \to \infty. \tag{5.147}$$

We now examine three choices of α_h: $\alpha_h \to 0$, $\alpha_h = \hat{\alpha} \langle u \rangle_{L^p(\Omega)}$ for some positive constant $\hat{\alpha}$, and a choice such that $M_K^{ani,2}$ is invariant under a scaling transformation of u and $\sigma_h = \sum_K |K| \rho_K^{ani,2} \le C$ for some constant C.

Case 1. Take $\alpha_h \to 0$. Since the equidistribution and alignment conditions are invariant under a scaling transformation of the monitor function (cf. Theorem 4.1.2), we can redefine the monitor function and the mesh density function as

$$M_K = \alpha_h^{\frac{2q}{d+q(2-m)}} M_K^{ani,2}, \quad \rho_K = \alpha_h^{\frac{dq}{d+q(2-m)}} \rho_K^{ani,2}.$$

Taking the limit $\alpha_h \to 0$ and noticing that $\| \, |H_K(u)| \, \| = \|H_K(u)\|$, we get

$$M_K = \det(|H_K(u)|)^{-\frac{1}{d+q(2-m)}} \|H_K(u)\|^{\frac{mq}{d+q(2-m)}} |H_K(u)|, \tag{5.148}$$

$$\rho_K = \det(|H_K(u)|)^{\frac{q(2-m)}{2(d+q(2-m))}} \|H_K(u)\|^{\frac{mqd}{2(d+q(2-m))}}, \tag{5.149}$$

which correspond to the non-regularized situation. To ensure M_K to be positive definite, we assume that there exists a positive constant β such that

$$|H(u)(\boldsymbol{x})| \ge \beta I, \quad \text{a.e. in } \Omega. \tag{5.150}$$

Here, the inequality sign "$\ge$" means that the difference between the left- and right-hand sides is positive semi-definite. Assuming that $u \in W^{2,\infty}(\Omega)$, from (5.148) and Theorem 4.2.3 we have

$$
\begin{aligned}
h_K &\le \frac{\hat{h}}{\sqrt{\lambda_{min}(M_K)}} N^{-\frac{1}{d}} \sigma_h^{\frac{1}{d}} \\
&\le C \frac{\det\left(|H_K(u)|\right)^{\frac{1}{2(d+q(2-m))}}}{\|H_K(u)\|^{\frac{mq}{2(d+q(2-m))}} \sqrt{\lambda_{min}(|H_K(u)|)}} N^{-\frac{1}{d}} \left(\sum_K |K| \rho_K\right)^{\frac{1}{d}} \\
&\le C |u|_{W^{2,\infty}(\Omega)}^{\frac{d}{2(d+q(2-m))}} \beta^{-\frac{d+2q}{2(d+q(2-m))}} N^{-\frac{1}{d}} \left(\sum_K |K| \rho_K\right)^{\frac{1}{d}} \\
&\le C |u|_{W^{2,\infty}(\Omega)}^{\frac{d}{2(d+q(2-m))}} \beta^{-\frac{d+2q}{2(d+q(2-m))}} N^{-\frac{1}{d}} \\
&\quad \times \left(\sum_K |K| \det(|H_K(u)|)^{\frac{q(2-m)}{2(d+q(2-m))}} \|H_K(u)\|^{\frac{mqd}{2(d+q(2-m))}}\right)^{\frac{1}{d}} \\
&\le C |u|_{W^{2,\infty}(\Omega)}^{\frac{d}{2(d+q(2-m))}} \beta^{-\frac{d+2q}{2(d+q(2-m))}} N^{-\frac{1}{d}} \left(\sum_K |K| \cdot \|H_K(u)\|^{\frac{dq(2-m)}{2(d+q(2-m))} + \frac{mqd}{2(d+q(2-m))}}\right)^{\frac{1}{d}} \\
&= C |u|_{W^{2,\infty}(\Omega)}^{\frac{d}{2(d+q(2-m))}} \beta^{-\frac{d+2q}{2(d+q(2-m))}} N^{-\frac{1}{d}} \left(\sum_K |K| \cdot \|H_K(u)\|^{\frac{dq}{d+q(2-m)}}\right)^{\frac{1}{d}}.
\end{aligned}
$$

Assuming $q \le p$ and applying Lemma 5.2.3 we get

$$
\begin{aligned}
h &\le C |u|_{W^{2,\infty}(\Omega)}^{\frac{d}{2(d+q(2-m))}} \beta^{-\frac{d+2q}{2(d+q(2-m))}} N^{-\frac{1}{d}} \|H(u)\|_{L^p(\Omega)}^{\frac{q}{d+q(2-m)}} \\
&\le C |u|_{W^{2,\infty}(\Omega)}^{\frac{d}{2(d+q(2-m))}} \beta^{-\frac{d+2q}{2(d+q(2-m))}} N^{-\frac{1}{d}} |u|_{W^{2,p}(\Omega)}^{\frac{q}{d+q(2-m)}}.
\end{aligned} \tag{5.151}
$$

Thus, the mesh condition (5.147) is satisfied.

The quantities $E(\mathscr{T}_h)$ and $E_{h.o.t.}(\mathscr{T}_h)$ can be estimated directly from the definition (5.137). A similar procedure as for deriving (5.144) and (5.146) shows that

$$
E(\mathscr{T}_h) \le C N^{-\frac{q(2-m)}{d}} \sigma_h^{\frac{d+q(2-m)}{d}}; \tag{5.152}
$$

furthermore, assuming that $u \in W^{3,\infty}(\Omega)$,

$$
\begin{aligned}
E_{h.o.t.}(\mathscr{T}_h) &\le C \sum_K |K| \cdot \left[\frac{1}{d} \mathrm{tr}\left((F_K')^{-1} (F_K')^{-T}\right)\right]^{\frac{mq}{2}} \\
&\qquad \times \left[\frac{1}{d} \mathrm{tr}\left((F_K')^T |H_K(u)| F_K'\right)\right]^q \|H_K^{-1}(u)\|^q \langle H(u) - H_K(u) \rangle_{L^p(K)}^q \\
&\le C N^{-\frac{q(2-m)}{d}} \sigma_h^{\frac{d+q(2-m)}{d}} \cdot \beta^{-q} h^q |u|_{W^{3,\infty}(\Omega)}^q,
\end{aligned}
$$

where

$$\sigma_h = \sum_K |K| \rho_K \to \int_\Omega \det(|H(u)|)^{\frac{q(2-m)}{2(d+q(2-m))}} \|H(u)\|^{\frac{mqd}{2(d+q(2-m))}} d\boldsymbol{x} \tag{5.153}$$

as $N \to \infty$. The interpolation error is thus bounded by

$$|u - \Pi_k u|_{W^{m,q}(\Omega)} \le C N^{-\frac{(2-m)}{d}} \sigma_h^{\frac{d+q(2-m)}{dq}} \left(1 + h\beta^{-1} |u|_{W^{3,\infty}(\Omega)}\right). \tag{5.154}$$

Case 2. Take $\alpha_h = \hat{\alpha} \langle u \rangle_{L^p(\Omega)}$ for a constant $\hat{\alpha} > 0$. Then

$$\begin{aligned}
\sigma_h &= \sum_K |K| \rho_K^{ani,2} \\
&\le C \sum_K |K| \, \|H_{K,\alpha}\|^{\frac{dq}{d+q(2-m)}} \\
&\le C \sum_K |K| \left(1 + \left(\hat{\alpha} \langle u \rangle_{L^p(\Omega)}\right)^{-\frac{dq}{d+q(2-m)}} \|H_K(u)\|^{\frac{dq}{d+q(2-m)}}\right) \\
&\le C \left(1 + \left(\hat{\alpha} \langle u \rangle_{L^p(\Omega)}\right)^{-\frac{dq}{d+q(2-m)}} \sum_K |K| \, \|H_K(u)\|^{\frac{dq}{d+q(2-m)}}\right) \\
&\le C \left(1 + \left(\hat{\alpha} \langle u \rangle_{L^p(\Omega)}\right)^{-\frac{dq}{d+q(2-m)}} \langle u \rangle_{W^{2,p}(\Omega)}^{\frac{dq}{d+q(2-m)}}\right),
\end{aligned}$$

where in the last step we have assumed $q \le p$ and used Lemma 5.2.3. Assume that $u \in W^{2,\infty}(\Omega)$. Noticing that $H_{K,\alpha} \ge I$, from Theorem 4.2.3 we obtain

$$\begin{aligned}
h \le C N^{-\frac{1}{d}} & \left(1 + \left(\frac{\langle u \rangle_{W^{2,p}(\Omega)}}{\hat{\alpha} \langle u \rangle_{L^p(\Omega)}}\right)^{\frac{q}{d+q(2-m)}}\right) \\
& \times \left(1 + \left(\frac{\langle u \rangle_{W^{2,\infty}(\Omega)}}{\hat{\alpha} \langle u \rangle_{L^p(\Omega)}}\right)^{\frac{d}{2(d+q(2-m))}}\right),
\end{aligned} \tag{5.155}$$

which implies that the mesh satisfies the property (5.147). From (5.137), (5.144), and (5.146) it is not difficult to obtain the error bound

$$|u - \Pi_k u|_{W^{m,q}(\Omega)} \le C N^{-\frac{(2-m)}{d}} \sigma_h^{\frac{d+q(2-m)}{dq}} \left(\hat{\alpha} \langle u \rangle_{L^p(\Omega)} + h |u|_{W^{3,\infty}(\Omega)}\right), \tag{5.156}$$

where as $N \to \infty$,

$$\sigma_h \to \int_\Omega \det\left(I + (\hat{\alpha}\,\langle u\rangle_{L^p(\Omega)})^{-1}|H(u)|\right)^{\frac{q(2-m)}{2(d+q(2-m))}} \times \|I + (\hat{\alpha}\,\langle u\rangle_{L^p(\Omega)})^{-1}|H(u)|\|^{\frac{mqd}{2(d+q(2-m))}}\, d\boldsymbol{x}. \tag{5.157}$$

Case 3. In this case α_h is chosen such that σ_h is bounded by a constant and $M_K^{ani,2}$ is invariant under a scaling transformation of u. Unlike in the cases considered in the previous subsections, α_h cannot be explicitly determined because it cannot be factored out of the matrix determinant and norm terms in (5.142). Indeed, we implicitly define $\alpha_h = \alpha_h^{ani,2}$, where $\alpha_h^{ani,2}$ is the solution of the algebraic equation

$$\sigma_h \equiv \sum_K |K|\rho_K^{ani,2} = 2c_{\frac{dq}{d+q(2-m)}}|\Omega|,$$

or from (5.142),

$$\sum_K |K| \det\left(I + \frac{1}{\alpha_h^{ani,2}}|H_K(u)|\right)^{\frac{q(2-m)}{2(d+q(2-m))}} \left\|I + \frac{1}{\alpha_h^{ani,2}}|H_K(u)|\right\|^{\frac{mqd}{2(d+q(2-m))}} = 2c_{\frac{dq}{d+q(2-m)}}|\Omega|. \tag{5.158}$$

The constant $c_{\frac{dq}{d+q(2-m)}}$ is defined in (B.4) and the reason for using the factor $2c_{\frac{dq}{d+q(2-m)}}$ will be clear from the derivation of an upper bound for α_h below (cf. (5.160)). Note that equation (5.158) has a unique solution provided $H_K(u)$ is not zero for all elements, since the left-hand side of (5.158) is monotonically decreasing with respect to $\alpha_h^{ani,2}$ and tends to $|\Omega|$ as $\alpha_h^{ani,2} \to \infty$ and $+\infty$ as $\alpha_h^{ani,2} \to 0$. As a consequence, (5.158) can be easily solved numerically for α_h, e.g., using a simple root finding algorithm such as the bisection method.

Upper and lower bounds for α_h can be obtained as follows. Since

$$\begin{aligned}
\rho_K^{ani,2} &= \det(H_{K,\alpha})^{\frac{q(2-m)}{2(d+q(2-m))}} \|H_{K,\alpha}\|^{\frac{mqd}{2(d+q(2-m))}} \\
&\ge \det(H_{K,\alpha})^{\frac{q}{d+q(2-m)}} \\
&\ge \left(1 + (\alpha_h^{ani,2})^{-d} \det(|H_K(u)|)\right)^{\frac{q}{d+q(2-m)}} \\
&\ge 2^{\frac{q}{d+q(2-m)}-1} \left(1 + (\alpha_h^{ani,2})^{-\frac{dq}{d+q(2-m)}} \det(|H_K(u)|)^{\frac{q}{d+q(2-m)}}\right),
\end{aligned}$$

it follows from (5.158) that

$$
\begin{aligned}
2c_{\frac{dq}{d+q(2-m)}} |\Omega| &= \sum_K |K| \rho_K^{ani,2} \\
&\geq 2^{\frac{q}{d+q(2-m)}-1} \left(|\Omega| + (\alpha_h^{ani,2})^{-\frac{dq}{d+q(2-m)}} \sum_K |K| \det(|H_K(u)|)^{\frac{q}{d+q(2-m)}} \right),
\end{aligned}
$$

or

$$
\begin{aligned}
\alpha_h^{ani,2} \geq \Bigg[& \left(2^{2-\frac{q}{d+q(2-m)}} c_{\frac{dq}{d+q(2-m)}} - 1 \right)^{-1} |\Omega|^{-1} \\
& \times \sum_K |K| \det(|H_K(u)|)^{\frac{q}{d+q(2-m)}} \Bigg]^{\frac{d+q(2-m)}{dq}} .
\end{aligned}
\tag{5.159}
$$

From Corollary B.0.1,

$$
\begin{aligned}
\rho_K^{ani,2} &\leq \|H_{K,\alpha}\|^{\frac{dq}{d+q(2-m)}} \\
&\leq \left(1 + (\alpha_h^{ani,2})^{-1} \|H_K(u)\| \right)^{\frac{dq}{d+q(2-m)}} \\
&\leq c_{\frac{dq}{d+q(2-m)}} \left(1 + (\alpha_h^{ani,2})^{-\frac{dq}{d+q(2-m)}} \|H_K(u)\|^{\frac{dq}{d+q(2-m)}} \right),
\end{aligned}
$$

so

$$
\begin{aligned}
2c_{\frac{dq}{d+q(2-m)}} |\Omega| &= \sum_K |K| \rho_K^{ani,2} \\
&\leq c_{\frac{dq}{d+q(2-m)}} \left(|\Omega| + (\alpha_h^{ani,2})^{-\frac{dq}{d+q(2-m)}} \sum_K |K| \cdot \|H_K(u)\|^{\frac{dq}{d+q(2-m)}} \right),
\end{aligned}
\tag{5.160}
$$

which gives

$$
\alpha_h^{ani,2} \leq \left[\frac{1}{|\Omega|} \sum_K |K| \cdot \|H_K(u)\|^{\frac{dq}{d+q(2-m)}} \right]^{\frac{d+q(2-m)}{dq}} .
\tag{5.161}
$$

Noticing that $\sigma_h \leq C$, we get from (5.137), (5.144), and (5.146) that

$$
|u - \Pi_k u|_{W^{m,q}(\Omega)} \leq C N^{-\frac{(2-m)}{d}} \left(\alpha_h^{ani,2} + h |u|_{W^{3,\infty}(\Omega)} \right),
\tag{5.162}
$$

where $\alpha_h^{ani,2}$ is bounded from below and above in (5.159) and (5.161). If the mesh satisfies the property (5.147), it is obvious from (5.159) and (5.161) that $\alpha_h^{ani,2} \to \alpha^{ani,2}$ as $N \to \infty$, where $\alpha^{ani,2}$ is a number bounded by

$$
C_1 \| \sqrt[d]{\det(|H(u)|)} \|_{L^{\frac{dq}{d+q(2-m)}}(\Omega)} \leq \alpha^{ani,2} \leq C_2 |u|_{W^{2,\frac{dq}{d+q(2-m)}}(\Omega)} .
\tag{5.163}
$$

To see if the mesh satisfies the property (5.147), note that σ_h is bounded for the current choice of $\alpha_h^{ani,2}$. From Theorem 4.2.3 we obtain

$$h \le CN^{-\frac{1}{d}} \left(1 + \left(\frac{\langle u \rangle_{W^{2,\infty}(\Omega)}}{\alpha_h^{ani,2}} \right)^{\frac{d}{2(d+q(2-m))}} \right). \tag{5.164}$$

As in Case 2 of the previous subsection, $\alpha_h^{ani,2}$ defined through (5.158) is unfortunately not guaranteed to have a positive lower bound. To avoid this difficulty, we define a new α_h as

$$\alpha_h = \max\{\alpha_h^{ani,2}, \hat{\alpha} \langle u \rangle_{L^p(\Omega)}\}, \tag{5.165}$$

where $\hat{\alpha}$ is a fixed positive constant. Then (5.164) reduces to

$$h \le CN^{-\frac{1}{d}} \left(1 + \left(\frac{\langle u \rangle_{W^{2,\infty}(\Omega)}}{\hat{\alpha} \langle u \rangle_{L^p(\Omega)}} \right)^{\frac{d}{2(d+q(2-m))}} \right), \tag{5.166}$$

which implies that the mesh satisfies (5.147). When $\alpha_h^{ani,2} \ge \hat{\alpha} \langle u \rangle_{L^p(\Omega)}$, the interpolation error is bounded as in (5.162); otherwise, it is bounded as in (5.156).

These results are summarized in the following theorem.

Theorem 5.2.4 *Suppose that $q \le p$, $u \in W^{3,\infty}(\Omega)$, and the condition (5.22) is satisfied with $l = 2$ and $m = 0$ or $m = 1$. Suppose also that the mesh density function and the monitor function are chosen as in (5.142) and (5.143), respectively. Consider a family of corresponding M-uniform meshes.*

(a) If the Hessian satisfies (5.150) and the regularization parameter is taken as $\alpha_h \to 0$, then

$$|u - \Pi_k u|_{W^{m,q}(\Omega)} \le CN^{-\frac{(2-m)}{d}} \sigma_h^{\frac{d+q(2-m)}{dq}} \left(1 + h\beta^{-1} |u|_{W^{3,\infty}(\Omega)} \right), \tag{5.167}$$

where

$$h \le CN^{-\frac{1}{d}} \beta^{-\frac{d+2q}{2(d+q(2-m))}} |u|_{W^{2,\infty}(\Omega)}^{\frac{d}{2(d+q(2-m))}} |u|_{W^{2,p}(\Omega)}^{\frac{q}{d+q(2-m)}}, \tag{5.168}$$

and

$$\sigma_h \to \int_\Omega det(|H(u)|)^{\frac{q(2-m)}{2(d+q(2-m))}} \|H(u)\|^{\frac{mqd}{2(d+q(2-m))}} d\boldsymbol{x} \quad as \quad N \to \infty. \tag{5.169}$$

(b) If the regularization parameter is taken as $\alpha_h = \hat{\alpha} \langle u \rangle_{L^p(\Omega)}$ for some positive constant $\hat{\alpha}$, then

$$|u - \Pi_k u|_{W^{m,q}(\Omega)} \le CN^{-\frac{(2-m)}{d}} \sigma_h^{\frac{d+q(2-m)}{dq}} \left(\hat{\alpha} \langle u \rangle_{L^p(\Omega)} + h|u|_{W^{3,\infty}(\Omega)} \right), \tag{5.170}$$

where

$$h \le CN^{-\frac{1}{d}} \left(1 + \left(\frac{\langle u \rangle_{W^{2,p}(\Omega)}}{\hat{\alpha} \langle u \rangle_{L^p(\Omega)}} \right)^{\frac{q}{d+q(2-m)}} \right) \times \left(1 + \left(\frac{\langle u \rangle_{W^{2,\infty}(\Omega)}}{\hat{\alpha} \langle u \rangle_{L^p(\Omega)}} \right)^{\frac{d}{2(d+q(2-m))}} \right), \tag{5.171}$$

and as $N \to \infty$,

$$\sigma_h \to \int_\Omega det\left(I + (\hat{\alpha} \langle u \rangle_{L^p(\Omega)})^{-1} |H(u)|\right)^{\frac{q(2-m)}{2(d+q(2-m))}} \times \|I + (\hat{\alpha} \langle u \rangle_{L^p(\Omega)})^{-1} |H(u)|\|^{\frac{mqd}{2(d+q(2-m))}} d\boldsymbol{x}. \tag{5.172}$$

(c) If α_h is taken as

$$\alpha_h = \max\{\alpha_h^{ani,2}, \hat{\alpha} \langle u \rangle_{L^p(\Omega)}\}, \tag{5.173}$$

where $\alpha_h^{ani,2}$ is defined in (5.158) and $\hat{\alpha}_h$ is a positive constant, then when $\alpha_h^{ani,2} \ge \hat{\alpha} \langle u \rangle_{L^p(\Omega)}$ the interpolation error is bounded by

$$|u - \Pi_k u|_{W^{m,q}(\Omega)} \le CN^{-\frac{(2-m)}{d}} \left(\alpha_h^{ani,2} + h|u|_{W^{3,\infty}(\Omega)} \right). \tag{5.174}$$

Here,

$$h \le CN^{-\frac{1}{d}} \left(1 + \left(\frac{\langle u \rangle_{W^{2,\infty}(\Omega)}}{\hat{\alpha} \langle u \rangle_{L^p(\Omega)}} \right)^{\frac{d}{2(d+q(2-m))}} \right) \tag{5.175}$$

and

$$\alpha_h^{ani,2} \to \alpha^{ani,2}, \quad as \quad N \to \infty$$

with $\alpha^{ani,2}$ being bounded by

$$C_1 \| \sqrt[d]{det(|H(u)|)} \|_{L^{\frac{dq}{d+q(2-m)}}(\Omega)} \le \alpha^{ani,2} \le C_2 |u|_{W^{2,\frac{dq}{d+q(2-m)}}(\Omega)}. \tag{5.176}$$

When $\alpha_h^{ani,2} \le \hat{\alpha} \langle u \rangle_{L^p(\Omega)}$, the interpolation error is bounded by (5.170).

For $m = 0$, (5.167) reduces to

$$\|u - \Pi_k u\|_{L^q(\Omega)} \le CN^{-\frac{2}{d}} \| \sqrt[d]{\det(|H(u)|)} \|_{L^{\frac{dq}{d+2q}}(\Omega)} + h.o.t. \tag{5.177}$$

This bound has been obtained by Huang and Sun [193] for $q = 2$ and by Chen et al. [99] for general $q \ge 1$. Chen et al. also show that the bound is optimal in the sense that it is a lower bound if u is strictly convex or concave. For $m = 1$, (5.167) yields

$$|u-\Pi_k u|_{W^{1,q}(\Omega)} \le CN^{-\frac{1}{d}} \left[\int_\Omega \det(|H(u)|)^{\frac{q}{2(d+q)}} \|H(u)\|^{\frac{dq}{2(d+q)}} dx \right]^{\frac{d+q}{dq}} + h.o.t. \tag{5.178}$$

It is unclear whether or not this bound is optimal (the smallest achievable) among all possible monitor functions. However, as we see in the next subsection, it is smaller (in terms of the solution-dependent factor) than that resulting from the Hessian monitor function.

It is useful to compare these bounds with those for isotropic meshes. We recall from (5.92) that the error bound for the isotropic case with $l = 2$ and $m = 0$ or $m = 1$ is given by

$$|u-\Pi_k u|_{W^{m,q}(\Omega)} \le CN^{-\frac{(2-m)}{d}} |u|_{W^{2,\frac{dq}{d+q(2-m)}}(\Omega)}. \tag{5.179}$$

By comparing this with (5.177) and (5.178) one can conclude that an anisotropic mesh generally leads to a lower asymptotic error bound than an isotropic one for the same number of mesh elements.

While the regularization situation is more complicated, the same conclusion can be reached. For example, for the choice $\alpha_h = \alpha_h^{ani,2}$, (5.174) leads to

$$|u-\Pi_k u|_{W^{m,q}(\Omega)} \le CN^{-\frac{(2-m)}{d}} \alpha^{ani,2} + h.o.t. \tag{5.180}$$

where $\alpha^{ani,2}$ is the solution of

$$\int_\Omega \det\left(I + \frac{1}{\alpha^{ani,2}}|H(u)|\right)^{\frac{q(2-m)}{2(d+(2-m)q)}} \left\|I + \frac{1}{\alpha^{ani,2}}|H(u)|\right\|^{\frac{mqd}{2(d+q(2-m))}} dx = 2c_{\frac{dq}{d+q(2-m)}} |\Omega|, \tag{5.181}$$

and bounded below and above as in (5.176). (Equation (5.181) is actually the continuous version of (5.158).) Inequality (5.176) implies that up to a multiplicative constant, the bound in (5.180) is smaller than that in (5.179).

5.2.4 The Hessian as the monitor function

A popular choice for the monitor function (or the metric tensor) in practice (e.g., see [54, 55, 93]) has been the Hessian of u,

$$M_K = |H_K(u)|. \tag{5.182}$$

This is largely motivated by the results of D'Azevedo [111] and D'Azevedo and Simpson [112] on linear interpolation for quadratic functions on triangles. It is instructive to examine the corresponding interpolation error bound and compare it with that for the optimal monitor function obtained in the previous subsection. For simplicity we only consider this for the non-regularized case $\alpha_h \to 0$. To ensure the positive definiteness of the monitor function we assume that u satisfies the condition (5.150). Then for an associated M-uniform mesh, from (5.137) and the alignment and equidistribution conditions we have

$$
\begin{aligned}
E(\mathscr{T}_h) &= C\sum_K |K| \cdot \left[\frac{1}{d}\mathrm{tr}\left((F_K')^{-1}(F_K')^{-T}\right)\right]^{\frac{mq}{2}} \left[\frac{1}{d}\mathrm{tr}\left((F_K')^T|H_K(u)|F_K'\right)\right]^q \\
&\le \sum_K |K| \cdot \left[\frac{1}{d}\mathrm{tr}\left((F_K')^{-1}|H_K(u)|^{-1}(F_K')^{-T}\right)\right]^{\frac{mq}{2}} \\
&\qquad \times \left[\frac{1}{d}\mathrm{tr}\left((F_K')^T|H_K(u)|F_K'\right)\right]^q \|H_K(u)\|^{\frac{mq}{2}} \\
&\le C\sum_K |K| \left(|K|\det(|H_K(u)|)^{\frac{1}{2}}\right)^{\frac{q(2-m)}{d}} \|H_K(u)\|^{\frac{mq}{2}} \\
&= CN^{-\frac{q(2-m)}{2}} \left(\sum_K |K|\det(|H_K(u)|)^{\frac{1}{2}}\right)^{\frac{q(2-m)}{d}} \left(\sum_K |K|\cdot\|H_K(u)\|^{\frac{mq}{2}}\right) \qquad (5.183)
\end{aligned}
$$

and

$$
\begin{aligned}
E_{h.o.t.}(\mathscr{T}_h) \le CN^{-\frac{q(2-m)}{2}} &\left(\sum_K |K|\det(|H_K(u)|)^{\frac{1}{2}}\right)^{\frac{q(2-m)}{d}} \\
&\times \left(\sum_K |K| \|H_K(u)\|^{\frac{mq}{2}} \|H_K^{-1}(u)\|^q\right) \langle H(u)-H_K(u)\rangle_{L^p(K)}^q . \qquad (5.184)
\end{aligned}
$$

Thus,

$$
\begin{aligned}
&|u-\Pi_k u|_{W^{m,q}(\Omega)} \\
&\quad \le CN^{-\frac{(2-m)}{2}} \left(\sum_K |K|\det(|H_K(u)|)^{\frac{1}{2}}\right)^{\frac{(2-m)}{d}} \left[\left(\sum_K |K|\cdot\|H_K(u)\|^{\frac{mq}{2}}\right)^{\frac{1}{q}} \right. \\
&\qquad \left. + \left(\sum_K |K| \|H_K(u)\|^{\frac{mq}{2}} \|H_K^{-1}(u)\|^q\right)^{\frac{1}{q}} \langle H(u)-H_K(u)\rangle_{L^p(K)} \right] . \qquad (5.185)
\end{aligned}
$$

The $N \to \infty$ asymptotic behavior of this error bound has a similar analysis to that in the previous subsection and is not given here.

We can easily compare the bound (5.185) with the bound (5.152) for $E(\mathscr{T}_h)$ for the optimal monitor function defined in §5.2.3. For an M-uniform mesh associated with the optimal monitor function (5.143) or (5.148) for the case $\alpha_h \to 0$, $E(\mathscr{T}_h)$ is bounded by

$$E(\mathscr{T}_h) \le CN^{-\frac{q(2-m)}{d}} \left(\sum_K |K| \det(|H_K(u)|)^{\frac{q(2-m)}{2(d+q(2-m))}} \|H_K(u)\|^{\frac{mqd}{2(d+q(2-m))}} \right)^{\frac{d+q(2-m)}{d}} \tag{5.186}$$

From Hölder's inequality,

$$\left(\sum_K |K| \det(|H_K(u)|)^{\frac{q(2-m)}{2(d+q(2-m))}} \|H_K(u)\|^{\frac{mqd}{2(d+q(2-m))}} \right)^{\frac{d+q(2-m)}{d}} \le \left(\sum_K |K| \det(|H_K(u)|)^{\frac{1}{2}} \right)^{\frac{q(2-m)}{d}} \left(\sum_K |K| \cdot \|H_K(u)\|^{\frac{mq}{2}} \right). \tag{5.187}$$

Thus, the solution-dependent factor in bound (5.186) is smaller than that in bound (5.183). This can be seen more clearly for the case $m = 0$. In fact, for this case, (5.186) becomes

$$E(\mathscr{T}_h) \le CN^{-\frac{2q}{d}} \left(\sum_K |K| \det(|H_K(u)|)^{\frac{q}{d+2q}} \right)^{\frac{d+2q}{d}} \tag{5.188}$$

and (5.183) reduces to

$$E(\mathscr{T}_h) \le CN^{-\frac{2q}{d}} \left(\sum_K |K| \det(|H_K(u)|)^{\frac{1}{2}} \right)^{\frac{2q}{d}}, \tag{5.189}$$

with the latter involving a greater solution-dependent factor. This observation shows that the bound associated with the Hessian monitor function (5.182) contains a greater solution-dependent factor than that associated with the optimal monitor function (5.148). So while the Hessian monitor function provides the information necessary for specifying the shape and orientation of mesh elements, (5.182) is in this sense not optimal.

5.2.5 Summary of formulas – continuous form

For convenience we summarize the continuous form for the formulas for the optimal monitor function developed in this section. This form proves especially handy in the next chapter when we study the variational approach for generating adaptive meshes. The reader is referred back to Table 5.1 for the physical meanings of the parameters k, l, m, p, and q. For simplicity, the subscripts "*iso*", "*ani*, 1", and "*ani*, 2" are suppressed in the formulas.

The error bounds obtained in the previous subsections and adaptive meshes for these formulations of the monitor function are examined for a number of numerical examples in Chapter 6; e.g., see Examples 6.4.2, 6.4.3, 6.5.2, 6.5.3, 6.5.6, and 6.5.7.

(a) **The isotropic case.**

$$\begin{cases} M &= \left(1+\frac{1}{\alpha}\|D^l u\|_{l_p}\right)^{\frac{2q}{d+q(l-m)}} I, \\ \rho &= \left(1+\frac{1}{\alpha}\|D^l u\|_{l_p}\right)^{\frac{dq}{d+q(l-m)}}, \\ \alpha &= \left(\frac{1}{|\Omega|}\int_\Omega \|D^l u\|_{l_p}^{\frac{dq}{d+q(l-m)}} dx\right)^{\frac{d+q(l-m)}{dq}}. \end{cases} \tag{5.190}$$

(b) **The anisotropic case with $l = 1$ and $m = 0$.** A special case is piecewise constant interpolation with the error measured in the L^q norm.

$$\begin{cases} M &= \left(1+\frac{1}{\alpha}\|\nabla u\|^2\right)^{-\frac{1}{d+q}} \left[I+\frac{1}{\alpha}\nabla u \nabla u^T\right], \\ \rho &= \left(1+\frac{1}{\alpha}\|\nabla u\|^2\right)^{\frac{q}{2(d+q)}}, \\ \alpha &= \max\left\{\hat{\alpha}^2 \langle u\rangle^2_{L^p(\Omega)},\ \left(\frac{1}{|\Omega|}\int_\Omega \|\nabla u\|^{\frac{q}{d+q}} dx\right)^{\frac{2(d+q)}{q}}\right\}, \end{cases} \tag{5.191}$$

where $\hat{\alpha}$ is a given positive constant.

(c) **The anisotropic case with $l = 2$ and $m = 0$ or 1.** A special case is piecewise linear interpolation with the error measured in the L^q norm or H^1 semi-norm.

$$\begin{cases} M &= \det\left(I+\frac{1}{\alpha}|H(u)|\right)^{-\frac{1}{d+q(2-m)}} \left\|I+\frac{1}{\alpha}|H(u)|\right\|^{\frac{mq}{d+q(2-m)}} \left[I+\frac{1}{\alpha}|H(u)|\right], \\ \rho(\boldsymbol{x};\alpha) &= \det\left(I+\frac{1}{\alpha}|H(u)|\right)^{\frac{q(2-m)}{2(d+(2-m)q)}} \left\|I+\frac{1}{\alpha}|H(u)|\right\|^{\frac{mqd}{2(d+q(2-m))}}, \\ \tilde{\alpha}: & \int_\Omega \rho(\boldsymbol{x};\tilde{\alpha}) d\boldsymbol{x} = 2c_{\frac{dq}{d+q(2-m)}} |\Omega|, \end{cases} \tag{5.192}$$

$$\alpha = \max\left\{\hat{\alpha}\langle u\rangle_{L^p(\Omega)},\ \tilde{\alpha}\right\}, \tag{5.193}$$

where the constant $c_{\frac{dq}{d+q(2-m)}}$ is defined in (B.4) and $\hat{\alpha}$ is a given positive constant. Recall that in this anisotropic case, $\tilde{\alpha}$ is defined implicitly through an algebraic equation (cf. (5.192)).

(d) **A (non-optimal) anisotropic case.**

$$M = |H(u)|. \tag{5.194}$$

5.3 Computation of monitor functions

5.3.1 Recovery of solution derivatives

The monitor functions defined in the previous section involve solution derivatives. However, in most practical applications only approximations to the nodal values of the solution are known, so the problem arises of how to approximate derivatives in terms of these nodal values. In this subsection we study two Hessian (and gradient) recovery techniques, one based on least squares fitting (Zhang and Naga [353] and §2.5.1) and the other based on a Galerkin formulation (Dolejší [122]). Although the techniques can work in any d-dimensions, for simplicity we restrict our discussion to 2D. We further note that the hierarchical basis strategy discussed in §5.4.2 can also be used for recovery of the gradient and Hessian of the solution.

The least squares fitting method. Suppose that the solution values u_j, defined at the vertices of the mesh (or triangularization) $\mathscr{T}_h$, are known. For a vertex $\boldsymbol{x}_j = (x_j, y_j)$, let $\boldsymbol{x}_{j_i}$, $i = 1, ..., N_j$ (with $N_j \geq 6$ for the 6 basis functions for 2D quadratic polynomials) be the N_j neighboring vertices closest (in a connectivity sense) to $\boldsymbol{x}_j$, including $\boldsymbol{x}_j$ itself. Let $\hat{\boldsymbol{x}}_j$ be the center of these points, and define

$$H_j^x = \max_{i=1,...,N_j} |x_{j_i} - \hat{x}_j|, \quad H_j^y = \max_{i=1,...,N_j} |y_{j_i} - \hat{y}_j|.$$

Denote the first three Legendre polynomials by $P_0(x) = 1$, $P_1(x) = x$, and $P_2(x) = (2x^2 - 1)/2$. Then a quadratic polynomial in the form

$$q(x,y) = \sum_{k=0}^{2} \sum_{l=0}^{2-k} a_{k,l} P_k \left(\frac{x - \hat{x}_j}{H_j^x} \right) P_l \left(\frac{y - \hat{y}_j}{H_j^y} \right)$$

is determined by least squares fitting, i.e.,

$$\min_{a_{0,0},...,a_{1,1},...,a_{2,0}} \sum_{i=1}^{N_j} (q(x_{j_i}, y_{j_i}) - u_{j_i})^2.$$

The values of the first and second derivatives of this quadratic polynomial at $\boldsymbol{x}_j$ are used as approximations to those of u, i.e., $\nabla u(\boldsymbol{x}_j) \approx \nabla q(\boldsymbol{x}_j)$ and $H(u)(\boldsymbol{x}_j) \approx H(q)(\boldsymbol{x}_j)$.

The Galerkin formulation method. Let u^h be a piecewise linear approximation to u on $\mathscr{T}_h$. Consider an interior vertex $\boldsymbol{x}_j$ and denote by ϕ_j the linear basis function associated with $\boldsymbol{x}_j$. Using a lump sum and the divergence theorem, we have

$$\frac{\partial^2 u^h}{\partial x^2}(\boldsymbol{x}_j)\int_\Omega \phi_j d\boldsymbol{x} \approx \int_\Omega \frac{\partial^2 u^h}{\partial x^2}\phi_j d\boldsymbol{x} = -\int_\Omega \frac{\partial u^h}{\partial x}\frac{\partial \phi_j}{\partial x} d\boldsymbol{x},$$

so

$$\frac{\partial^2 u}{\partial x^2}(\boldsymbol{x}_j) \approx \frac{\partial^2 u^h}{\partial x^2}(\boldsymbol{x}_j) \approx -\frac{1}{\int_\Omega \phi_j d\boldsymbol{x}}\int_\Omega \frac{\partial u^h}{\partial x}\frac{\partial \phi_j}{\partial x} d\boldsymbol{x}. \tag{5.195}$$

Notice that the integrals on the right-hand side are well defined and easily computed. The other second derivatives can be computed similarly. Moreover, second derivatives at a boundary vertex can be computed similarly using integration by parts although the fact that ϕ_j does not vanish on $\partial\Omega$ must be taken into account.

It is easy to show that in 1D the approximation (5.195) reduces to

$$\frac{d^2 u}{dx^2}(x_j) \approx \frac{2}{x_{j+1}-x_{j-1}}\left(\frac{u_{j+1}-u_j}{x_{j+1}-x_j} - \frac{u_j - u_{j-1}}{x_j - x_{j-1}}\right). \tag{5.196}$$

In this sense the Galerkin formulation approximation to the Hessian matrix can be viewed as a multidimensional generalization of the 1D central finite difference approximation.

5.3.2 Computation of the absolute value of Hessian matrix

As seen in §5.2.5, the absolute value of the Hessian of the function u, $|H|$, is involved in the definition of the monitor function for the case $l = 2$. Recall from (5.6) and (5.7) that $|H|$ is defined as

$$|H| = Q\,\mathrm{diag}(|\lambda_1|,\ldots,|\lambda_d|)\,Q^T \tag{5.197}$$

provided that the eigen-decomposition of H is

$$H = Q\,\mathrm{diag}(\lambda_1,\ldots,\lambda_d)\,Q^T, \tag{5.198}$$

where Q is an orthogonal matrix and λ_i's are the eigenvalues of H. The objective of this subsection is to see how to compute $|H|$ from a given symmetric matrix

$H \in \mathbb{R}^{d\times d}$ through its eigen-decomposition (5.198). We restrict our discussion to the cases of mesh adaptation in two and three dimensions, i.e., $d = 2$ and $d = 3$.

For $d = 2$, write

$$H = \begin{bmatrix} h_{11} & h_{12} \\ h_{12} & h_{22} \end{bmatrix}. \tag{5.199}$$

When $h_{12} = 0$, H becomes diagonal and

$$|H| = \begin{bmatrix} |h_{11}| & 0 \\ 0 & |h_{22}| \end{bmatrix}. \tag{5.200}$$

For $h_{12} \neq 0$, the characteristic equation of H is

$$\lambda^2 - (h_{11}+h_{22})\lambda + (h_{11}h_{22} - h_{12}^2) = 0,$$

the eigenvalues are

$$\lambda_{1,2} = \frac{1}{2}\left(h_{11}+h_{22} \pm \sqrt{(h_{11}-h_{22})^2 + 4h_{12}^2}\right), \tag{5.201}$$

and the corresponding eigenvectors are given by

$$Q = \begin{bmatrix} \frac{h_{12}}{\sqrt{h_{12}^2+(h_{11}-\lambda_1)^2}} & \frac{\lambda_2-h_{22}}{\sqrt{h_{12}^2+(h_{22}-\lambda_2)^2}} \\ \frac{\lambda_1-h_{11}}{\sqrt{h_{12}^2+(h_{11}-\lambda_1)^2}} & \frac{h_{12}}{\sqrt{h_{12}^2+(h_{22}-\lambda_2)^2}} \end{bmatrix}. \tag{5.202}$$

Having obtained the eigenvalues and eigenvectors, we can easily compute $|H|$ from (5.197).

The situation for $d = 3$ and

$$H = \begin{bmatrix} h_{11} & h_{12} & h_{13} \\ h_{12} & h_{22} & h_{23} \\ h_{13} & h_{23} & h_{33} \end{bmatrix} \tag{5.203}$$

is considerably more complicated. If $h_{12}^2 + h_{13}^2 + h_{23}^2 = 0$, H becomes diagonal and $|H|$ is obtained by simply taking the absolute value of the diagonal entries, so we consider the case when $h_{12}^2 + h_{13}^2 + h_{23}^2 > 0$. It is not difficult to obtain the characteristic equation as

$$\lambda^3 + a\lambda^2 + b\lambda + c = 0, \tag{5.204}$$

where

$$\begin{cases} a = -(h_{11}+h_{22}+h_{33}), \\ b = \begin{vmatrix} h_{11} & h_{12} \\ h_{12} & h_{22} \end{vmatrix} + \begin{vmatrix} h_{11} & h_{13} \\ h_{13} & h_{33} \end{vmatrix} + \begin{vmatrix} h_{22} & h_{23} \\ h_{23} & h_{33} \end{vmatrix}, \\ c = -\det(H). \end{cases} \tag{5.205}$$

Using Cardano's transformation $\lambda = \mu - a/3$, we rewrite (5.204) as

$$\mu^3 + (b - \frac{1}{3}a^2)\mu + (\frac{2}{27}a^3 - \frac{1}{3}ab + c) = 0. \tag{5.206}$$

Note that

$$\begin{aligned} b - \frac{1}{3}a^2 &= \frac{1}{3}(h_{11}h_{22} + h_{11}h_{33} + h_{22}h_{33} - h_{11}^2 - h_{22}^2 - h_{33}^2) - h_{12}^2 - h_{13}^2 - h_{23}^2 \\ &\le -h_{12}^2 - h_{13}^2 - h_{23}^2 \\ &< 0. \end{aligned}$$

The roots of (5.206) have relatively simple form, which gives the eigenvalues

$$\begin{cases} \lambda_1 = 2\sqrt{-\frac{1}{3}(b - \frac{1}{3}a^2)}\cos(\frac{\theta}{3}) - \frac{1}{3}a, \\ \lambda_2 = 2\sqrt{-\frac{1}{3}(b - \frac{1}{3}a^2)}\cos(\frac{\theta+2\pi}{3}) - \frac{1}{3}a, \\ \lambda_3 = 2\sqrt{-\frac{1}{3}(b - \frac{1}{3}a^2)}\cos(\frac{\theta+4\pi}{3}) - \frac{1}{3}a, \end{cases} \tag{5.207}$$

where

$$\theta = \arccos\left(\frac{-\frac{1}{2}(\frac{2}{27}a^3 - \frac{1}{3}ab + c)}{\sqrt{-\left(\frac{1}{3}(b - \frac{1}{3}a^2)\right)^3}}\right).$$

To now compute Q, let

$$\boldsymbol{v} = \begin{bmatrix} h_{11} - \lambda_1 \\ h_{12} \\ h_{13} \end{bmatrix}.$$

If $\|\boldsymbol{v}\| = 0$, then

$$H - \lambda_1 I = \begin{bmatrix} 0 & 0 & 0 \\ 0 & h_{22} - \lambda_1 & h_{23} \\ 0 & h_{23} & h_{33} - \lambda_1 \end{bmatrix},$$

which is essentially the same as (5.199) and has eigenvalues 0, $\lambda_2 - \lambda_1$, and $\lambda_3 - \lambda_1$. Like the 2D case, we can find an orthogonal matrix Q such that

$$H - \lambda_1 I = Q \begin{bmatrix} 0 & 0 & 0 \\ 0 & \lambda_2 - \lambda_1 & 0 \\ 0 & 0 & \lambda_3 - \lambda_1 \end{bmatrix} Q^T,$$

and then

$$H = Q \begin{bmatrix} \lambda_1 & 0 & 0 \\ 0 & \lambda_2 & 0 \\ 0 & 0 & \lambda_3 \end{bmatrix} Q^T. \tag{5.208}$$

Otherwise, for the case $\|\boldsymbol{v}\| \neq 0$, we define a Householder matrix

$$Q_1 = I - \frac{2}{\|\boldsymbol{u}\|^2}\boldsymbol{u}\boldsymbol{u}^T,$$

where

$$\boldsymbol{u} = \boldsymbol{v} + \|\boldsymbol{v}\| \operatorname{sign}(h_{11} - \lambda_1) \begin{bmatrix} 1 \\ 0 \\ 0 \end{bmatrix}$$

and $\operatorname{sign}(x) = 1$ for $x \geq 0$ and -1 for $x < 0$. It follows that

$$Q_1\boldsymbol{v} = \boldsymbol{v} - \boldsymbol{u} = -\|\boldsymbol{v}\| \operatorname{sign}(h_{11} - \lambda_1) \begin{bmatrix} 1 \\ 0 \\ 0 \end{bmatrix}.$$

Then $Q_1(H - \lambda_1 I)$ has the form

$$Q_1(H - \lambda_1 I) = \begin{bmatrix} a_{11} & a_{12} & a_{13} \\ 0 & a_{22} & a_{23} \\ 0 & a_{32} & a_{33} \end{bmatrix},$$

where $a_{11} = -\|\boldsymbol{v}\| \operatorname{sign}(h_{11} - \lambda_1) \neq 0$. We now define

$$Q_2 = \begin{bmatrix} 1 & 0 & 0 \\ 0 & c_2 & s_2 \\ 0 & -s_2 & c_2 \end{bmatrix}$$

as follows: If $a_{22}^2 + a_{32}^2 > 0$, then

$$c_2 = \frac{a_{22}}{\sqrt{a_{22}^2 + a_{32}^2}}, \quad s_2 = \frac{a_{32}}{\sqrt{a_{22}^2 + a_{32}^2}},$$

and we have

$$Q_2Q_1(H - \lambda_1 I) = \begin{bmatrix} a_{11} & a_{12} & a_{13} \\ 0 & \sqrt{a_{22}^2 + a_{32}^2} & c_2a_{23} + s_2a_{33} \\ 0 & 0 & -s_2a_{23} + c_2a_{33} \end{bmatrix}.$$

Since $Q_2Q_1(H - \lambda_1 I)$ is singular, it must have the form

$$Q_2Q_1(H - \lambda_1 I) = \begin{bmatrix} a_{11} & a_{12} & a_{13} \\ 0 & \sqrt{a_{22}^2 + a_{32}^2} & c_2a_{23} + s_2a_{33} \\ 0 & 0 & 0 \end{bmatrix}.$$

If $a_{22}^2 + a_{32}^2 = 0$ but $a_{23}^2 + a_{33}^2 > 0$, we define

$$c_2 = \frac{a_{23}}{\sqrt{a_{23}^2 + a_{33}^2}}, \quad s_2 = \frac{a_{33}}{\sqrt{a_{23}^2 + a_{33}^2}}$$

and obtain

$$Q_2 Q_1 (H - \lambda_1 I) = \begin{bmatrix} a_{11} & a_{12} & a_{13} \\ 0 & 0 & \sqrt{a_{22}^2 + a_{32}^2} \\ 0 & 0 & 0 \end{bmatrix}.$$

If both $a_{22}^2 + a_{32}^2 = 0$ and $a_{23}^2 + a_{33}^2 = 0$, then we define

$$c_2 = 1, \quad s_2 = 0$$

and have

$$Q_2 Q_1 (H - \lambda_1 I) = \begin{bmatrix} a_{11} & a_{12} & a_{13} \\ 0 & 0 & 0 \\ 0 & 0 & 0 \end{bmatrix}.$$

Thus, for all three cases $Q_2 Q_1 (H - \lambda_1 I)$ has the form

$$Q_2 Q_1 (H - \lambda_1 I) = \begin{bmatrix} b_{11} & b_{12} & b_{13} \\ b_{21} & b_{22} & b_{23} \\ 0 & 0 & 0 \end{bmatrix}.$$

Direct calculation verifies that $Q_2 Q_1 (H - \lambda_1 I) Q_1^T Q_2^T$ has the same form, i.e.,

$$Q_2 Q_1 (H - \lambda_1 I) Q_1^T Q_2^T = \begin{bmatrix} c_{11} & c_{12} & c_{13} \\ c_{21} & c_{22} & c_{23} \\ 0 & 0 & 0 \end{bmatrix}.$$

By symmetry it follows that

$$Q_2 Q_1 (H - \lambda_1 I) Q_1^T Q_2^T = \begin{bmatrix} c_{11} & c_{12} & 0 \\ c_{21} & c_{22} & 0 \\ 0 & 0 & 0 \end{bmatrix},$$

which has eigenvalues 0, $\lambda_2 - \lambda_1$, and $\lambda_3 - \lambda_1$. Finally, choosing Q_3 such that

$$Q_3^T Q_2 Q_1 (H - \lambda_1 I) Q_1^T Q_2^T Q_3 = \begin{bmatrix} \lambda_2 - \lambda_1 & 0 & 0 \\ 0 & \lambda_3 - \lambda_1 & 0 \\ 0 & 0 & 0 \end{bmatrix}$$

and letting $Q = Q_1^T Q_2^T Q_3$, we obtain the eigen-decomposition (5.208), and then compute $|H|$ using (5.197).

5.3.3 Smoothing

As in 1D, a smoother monitor function in multidimensions generally leads to a smoother mesh. Once again, direct smoothing of the monitor function can be based on use of the Laplace operator in the computational coordinate (cf. §2.5.2), i.e.,

$$\begin{cases} (I-\beta^{-2}\Delta_{\boldsymbol{\xi}})\tilde{M} = M, & \text{in } \Omega_c \\ \frac{\partial M}{\partial \boldsymbol{n}} = 0, & \text{on } \partial\Omega_c \end{cases} \tag{5.209}$$

where β is a positive parameter, $\Delta_{\boldsymbol{\xi}}$ is the Laplace operator in $\boldsymbol{\xi}$, and M and $\tilde{M}$ are viewed as functions on the computational domain Ω_c under a continuous, global coordinate transformation $\boldsymbol{x} = \boldsymbol{x}(\boldsymbol{\xi}) : \Omega_c \to \Omega$ (cf. §4.3). Generally speaking, the scheme (5.209) is not economical, and a local approximation is often used instead. For example, for a 2D rectangular mesh, we can use (cf. (2.138))

$$\tilde{M}_{i,j} = \frac{\sum\limits_{k=i-p}^{i+p} \sum\limits_{l=j-p}^{j+p} \gamma^{|i-k|+|j-l|} M_{k,l}}{\sum\limits_{k=i-p}^{i+p} \sum\limits_{l=j-p}^{j+p} \gamma^{|i-k|+|j-l|}} \tag{5.210}$$

where $\gamma \in (0,1)$ and p is a given integer. A similar scheme can be defined for a general mesh $\mathscr{T}_h$. If ω_j is the patch of elements which contain the j-th vertex as a vertex, then the smoothed monitor function can be defined as

$$\tilde{M}(\boldsymbol{x}_j) = \frac{1}{|\omega_j|} \sum_{K\in\omega_j} \frac{|K|}{d+1} \sum_{i=1}^{d+1} M(\boldsymbol{x}_{K,i}), \tag{5.211}$$

where $\boldsymbol{x}_{K,i}$, $i = 1,...,d$ denote the vertices of K. A successful smoothing strategy can be to consecutively repeat scheme (5.210) or (5.211) three or four times for each step in actual computation.

5.3.4 Monitor functions for multicomponent solutions

The monitor function has thus far been considered for solutions having a single component. When the solution has multiple components, one can construct a monitor function based on some weighted norm of the solution vector, for instance, $\sum_i w_i u_i^2$ where w_i's are weights and u_i's are the solution components. Another simple but conservative way is to define $M = M(\boldsymbol{x})$ as an intersection of the monitor functions associated with the individual solution components. This idea is similar to that used in §2.5.3 for the 1D case (cf. (2.150)), but the calculation of an intersection of matrices is more complicated. A numerical method for approximating the intersection is given by Borouchaki et al. [54]. The interested reader is also referred to Van Dam [332] for discussion on this issue.

5.4 Monitor functions based on semi–a posteriori and a posteriori error estimates

Recall that the definition of the monitor functions in §5.2 involves derivatives of the exact solution. In practice, these solution derivatives are approximated from a computed solution (see §5.3), and in this sense the implementation of a mesh adaptation technique associated with the monitor functions is a posteriori. Still, it is often desirable to develop monitor functions directly based on a posteriori error estimates since special features of the underlying PDE can be built into the mechanism of mesh adaptation. A posteriori error estimation for PDEs is a large area of research, and in this section we study two such methods suitable for mesh adaptation, one based on residuals and edge jumps and the other based on hierarchical bases of a finite element solution.

For simplicity, the methods are described for a 2D model problem

$$\begin{cases} -\Delta u = f, & \text{in } \Omega \\ u = 0, & \text{on } \partial\Omega \end{cases} \tag{5.212}$$

where Ω is a polygonal domain and f is a given function, but it should be emphasized that they work for other elliptic PDEs in any dimensions. Recall that the weak formulation of (5.212) is to find $u \in H_0^1(\Omega)$ such that

$$B(u,v) = (f,v), \qquad \forall v \in H_0^1(\Omega) \tag{5.213}$$

where

$$B(u,v) = \int_\Omega \nabla u \cdot \nabla v d\boldsymbol{x}, \quad (f,v) = \int_\Omega f v d\boldsymbol{x}.$$

We consider the linear finite element solution of (5.212). Let $\{\mathscr{T}_h\}$ be an affine family of triangular meshes for Ω. A linear finite element approximation u^h on $\mathscr{T}_h$ is then defined as $u^h \in \mathscr{S}^h$ such that

$$B(u^h,v^h) = (f,v^h), \qquad \forall v^h \in \mathscr{S}^h \tag{5.214}$$

where the linear finite element space is defined as

$$\mathscr{S}^h = \left\{ v \in H_0^1(\Omega) \mid v|_K \circ F_K \in \hat{P} \equiv P_1, \ \forall K \in \mathscr{T}_h \right\}. \tag{5.215}$$

It is easy to show that the error $e^h = u - u^h$ satisfies the orthogonality condition

$$B(e^h, v^h) = 0 \quad \forall v^h \in \mathscr{S}^h \tag{5.216}$$

and the error equation

$$B(e^h, v) = (f, v) - B(u^h, v), \quad \forall v \in H_0^1(\Omega). \tag{5.217}$$

5.4.1 A semi–a posteriori method

The first method is based on a semi–a posteriori error bound. The derivation of the bound is similar to that for the a posteriori bound (2.264) in §2.9, except the current situation is more complicated. Let Π_1 be the P_K-interpolation operator associated with $\mathscr{S}^h$. From (5.216) and (5.217) it follows that, for any $v \in H_0^1(\Omega)$,

$$\begin{aligned} B(e^h, v) &= B(e^h, v - \Pi_1 v) \\ &= (f, v - \Pi_1 v) - B(u^h, v - \Pi_1 v) \\ &= \sum_K \int_K \Big(f(v - \Pi_1 v) - \nabla u^h \cdot \nabla (v - \Pi_1 v) \Big)\, d\boldsymbol{x}. \end{aligned}$$

Integrating by parts for the second term in the integral, we get

$$B(e^h, v) = \sum_K \int_K \Big(f + \Delta u^h \Big) (v - \Pi_1 v) d\boldsymbol{x} - \sum_K \int_{\partial K} \nabla u^h \cdot \boldsymbol{n}_K (v - \Pi_1 v) dS, \tag{5.218}$$

where $\boldsymbol{n}_K$ denotes the unit outward normal to ∂K. Define the residual r^h and edge jump R^h as

$$r^h(\boldsymbol{x}) = f + \Delta u^h, \quad \forall \boldsymbol{x} \in K,\ \forall K \in \mathscr{T}_h \tag{5.219}$$

$$R^h(\boldsymbol{x}) = \begin{cases} \nabla u^h|_K \cdot \boldsymbol{n}_K + u^h|_{K'} \cdot \boldsymbol{n}_{K'}, & \forall \boldsymbol{x} \in \gamma,\ \forall \gamma \in \partial \mathscr{T}_h \backslash \partial \Omega \\ 0 & \text{otherwise} \end{cases} \tag{5.220}$$

where $\partial \mathscr{T}_h \backslash \partial \Omega$ denotes the collection of all internal edges of $\mathscr{T}_h$ and K and K' are the elements sharing common edge γ. Then (5.218) can be written as

$$\begin{aligned} B(e^h, v) &= \sum_K \int_K r^h (v - \Pi_1 v) d\boldsymbol{x} - \sum_{\gamma \in \partial \mathscr{T}_h} \int_\gamma R^h (v - \Pi_1 v) dS \\ &= \sum_K \left(\int_K r^h (v - \Pi_1 v) d\boldsymbol{x} - \frac{1}{2} \sum_{\gamma \in \partial K} \int_\gamma R^h (v - \Pi_1 v) dS \right). \end{aligned}$$

Taking $v = e^h$ in the above equation, using Schwarz's inequality, and noticing that $e^h - \Pi_1 e^h = u - \Pi_1 u$, we obtain

$$
\begin{aligned}
|e^h|^2_{H^1(\Omega)} &= B(e^h, e^h) \\
&\le \sum_K \left(\|r^h\|_{L^2(K)} \|u - \Pi_1 u\|_{L^2(K)} + \frac{1}{2} \sum_{\gamma \in \partial K} \|R^h\|_{L^2(\gamma)} \|u - \Pi_1 u\|_{L^2(\gamma)} \right). \quad (5.221)
\end{aligned}
$$

Note that this bound is semi–a posteriori since it involves the residual r^h, edge jump R^h, and interpolation error $(u - \Pi_1 u)$ of the solution u. Once again, from Schwarz's inequality,

$$
\begin{aligned}
&\sum_{\gamma \in \partial K} \|R^h\|_{L^2(\gamma)} \|u - \Pi_1 u\|_{L^2(\gamma)} \\
&\qquad \le \left(\sum_{\gamma \in \partial K} |\gamma| \|R^h\|^2_{L^2(\gamma)} \right)^{\frac{1}{2}} \left(\sum_{\gamma \in \partial K} \frac{1}{|\gamma|} \|u - \Pi_1 u\|^2_{L^2(\gamma)} \right)^{\frac{1}{2}} \\
&\qquad \le \left(\sum_{\gamma \in \partial K} |\gamma|^{\frac{1}{2}} \|R^h\|_{L^2(\gamma)} \right) \left(\sum_{\gamma \in \partial K} \frac{1}{|\gamma|} \|u - \Pi_1 u\|^2_{L^2(\gamma)} \right)^{\frac{1}{2}}.
\end{aligned}
$$

Combining this with (5.221) and applying Theorems 5.1.5 and 5.1.6, we obtain

$$
\begin{aligned}
&|e^h|^2_{H^1(\Omega)} \\
&\le C \sum_K \left(\frac{1}{|K|^{\frac{1}{2}}} \|r^h\|_{L^2(K)} + \frac{1}{|K|} \sum_{\gamma \in \partial K} |\gamma|^{\frac{1}{2}} \|R^h\|_{L^2(\gamma)} \right) \\
&\qquad \times |K| \cdot \left[\frac{1}{|K|} \int_K \left(\mathrm{tr}\left((F_K')^T |H(u)| F_K' \right) \right)^2 d\boldsymbol{x} \right]^{\frac{1}{2}} \\
&\approx C \sum_K \left(\frac{1}{|K|^{\frac{1}{2}}} \|r^h\|_{L^2(K)} + \frac{1}{|K|} \sum_{\gamma \in \partial K} |\gamma|^{\frac{1}{2}} \|R^h\|_{L^2(\gamma)} \right) |K| \cdot \mathrm{tr}\left((F_K')^T |H_K(u)| F_K' \right).
\end{aligned}
$$

A regularization of this bound with a constant $\alpha_h > 0$ gives

$$
\begin{aligned}
|e^h|^2_{H^1(\Omega)} \lesssim C &\left(1 + \frac{1}{\alpha_h |K|^{\frac{1}{2}}} \|r^h\|_{L^2(K)} + \frac{1}{\alpha_h |K|} \sum_{\gamma \in \partial K} |\gamma|^{\frac{1}{2}} \|R^h\|_{L^2(\gamma)} \right) \\
&\times |K| \cdot \mathrm{tr}\left((F_K')^T \left[I + \frac{1}{\alpha_h} |H_K(u)| \right] F_K' \right). \quad (5.222)
\end{aligned}
$$

Following the same procedure used in §5.2, by minimizing this bound we obtain the optimal monitor function as

$$M_K = \left(1 + \frac{1}{\alpha_h |K|^{\frac{1}{2}}} \|r^h\|_{L^2(K)} + \frac{1}{\alpha_h |K|} \sum_{\gamma \in \partial K} |\gamma|^{\frac{1}{2}} \|R^h\|_{L^2(\gamma)} \right)^{\frac{2}{d+2}}$$
$$\times \det \left(I + \frac{1}{\alpha_h} |H_K(u)| \right)^{-\frac{1}{d+2}} \left[I + \frac{1}{\alpha_h} |H_K(u)| \right]. \tag{5.223}$$

The regularization parameter α_h can be chosen in the three ways discussed in §5.2.3.

This semi–a posteriori monitor function involves the Hessian matrix of the exact solution, although its actual computation requires recovery of the Hessian matrix from solution nodal approximations. The residual and the edge jump terms provide a mechanism to incorporate the structure of the underlying PDE into the mesh adaptation.

It is instructive to compare this monitor function with that based on interpolation error. Since $m = 1$ and $q = 2$, (5.143) gives

$$M_K^{ani,2} = \left\| I + \frac{1}{\alpha_h} |H_K(u)| \right\|^{\frac{2}{d+2}} \det \left(I + \frac{1}{\alpha_h} |H_K(u)| \right)^{-\frac{1}{d+2}}$$
$$\times \left[I + \frac{1}{\alpha_h} |H_K(u)| \right]. \tag{5.224}$$

Thus, the main difference lies in the first factor, which is the l^2 norm of the regularized Hessian matrix for $M_K^{ani,2}$ and a term involving r^h and R^h in (5.223).

5.4.2 A hierarchical basis method

Assume that a reconstruction procedure R_h is applied to the linear finite element solution u^h (e.g., see Huang, Kamenski, and Lang [182]), and assume that it satisfies the following two conditions:

(a) *Saturation condition.* There exists a constant $\beta \in (0,1)$ such that

$$|u - R^h u^h|_{H^1(\Omega)} \leq \beta |u - u^h|_{H^1(\Omega)}. \tag{5.225}$$

(b) For the P_K-interpolation operator Π_1 associated with $\mathcal{S}^h$, it holds that

$$\Pi_1 R_h v^h = v^h, \quad \forall v^h \in \mathcal{S}^h. \tag{5.226}$$

The condition (5.226) can be satisfied relatively easily, and below we describe a reconstruction based on hierarchical bases that satisfies this condition. On the other hand, the saturation condition implies that the reconstructed solution, $R_h u^h$, is better than u^h. While this appears to be a natural condition, it is known to be notoriously difficult to prove rigorously. The condition has been used by a number of researchers in convergence analysis of various adaptive finite element approximations; e.g., see in Dörfler and Nochetto [126], Dörfler [125], and Achchab et al. [1].

Under these conditions, it is straightforward to show that

$$|u-u^h|_{H^1(\Omega)} \leq \frac{1}{1-\beta}|R_h u^h - u^h|_{H^1(\Omega)} = \frac{1}{1-\beta}|R_h u^h - \Pi_1 R_h u^h|_{H^1(\Omega)}. \quad (5.227)$$

That is, the finite element error is bounded by the computable interpolation error of the reconstructed solution. If further we assume that $R_h u^h$ is piecewise quadratic, from Theorem 5.1.5 we have

$$|u-u^h|_{H^1(\Omega)} \leq \frac{C}{1-\beta}\left[\sum_K |K| \left(\text{tr}\left((F_K')^T |H_K(R_h u^h)| F_K'\right)\right)^2\right]^{\frac{1}{2}}. \quad (5.228)$$

Once again, regularizing this bound with $\alpha_h > 0$ and minimizing it, we obtain

$$\begin{aligned} M_K = & \left\| I + \frac{1}{\alpha_h}|H_K(R_h u^h)| \right\|^{\frac{2}{d+2}} \det\left(I + \frac{1}{\alpha_h}|H_K(R_h u^h)|\right)^{-\frac{1}{d+2}} \\ & \times \left[I + \frac{1}{\alpha_h}|H_K(R_h u^h)| \right] \end{aligned} \quad (5.229)$$

as the optimal monitor function. Notice that this monitor function is almost the same as (5.224) except that the solution u is now replaced with the reconstructed solution $R_h u^h$.

We can now describe a reconstruction procedure based on hierarchical basis approximation. For a given edge $\gamma \in \partial \mathscr{T}_h$, let the indexes of its two vertices be i and j and the corresponding linear basis functions be ϕ_i and ϕ_j, and define the edge bubble function associated with γ by $\psi_l = 4\phi_i\phi_j$. Note that ψ_l is piecewise quadratic, its support is $K \cup K'$ where K and K' are the elements sharing common edge γ, and it vanishes on all edges but γ. Let $\mathscr{W}^h$ be the linear span of the bubble functions associated with all internal edges of $\mathscr{T}_h$, and define $z^h \in \mathscr{W}^h$ by

$$B(z^h, v^h) = (f, v^h) - B(u^h, v^h), \quad \forall v^h \in \mathscr{W}^h. \quad (5.230)$$

The so-defined z^h can be viewed as a projection of the true error e^h onto the subspace $\mathscr{W}^h$ and thus an estimate to e^h. Once z_h is obtained, $R_h u^h$ is defined as

$$R_h u^h = u^h + z^h. \quad (5.231)$$

Since z^h vanishes at vertices, we have $\Pi_1 z^h = 0$, and thus R_h satisfies condition (5.226).

Despite the fact that (5.230) defines a global system so its solution can be costly, numerical results show that a few symmetric Gauss-Seidel iterations can be sufficient for producing an approximation to z^h good enough for the purpose of mesh adaptation [182].

5.5 Additional considerations for defining monitor functions

In addition to error estimates, monitor functions can also be designed based on geometric and physical considerations to meet special needs in practical computation. In this section we discuss two such situations.

5.5.1 *Monitor functions based on distance to interfaces*

When applying the MMPDE moving mesh method to phase change problems, Mackenzie and his collaborators [47, 247] employ a scalar-matrix-type monitor function

$$M(\boldsymbol{x}) = \left(1 + \frac{\mu_1}{\sqrt{1+\mu_2^2|\boldsymbol{x}-\boldsymbol{x}^*|^2}}\right) I, \tag{5.232}$$

where μ_1 and μ_2 are user-prescribed positive parameters, and $\boldsymbol{x}^*$ is the point on the (numerical estimate of the) phase front that is closest to $\boldsymbol{x}$. The purpose of this monitor function is to concentrate mesh points around the phase front. It is numerically shown that μ_1 controls the minimum mesh spacing while μ_2 controls the rate at which mesh clustering occurs.

5.5.2 *Monitor functions based on a reference mesh*

There are a variety of situations where one needs to define a monitor function based on an existing mesh. For example, in shape design [344] the shape changes from time to time, and once modified, a new mesh consistent with the new boundary has to be generated. One often wants the new mesh to be as close as possible to the old mesh. One way to do this is to first compute the monitor function based on the old mesh and then generate the new one using the equidistribution and alignment conditions (4.74) and (4.75) for the computed monitor function. Another situation where one wishes to construct a monitor function on a given mesh arises in the numerical solution of PDEs. An adaptive mesh of possibly poor quality can often be generated using a simple method such as the method of characteristics (e.g., see Fletcher [148], Finlayson [145], and Anderson [14]). An adaptive mesh having higher quality can then be generated using a variational method (as discussed in Chapter 6) for the monitor function computed using the mesh of poorer quality. A similar idea for generating adaptive meshes is used in the so-called reference Jacobian method described in the next chapter.

Without loss of generality, we show here how to construct a monitor function for an unstructured, affine reference mesh $\mathscr{T}_h$ on a d-dimensional domain Ω. The development is based on the condition (4.35):

$$(F_K^{'})^{-T}(F_K^{'})^{-1} = \left(\frac{\sigma_h}{N}\right)^{-\frac{2}{d}} M_K. \tag{5.233}$$

Recall that this condition is used in §4.1 to determine an M-uniform mesh for a given monitor function. We use it now for the converse: to determine M_K where $F_K^{'}$, associated with the given reference mesh $\mathscr{T}_h$, is known. Since (5.233) is invariant under a scaling transformation of M, i.e., $M \to cM$ for any nonzero constant c (cf. Theorem 4.1.2), M can only be determined from it up to a multiplicative constant. In particular,

$$M_{ref,K} = \theta (F_K^{'})^{-T}(F_K^{'})^{-1}, \quad \forall K \in \mathscr{T}_h \tag{5.234}$$

where θ is an arbitrary non-zero constant. However, since our ultimate goal is to generate a new mesh satisfying (5.233), we can simply choose $\theta = 1$ in (5.234) and thus assume that

$$M_{ref,K} = (F_K^{'})^{-T}(F_K^{'})^{-1}, \quad \forall K \in \mathscr{T}_h. \tag{5.235}$$

The nodal values of M can be calculated using volume averaging. Specifically, if ω_j is the collection of the elements which have the j-th vertex as one of their vertices, i.e., each ω_j is an element patch associated with the j-th vertex, then let

$$M_{ref,j} = \frac{\sum_{K\in\omega_j} |K| M_K}{\sum_{K\in\omega_j} |K|}. \tag{5.236}$$

Finally, $M = M_{ref}(\boldsymbol{x})$ can be defined as a piecewise linear function.

The monitor function for a structured mesh can be defined similarly. In continuous form, we simply let

$$M_{ref}(\boldsymbol{x}) = \boldsymbol{J}^{-T}\boldsymbol{J}^{-1}. \tag{5.237}$$

In other words, the monitor function is defined as the metric of the inverse coordinate transformation.

5.6 Biographical notes

The estimates for interpolation error in §5.1 and the corresponding optimal monitor functions have first been developed in [193]; also see [178, 179].

Beckett and Mackenzie [42] appear to be the first to use a global or integral definition of the adaptation intensity parameter α. It is extensively studied and extended to multi-dimensions in [175, 195]. The approach used in §5.2 for defining α_h is adopted from [175].

The monitor function (5.223) has also been obtained in [183] for finite element solution of variational problems by minimizing a bound on the variation of the underlying functional. The hierarchical basis a posteriori approach in §5.4.2 was developed in [182]. Cao et al. [83] study two a posteriori strategies based on elements and hierarchical bases for computing monitor functions in the form of scalar matrices. Tang [318] uses edge jumps to define monitor functions also in a scalar matrix form. A newer method of computing the monitor function from an edge-based a posteriori error estimate proposed by Agouzal et al. [7] deserves special attention, although it is unclear whether or not the computed monitor function is optimal in some sense. A number of heuristic strategies for choosing the monitor function (or the metric tensor) are described by Frey and George [150].

5.7 Exercises

1. Verify Lemma 5.1.1 for any diagonal matrix $S = \text{diag}(\lambda_1, ..., \lambda_d)$.
2. Show that the higher order terms in (5.11) and (5.14) vanish for any quadratic polynomial u on K.
3. Derive in detail the inequality (5.38).
4. Prove the second inequality in (5.23).
5. Complete the details of the proof of Theorem 5.1.4 to show (5.43).
6. Prove (5.51).
7. Prove that equality in (5.59) holds if the mesh satisfies (5.64) and (5.65).
8. Complete the proof of Lemma 5.2.1.
9. Prove (5.108).
10. Derive the asymptotic bound (5.136).
11. Derive asymptotic bounds (5.177) and (5.178).
12. Prove inequality (5.187).
13. Show that (5.195) reduces to (5.196) in 1D.

Chapter 6
Variational Mesh Adaptation Methods

In this chapter and the following one, we discuss the general mesh generation problem. The first main class of methods we consider are variational methods. They are applicable for either nonadaptive or adaptive mesh generation, and natural relationships between these two different goals are examined. While mesh generation ideas generally apply for either the static or dynamic case, we often limit discussion to static mesh generation since extending it to compute a dynamic mesh is straightforward in principle using the MMPDE approach discussed in §6.1.2 (see also see §2.3). Although variational methods have most commonly been used for finite difference computations for structured meshes, they can also be employed for unstructured mesh generation and adaptation (e.g., see Cao, Huang, and Russell [81]) and for mesh smoothing (e.g., see Canann et al. [79] and Knupp [215]).

The variational approach is motivated by the fact that it is a very natural way to formulate an elliptic mesh generation system which can incorporate mesh quality control into the mesh adaptation. An elliptic system is advantageous for mesh generation because it generally produces a mesh with desirable smoothness properties while allowing for specification of a complete boundary correspondence. This is in sharp contrast to algebraic and hyperbolic mesh generators. An algebraic method uses transfinite interpolation and generates a mesh which often lacks in smoothness, while a hyperbolic system involves hyperbolic PDEs which only allow for specification of boundary conditions on inflow boundary segments and can produce a non-smooth mesh if the boundary point distribution is non-smooth.

Variational methods have historically been the primary ones used for mesh generation, and a great many have been developed using a multitude of error based, geometric, physical, and other considerations. A number of them have had good success in applications, and they can often appear to give similar results when they work. As pointed out by Brackbill [57], "the marginal utility of using an adaptive grid over using a uniform grid is much greater than the marginal utility of using one method for adaptive gridding over another. It is almost always better to use an

W. Huang and R.D. Russell, *Adaptive Moving Mesh Methods*, Applied Mathematical Sciences 174, DOI 10.1007/978-1-4419-7196-2_6, © Springer Science+Business Media, LLC 2011

adaptive grid of any kind than to use none." On the other hand, methods can differ significantly in reliability and robustness, depending largely upon whether or not one can choose an appropriate monitor function. Making a suitable choice relies on an understanding of how the monitor function influences the resulting mesh properties and how the underlying method is related to the solution error properties. A major objective of this chapter is to understand some of these important issues for the most commonly used methods.

With the variational approach for mesh adaptation, the adaptive meshes are generated as images of a computational mesh under a coordinate transformation between the computational and physical domains. This coordinate transformation is determined as the minimizer of a so-called adaptation functional which is commonly designed to measure the difficulty in the numerical approximation of the physical solution, although it is often designed to have certain geometric mesh properties as well. Generally, a monitor function is used to control the mesh concentration, so a major focus is on studying how a given monitor function will effect the mesh for the various methods.

Recall from Chapter 4 that the effects of the monitor function on the behavior of an M-uniform mesh are characterized by the equidistribution and alignment conditions (cf. Theorem 4.2.3). Moreover, it is shown in Chapter 5 that optimal monitor functions can be chosen for M-uniform meshes based on error estimates or other considerations. It is thus natural to consider algorithms for generating M-uniform meshes for a given monitor function. This is the motivation for the development of the equidistribution-and-alignment based method in §6.4. However, the equidistribution and alignment conditions (as the basic principles for general mesh adaptation), along with their invariances under a scaling transformation of M (cf. Theorem 4.1.2) and under rotation, translation, and dilation transformations of $\hat{K}$ (cf. Theorem 4.2.2) or $\boldsymbol{\xi}$ (cf. Theorem 4.3.2), provide basic tools for use in understanding other methods as well.

6.1 General framework for variational methods and MMPDEs

With a variational method, the coordinate transformation needed for adaptive mesh generation is determined as the minimizer of an adaptation functional. Although different methods use different functionals, most of them can be cast in the same general form (see (6.11) below). Consequently, it is convenient to discuss some common issues before individual methods are constructed. In this spirit, the next few sections are devoted to the study of some general topics, such as Euler-Lagrange equations, mesh equations, boundary conditions, moving mesh PDEs (MMPDEs), existence of minimizers, and discretization and solution of mesh equations.

Like in Chapter 3, most discussion in this chapter is given for the 3D case. Recall that the formulas in 2D can be obtained by setting the third base vector to be the unit vector $\boldsymbol{a}_3 = \boldsymbol{a}^3 = [0,0,1]^T$ and dropping third components in 3D formulas; cf. §3.1.4.

We choose the approach of solving Euler-Lagrange equations for the minimizer of an adaptation functional. Alternatively, one could find the minimizer by first discretizing the functional and then solving the minimization problem directly; e.g., see Castillo [90] and Azarenok [19]. Caution should be taken for the latter approach since many discretizations such as a central finite difference discretization can cause problems for minimization, such as strong decoupling or loss of integral constraints satisfied by the underlying functional [90]. (These problems typically do not arise with the Euler-Lagrange approach.) A way to avoid these problems is to directly form the discrete objective function by mimicking the formulation of the continuous adaptation functional [91]. Regardless, the minimization problem can be solved globally or locally using a Gauss-Seidel-type Newton iteration or a preconditioned Krylov subspace method [288]. Direct minimization has also been used with other objective functions or with error bounds (e.g., see [36, 28, 330, 331, 329]), but the existence of the minimizers and their convexity properties are generally more difficult to analyze than when working with the functionals directly.

6.1.1 General adaptation functional and mesh equations

The adaptation functional can be formulated in terms of either the coordinate transformation $\boldsymbol{x} = \boldsymbol{x}(\boldsymbol{\xi}) : \Omega_c \to \Omega$ or the inverse coordinate transformation $\boldsymbol{\xi} = \boldsymbol{\xi}(\boldsymbol{x}) : \Omega \to \Omega_c$, where as before Ω_c and Ω are the computational and physical domains, respectively. The functional for $\boldsymbol{\xi} = \boldsymbol{\xi}(\boldsymbol{x})$ has a general form

$$I[\boldsymbol{\xi}] = \int_{\Omega} F(\nabla\boldsymbol{\xi}, \boldsymbol{\xi}, \boldsymbol{x}) d\boldsymbol{x} \tag{6.1}$$

while that for $\boldsymbol{x} = \boldsymbol{x}(\boldsymbol{\xi})$ has a form

$$\hat{I}[\boldsymbol{x}] = \int_{\Omega_c} \hat{F}(\nabla_{\boldsymbol{\xi}}\boldsymbol{x}, \boldsymbol{x}, \boldsymbol{\xi}) d\boldsymbol{\xi}, \tag{6.2}$$

where the Jacobian matrices

$$\nabla\boldsymbol{\xi} = \boldsymbol{J}^{-1} = \begin{bmatrix} \frac{\partial \xi_1}{\partial x_1} & \frac{\partial \xi_1}{\partial x_2} & \frac{\partial \xi_1}{\partial x_3} \\ \frac{\partial \xi_2}{\partial x_1} & \frac{\partial \xi_2}{\partial x_2} & \frac{\partial \xi_3}{\partial x_3} \\ \frac{\partial \xi_3}{\partial x_1} & \frac{\partial \xi_3}{\partial x_2} & \frac{\partial \xi_3}{\partial x_3} \end{bmatrix}, \quad \nabla_{\boldsymbol{\xi}}\boldsymbol{x} = \boldsymbol{J} = \begin{bmatrix} \frac{\partial x_1}{\partial \xi_1} & \frac{\partial x_1}{\partial \xi_2} & \frac{\partial x_1}{\partial \xi_3} \\ \frac{\partial x_2}{\partial \xi_1} & \frac{\partial x_2}{\partial \xi_2} & \frac{\partial x_2}{\partial \xi_3} \\ \frac{\partial x_3}{\partial \xi_1} & \frac{\partial x_3}{\partial \xi_2} & \frac{\partial x_3}{\partial \xi_3} \end{bmatrix}. \tag{6.3}$$

The determinant of $\boldsymbol{J}$, or simply the Jacobian, is $J = \det(\boldsymbol{J})$.

It is important to emphasize that the functionals are mathematically equivalent for suitable choices of the integrands F and $\hat{F}$. For example, by interchanging the roles of independent and dependent variables $\boldsymbol{x}$ and $\boldsymbol{\xi}$ in (6.1), from (6.3) one gets

$$\hat{I}[\boldsymbol{x}] = \int_{\Omega_c} F\left((\nabla_{\boldsymbol{\xi}}\boldsymbol{x})^{-1}, \boldsymbol{\xi}, \boldsymbol{x}\right) J \, d\boldsymbol{\xi}. \tag{6.4}$$

Thus, the functional $\hat{I}[\boldsymbol{x}]$ is equivalent to $I[\boldsymbol{\xi}]$ if its integrand is chosen as

$$\hat{F} = F\left((\nabla_{\boldsymbol{\xi}}\boldsymbol{x})^{-1}, \boldsymbol{\xi}, \boldsymbol{x}\right) J. \tag{6.5}$$

Interestingly, most of the existing variational methods have been developed in terms of the inverse coordinate transformation in (6.1). This partly originates from Winslow's early idea [341] of defining mesh lines so as to play the role of equipotentials in a potential problem. For example, a 2D mesh is formed by intersecting the "equipotentials" ξ = constant and η = constant with "potentials" satisfying Laplace's equations

$$\frac{\partial^2 \xi}{\partial x^2} + \frac{\partial^2 \xi}{\partial y^2} = 0, \quad \frac{\partial^2 \eta}{\partial x^2} + \frac{\partial^2 \eta}{\partial y^2} = 0. \tag{6.6}$$

It is easy to verify that these equations constitute the Euler-Lagrange equations for a functional of the type (6.1) in the form

$$I[\xi, \eta] = \int_{\Omega} \left[\left(\frac{\partial \xi}{\partial x}\right)^2 + \left(\frac{\partial \xi}{\partial y}\right)^2 + \left(\frac{\partial \eta}{\partial x}\right)^2 + \left(\frac{\partial \eta}{\partial y}\right)^2 \right] dxdy. \tag{6.7}$$

In practice, the roles of the independent and dependent variables in (6.6), (x, y) and (ξ, η), need to be interchanged because the node location is given by the mapping $x = x(\xi, \eta)$ and $y = y(\xi, \eta)$ instead of the inverse mapping. This can be shown to give the system

$$\begin{aligned}
&(x_\eta^2 + y_\eta^2) x_{\xi\xi} - 2(x_\xi x_\eta + y_\xi y_\eta) x_{\xi\eta} + (x_\xi^2 + y_\xi^2) x_{\eta\eta} = 0, \\
&(x_\eta^2 + y_\eta^2) y_{\xi\xi} - 2(x_\xi x_\eta + y_\xi y_\eta) y_{\xi\eta} + (x_\xi^2 + y_\xi^2) y_{\eta\eta} = 0.
\end{aligned} \tag{6.8}$$

Another reason for the popularity of (6.1) is the fact that the alternative system

$$\frac{\partial^2 x}{\partial \xi^2} + \frac{\partial^2 x}{\partial \eta^2} = 0, \quad \frac{\partial^2 y}{\partial \xi^2} + \frac{\partial^2 y}{\partial \eta^2} = 0, \tag{6.9}$$

which corresponds to the functional

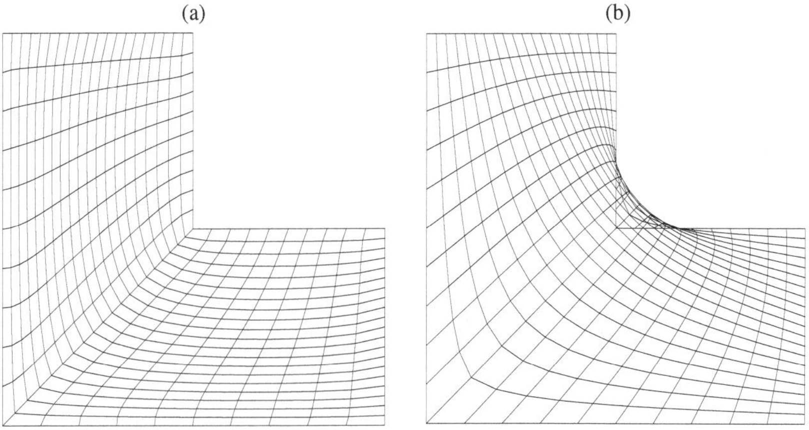

Fig. 6.1 Meshes generated using systems (a) (6.6) and (b) (6.9) for a given boundary correspondence. The computational domain Ω_c is taken as the unit square.

$$\hat{I}[x,y] = \int_{\Omega_c} \left[\left(\frac{\partial x}{\partial \xi}\right)^2 + \left(\frac{\partial x}{\partial \eta}\right)^2 + \left(\frac{\partial y}{\partial \xi}\right)^2 + \left(\frac{\partial y}{\partial \eta}\right)^2 \right] d\xi d\eta \tag{6.10}$$

of the type (6.2), can more easily result in a folded mesh than system (6.6) when the physical domain is concave (e.g., see Dvinsky [129] for explanation of this along with some specific examples). Figure 6.1 shows meshes generated using (6.6) (or (6.8)) and using (6.9). Note that the mesh generated using (6.9) is folded.

The functionals (6.7) and (6.10) are of course not equivalent, and the price paid for the more robust formulation is that (6.8) is much more complicated to solve than (6.9).

Following convention we restrict our discussion to functionals of the form (6.1), although functionals of the forms (6.1) and (6.2) can easily be transformed into each other. Rewrite (6.1) in the slightly different form

$$I[\boldsymbol{\xi}] = \int_{\Omega} F(\boldsymbol{a}^1, \boldsymbol{a}^2, \boldsymbol{a}^3, J, \boldsymbol{x}) d\boldsymbol{x}, \tag{6.11}$$

where the functional depends explicitly upon the base vectors $\boldsymbol{a}^i = \nabla \xi_i$, the Jacobian J, and the variable $\boldsymbol{x}$. As we shall see, the dependence upon $\boldsymbol{x}$ often arises through the monitor function $M = M(\boldsymbol{x})$.

To derive the Euler-Lagrange equation for the functional, recall that

$$J = \boldsymbol{a}_1 \cdot (\boldsymbol{a}_2 \times \boldsymbol{a}_3), \quad J^{-1} = \boldsymbol{a}^1 \cdot (\boldsymbol{a}^2 \times \boldsymbol{a}^3).$$

Letting δ be the variation operator and using the relation (3.7), we have

$$\begin{aligned}
\delta J^{-1} &= (\delta \boldsymbol{a}^1)\cdot(\boldsymbol{a}^2\times\boldsymbol{a}^3)+(\delta \boldsymbol{a}^2)\cdot(\boldsymbol{a}^3\times\boldsymbol{a}^1)+(\delta \boldsymbol{a}^3)\cdot(\boldsymbol{a}^1\times\boldsymbol{a}^2)\\
&= \sum_{i=1}^{3}(\delta \boldsymbol{a}^i)\cdot(\boldsymbol{a}^j\times\boldsymbol{a}^k) \qquad (i,j,k) \text{ cyclic}\\
&= \frac{1}{J}\sum_{i=1}^{3}\boldsymbol{a}_i\cdot\delta \boldsymbol{a}^i.
\end{aligned}$$

Thus, the first variation of the functional (6.11) has the form

$$\begin{aligned}
\delta I[\boldsymbol{\xi}] &= \int_\Omega \left[\sum_{i=1}^{3}\frac{\partial F}{\partial \boldsymbol{a}^i}\cdot\delta \boldsymbol{a}^i + \frac{\partial F}{\partial J}\delta J\right] d\boldsymbol{x}\\
&= \int_\Omega \left[\sum_{i=1}^{3}\frac{\partial F}{\partial \boldsymbol{a}^i}\cdot\delta \boldsymbol{a}^i - J^2\frac{\partial F}{\partial J}\delta J^{-1}\right] d\boldsymbol{x}\\
&= \int_\Omega \sum_{i=1}^{3}\left[\frac{\partial F}{\partial \boldsymbol{a}^i} - J\frac{\partial F}{\partial J}\boldsymbol{a}_i\right]\cdot\delta \boldsymbol{a}^i d\boldsymbol{x}\\
&= \int_\Omega \sum_{i=1}^{3}\left[\frac{\partial F}{\partial \boldsymbol{a}^i} - J\frac{\partial F}{\partial J}\boldsymbol{a}_i\right]\cdot\nabla\delta\xi_i d\boldsymbol{x}\\
&= -\int_\Omega \sum_{i=1}^{3}\nabla\cdot\left[\frac{\partial F}{\partial \boldsymbol{a}^i} - J\frac{\partial F}{\partial J}\boldsymbol{a}_i\right]\delta\xi_i d\boldsymbol{x}\\
&\quad + \int_{\partial\Omega} \sum_{i=1}^{3}\boldsymbol{n}\cdot\left[\frac{\partial F}{\partial \boldsymbol{a}^i} - J\frac{\partial F}{\partial J}\boldsymbol{a}_i\right]\delta\xi_i dS, \qquad (6.12)
\end{aligned}$$

where Gauss' theorem has been used in the final step. Setting the first variation to be zero for all admissible $\delta\xi_i$, we obtain the Euler-Lagrange equation as

$$-\nabla\cdot\left[\frac{\partial F}{\partial \boldsymbol{a}^i} - J\frac{\partial F}{\partial J}\boldsymbol{a}_i\right] = 0, \quad i=1,2,3. \qquad (6.13)$$

The unknown function in (6.13) is the inverse coordinate transformation $\boldsymbol{\xi} = \boldsymbol{\xi}(\boldsymbol{x})$, so to obtain a mesh equation in terms of the coordinate transformation $\boldsymbol{x} = \boldsymbol{x}(\boldsymbol{\xi})$, which gives the location of the mesh nodes directly (see also §3.1 and (3.38)), we need to interchange the roles of dependent and independent variables.

From (3.9), the mesh equation (6.13) in conservative form is

$$-\frac{1}{J}\sum_j\frac{\partial}{\partial\xi_j}\left[J\boldsymbol{a}^j\cdot\frac{\partial F}{\partial \boldsymbol{a}^i} - J^2\frac{\partial F}{\partial J}\boldsymbol{a}^j\cdot\boldsymbol{a}_i\right] = 0, \quad i=1,2,3 \qquad (6.14)$$

where $\boldsymbol{a}^i$'s are viewed as functions of $\boldsymbol{a}_j$'s through the relations (3.7). Since $\boldsymbol{a}^j\cdot\boldsymbol{a}_i = \delta_{ij}$ (again by (3.7)), this equation simplifies to

$$-\frac{1}{J}\sum_j \frac{\partial}{\partial \xi_j}\left(J\boldsymbol{a}^j \cdot \frac{\partial F}{\partial \boldsymbol{a}^i}\right) + \frac{1}{J}\frac{\partial}{\partial \xi_i}\left(J^2 \frac{\partial F}{\partial J}\right) = 0, \quad i = 1,2,3. \tag{6.15}$$

To derive the non-conservative form of the mesh equation, use (3.8) for ∇ in the first term and (3.9) in the second term of (6.13), and in a similar way as for (6.15) obtain that

$$-\sum_j \boldsymbol{a}^j \cdot \frac{\partial}{\partial \xi_j}\frac{\partial F}{\partial \boldsymbol{a}^i} + \frac{1}{J}\frac{\partial}{\partial \xi_i}\left(J^2 \frac{\partial F}{\partial J}\right) = 0, \quad i = 1,2,3. \tag{6.16}$$

Note that

$$\frac{\partial}{\partial \xi_j}\frac{\partial F}{\partial \boldsymbol{a}^i} = \sum_k \frac{\partial^2 F}{\partial \boldsymbol{a}^i \partial \boldsymbol{a}^k}\frac{\partial \boldsymbol{a}^k}{\partial \xi_j} + \frac{\partial^2 F}{\partial \boldsymbol{a}^i \partial J}\frac{\partial J}{\partial \xi_j} + \frac{\partial^2 F}{\partial \boldsymbol{a}^i \partial \boldsymbol{x}}\frac{\partial \boldsymbol{x}}{\partial \xi_j}, \tag{6.17}$$

where the 3×3 matrices $\frac{\partial^2 F}{\partial \boldsymbol{a}^i \partial \boldsymbol{a}^k}$ and $\frac{\partial^2 F}{\partial \boldsymbol{a}^i \partial \boldsymbol{x}}$, which are not necessarily symmetric, are defined as

$$\left(\frac{\partial^2 F}{\partial \boldsymbol{a}^i \partial \boldsymbol{a}^k}\right)_{(m,n)} = \frac{\partial^2 F}{\partial (\boldsymbol{a}^i)_m \partial (\boldsymbol{a}^k)_n}, \qquad \left(\frac{\partial^2 F}{\partial \boldsymbol{a}^i \partial \boldsymbol{x}}\right)_{(m,n)} = \frac{\partial^2 F}{\partial (\boldsymbol{a}^i)_m \partial x_n}. \tag{6.18}$$

Here, $(\boldsymbol{a}^i)_m$ denotes the m-th component of $\boldsymbol{a}^i$ and x_n the n-th component of $\boldsymbol{x}$.

Completing the derivation requires the two identities

$$\frac{\partial J}{\partial \xi_l} = \sum_i J\boldsymbol{a}^i \cdot \frac{\partial \boldsymbol{a}_i}{\partial \xi_l}, \tag{6.19}$$

$$\frac{\partial \boldsymbol{a}^i}{\partial \xi_l} = -\sum_s \left(\boldsymbol{a}^i \cdot \frac{\partial \boldsymbol{a}_s}{\partial \xi_l}\right)\boldsymbol{a}^s. \tag{6.20}$$

These follow from

$$\begin{aligned}
\frac{\partial J}{\partial \xi_l} &= \frac{\partial}{\partial \xi_l}\boldsymbol{a}_1 \cdot (\boldsymbol{a}_2 \times \boldsymbol{a}_3) \\
&= \sum_i \frac{\partial \boldsymbol{a}_i}{\partial \xi_l} \cdot (\boldsymbol{a}_j \times \boldsymbol{a}_k) \qquad (i,j,k) \text{ cyclic} \\
&= \sum_i J\boldsymbol{a}^i \cdot \frac{\partial \boldsymbol{a}_i}{\partial \xi_l}
\end{aligned}$$

and, from (6.19),

$$
\begin{aligned}
\frac{\partial \boldsymbol{a}^i}{\partial \xi_l} &= \frac{\partial}{\partial \xi_l}\left(\frac{1}{J}\boldsymbol{a}_j \times \boldsymbol{a}_k\right) \qquad (i,j,k) \text{ cyclic} \\
&= -\frac{1}{J^2}\left(\boldsymbol{a}_j \times \boldsymbol{a}_k\right)\frac{\partial J}{\partial \xi_l} + \frac{1}{J}\frac{\partial \boldsymbol{a}_j}{\partial \xi_l} \times \boldsymbol{a}_k + \frac{1}{J}\boldsymbol{a}_j \times \frac{\partial \boldsymbol{a}_k}{\partial \xi_l} \\
&= -\frac{1}{J}\boldsymbol{a}^i \frac{\partial J}{\partial \xi_l} + \frac{\partial \boldsymbol{a}_j}{\partial \xi_l} \times (\boldsymbol{a}^i \times \boldsymbol{a}^j) + \left(\boldsymbol{a}^k \times \boldsymbol{a}^i\right) \times \frac{\partial \boldsymbol{a}_k}{\partial \xi_l} \\
&= -\boldsymbol{a}^i \left[\left(\boldsymbol{a}^i \cdot \frac{\partial \boldsymbol{a}_i}{\partial \xi_l}\right) + \left(\boldsymbol{a}^j \cdot \frac{\partial \boldsymbol{a}_j}{\partial \xi_l}\right) + \left(\boldsymbol{a}^k \cdot \frac{\partial \boldsymbol{a}_k}{\partial \xi_l}\right)\right] \\
&\quad + \left[\left(\boldsymbol{a}^j \cdot \frac{\partial \boldsymbol{a}_j}{\partial \xi_l}\right)\boldsymbol{a}^i - \left(\boldsymbol{a}^i \cdot \frac{\partial \boldsymbol{a}_j}{\partial \xi_l}\right)\boldsymbol{a}^j\right] \\
&\quad + \left[\left(\boldsymbol{a}^k \cdot \frac{\partial \boldsymbol{a}_k}{\partial \xi_l}\right)\boldsymbol{a}^i - \left(\boldsymbol{a}^i \cdot \frac{\partial \boldsymbol{a}_k}{\partial \xi_l}\right)\boldsymbol{a}^k\right] \\
&= -\left(\boldsymbol{a}^i \cdot \frac{\partial \boldsymbol{a}_i}{\partial \xi_l}\right)\boldsymbol{a}^i - \left(\boldsymbol{a}^i \cdot \frac{\partial \boldsymbol{a}_j}{\partial \xi_l}\right)\boldsymbol{a}^j - \left(\boldsymbol{a}^i \cdot \frac{\partial \boldsymbol{a}_k}{\partial \xi_l}\right)\boldsymbol{a}^k \\
&= -\sum_s \left(\boldsymbol{a}^i \cdot \frac{\partial \boldsymbol{a}_s}{\partial \xi_l}\right)\boldsymbol{a}^s.
\end{aligned}
$$

Substituting (6.19) and (6.20) into (6.17) gives

$$
\begin{aligned}
\frac{\partial}{\partial \xi_j}\frac{\partial F}{\partial \boldsymbol{a}^i} &= -\sum_{k,s} \frac{\partial^2 F}{\partial \boldsymbol{a}^i \partial \boldsymbol{a}^k}\boldsymbol{a}^s \left(\boldsymbol{a}^k \cdot \frac{\partial \boldsymbol{a}_s}{\partial \xi_j}\right) \\
&\quad + J\sum_s \frac{\partial^2 F}{\partial \boldsymbol{a}^i \partial J}\left(\boldsymbol{a}^s \cdot \frac{\partial \boldsymbol{a}_s}{\partial \xi_j}\right) + \frac{\partial^2 F}{\partial \boldsymbol{a}^i \partial \boldsymbol{x}}\frac{\partial \boldsymbol{x}}{\partial \xi_j}.
\end{aligned} \tag{6.21}
$$

Similarly, for the second term in (6.16) we have

$$
\begin{aligned}
\frac{1}{J}\frac{\partial}{\partial \xi_i}\left(J^2 \frac{\partial F}{\partial J}\right) &= J^2\left(\frac{2}{J}\frac{\partial F}{\partial J} + \frac{\partial^2 F}{\partial J^2}\right)\sum_s \left(\boldsymbol{a}^s \cdot \frac{\partial \boldsymbol{a}_s}{\partial \xi_i}\right) \\
&\quad - J\sum_{k,s}\left(\frac{\partial^2 F}{\partial J \partial \boldsymbol{a}^k} \cdot \boldsymbol{a}^s\right)\left(\boldsymbol{a}^k \cdot \frac{\partial \boldsymbol{a}_s}{\partial \xi_i}\right) + J\frac{\partial^2 F}{\partial J \partial \boldsymbol{x}} \cdot \frac{\partial \boldsymbol{x}}{\partial \xi_i}.
\end{aligned} \tag{6.22}
$$

Inserting (6.21) and (6.22) into (6.16) and recalling that $\boldsymbol{a}_s = \partial \boldsymbol{x}/\partial \xi_s$, we obtain

$$
\begin{aligned}
&\sum_{j,s}\left[\sum_k\left((\boldsymbol{a}^j)^T\frac{\partial^2 F}{\partial \boldsymbol{a}^i\partial \boldsymbol{a}^k}\boldsymbol{a}^s\right)(\boldsymbol{a}^k)^T - J\left((\boldsymbol{a}^j)^T\frac{\partial^2 F}{\partial \boldsymbol{a}^i\partial J}\right)(\boldsymbol{a}^s)^T\right]\frac{\partial^2 \boldsymbol{x}}{\partial \xi_j\partial \xi_s}\\
&\quad -\sum_j\left((\boldsymbol{a}^j)^T\frac{\partial^2 F}{\partial \boldsymbol{a}^i\partial \boldsymbol{x}}\right)\frac{\partial \boldsymbol{x}}{\partial \xi_j}\\
&\quad +\sum_s\left[-J\sum_k\left((\boldsymbol{a}^s)^T\frac{\partial^2 F}{\partial J\partial \boldsymbol{a}^k}\right)(\boldsymbol{a}^k)^T + J^2\left(\frac{2}{J}\frac{\partial F}{\partial J}+\frac{\partial^2 F}{\partial J^2}\right)(\boldsymbol{a}^s)^T\right]\frac{\partial^2 \boldsymbol{x}}{\partial \xi_i\partial \xi_s}\\
&\quad +J\left(\frac{\partial^2 F}{\partial J\partial \boldsymbol{x}}\right)^T\frac{\partial \boldsymbol{x}}{\partial \xi_i} = 0, \qquad i=1,2,3. \qquad (6.23)
\end{aligned}
$$

These equations can be combined into a single vector equation by multiplying (6.23) by $\boldsymbol{a}_i$ and summing over i, giving

$$
\begin{aligned}
&\sum_{j,s}\left[\sum_{k,i}\left((\boldsymbol{a}^j)^T\frac{\partial^2 F}{\partial \boldsymbol{a}^i\partial \boldsymbol{a}^k}\boldsymbol{a}^s\right)\boldsymbol{a}_i(\boldsymbol{a}^k)^T - J\sum_i\left((\boldsymbol{a}^j)^T\frac{\partial^2 F}{\partial \boldsymbol{a}^i\partial J}\right)\boldsymbol{a}_i(\boldsymbol{a}^s)^T\right]\frac{\partial^2 \boldsymbol{x}}{\partial \xi_j\partial \xi_s}\\
&\quad -\sum_j\left[\sum_i \boldsymbol{a}_i(\boldsymbol{a}^j)^T\frac{\partial^2 F}{\partial \boldsymbol{a}^i\partial \boldsymbol{x}}\right]\frac{\partial \boldsymbol{x}}{\partial \xi_j}\\
&\quad +\sum_{i,s}\left[-J\sum_k\left((\boldsymbol{a}^s)^T\frac{\partial^2 F}{\partial J\partial \boldsymbol{a}^k}\right)\boldsymbol{a}_i(\boldsymbol{a}^k)^T + J^2\left(\frac{2}{J}\frac{\partial F}{\partial J}+\frac{\partial^2 F}{\partial J^2}\right)\boldsymbol{a}_i(\boldsymbol{a}^s)^T\right]\frac{\partial^2 \boldsymbol{x}}{\partial \xi_i\partial \xi_s}\\
&\quad +J\sum_i \boldsymbol{a}_i\left(\frac{\partial^2 F}{\partial J\partial \boldsymbol{x}}\right)^T\frac{\partial \boldsymbol{x}}{\partial \xi_i} = 0. \qquad (6.24)
\end{aligned}
$$

Finally, since

$$
\begin{aligned}
\sum_j\left[\sum_i \boldsymbol{a}_i(\boldsymbol{a}^j)^T\frac{\partial^2 F}{\partial \boldsymbol{a}^i\partial \boldsymbol{x}}\right]\frac{\partial \boldsymbol{x}}{\partial \xi_j} &= \sum_{j,i}\boldsymbol{a}_i(\boldsymbol{a}^j)^T\frac{\partial^2 F}{\partial \boldsymbol{a}^i\partial \boldsymbol{x}}\boldsymbol{a}_j\\
&= \sum_{j,i}\boldsymbol{a}_i\left((\boldsymbol{a}^j)^T\frac{\partial^2 F}{\partial \boldsymbol{a}^i\partial \boldsymbol{x}}\boldsymbol{a}_j\right)\\
&= \sum_i\sum_j\left((\boldsymbol{a}^j)^T\frac{\partial^2 F}{\partial \boldsymbol{a}^i\partial \boldsymbol{x}}\boldsymbol{a}_j\right)\frac{\partial \boldsymbol{x}}{\partial \xi_i},
\end{aligned}
$$

by rearranging the indices we can simplify (6.24) to obtain the mesh equation

$$
\begin{aligned}
&\sum_{i,j}\left[\sum_{k,s}\left((\boldsymbol{a}^i)^T\frac{\partial^2 F}{\partial \boldsymbol{a}^s \partial \boldsymbol{a}^k}\boldsymbol{a}^j\right)\boldsymbol{a}_s(\boldsymbol{a}^k)^T - J\sum_s\left((\boldsymbol{a}^i)^T\frac{\partial^2 F}{\partial \boldsymbol{a}^s \partial J}\right)\boldsymbol{a}_s(\boldsymbol{a}^j)^T\right]\frac{\partial^2 \boldsymbol{x}}{\partial \xi_i \partial \xi_j}\\
&\quad -\sum_i\sum_k\left((\boldsymbol{a}^k)^T\frac{\partial^2 F}{\partial \boldsymbol{a}^i \partial \boldsymbol{x}}\boldsymbol{a}_k\right)\frac{\partial \boldsymbol{x}}{\partial \xi_i}\\
&\quad +\sum_{i,j}\left[-J\sum_k\left((\boldsymbol{a}^j)^T\frac{\partial^2 F}{\partial J \partial \boldsymbol{a}^k}\right)\boldsymbol{a}_i(\boldsymbol{a}^k)^T + J^2\left(\frac{2}{J}\frac{\partial F}{\partial J}+\frac{\partial^2 F}{\partial J^2}\right)\boldsymbol{a}_i(\boldsymbol{a}^j)^T\right]\frac{\partial^2 \boldsymbol{x}}{\partial \xi_i \partial \xi_j}\\
&\quad +J\sum_i \boldsymbol{a}_i\left(\frac{\partial^2 F}{\partial J \partial \boldsymbol{x}}\right)^T\frac{\partial \boldsymbol{x}}{\partial \xi_i} = 0.
\end{aligned}
\tag{6.25}
$$

This non-conservative equation can be written in the more concise form

$$
\sum_{i,j} A_{i,j}\frac{\partial^2 \boldsymbol{x}}{\partial \xi_i \partial \xi_j} + \sum_i B_i \frac{\partial \boldsymbol{x}}{\partial \xi_i} = 0, \tag{6.26}
$$

where the coefficient matrices

$$
\begin{aligned}
A_{i,j} &= \sum_{k,s}\left((\boldsymbol{a}^i)^T\frac{\partial^2 F}{\partial \boldsymbol{a}^s \partial \boldsymbol{a}^k}\boldsymbol{a}^j\right)\boldsymbol{a}_s(\boldsymbol{a}^k)^T - J\sum_s\left((\boldsymbol{a}^i)^T\frac{\partial^2 F}{\partial \boldsymbol{a}^s \partial J}\right)\boldsymbol{a}_s(\boldsymbol{a}^j)^T\\
&\quad -J\sum_k\left((\boldsymbol{a}^j)^T\frac{\partial^2 F}{\partial J \partial \boldsymbol{a}^k}\right)\boldsymbol{a}_i(\boldsymbol{a}^k)^T + J^2\left(\frac{2}{J}\frac{\partial F}{\partial J}+\frac{\partial^2 F}{\partial J^2}\right)\boldsymbol{a}_i(\boldsymbol{a}^j)^T,\\
B_i &= -\sum_k\left((\boldsymbol{a}^k)^T\frac{\partial^2 F}{\partial \boldsymbol{a}^i \partial \boldsymbol{x}}\boldsymbol{a}_k\right) + J\boldsymbol{a}_i\left(\frac{\partial^2 F}{\partial J \partial \boldsymbol{x}}\right)^T.
\end{aligned}
\tag{6.27}
$$

A Special Case: We derive here the form of this mesh equation for the important special case where F has the simpler form

$$
F(\boldsymbol{a}^1,\boldsymbol{a}^2,\boldsymbol{a}^3,J,\boldsymbol{x}) = F_1(\rho,\beta)+F_2(\rho,J), \tag{6.28}
$$

where $\rho(\boldsymbol{x}) = \sqrt{\det(M(\boldsymbol{x}))}$ is the mesh density function associated with a given monitor function $M = M(\boldsymbol{x})$ and

$$
\beta = \sum_i (\boldsymbol{a}^i)^T M^{-1}\boldsymbol{a}^i = \sum_i (\nabla \xi_i)^T M^{-1}\nabla \xi_i. \tag{6.29}
$$

As we see later in this chapter, this form characterizes many of the popular adaptive variational methods, with these terms naturally appearing in a functional to represent the equidistribution and alignment conditions (4.74) and (4.75). Since

$$
\frac{\partial \beta}{\partial \boldsymbol{a}^s} = 2M^{-1}\boldsymbol{a}^s,
$$

we have

$$\frac{\partial F}{\partial \boldsymbol{a}^s} = 2\frac{\partial F_1}{\partial \beta} M^{-1}\boldsymbol{a}^s,$$
$$\frac{\partial^2 F}{\partial \boldsymbol{a}^s \partial \boldsymbol{a}^k} = 4\frac{\partial^2 F_1}{\partial \beta^2}\left(M^{-1}\boldsymbol{a}^s\right)\left(M^{-1}\boldsymbol{a}^k\right)^T + 2\frac{\partial F_1}{\partial \beta} M^{-1}\delta_{sk}.$$

Thus, the first term of $A_{i,j}$ is

$$\begin{aligned}
&\sum_{k,s}\left((\boldsymbol{a}^i)^T \frac{\partial^2 F}{\partial \boldsymbol{a}^s \partial \boldsymbol{a}^k}\boldsymbol{a}^j\right)\boldsymbol{a}_s(\boldsymbol{a}^k)^T \\
&= 4\frac{\partial^2 F_1}{\partial \beta^2}\sum_{k,s}\left((\boldsymbol{a}^i)^T\left(M^{-1}\boldsymbol{a}^s\right)\left(M^{-1}\boldsymbol{a}^k\right)^T\boldsymbol{a}^j\right)\boldsymbol{a}_s(\boldsymbol{a}^k)^T \\
&\quad + 2\frac{\partial F_1}{\partial \beta}\sum_{k,s}\left((\boldsymbol{a}^i)^T M^{-1}\delta_{sk}\boldsymbol{a}^j\right)\boldsymbol{a}_s(\boldsymbol{a}^k)^T \\
&= 4\frac{\partial^2 F_1}{\partial \beta^2}\sum_{k,s}\boldsymbol{a}_s\left((\boldsymbol{a}^s)^T M^{-1}\boldsymbol{a}^i\right)\left((\boldsymbol{a}^j)^T M^{-1}\boldsymbol{a}^k\right)(\boldsymbol{a}^k)^T \\
&\quad + 2\frac{\partial F_1}{\partial \beta}\sum_{k}\left((\boldsymbol{a}^i)^T M^{-1}\boldsymbol{a}^j\right)\boldsymbol{a}_k(\boldsymbol{a}^k)^T \\
&= 4\frac{\partial^2 F_1}{\partial \beta^2}\sum_{k,s}\boldsymbol{a}_s(\boldsymbol{a}^s)^T\left(M^{-1}\boldsymbol{a}^i\right)\left(M^{-1}\boldsymbol{a}^j\right)^T\boldsymbol{a}^k(\boldsymbol{a}^k)^T \\
&\quad + 2\frac{\partial F_1}{\partial \beta}\left((\boldsymbol{a}^i)^T M^{-1}\boldsymbol{a}^j\right)\sum_{k}\boldsymbol{a}_k(\boldsymbol{a}^k)^T \\
&= 4\frac{\partial^2 F_1}{\partial \beta^2}\left(M^{-1}\boldsymbol{a}^i\right)\left(M^{-1}\boldsymbol{a}^j\right)^T\sum_{k}\boldsymbol{a}^k(\boldsymbol{a}^k)^T + 2\frac{\partial F_1}{\partial \beta}\left((\boldsymbol{a}^i)^T M^{-1}\boldsymbol{a}^j\right)I, \qquad (6.30)
\end{aligned}$$

where the fact that $\sum_k \boldsymbol{a}_k(\boldsymbol{a}^k)^T = I$ is used in the last step.

Next, we calculate $\frac{\partial^2 F}{\partial \boldsymbol{a}^i \partial \boldsymbol{x}}$. Since

$$\nabla = \sum_s \boldsymbol{a}^s \frac{\partial}{\partial \xi_s} \quad \text{and} \quad \frac{\partial}{\partial x_n} = \sum_s (\boldsymbol{a}^s)_n \frac{\partial}{\partial \xi_s},$$

we have

$$
\begin{aligned}
\frac{\partial^2 F}{\partial \boldsymbol{a}^i \partial x_n} &= \frac{\partial}{\partial x_n}\frac{\partial F}{\partial \boldsymbol{a}^i} \\
&= \frac{\partial}{\partial x_n}\left(2\frac{\partial F_1}{\partial \beta} M^{-1}\boldsymbol{a}^i\right) \\
&= 2\frac{\partial^2 F_1}{\partial \beta^2} M^{-1}\boldsymbol{a}^i \frac{\partial \beta}{\partial x_n} + 2\frac{\partial^2 F_1}{\partial \beta \partial \rho} M^{-1}\boldsymbol{a}^i \frac{\partial \rho}{\partial x_n} + 2\frac{\partial F_1}{\partial \beta}\frac{\partial M^{-1}}{\partial x_n}\boldsymbol{a}^i \\
&= 2\frac{\partial^2 F_1}{\partial \beta^2} M^{-1}\boldsymbol{a}^i \sum_l (\boldsymbol{a}^l)^T \frac{\partial M^{-1}}{\partial x_n}\boldsymbol{a}^l + 2\frac{\partial^2 F_1}{\partial \beta \partial \rho} M^{-1}\boldsymbol{a}^i \frac{\partial \rho}{\partial x_n} + 2\frac{\partial F_1}{\partial \beta}\frac{\partial M^{-1}}{\partial x_n}\boldsymbol{a}^i \\
&= 2\frac{\partial^2 F_1}{\partial \beta^2} M^{-1}\boldsymbol{a}^i \sum_l (\boldsymbol{a}^l)^T \sum_s (\boldsymbol{a}^s)_n \frac{\partial M^{-1}}{\partial \xi_s}\boldsymbol{a}^l + 2\frac{\partial^2 F_1}{\partial \beta \partial \rho} M^{-1}\boldsymbol{a}^i \sum_s (\boldsymbol{a}^s)_n \frac{\partial \rho}{\partial \xi_s} \\
&\quad + 2\frac{\partial F_1}{\partial \beta}\sum_s (\boldsymbol{a}^s)_n \frac{\partial M^{-1}}{\partial \xi_s}\boldsymbol{a}^i \\
&= \sum_s \Bigg[2\frac{\partial^2 F_1}{\partial \beta^2}\left(M^{-1}\boldsymbol{a}^i\right)(\boldsymbol{a}^s)_n \sum_l (\boldsymbol{a}^l)^T \frac{\partial M^{-1}}{\partial \xi_s}\boldsymbol{a}^l \\
&\qquad + 2\frac{\partial^2 F_1}{\partial \beta \partial \rho}\left(M^{-1}\boldsymbol{a}^i\right)(\boldsymbol{a}^s)_n \frac{\partial \rho}{\partial \xi_s} + 2\frac{\partial F_1}{\partial \beta}\left(\frac{\partial M^{-1}}{\partial \xi_s}\boldsymbol{a}^i\right)(\boldsymbol{a}^s)_n \Bigg]
\end{aligned}
$$

and

$$
\begin{aligned}
\frac{\partial^2 F}{\partial \boldsymbol{a}^i \partial \boldsymbol{x}} = \sum_s \Bigg[& 2\frac{\partial^2 F_1}{\partial \beta^2}\left(M^{-1}\boldsymbol{a}^i\right)(\boldsymbol{a}^s)^T \sum_l (\boldsymbol{a}^l)^T \frac{\partial M^{-1}}{\partial \xi_s}\boldsymbol{a}^l \\
& + 2\frac{\partial^2 F_1}{\partial \beta \partial \rho}\left(M^{-1}\boldsymbol{a}^i\right)(\boldsymbol{a}^s)^T \frac{\partial \rho}{\partial \xi_s} + 2\frac{\partial F_1}{\partial \beta}\left(\frac{\partial M^{-1}}{\partial \xi_s}\boldsymbol{a}^i\right)(\boldsymbol{a}^s)^T \Bigg].
\end{aligned}
$$

Since $(\boldsymbol{a}^s)^T \boldsymbol{a}_k = \delta_{sk}$,

$$
\begin{aligned}
&\sum_k \left((\boldsymbol{a}^k)^T \frac{\partial^2 F}{\partial \boldsymbol{a}^i \partial \boldsymbol{x}} \boldsymbol{a}_k \right) \\
&= \sum_{k,s} \Bigg[2 \frac{\partial^2 F_1}{\partial \beta^2} (\boldsymbol{a}^k)^T \left(M^{-1} \boldsymbol{a}^i\right) (\boldsymbol{a}^s)^T \boldsymbol{a}_k \sum_l (\boldsymbol{a}^l)^T \frac{\partial M^{-1}}{\partial \xi_s} \boldsymbol{a}^l \\
&\quad + 2 \frac{\partial^2 F_1}{\partial \beta \partial \rho} (\boldsymbol{a}^k)^T \left(M^{-1} \boldsymbol{a}^i\right) (\boldsymbol{a}^s)^T \boldsymbol{a}_k \frac{\partial \rho}{\partial \xi_s} + 2 \frac{\partial F_1}{\partial \beta} (\boldsymbol{a}^k)^T \left(\frac{\partial M^{-1}}{\partial \xi_s} \boldsymbol{a}^i \right) (\boldsymbol{a}^s)^T \boldsymbol{a}_k \Bigg] \\
&= \sum_k \Bigg[2 \frac{\partial^2 F_1}{\partial \beta^2} (\boldsymbol{a}^k)^T \left(M^{-1} \boldsymbol{a}^i\right) \sum_l (\boldsymbol{a}^l)^T \frac{\partial M^{-1}}{\partial \xi_k} \boldsymbol{a}^l \\
&\quad + 2 \frac{\partial^2 F_1}{\partial \beta \partial \rho} (\boldsymbol{a}^k)^T \left(M^{-1} \boldsymbol{a}^i\right) \frac{\partial \rho}{\partial \xi_k} + 2 \frac{\partial F_1}{\partial \beta} (\boldsymbol{a}^k)^T \left(\frac{\partial M^{-1}}{\partial \xi_k} \boldsymbol{a}^i \right) \Bigg] \\
&= \sum_k \Bigg[2 \frac{\partial^2 F_1}{\partial \beta^2} \left((\boldsymbol{a}^k)^T M^{-1} \boldsymbol{a}^i \right) \left(\sum_l (\boldsymbol{a}^l)^T \frac{\partial M^{-1}}{\partial \xi_k} \boldsymbol{a}^l \right) \\
&\quad + 2 \frac{\partial^2 F_1}{\partial \beta \partial \rho} \left((\boldsymbol{a}^k)^T M^{-1} \boldsymbol{a}^i \right) \frac{\partial \rho}{\partial \xi_k} + 2 \frac{\partial F_1}{\partial \beta} \left((\boldsymbol{a}^k)^T \frac{\partial M^{-1}}{\partial \xi_k} \boldsymbol{a}^i \right) \Bigg]. \qquad (6.31)
\end{aligned}
$$

For the term $\frac{\partial^2 F}{\partial J \partial \boldsymbol{x}}$, we have

$$
\frac{\partial^2 F}{\partial J \partial \boldsymbol{x}} = \frac{\partial^2 F_2}{\partial J \partial \rho} \frac{\partial \rho}{\partial \boldsymbol{x}} = \frac{\partial^2 F_2}{\partial J \partial \rho} \sum_s \boldsymbol{a}^s \frac{\partial \rho}{\partial \xi_s},
$$

and therefore,

$$
\begin{aligned}
\sum_i \boldsymbol{a}_i \left(\frac{\partial^2 F}{\partial J \partial \boldsymbol{x}} \right)^T \frac{\partial \boldsymbol{x}}{\partial \xi_i} &= \frac{\partial^2 F_2}{\partial J \partial \rho} \sum_{i,s} \boldsymbol{a}_i (\boldsymbol{a}^s)^T \frac{\partial \rho}{\partial \xi_s} \boldsymbol{a}_i \\
&= \frac{\partial^2 F_2}{\partial J \partial \rho} \sum_i \boldsymbol{a}_i \frac{\partial \rho}{\partial \xi_i} \\
&= \frac{\partial^2 F_2}{\partial J \partial \rho} \sum_i \frac{\partial \rho}{\partial \xi_i} \frac{\partial \boldsymbol{x}}{\partial \xi_i}. \qquad (6.32)
\end{aligned}
$$

Using (6.30)–(6.32) and the fact that $\frac{\partial^2 F}{\partial \boldsymbol{a}^s \partial J} = \boldsymbol{0}$ and $\frac{\partial^2 F}{\partial J \partial \boldsymbol{a}^k} = \boldsymbol{0}$ for the special functional form (6.28) under consideration, the mesh equation in non-conservative form is

$$\sum_{i,j}\Bigg[4\frac{\partial^2 F_1}{\partial \beta^2}\left(M^{-1}\boldsymbol{a}^i\right)\left(M^{-1}\boldsymbol{a}^j\right)^T\sum_k \boldsymbol{a}^k(\boldsymbol{a}^k)^T+2\frac{\partial F_1}{\partial \beta}\left((\boldsymbol{a}^i)^T M^{-1}\boldsymbol{a}^j\right)I$$
$$+J^2\left(\frac{2}{J}\frac{\partial F_2}{\partial J}+\frac{\partial^2 F_2}{\partial J^2}\right)\boldsymbol{a}_i(\boldsymbol{a}^j)^T\Bigg]\frac{\partial^2 \boldsymbol{x}}{\partial \xi_i \partial \xi_j}$$
$$-\sum_i\sum_k\Bigg[2\frac{\partial^2 F_1}{\partial \beta^2}\left((\boldsymbol{a}^k)^T M^{-1}\boldsymbol{a}^i\right)\left(\sum_l (\boldsymbol{a}^l)^T\frac{\partial M^{-1}}{\partial \xi_k}\boldsymbol{a}^l\right)$$
$$+2\frac{\partial^2 F_1}{\partial \beta \partial \rho}\left((\boldsymbol{a}^k)^T M^{-1}\boldsymbol{a}^i\right)\frac{\partial \rho}{\partial \xi_k}+2\frac{\partial F_1}{\partial \beta}\left((\boldsymbol{a}^k)^T\frac{\partial M^{-1}}{\partial \xi_k}\boldsymbol{a}^i\right)\Bigg]\frac{\partial \boldsymbol{x}}{\partial \xi_i}$$
$$+J\frac{\partial^2 F_2}{\partial J \partial \rho}\sum_i \frac{\partial \rho}{\partial \xi_i}\frac{\partial \boldsymbol{x}}{\partial \xi_i}=0. \tag{6.33}$$

Writing this in the concise form (6.26), the coefficient matrices for (6.33) are

$$A_{ij}=4\frac{\partial^2 F_1}{\partial \beta^2}\left(M^{-1}\boldsymbol{a}^i\right)\left(M^{-1}\boldsymbol{a}^j\right)^T\sum_k \boldsymbol{a}^k(\boldsymbol{a}^k)^T+2\frac{\partial F_1}{\partial \beta}\left((\boldsymbol{a}^i)^T M^{-1}\boldsymbol{a}^j\right)I$$
$$+J^2\left(\frac{2}{J}\frac{\partial F_2}{\partial J}+\frac{\partial^2 F_2}{\partial J^2}\right)\boldsymbol{a}_i(\boldsymbol{a}^j)^T,$$
$$B_i=-\sum_k\Bigg[2\frac{\partial^2 F_1}{\partial \beta^2}\left((\boldsymbol{a}^k)^T M^{-1}\boldsymbol{a}^i\right)\left(\sum_l (\boldsymbol{a}^l)^T\frac{\partial M^{-1}}{\partial \xi_k}\boldsymbol{a}^l\right)$$
$$+2\frac{\partial^2 F_1}{\partial \beta \partial \rho}\left((\boldsymbol{a}^k)^T M^{-1}\boldsymbol{a}^i\right)\frac{\partial \rho}{\partial \xi_k}+2\,\frac{\partial F_1}{\partial \beta}\left((\boldsymbol{a}^k)^T\frac{\partial M^{-1}}{\partial \xi_k}\boldsymbol{a}^i\right)\Bigg]I$$
$$+J\frac{\partial^2 F_2}{\partial J \partial \rho}\frac{\partial \rho}{\partial \xi_i}I, \tag{6.34}$$

where I is the 3×3 identity matrix.

6.1.2 Moving mesh PDEs

In this subsection we extend the above framework to the case of a time-dependent coordinate transformation $\boldsymbol{x}=\boldsymbol{x}(\boldsymbol{\xi},t)$ having an inverse $\boldsymbol{\xi}=\boldsymbol{\xi}(\boldsymbol{x},t)$, which is the typical situation arising when solving a time-dependent physical PDE. The coordinate transformation can again be determined through one of the mesh equations (6.13), (6.15), (6.25), or (6.33), but with the monitor function now time-dependent, i.e., $M=M(\boldsymbol{x},t)$. When discretized in space, these mesh equations lead to a system of algebraic equations and form a DAE (Differential-Algebraic Equation) system when combined with the physical PDE. As previously discussed in §2.3.1, such a DAE system is often difficult to integrate, making it attractive to use a modified mesh equation which involves the mesh speed, giving a system of differential equations to integrate after discretization in space. This motivation has led to the development of so-called moving mesh PDEs or MMPDEs. Recall that another reason for this approach is that the mesh equations are highly nonlinear, and a direct application of Newton's iteration on an algebraic system of discretized mesh equations may encounter difficulty with convergence, especially when the monitor function has large variations in space. An effective way to modify the mesh equation is often to introduce a pseudo-time variable and use the continuation method on it. Thus, in either case one encounters a mesh equation involving mesh speeds.

In §2.3, 1D MMPDEs are defined as modified gradient flow equations for the adaptation functionals. This is also straightforward to do in the multidimensional case. Specifically, we define an MMPDE

$$\frac{\partial \boldsymbol{\xi}}{\partial t} = -\frac{1}{\tau p(\boldsymbol{x},t)} \frac{\delta I}{\delta \boldsymbol{\xi}}, \tag{6.35}$$

where $p = p(\boldsymbol{x},t)$ is a balancing function (discussed below), $\tau > 0$ is a user specified parameter for adjusting the time scale of mesh movement, and $\frac{\delta I}{\delta \boldsymbol{\xi}}$ is the functional derivative of I with respect to the unknown function $\boldsymbol{\xi}$. For the general functional defined in (6.11), its functional derivative equals the left-hand side of (6.13), i.e.,

$$\frac{\delta I}{\delta \xi_i} = -\nabla \cdot \left[\frac{\partial F}{\partial \boldsymbol{a}^i} - J \frac{\partial F}{\partial J} \boldsymbol{a}_i \right].$$

Thus, the MMPDE (6.35) is

$$\frac{\partial \xi_i}{\partial t} = \frac{1}{\tau p(\boldsymbol{x},t)} \nabla \cdot \left[\frac{\partial F}{\partial \boldsymbol{a}^i} - J \frac{\partial F}{\partial J} \boldsymbol{a}_i \right], \quad i = 1,2,3. \tag{6.36}$$

Next, the dependent and independent variables in (6.36) are interchanged to obtain an MMPDE in terms of the coordinate transformation $\boldsymbol{x} = \boldsymbol{x}(\boldsymbol{\xi},t)$. Recalling that the mesh speeds $\dot{\boldsymbol{x}}$ and $\boldsymbol{\xi}_t$ are related through equation (3.16), which is

$$\dot{\boldsymbol{x}} = -\boldsymbol{J}\boldsymbol{\xi}_t = -\sum_i \boldsymbol{a}_i \frac{\partial \xi_i}{\partial t},$$

we obtain the semi-conservative form of the MMPDE

$$\dot{\boldsymbol{x}} = -\frac{1}{\tau p(\boldsymbol{x},t)} \sum_i \boldsymbol{a}_i \nabla \cdot \left[\frac{\partial F}{\partial \boldsymbol{a}^i} - J \frac{\partial F}{\partial J} \boldsymbol{a}_i \right], \tag{6.37}$$

or from (6.15),

$$\dot{\boldsymbol{x}} = -\frac{1}{\tau J p(\boldsymbol{x},t)} \sum_i \boldsymbol{a}_i \left[\sum_j \frac{\partial}{\partial \xi_j} \left(J \boldsymbol{a}^j \cdot \frac{\partial F}{\partial \boldsymbol{a}^i} \right) - \frac{\partial}{\partial \xi_i} \left(J^2 \frac{\partial F}{\partial J} \right) \right]. \tag{6.38}$$

The MMPDE can also be cast in the non-conservative form

$$\dot{\boldsymbol{x}} = \frac{1}{\tau p(\boldsymbol{x},t)} \left[\sum_{i,j} A_{i,j} \frac{\partial^2 \boldsymbol{x}}{\partial \xi_i \partial \xi_j} + \sum_i B_i \frac{\partial \boldsymbol{x}}{\partial \xi_i} \right], \tag{6.39}$$

where the coefficient matrices are given in (6.27) for the general functional (6.11) or in (6.34) for the special functional (6.28). Note that (6.37) or (6.39) can be regarded as a multi-dimensional analogue of MMPDE5, and when it can be done without confusion, we shall refer to this form of the multi-dimensional MMPDE as simply MMPDE5.

Choice of the balancing function $p = p(\boldsymbol{x},t)$**.** Ideally the function $p = p(\boldsymbol{x},t)$ should be chosen so that all the mesh points move with a uniform time scale because an MMPDE having this time scale could be integrated numerically more easily and more reliably with a constant value of τ. Unfortunately, it is unclear mathematically how to make a PDE have a uniform time scale. We use here a heuristic, spatial balance criterion from [175]; namely, p is chosen such that the coefficients in the mesh equation, especially those of the second order derivatives, change evenly over the spatial domain. In this way, the MMPDE behaves more like a diffusion equation with an almost constant diffusion coefficient. For example, to spatially balance the MMPDE (6.39) (or (6.38)), we can choose

$$p(\boldsymbol{x},t) = \sqrt{\sum_{i,j} \|A_{i,j}\|_F^2}, \tag{6.40}$$

where $\|\cdot\|_F$ denotes the Frobenius matrix norm. This choice emphasizes exclusively the second order derivative terms of the coordinate transformation. An alternative is

$$p(\boldsymbol{x},t) = \sqrt{\sum_{i,j} \|A_{i,j}\|_F^2 + \sum_i \|B_i\|_F^2}. \tag{6.41}$$

However, since $A_{i,j}$ and B_i have different dimensions, the roles they play in (6.41) can differ significantly.

Choice of the parameter τ**.** The parameter τ provides a mechanism for the user to adjust the time scale of mesh movement. The smaller τ is, the faster the mesh responds to changes in the monitor function (and therefore, to the physical solution), and the stiffer the MMPDE becomes. On the other hand, for a very large value of τ, the mesh will change slowly. An optimal choice of τ often requires some tuning. Fortunately, numerical experience shows that mesh movement is not very sensitive to this parameter, and a value in the range $[10^{-3}, 10^{-1}]$ works well for most problems. Moreover, as in the case of the numerical solution of problems having blowup solutions, the parameter τ can be chosen to have a special dimension or to be dimensionless using a dimensional analysis of the underlying PDE and MMPDE (cf. §2.8). In the latter case, a special form of MMPDE and monitor function is taken such that choosing a proper value for the dimensionless parameter is relatively easy and robust.

6.1.3 Boundary conditions for coordinate transformation

A complete specification of the coordinate transformation requires supplementing the mesh equations with suitable boundary conditions. The simplest boundary conditions are of Dirichlet type. If Γ is either the boundary of the entire physical domain Ω or a portion of it, with Γ_c its counterpart for the computational domain Ω_c, then a Dirichlet boundary condition for Γ has the form

$$\boldsymbol{x} = \boldsymbol{g}(\boldsymbol{\xi}, t), \qquad \text{on } \Gamma_c \subseteq \partial \Omega_c \tag{6.42}$$

where $\boldsymbol{g}(\boldsymbol{\xi}, t)$ is a given function. Boundary points specified by (6.42) are generally not adaptive to the physical solution, and they stay fixed when $\boldsymbol{g}$ is independent of t.

Boundary points can also be specified through the natural boundary conditions for the functional (6.11). Specifically, if

$$\psi(\boldsymbol{x}, t) = 0 \tag{6.43}$$

denotes the equation implicitly defining the boundary surface Γ, then applying the variation operator to it gives

$$\nabla \psi(\boldsymbol{x}, t) \cdot \delta \boldsymbol{x} = 0, \quad \forall \boldsymbol{x} \in \Gamma.$$

Since $\delta \boldsymbol{x} = \sum_i \frac{\partial \boldsymbol{x}}{\partial \xi_i} = \sum_i \boldsymbol{a}_i \delta \xi_i$ and $\nabla \psi \propto \boldsymbol{n}$, where $\boldsymbol{n}$ is the outward normal to the surface,

$$\sum_i (\boldsymbol{a}_i \cdot \boldsymbol{n}) \delta \xi_i = 0, \quad \forall \boldsymbol{x} \in \Gamma. \tag{6.44}$$

Setting $\delta I[\boldsymbol{\xi}] = 0$, (6.12) and (6.13) imply

$$\int_{\partial \Omega} \sum_{i=1}^{3} \boldsymbol{n} \cdot \left[\frac{\partial F}{\partial \boldsymbol{a}^i} - J \frac{\partial F}{\partial J} \boldsymbol{a}_i \right] \delta \xi_i dS = 0 \tag{6.45}$$

for all admissible functions $\delta \xi_i$ satisfying the constraint (6.44). This can be shown to imply that

$$(\boldsymbol{a}_1 \cdot \boldsymbol{n}) \left(\frac{\partial F}{\partial \boldsymbol{a}^2} \cdot \boldsymbol{n} \right) = (\boldsymbol{a}_2 \cdot \boldsymbol{n}) \left(\frac{\partial F}{\partial \boldsymbol{a}^1} \cdot \boldsymbol{n} \right), \quad \text{on } \Gamma \tag{6.46}$$

$$(\boldsymbol{a}_1 \cdot \boldsymbol{n}) \left(\frac{\partial F}{\partial \boldsymbol{a}^3} \cdot \boldsymbol{n} \right) = (\boldsymbol{a}_3 \cdot \boldsymbol{n}) \left(\frac{\partial F}{\partial \boldsymbol{a}^1} \cdot \boldsymbol{n} \right), \quad \text{on } \Gamma. \tag{6.47}$$

The boundary point distribution on Γ is then determined from (6.46) and (6.47) and the boundary equation (6.43).

Orthogonal boundary conditions are useful in some situations. Without loss of generality, we assume that Γ coincides with a coordinate plane in the computational coordinates, i.e.,

$$\xi_1(\boldsymbol{x},t) = 0. \tag{6.48}$$

Then the orthogonality conditions require that the two sets of coordinate surfaces

$$\xi_2(\boldsymbol{x},t) = \text{ constant} \qquad \text{and} \qquad \xi_3(\boldsymbol{x},t) = \text{ constant}$$

be orthogonal to the physical boundary represented by (6.48). This leads to

$$\nabla \xi_1 \cdot \nabla \xi_2 = 0, \quad \nabla \xi_1 \cdot \nabla \xi_3 = 0, \quad \text{on } \Gamma$$

or

$$\boldsymbol{a}^1 \cdot \boldsymbol{a}^2 = 0, \quad \boldsymbol{a}^1 \cdot \boldsymbol{a}^3 = 0 \quad \text{on } \Gamma. \tag{6.49}$$

An alternative way to implement orthogonal boundary conditions is to use a projection method. With this method, the locations of boundary points are obtained by projecting the mesh points which are inside the domain but next to the physical boundary onto the boundary. This method is easy to implement and often very effective.

Finally, boundary points can be distributed according to a lower dimensional mesh equation or MMPDE, e.g., see [189]. To illustrate, consider the problem of distributing the boundary points on a boundary surface Γ in 3D. A 2D mesh equation or MMPDE can be used for Γ. The monitor function needed for the 2D mesh equation can be defined by projecting the 3D monitor function $M = M(\boldsymbol{x})$ onto Γ, i.e.,

$$M_\Gamma = \begin{bmatrix} \boldsymbol{t}_1^T \\ \boldsymbol{t}_2^T \end{bmatrix} M(\boldsymbol{x}) \left[\boldsymbol{t}_1, \boldsymbol{t}_2 \right], \quad \forall \boldsymbol{x} \in \Gamma \tag{6.50}$$

where $\boldsymbol{t}_1$ and $\boldsymbol{t}_t$ are two normalized, orthogonal directions tangent to Γ.

6.2 Existence of minimizer

Aside from the obvious practical desire to properly concentrate the mesh while controlling the mesh quality, there is the fundamental theoretical issue of how to select an adaptation functional for which the existence of a minimizer is ensured. As we discuss in this section, this is a major reason why one cannot simply choose any error bound for an adaptation functional. Fortuitously, however, the existence of a minimizer for many adaptation functionals can be proven from standard theory in calculus of variations and mathematical elasticity.

6.2.1 Convex functionals

Developing the basic existence theory requires some theoretical tools, beginning with convex functionals for systems. Assume that $\Omega \subset \mathbb{R}^3$ is a bounded, open set with Lipschitz-continuous boundary $\partial\Omega$. Consider an integral functional of the form

$$I[\boldsymbol{w}] = \int_{\Omega} L(\nabla \boldsymbol{w}, \boldsymbol{x}) d\boldsymbol{x} \tag{6.51}$$

for function $\boldsymbol{w} : \Omega \to \mathbb{R}^3$, where $L : \mathbb{R}^{3\times 3} \times \mathbb{R}^3 \to \mathbb{R}$ is a smooth function and the gradient

$$\nabla \boldsymbol{w} = \begin{bmatrix} \frac{\partial w_1}{\partial x_1} & \frac{\partial w_1}{\partial x_2} & \frac{\partial w_1}{\partial x_3} \\ \frac{\partial w_2}{\partial x_1} & \frac{\partial w_2}{\partial x_2} & \frac{\partial w_2}{\partial x_3} \\ \frac{\partial w_3}{\partial x_1} & \frac{\partial w_3}{\partial x_2} & \frac{\partial w_3}{\partial x_3} \end{bmatrix}.$$

Note that we assume here that L does not depend upon $\boldsymbol{w}$ explicitly. Moreover, while the functional is expressed explicitly in terms of the independent variable $\boldsymbol{x}$, the function $\boldsymbol{w} = \boldsymbol{w}(\boldsymbol{x})$ can be given either in terms of the inverse coordinate transformation $\boldsymbol{\xi} = \boldsymbol{\xi}(\boldsymbol{x})$ or the coordinate transformation $\boldsymbol{x} = \boldsymbol{x}(\boldsymbol{\xi})$ (where $\boldsymbol{\xi}$ is the independent variable) – cf. §6.1.1.

Ensuring existence of a minimizer can be closely associated with the concepts of *coercivity* and *convexity* of the functional. Loosely speaking, coercivity ensures that the functional grows rapidly as $\|\nabla \boldsymbol{w}\| \to \infty$, so there exists a bounded minimizing sequence of functions. Convexity provides a kind of compactness guaranteeing that the minimizing sequence has a convergent subsequence in an infinite dimensional function space.

More specifically, for a given number p with

$$1 < p < \infty,$$

the function L (and functional $I[\boldsymbol{w}]$ in (6.51))[1] is said to be *coercive* if there exist constants $\alpha > 0$ and $\beta \geq 0$ such that

[1] Following convention, the definitions of coercivity and of convexity for the function L and for the corresponding functional are synonymic.

$$L(P,\boldsymbol{x}) \geq \alpha \|P\|^p - \beta, \quad \forall P \in \mathbb{R}^{3\times 3}, \boldsymbol{x} \in \Omega, \tag{6.52}$$

where $\|\cdot\|$ denotes a matrix norm. The function L is said to be *convex* if

$$\sum_{i,j,m,n} \frac{\partial^2 L}{\partial p_{i,j} \partial p_{m,n}}(P,\boldsymbol{x}) \xi_{i,j} \xi_{m,n} \geq 0,$$
$$\forall P \equiv (p_{i,j}),\ \Xi \equiv (\xi_{ij}) \in \mathbb{R}^{3\times 3}, \boldsymbol{x} \in \mathbb{R}^3. \tag{6.53}$$

By convention, we call (6.52) and (6.53) the coercivity and convexity conditions, respectively.

The coercivity condition implies that $\boldsymbol{w} \in W^{1,p}(\Omega;\mathbb{R}^3)$ when $I[\boldsymbol{w}] < \infty$. Thus, the admissible set for a Dirichlet boundary condition can be defined as

$$\mathscr{A} = \{\boldsymbol{w} \in W^{1,p}(\Omega;\mathbb{R}^3) \,|\, \boldsymbol{w} = \boldsymbol{g} \text{ on } \partial\Omega\}, \tag{6.54}$$

where $\boldsymbol{g} \in W^{1,p}(\Omega;\mathbb{R}^3)$ is a given function.

Theorem 6.2.1 (Existence of minimizers of convex functionals) *Assume that $L = L(P,\boldsymbol{x})$ satisfies the coercivity condition (6.52) and convexity condition (6.53), and assume further that the admissible set $\mathscr{A}$ is nonempty. Then there exists a minimizer $\mathbf{w}^* \in \mathscr{A}$ satisfying*

$$I[\mathbf{w}^*] = \inf_{\mathbf{w} \in \mathscr{A}} I[\mathbf{w}].$$

Proof. See Evans [137]. □

A uniqueness result for a minimizer, which requires a stronger convexity assumption and applies for a smaller class of functionals, is given in the following theorem.

Theorem 6.2.2 (Uniqueness of minimizers of uniformly convex functionals) *Assume that $L = L(P,\boldsymbol{x})$ is uniformly convex in the variable P, i.e., there exists a constant $\theta > 0$ such that*

$$\sum_{i,j,m,n} \frac{\partial^2 L}{\partial p_{i,j} \partial p_{m,n}}(P,\boldsymbol{x}) \xi_{i,j} \xi_{m,n} \geq \theta \sum_{i,j} \xi_{i,j}^2,$$
$$\forall P \equiv (p_{i,j}),\ \Xi \equiv (\xi_{ij}) \in \mathbb{R}^{3\times 3}, \boldsymbol{x} \in \mathbb{R}^3. \tag{6.55}$$

Then if the admissible set $\mathscr{A}$ is nonempty, a minimizer $\mathbf{w}^* \in \mathscr{A}$ of the functional $I[\mathbf{w}]$ in (6.51) is unique.

Proof. See Evans [137]. □

The following lemma is handy for proving the convexity of a functional.

Lemma 6.2.1 *Suppose that a given function $g(P,\boldsymbol{x}) \in C^2(\mathbb{R}^{3\times 3} \times \mathbb{R};\mathbb{R})$ is convex in P and a function $f \in C^2(\mathbb{R} \times \mathbb{R};\mathbb{R})$ satisfies*

$$\frac{\partial f}{\partial g}(g(P,\boldsymbol{x}),\boldsymbol{x}) \geq 0, \quad \frac{\partial^2 f}{\partial g^2}(g(P,\boldsymbol{x}),\boldsymbol{x}) \geq 0, \quad \forall P \in \mathbb{R}^{3\times 3}, \boldsymbol{x} \in \mathbb{R}. \tag{6.56}$$

Then the composite function $f \circ g \equiv f(g(P,\boldsymbol{x}),\boldsymbol{x}) \in C^2(\mathbb{R}^{3\times 3};\mathbb{R})$ is convex in P.

Proof. From the chain rule,

$$\frac{\partial (f \circ g)}{\partial p_{i,j}} = \frac{\partial f}{\partial g}(g(P,\boldsymbol{x}),\boldsymbol{x})\frac{\partial g}{\partial p_{i,j}}$$

and

$$\frac{\partial^2 (f \circ g)}{\partial p_{i,j}\partial p_{m,n}} = \frac{\partial^2 f}{\partial g^2}(g(P,\boldsymbol{x}),\boldsymbol{x})\frac{\partial g}{\partial p_{i,j}}\frac{\partial g}{\partial p_{m,n}} + \frac{\partial f}{\partial g}(g(P,\boldsymbol{x}),\boldsymbol{x})\frac{\partial^2 g}{\partial p_{i,j}\partial p_{m,n}}.$$

It follows that, for any $\Xi = (\xi_{i,j}) \in \mathbb{R}^{3\times 3}$,

$$\begin{aligned}
&\sum_{i,j,m,n} \frac{\partial^2 (f \circ g)}{\partial p_{i,j}\partial p_{m,n}}(P,\boldsymbol{x})\xi_{i,j}\xi_{m,n} \\
&= \sum_{i,j,m,n} \frac{\partial^2 f}{\partial g^2}(g(P,\boldsymbol{x}),\boldsymbol{x})\frac{\partial g}{\partial p_{i,j}}(P,\boldsymbol{x})\frac{\partial g}{\partial p_{m,n}}(P,\boldsymbol{x})\xi_{i,j}\xi_{m,n} \\
&\quad + \sum_{i,j,m,n} \frac{\partial f}{\partial g}(g(P,\boldsymbol{x}),\boldsymbol{x})\frac{\partial^2 g}{\partial p_{i,j}\partial p_{m,n}}(P,\boldsymbol{x})\xi_{i,j}\xi_{m,n} \\
&= \frac{\partial^2 f}{\partial g^2}(g(P,\boldsymbol{x}),\boldsymbol{x})\left(\sum_{i,j}\frac{\partial g}{\partial p_{i,j}}(P,\boldsymbol{x})\xi_{i,j}\right)^2 \\
&\quad + \frac{\partial f}{\partial g}(g(P,\boldsymbol{x}),\boldsymbol{x})\sum_{i,j,m,n}\frac{\partial^2 g}{\partial p_{i,j}\partial p_{m,n}}(P,\boldsymbol{x})\xi_{i,j}\xi_{m,n} \\
&\geq 0,
\end{aligned}$$

where in the last step we have used (6.56) and the convexity condition on g. Hence, $f \circ g$ is convex in P. □

6.2.2 Polyconvex functionals

A number of adaptation functionals are of a type which, though not convex, can nevertheless be shown to have a minimizer. The distinct features of this type of functional are that they explicitly involve the determinant of the gradient and are polyconvex.

We first consider functionals whose integrand is of the form

$$L(P,\boldsymbol{x}) = F(P,\det(P),\boldsymbol{x}), \quad \forall P \in \mathbb{R}^{3\times 3},\ \boldsymbol{x} \in \Omega \tag{6.57}$$

where $F : \mathbb{R}^{3\times 3} \times \mathbb{R} \times \Omega \rightarrow \mathbb{R}$ is a smooth function. The function L is called *polyconvex* if $F(P,r,\boldsymbol{x})$ is convex in variables P and r, i.e.,

$$\sum_{i,j,m,n} \frac{\partial^2 F}{\partial p_{i,j}\partial p_{m,n}}(P,r,\boldsymbol{x})\xi_{i,j}\xi_{m,n} + 2\sum_{i,j} \frac{\partial^2 F}{\partial p_{i,j}\partial r}(P,r,\boldsymbol{x})\xi_{i,j}\eta + \frac{\partial^2 F}{\partial r^2}(P,r,\boldsymbol{x})\eta^2$$
$$\geq 0, \quad \forall P \equiv (p_{i,j}),\ \Xi \equiv (\xi_{ij}) \in \mathbb{R}^{3\times 3},\ \boldsymbol{x} \in \mathbb{R}^3,\ r,\ \eta \in \mathbb{R}. \tag{6.58}$$

Theorem 6.2.3 (Existence of minimizers of polyconvex functionals) *Assume that $3 < p < \infty$ ($2 < p < \infty$ in 2D) and F in (6.57) satisfies the coercivity condition (6.52) and is polyconvex. If the admissible set $\mathscr{A}$ is nonempty, then there exists a minimizer $\mathbf{w}^* \in \mathscr{A}$ satisfying*

$$I[\mathbf{w}^*] = \inf_{\mathbf{w}\in\mathscr{A}} I[\mathbf{w}].$$

Proof. See Evans [137].

Next, consider functionals with integrand of the form

$$L(P,\boldsymbol{x}) = F(P,\mathrm{Cof}(P),\det(P),\boldsymbol{x}), \quad \forall P \in \mathbb{R}_+^{3\times 3},\ \boldsymbol{x} \in \Omega \tag{6.59}$$

where $F : \mathbb{R}^{3\times 3} \times \mathbb{R}^{3\times 3} \times \mathbb{R}_+ \times \Omega \rightarrow \mathbb{R}$ is a smooth function, $\mathbb{R}_+ = (0,+\infty)$, $\mathbb{R}_+^{3\times 3} = \{P \in \mathbb{R}^{3\times 3};\ \det(P) > 0\}$, and $\mathrm{Cof}(P)$ is the cofactor of P, which is defined as $\mathrm{Cof}(P) = \det(P)P^{-T}$ if P is invertible. The motivation for considering this type of functional is that one can show that they have a minimizer with the desired property $\det(\nabla \boldsymbol{w}) > 0$, which corresponds to the (local) non-singularity of the coordinate transformation when used in the next subsection to study mesh adaptation.

The definitions of coercivity, polyconvexity, and the admissible set are extended to this type of functional as follows: A functional in (6.51) with L in (6.59) is *coercive* if there exist constants $\alpha > 0$, β, $p > 1$, $q > 1$, and $s > 1$ such that

$$L(P,\boldsymbol{x}) \geq \alpha\left(\|P\|^p + \|\mathrm{Cof}(P)\|^q + \det(P)^s\right) + \beta, \quad \forall P \in \mathbb{R}_+^{3\times 3},\ \boldsymbol{x} \in \Omega. \tag{6.60}$$

The function L is *polyconvex* if $F(P,Q,r,\boldsymbol{x})$ is convex in the variables $P \in \mathbb{R}^{3\times 3}$, $Q \in \mathbb{R}^{3\times 3}$, and $r \in \mathbb{R}_+$. For $\partial\Omega = \Gamma_D \cup \Gamma_N$ with $|\Gamma_D| > 0$, from (6.60) we define the admissible set as

$$\mathscr{A} = \Big\{ \quad \boldsymbol{w}^{1,p}(\Omega;\mathbb{R}^3): \quad \mathrm{Cof}(\nabla \boldsymbol{w}) \in L^q(\Omega); \quad \det(\nabla \boldsymbol{w}) \in L^s(\Omega);$$
$$\boldsymbol{w} = g \text{ on } \Gamma_D; \quad \det(\nabla \boldsymbol{w}) > 0 \text{ a.e. in } \Omega \quad \Big\}. \tag{6.61}$$

Theorem 6.2.4 (Existence of minimizers of polyconvex functionals) *Assume that L has the form (6.59), satisfies the coercivity condition (6.60) with* $p \geq 2$, $q \geq \frac{p}{p-1}$, *and* $s > 1$, *and is polyconvex. Assume further that L satisfies*

$$\lim_{det(P)\to 0^+} L(P,\boldsymbol{x}) = +\infty. \tag{6.62}$$

If the admissible set $\mathscr{A}$ *defined in (6.61) is nonempty, then there exists a minimizer* $\boldsymbol{w}^* \in \mathscr{A}$ *satisfying*

$$I[\boldsymbol{w}^*] = \inf_{\boldsymbol{w}\in\mathscr{A}} I[\boldsymbol{w}].$$

Proof. See Ciarlet [105]. □

6.2.3 Examples of convex and polyconvex mesh adaptation functionals

In this subsection we study several mesh adaptation functionals which are coercive and convex or polyconvex. These functionals are formulated in terms of the inverse coordinate transformation. The correspondences between the general notation in the two previous subsections and that for the examples here are as follows:

$$\begin{array}{lllll} \boldsymbol{w} = \boldsymbol{w}(\boldsymbol{x}) & \longleftrightarrow & \boldsymbol{\xi} = \boldsymbol{\xi}(\boldsymbol{x}) & & \\ \nabla \boldsymbol{w} & \longleftrightarrow & \nabla \boldsymbol{\xi} = \boldsymbol{J}^{-1} & \longleftrightarrow & P \\ \mathrm{Cof}(\nabla \boldsymbol{w}) & \longleftrightarrow & \mathrm{Cof}(\boldsymbol{J}^{-1}) = \frac{1}{J}\boldsymbol{J}^T & \longleftrightarrow & Q = \mathrm{Cof}(P) \\ \det(\nabla \boldsymbol{w}) & \longleftrightarrow & \det(\boldsymbol{J}^{-1}) = J^{-1} & \longleftrightarrow & r. \end{array} \tag{6.63}$$

For theoretical purpose we assume here that the monitor function is chosen such that

$$\underline{m}I \leq M(\boldsymbol{x}) \leq \overline{m}I, \quad \forall \boldsymbol{x} \in \Omega \tag{6.64}$$

for two positive constants $\underline{m}$ and $\overline{m}$.

Example 6.2.1 Suppose that L is given by

$$L(\nabla \boldsymbol{\xi}, \boldsymbol{x}) = w(\boldsymbol{x}) \left(\mathrm{tr}\left(\boldsymbol{J}^{-1} M^{-1} \boldsymbol{J}^{-T}\right)\right)^{\frac{p}{2}}, \tag{6.65}$$

where $p \geq 2$ and $w = w(\boldsymbol{x})$ is a given strictly positive function, i.e., $w(\boldsymbol{x}) \geq \underline{w} > 0$. Recall that

$$J^{-T} = \nabla \boldsymbol{\xi}^T = [\nabla \xi_1, \nabla \xi_2, \nabla \xi_3] .$$

From the definition of β in (6.29),

$$\operatorname{tr}\left(J^{-1} M^{-1} J^{-T}\right) = \sum_i \nabla \xi_i^T M^{-1} \nabla \xi_i \equiv \beta(J^{-1}), \tag{6.66}$$

so the corresponding adaptation functional can be expressed by

$$I[\boldsymbol{\xi}] = \int_\Omega w(\boldsymbol{x}) \left(\sum_i \nabla \xi_i^T M^{-1} \nabla \xi_i \right)^{\frac{p}{2}} d\boldsymbol{x}. \tag{6.67}$$

Since

$$\beta(P) = \operatorname{tr}\left(P M^{-1} P^T\right) = \sum_i [p_{i,1},\ p_{i,2},\ p_{i,3}] M^{-1} \begin{bmatrix} p_{i,1} \\ p_{i,2} \\ p_{i,3} \end{bmatrix} \tag{6.68}$$

for any $P \in \mathbb{R}^{3\times 3}$, it follows from (6.64) that

$$\begin{aligned} L(P, \boldsymbol{x}) &= w(\boldsymbol{x}) \left(\beta(P)\right)^{\frac{p}{2}} \\ &\geq \underline{w} \overline{m}^{-\frac{p}{2}} \left(\sum_i [p_{i,1},\ p_{i,2},\ p_{i,3}] \begin{bmatrix} p_{i,1} \\ p_{i,2} \\ p_{i,3} \end{bmatrix} \right)^{\frac{p}{2}} \\ &= \underline{w} \overline{m}^{-\frac{p}{2}} \|P\|_F^p. \end{aligned}$$

Thus, L satisfies the coercivity condition (6.52).

Moreover, for any $\Xi = (\xi_{i,j}) \in \mathbb{R}^{3\times 3}$,

$$\sum_{i,j,k,l} \frac{\partial^2 \beta(P)}{\partial p_{i,j} \partial p_{k,l}} \xi_{i,j} \xi_{k,l} = 2 \sum_i [\xi_{i,1}\ \xi_{i,2}\ \xi_{i,3}] M^{-1} \begin{bmatrix} \xi_{i,1} \\ \xi_{i,2} \\ \xi_{i,3} \end{bmatrix} \geq 2 \overline{m}^{-1} \|\Xi\|_F^2,$$

which implies that $\beta(P)$ is uniformly convex. The convexity of L follows from Lemma 6.2.1.

Hence, the mesh adaptation functional (6.67) is coercive and convex. From Theorem 6.2.1, if the admissible set $\mathscr{A}$ defined in (6.54) is nonempty, then the functional has a minimizer in $\mathscr{A}$. □

Example 6.2.2 Consider the integrand L for $I[\boldsymbol{\xi}] = \int_\Omega L(\nabla \boldsymbol{\xi}, \boldsymbol{x}) d\boldsymbol{x}$ now of the special form (6.28), i.e.,

$$L(\nabla\boldsymbol{\xi},\boldsymbol{x}) = F_1(\rho,\beta(\boldsymbol{J}^{-1})) + F_2(\rho,J), \tag{6.69}$$

where $\rho(\boldsymbol{x}) = \sqrt{\det(M)}$ and $\beta(\boldsymbol{J}^{-1}) = \sum_i (\nabla\xi_i)^T M^{-1} \nabla\xi_i$ (cf. (6.29)).

From the previous example, $\beta(P)$ is uniformly convex. Thus, if

$$\begin{aligned} &\frac{\partial F_1}{\partial \beta}(\rho,\beta) \geq 0, \quad \frac{\partial^2 F_1}{\partial \beta^2}(\rho,\beta) \geq 0, \quad \frac{\partial^2}{\partial r^2} F_2(\rho,\frac{1}{r}) \geq 0, \\ &\qquad \forall \rho,\, \beta \in \mathbb{R}_+,\, r \in \mathbb{R} \end{aligned} \tag{6.70}$$

then Lemma 6.2.1 and (6.58) imply that the function (6.69) is polyconvex. If, for some constants $C > 0$ and $p > d$ (where $d = 2$ for 2D and $d = 3$ for 3D), F_1 and F_2 further satisfy

$$F_1(\rho,\beta) \geq C\beta^{\frac{p}{2}}, \quad F_2(\rho,\frac{1}{r}) \geq 0, \qquad \forall \rho,\, \beta \in \mathbb{R}_+,\, r \in \mathbb{R} \tag{6.71}$$

then L satisfies the coercivity condition (6.52). From Theorem 6.2.3, the functional $I[\boldsymbol{\xi}]$ associated with (6.69) has a minimizer provided that the admissible set $\mathscr{A}$ is nonempty.

An important case is the functional

$$I[\boldsymbol{\xi}] = \int_\Omega \left[\rho \left(\sum_i (\nabla\xi_i)^T M^{-1} \nabla\xi_i \right)^d + \frac{\rho}{(J\rho)^2} \right] d\boldsymbol{x}, \tag{6.72}$$

which is an adaptation functional developed directly based on equidistribution and alignment – see (6.113) in §6.4. It can be cast in the form (6.69) with

$$F_1(\rho,\beta) = \rho\beta^{\frac{p}{2}}, \quad F_2(\rho,\frac{1}{r}) = \frac{r^2}{\rho}, \quad p = 2d.$$

It is easily verified that they satisfy (6.70) and (6.71), so the functional (6.72) is coercive and polyconvex. □

Example 6.2.3 The adaptation functional motivated by stored energy functions for hyperelastic materials [105] has the integrand

$$\begin{aligned} L(P,\boldsymbol{x}) &= F(P,\mathrm{Cof}(P),\det(P),\boldsymbol{x}) && (6.73)\\ &= w_1(\boldsymbol{x}) \left(\mathrm{tr}\left(PM^{-1}P^T\right)\right)^{\frac{p}{2}} + w_2(\boldsymbol{x}) \left(\mathrm{tr}\left(\mathrm{Cof}(PM^{-1}P^T)\right)\right)^{\frac{q}{2}} \\ &\quad + w_3(\boldsymbol{x})\det(P)^s + w_4(\boldsymbol{x})\det(P)^{-t}, \qquad \forall P \in \mathbb{R}_+^{3\times 3},\, \boldsymbol{x} \in \Omega. && (6.74) \end{aligned}$$

Here, $w_i(\boldsymbol{x}) \geq \underline{w_i} > 0$, $i = 1,\ldots,4$ are given strictly positive functions and p, q, s, and t are constants satisfying

$$p \geq 2, \quad q \geq \frac{p}{p-1}, \quad s > 1, \quad t > 0. \tag{6.75}$$

Note that

$$\lim_{\det(P) \to 0^+} L(P,\boldsymbol{x}) = +\infty.$$

Using the same procedure for proving the coercivity of $\beta(P)$ as in Example 6.2.1 we can show that $L(P,\boldsymbol{x})$ satisfies the coercivity condition (6.60). Furthermore, given the function $F = F(P,Q,r,\boldsymbol{x}) : \mathbb{R}^{3\times 3} \times \mathbb{R}^{3\times 3} \times \mathbb{R}_+ \times \Omega \longrightarrow \mathbb{R}$ in (6.73), for any $\Xi = (\xi_{i,j}), \Phi = (\phi_{i,j}) \in \mathbb{R}^{3\times 3}$, $\eta \in \mathbb{R}$, and $\boldsymbol{x} \in \mathbb{R}^3$ we can show

$$\begin{aligned}
&\sum_{i,j,m,n} \frac{\partial^2 F}{\partial p_{i,j} \partial p_{m,n}} \xi_{i,j} \xi_{m,n} + \sum_{i,j,m,n} \frac{\partial^2 F}{\partial q_{i,j} \partial q_{m,n}} \phi_{i,j} \phi_{m,n} + \frac{\partial^2 F}{\partial r^2} \eta^2 \\
&+ \sum_{i,j,m,n} \frac{\partial^2 F}{\partial p_{i,j} \partial q_{m,n}} \xi_{i,j} \phi_{m,n} + \sum_{i,j} \frac{\partial^2 F}{\partial p_{i,j} \partial r} \xi_{i,j} \eta + \sum_{i,j} \frac{\partial^2 F}{\partial q_{i,j} \partial r} \phi_{i,j} \eta \geq 0.
\end{aligned} \tag{6.76}$$

This is because the last three terms are zero, the first two terms are greater than zero (due to the fact that F is convex in P and in Q), and

$$\frac{\partial^2 F}{\partial r^2} \eta^2 = \left[s(s-1) w_3(\boldsymbol{x}) r^{s-2} + t(t+1) w_4(\boldsymbol{x}) r^{-t-2} \right] \eta^2 \geq 0,$$
$$\forall r \in \mathbb{R}_+, \, \boldsymbol{x} \in \mathbb{R}^3, \, \eta \in \mathbb{R}.$$

Inequality (6.76) implies that L is polyconvex.

Combining the above results, we conclude from Theorem 6.2.4 that the functional for (6.73) has a minimizer in the admissible set $\mathscr{A}$ defined in (6.61) as long as the set is nonempty. It is emphasized that such a minimizer satisfies $\det(\nabla \boldsymbol{w}) > 0$ for almost every point in Ω, meaning that the coordinate transformation is nonsingular locally for all points in Ω except for a zero-measure subset. □

6.3 Discretization and solution procedures

In this section, discretizations of the MMPDEs given in §5.1 and some solution procedures are studied. Those for the time-independent mesh equations (6.15) and (6.26) will not be considered separately since their solutions can be obtained as steady-state solutions of the corresponding MMPDEs. Moreover, as previously mentioned, Newton's method to solve the highly nonlinear mesh equations often fails to converge. A better solution approach is generally to use a continuation method, which amounts to simply solving the MMPDEs, where time serves as a natural choice for the continuation parameter.

Both finite difference and finite element methods are considered for spatial discretization of the MMPDEs. For temporal discretization, only the two simplest integration schemes, Euler's method and the backward Euler method, are considered. This is in part because the MMPDEs are only auxiliary equations for which it is normally unnecessary to integrate to high accuracy. Euler's method and other explicit schemes have the advantage that only the evaluation of the residual is required. On the other hand, the backward Euler method and other implicit schemes, which require solving systems of nonlinear equations, have better stability properties and allow larger time steps.

It is assumed that a suitable underlying monitor function has been chosen. For simplicity, only the two-dimensional case is considered, but extension to three dimensions is straightforward.

6.3.1 Finite difference methods

Finite difference discretization of the MMPDEs bears much similarity to that for the physical PDEs discussed in §3.2. Specifically, we assume without loss of generality that the computational domain Ω_c is taken as the unit square and a rectangular mesh is given thereon, i.e.,

$$\mathscr{T}_h^c : \quad \xi_j = (j-1)\Delta\xi,\, \eta_k = (k-1)\Delta\eta, \quad j = 1,\dots,J,\, k = 1,\dots,K,$$

where $\Delta\xi = 1/(J-1)$, $\Delta\eta = 1/(K-1)$ for given positive integers J and K. The goal is to compute the corresponding structured moving mesh

$$\mathscr{T}_h(t): \quad (x_{j,k}(t), y_{j,k}(t)) = (x(\xi_j,\eta_k,t), y(\xi_j,\eta_k,t)), \quad j = 1,\dots,J,\, k = 1,\dots,K, \tag{6.77}$$

expressed in terms of a coordinate transformation $(x,y) = (x(\xi,\eta,t), y(\xi,\eta,t)) : \Omega_c \to \Omega$ at time step $t = t_{n+1}$, assuming that the mesh $\mathscr{T}_h(t_n) = \{(x_{j,k}(t_n), y_{j,k}(t_n))\}$ at the previous time step $t = t_n$ and the monitor function $M = M(x,y,t)$ are known.

Spatial finite difference discretization can be based on the MMPDE in the semi-conservative form (6.38) or in the non-conservative form (6.39). The MMPDE (6.38) has the 2D form

$$\begin{aligned}
&-\tau J p(x,y,t) \begin{bmatrix} \dot{x} \\ \dot{y} \end{bmatrix} \\
&= \boldsymbol{a}_1 \left[\frac{\partial}{\partial \xi}\left(J\boldsymbol{a}^1 \cdot \frac{\partial F}{\partial \boldsymbol{a}^1}\right) + \frac{\partial}{\partial \eta}\left(J\boldsymbol{a}^2 \cdot \frac{\partial F}{\partial \boldsymbol{a}^1}\right) - \frac{\partial}{\partial \xi}\left(J^2 \frac{\partial F}{\partial J}\right)\right] \\
&+\boldsymbol{a}_2 \left[\frac{\partial}{\partial \xi}\left(J\boldsymbol{a}^1 \cdot \frac{\partial F}{\partial \boldsymbol{a}^2}\right) + \frac{\partial}{\partial \eta}\left(J\boldsymbol{a}^2 \cdot \frac{\partial F}{\partial \boldsymbol{a}^2}\right) - \frac{\partial}{\partial \eta}\left(J^2 \frac{\partial F}{\partial J}\right)\right],
\end{aligned} \tag{6.78}$$

while the 2D form of (6.39) is

$$\tau p(x,y,t)\begin{bmatrix}\dot{x}\\ \dot{y}\end{bmatrix} = A\begin{bmatrix}x\\ y\end{bmatrix}. \tag{6.79}$$

Here the differentiation operator

$$A = A_{1,1}\frac{\partial^2}{\partial\xi^2} + (A_{1,2}+A_{2,1})\frac{\partial^2}{\partial\xi\partial\eta} + A_{2,2}\frac{\partial^2}{\partial\eta^2} + B_1\frac{\partial}{\partial\xi} + B_2\frac{\partial}{\partial\eta} \tag{6.80}$$

and the coefficient matrices are given in (6.27), or in (6.34) for the special case of (6.28). Since the spatial finite difference discretization of (6.78) and (6.79) can be done as in §3.2, we for brevity only outline what is involved.

First, the base vectors $\boldsymbol{a}_1$, $\boldsymbol{a}_2$, $\boldsymbol{a}^1$, and $\boldsymbol{a}^2$ can be discretized as in (3.56) and through the relations (3.28) and (3.29). Second, to avoid wide finite difference stencils, the outer differentiation operators in (6.78) can be approximated using central finite differences based on half points. For instance,

$$\frac{\partial}{\partial\xi}\left(J\boldsymbol{a}^1\cdot\frac{\partial F}{\partial\boldsymbol{a}^1}\right)\Bigg|_{j,k} \approx \frac{1}{\Delta\xi}\left[\left(J\boldsymbol{a}^1\cdot\frac{\partial F}{\partial\boldsymbol{a}^1}\right)\Bigg|_{j+\frac{1}{2},k} - \left(J\boldsymbol{a}^1\cdot\frac{\partial F}{\partial\boldsymbol{a}^1}\right)\Bigg|_{j-\frac{1}{2},k}\right].$$

Third, the balancing function $p(x,y,t)$ for (6.79) can be chosen as in (6.40) or (6.41). For (6.78), computational experience has indicated that it can often suffice to instead only calculate some of the diagonal entries of the coefficient matrices. For instance, in the case of F in (6.28), a reasonable choice (cf. (6.34)) is

$$p(x,y,t) = 2\left|\frac{\partial F_1}{\partial\beta}\right|\sqrt{((\boldsymbol{a}^1)^T M^{-1}\boldsymbol{a}^1)^2 + (\boldsymbol{a}^2)^T M^{-1}\boldsymbol{a}^2)^2}. \tag{6.81}$$

For time discretizations of the MMPDEs (6.78) and (6.79), Euler's method is simple and easy to implement, requiring only the evaluation of the right-hand-side terms. It is easiest for the semi-conservative form (6.78) because of the simpler formulation.

For the backward Euler method, Newton's iteration can in theory be used for solving the resulting nonlinear system of equations. However, the computation of the Jacobian matrix of the algebraic system can be quite time consuming. One remedy is to use a different linearization. For example, for the MMPDE (6.79), the coefficient matrices can be calculated at the current time $t = t_n$ and the backward Euler method applied only to the linear part. This leads to

$$\frac{\tau p(x^n,y^n,t_n)}{\Delta t_n}\begin{bmatrix}x^{n+1}-x^n\\ y^{n+1}-y^n\end{bmatrix} = A^n\begin{bmatrix}x^{n+1}\\ y^{n+1}\end{bmatrix}, \tag{6.82}$$

for $\Delta t_n = t_{n+1} - t_n$ and

$$A^n = A^n_{1,1}\frac{\partial^2}{\partial \xi^2} + (A^n_{1,2}+A^n_{2,1})\frac{\partial^2}{\partial \xi \partial \eta} + A^n_{2,2}\frac{\partial^2}{\partial \eta^2} + B^n_1 \frac{\partial}{\partial \xi} + B^n_2 \frac{\partial}{\partial \eta}, \tag{6.83}$$

where superscript n (or $n+1$) indicates a quantity evaluated at $t = t_n$ (or $t = t_{n+1}$).

The situation for the MMPDE (6.78) is slightly more complicated. Denoting the right-hand-side term of (6.78) by rhs, after dividing by J the discretization corresponding to the above is

$$\frac{\tau p(x^n, y^n, t_n)}{\Delta t_n}\begin{bmatrix} x^{n+1} - x^n \\ y^{n+1} - y^n \end{bmatrix} = -\left.\frac{rhs}{J}\right|^{n+1}, \tag{6.84}$$

where for simplicity the balancing factor is calculated at $t = t_n$. Linearizing the right-hand side about $(x,y) = (x^n, y^n)$ yields

$$-\left.\frac{rhs}{J}\right|^{n+1} \approx -\left.\frac{rhs}{J}\right|^{n} + \mathscr{J}\left(-\frac{rhs}{J}\right)\Big|^{n}\begin{bmatrix} x^{n+1} - x^n \\ y^{n+1} - y^n \end{bmatrix},$$

where $\mathscr{J}(-rhs/J)$ denotes the Jacobian matrix of the function $-rhs/J$. Inserting this into (6.84), one obtains a linearly implicit scheme

$$\frac{\tau p(x^n, y^n, t_n)}{\Delta t_n}\begin{bmatrix} x^{n+1} - x^n \\ y^{n+1} - y^n \end{bmatrix} = -\left.\frac{rhs}{J}\right|^{n} + \mathscr{J}\left(-\frac{rhs}{J}\right)\Big|^{n}\begin{bmatrix} x^{n+1} - x^n \\ y^{n+1} - y^n \end{bmatrix}. \tag{6.85}$$

Compared to the full backward Euler scheme, this scheme has the same accuracy and only slightly worse stability. Most importantly, the Jacobian matrix can often be replaced with a simple approximation and still maintain a suitable level of accuracy and stability. For example, comparing (6.78) and (6.79) one sees that

$$-\frac{rhs}{J} = A\begin{bmatrix} x \\ y \end{bmatrix}.$$

If the Jacobian matrix is approximated by

$$\mathscr{J}\left(-\frac{rhs}{J}\right)\Big|^{n} \approx A^n, \tag{6.86}$$

then the scheme becomes

$$\frac{\tau p(x^n, y^n, t_n)}{\Delta t_n}\begin{bmatrix} x^{n+1} - x^n \\ y^{n+1} - y^n \end{bmatrix} = -\left.\frac{rhs}{J}\right|^{n} + A^n\begin{bmatrix} x^{n+1} - x^n \\ y^{n+1} - y^n \end{bmatrix}. \tag{6.87}$$

Further simplifications are possible, such as replacing A with only the second-order terms

$$A := A_{1,1}\frac{\partial^2}{\partial \xi^2} + (A_{1,2}+A_{2,1})\frac{\partial^2}{\partial \xi \partial \eta} + A_{2,2}\frac{\partial^2}{\partial \eta^2}, \tag{6.88}$$

or for (6.28), by only some of the second-order terms, such as (cf. (6.34))

$$A := 2\frac{\partial F_1}{\partial \beta} I \Big[\left((\boldsymbol{a}^1)^T M^{-1} \boldsymbol{a}^1\right) \frac{\partial^2}{\partial \xi^2} + 2\left((\boldsymbol{a}^1)^T M^{-1} \boldsymbol{a}^2\right) \frac{\partial^2}{\partial \xi \partial \eta} + \left((\boldsymbol{a}^2)^T M^{-1} \boldsymbol{a}^2\right) \frac{\partial^2}{\partial \eta^2} \Big]. \tag{6.89}$$

Interestingly, the linearly implicit scheme (6.87) can be viewed as an explicit Euler discretization of the MMPDE

$$\begin{aligned} & -J\left(\tau p(x,y,t) I - \Delta t_n A \right) \begin{bmatrix} \dot{x} \\ \dot{y} \end{bmatrix} \\ = \; & \boldsymbol{a}_1 \left[\frac{\partial}{\partial \xi}\left(J\boldsymbol{a}^1 \cdot \frac{\partial F}{\partial \boldsymbol{a}^1}\right) + \frac{\partial}{\partial \eta}\left(J\boldsymbol{a}^2 \cdot \frac{\partial F}{\partial \boldsymbol{a}^1}\right) - \frac{\partial}{\partial \xi}\left(J^2 \frac{\partial F}{\partial J}\right) \right] \\ & + \boldsymbol{a}_2 \left[\frac{\partial}{\partial \xi}\left(J\boldsymbol{a}^1 \cdot \frac{\partial F}{\partial \boldsymbol{a}^2}\right) + \frac{\partial}{\partial \eta}\left(J\boldsymbol{a}^2 \cdot \frac{\partial F}{\partial \boldsymbol{a}^2}\right) - \frac{\partial}{\partial \eta}\left(J^2 \frac{\partial F}{\partial J}\right) \right], \end{aligned} \tag{6.90}$$

where the differentiation operator A is given in (6.80) (or (6.88) or (6.89)). This MMPDE has a similar structure to MMPDE4 in 1D (see (2.54)).

We conclude this subsection with brief comments about two practical issues. First, the system of algebraic equations resulting from an implicit discretization of the MMPDE can generally be solved efficiently using an iterative method such as GMRES (Generalized Minimal RESidual method) [289] or BiCGStab (BiConjugate Gradients STABilized method) [161, 334], together with preconditioning. A commonly used preconditioner is an ILU (Incomplete Lower-Upper triangular matrix decomposition) preconditioner – e.g., see [288]. Second, there are a variety of possible implementations of boundary conditions, as discussed in §6.1.3. Generally speaking, Dirichlet boundary conditions are the easiest to implement. In contrast, difficulty often occurs with a direct implementation of Neumann or orthogonal boundary conditions. A major problem is that boundary points move out of the boundary of the domain, partly because of the fact that corner points have to be fixed during the calculation. Since Dirichlet boundary conditions are not adaptive in general, they are not suitable for problems needing boundary adaptation. Numerical experience has shown that it can be preferable to implement adaptive boundary conditions in two steps. In the first step, the boundary points are redistributed using a low-dimensional MMPDE or an orthogonal projection method (for which boundary points are obtained by projecting the adjacent inside mesh points to the boundary). In the next step, the mesh equation supplemented with Dirichlet boundary conditions specified at these new boundary points is solved for interior mesh points.

6.3.2 Finite element methods

In this subsection, we consider finite element discretizations of the MMPDEs. Like the finite difference discretizations, they can be based on either the semi-conservative form (6.37) or the non-conservative form (6.39). Unlike for finite differences, the finite element formulations incorporate the boundary data in the (integral) formulations. For simplicity, we shall only consider the case of Dirichlet boundary conditions. If $\boldsymbol{v}(x,y)$ is an arbitrary admissible, vector-valued test function which vanishes on the boundary of Ω, then the weak formulations of the MMPDEs are given by

$$\int_{\Omega} \tau p(\boldsymbol{x},t)\,(\dot{\boldsymbol{x}}\cdot\boldsymbol{v})\,d\boldsymbol{x} = \sum_i \int_{\Omega} \left[\frac{\partial F}{\partial \boldsymbol{a}^i} - J\frac{\partial F}{\partial J}\boldsymbol{a}_i\right]\cdot\nabla\left(\boldsymbol{a}_i\cdot\boldsymbol{v}\right)d\boldsymbol{x} \tag{6.91}$$

and

$$\int_{\Omega_c} \tau p(\boldsymbol{x},t)\,(\dot{\boldsymbol{x}}\cdot\boldsymbol{v})\,d\boldsymbol{\xi}$$
$$= -\sum_{i,j}\int_{\Omega_c}\frac{\partial \boldsymbol{x}}{\partial \xi_i}\cdot\frac{\partial}{\partial \xi_j}(A_{i,j}\boldsymbol{v})\,d\boldsymbol{\xi} + \sum_i\int_{\Omega_c}\frac{\partial \boldsymbol{x}}{\partial \xi_i}\cdot(B_i\boldsymbol{v})\,d\boldsymbol{\xi}. \tag{6.92}$$

Finite element spatial discretizations for (6.91) and (6.92) can be done in the standard way (cf. §3.3). Specifically, if $\mathscr{T}_h^c$ is an affine family of quasi-uniform triangulations of the computational domain Ω_c and $\mathscr{T}_h(t)$ is the corresponding time-dependent triangulation of Ω having the same connectivity as $\mathscr{T}_h^c$, then for a given approximation space $S^{c,h}$ on $\mathscr{T}_h^c$ with basis $\{\psi_{c,j}\}$, the finite element approximation to the coordinate transformation has the representation

$$\boldsymbol{x}^h = \sum_j \boldsymbol{x}_j(t)\psi_{c,j}(\boldsymbol{\xi}), \qquad \dot{\boldsymbol{x}}^h = \sum_j \dot{\boldsymbol{x}}_j(t)\psi_{c,j}(\boldsymbol{\xi}), \tag{6.93}$$

where $\boldsymbol{x}_j$ and $\dot{\boldsymbol{x}}_j$ are the node location and corresponding mesh speed. For the approximation space $S_0^{c,h} = \{v \mid v\in S^{c,h}, \quad v|_{\partial\Omega_c} = 0\}$ defined on the computational domain, the finite element solution of (6.91) is determined by suitable boundary conditions (resulting from the boundary correspondence between Ω and Ω_c) and

$$\int_{\Omega} \tau p(\boldsymbol{x},t)\left(\dot{\boldsymbol{x}}^h\cdot\boldsymbol{v}\right)d\boldsymbol{x}$$
$$= \sum_i\int_{\Omega}\left[\frac{\partial F}{\partial \boldsymbol{a}^i} - J\frac{\partial F}{\partial J}\boldsymbol{a}_i\right]\cdot\nabla\left(\boldsymbol{a}_i\cdot\boldsymbol{v}\right)d\boldsymbol{x}, \quad \forall \boldsymbol{v}\in S_0^{c,h}\times\cdots\times S_0^{c,h}, \tag{6.94}$$

where the base vectors $\boldsymbol{a}_i$ and $\boldsymbol{a}^i$ are calculated using (6.93) (i.e., $\boldsymbol{a}_i \approx \frac{\partial \boldsymbol{x}^h}{\partial \xi_i}$) and (3.7). The finite element solution of (6.92) can be defined similarly.

The finite element discretizations (6.94) can be straightforwardly integrated in time using Euler's scheme or the backward Euler scheme for the implicit temporal discretization, and any of the approximate Jacobian matrices discussed in the previous subsection can be used in the iterative solution of the resulting algebraic systems.

6.4 Methods based on equidistribution and alignment conditions

Recall from Chapter 4 that the basic properties of an M-uniform mesh – the size, shape, and orientation of mesh elements – can be specified through a monitor function (cf. Theorem 4.2.3). Moreover, it is shown in Chapter 5 that the optimal monitor function can be defined for M-uniform meshes based on an interpolation or solution error. It is thus natural to formulate a variational method for generating M-uniform meshes, with the monitor function chosen based on the error analysis. We use here the approach of [176] for developing such a method. Specifically, two functionals are developed from the equidistribution and alignment conditions for an M-uniform mesh, and the final adaptation functional for the method is defined as a linear combination of them.

6.4.1 Functional for mesh alignment

The functional is developed based on (4.80), a continuous form of the alignment condition. Denote the eigenvalues of $\boldsymbol{J}^{-1}M^{-1}\boldsymbol{J}^{-T}$ by $\lambda_1, ..., \lambda_d$. Recalling that

$$\operatorname{tr}\left(\boldsymbol{J}^{-1}M^{-1}\boldsymbol{J}^{-T}\right) = \sum_i \lambda_i, \quad \det\left(\boldsymbol{J}^{-1}M^{-1}\boldsymbol{J}^{-T}\right) = \prod_i \lambda_i, \tag{6.95}$$

we can rewrite (4.80) as

$$\frac{1}{d}\sum_i \lambda_i = \left(\prod_i \lambda_i\right)^{\frac{1}{d}}. \tag{6.96}$$

From the arithmetic-mean geometric-mean inequality (cf. Theorem B.0.11), the geometric mean of the eigenvalues is less than or equal to their geometric mean, i.e.,

$$\left(\prod_i \lambda_i\right)^{\frac{1}{d}} \le \frac{1}{d}\sum_i \lambda_i. \tag{6.97}$$

This implies that the coordinate transformation satisfying the alignment condition (6.96) can be obtained by minimizing the difference between the two sides of inequality (6.97). Since $\nabla\boldsymbol{\xi} = \boldsymbol{J}^{-1}$,

$$\sum_i \lambda_i = \mathrm{tr}(\boldsymbol{J}^{-1}M^{-1}\boldsymbol{J}^{-T}) = \sum_i (\nabla\xi_i)^T M^{-1}\nabla\xi_i, \tag{6.98}$$

and

$$\prod_i \lambda_i = \det(\boldsymbol{J}^{-1}M^{-1}\boldsymbol{J}^{-T}) = \frac{1}{(J\rho)^2}, \tag{6.99}$$

where $J = \det(\boldsymbol{J})$ and $\rho = \sqrt{\det(M)}$. Thus, (6.97) becomes

$$\left(\frac{1}{(J\rho)^2}\right)^{\frac{1}{d}} \le \frac{1}{d}\sum_i (\nabla\xi_i)^T M^{-1}\nabla\xi_i,$$

or equivalently

$$\frac{d^{\frac{d}{2}}}{J} \le \rho\left(\sum_i (\nabla\xi_i)^T M^{-1}\nabla\xi_i\right)^{\frac{d}{2}}. \tag{6.100}$$

Integrating over the physical domain yields

$$d^{\frac{d}{2}}\int_{\Omega_c} d\boldsymbol{\xi} \le \int_\Omega \rho\left(\sum_i (\nabla\xi_i)^T M^{-1}\nabla\xi_i\right)^{\frac{d}{2}} d\boldsymbol{x}.$$

Hence, the adaptation functional associated with mesh alignment for the inverse coordinate transformation $\boldsymbol{\xi} = \boldsymbol{\xi}(\boldsymbol{x})$ can be defined as

$$I_{ali}[\boldsymbol{\xi}] = \frac{1}{2}\int_\Omega \rho\left(\sum_i (\nabla\xi_i)^T M^{-1}\nabla\xi_i\right)^{\frac{d}{2}} d\boldsymbol{x}. \tag{6.101}$$

The functional in (6.101) can also be derived using the concept of conformal norm in differential geometry, and for this reason the alignment is called the isotropy or conformity condition

in [176]. Moreover, in two dimensions ($d = 2$), (6.101) gives the energy of a harmonic mapping – see §6.5.2 and [129] .

6.4.2 Functional for equidistribution

Consider now the equidistribution condition (4.74). From Theorem A.0.3 one obtains that for any $\gamma > 1$,

$$\left(\frac{1}{\sigma}\int_\Omega \frac{\rho}{(J\rho)^\gamma}d\boldsymbol{x}\right)^{1/\gamma} = \left(\frac{1}{\sigma}\int_\Omega \rho\left(\frac{1}{J\rho}\right)^\gamma d\boldsymbol{x}\right)^{1/\gamma}$$
$$\geq \frac{1}{\sigma}\int_\Omega \rho\left(\frac{1}{J\rho}\right)d\boldsymbol{x} = \frac{1}{\sigma}\int_{\Omega_c} d\xi, \qquad (6.102)$$

with equality if and only if (4.74) holds. (Recall that $\sigma = \int_\Omega \rho d\boldsymbol{x}$.) Thus, the adaptation functional defined using the equidistribution condition (4.74) has the general form

$$I_{eq}[\boldsymbol{\xi}] = \int_\Omega \frac{\rho}{(J\rho)^\gamma}d\boldsymbol{x}, \qquad (6.103)$$

where the parameter $\gamma > 1$.

6.4.3 Mesh adaptation functional

Recall that M-uniform meshes satisfy both the equidistribution and alignment conditions (4.74) and (4.75). To generate such meshes, it is thus natural to combine the functionals developed in the two previous subsections. To this end, taking the γ-th power on both sides of (6.100), dividing by $\rho^{\gamma-1}$, and integrating over Ω, one gets

$$d^{\frac{d\gamma}{2}}\int_\Omega \frac{\rho}{(J\rho)^\gamma}d\boldsymbol{x} \leq \int_\Omega \rho\left(\sum_i (\nabla\xi_i)^T M^{-1}\nabla\xi_i\right)^{\frac{d\gamma}{2}} d\boldsymbol{x}. \qquad (6.104)$$

Taking the γ-th power in (6.102), it follows that

$$\sigma^{1-\gamma}\left(\int_{\Omega_c} d\boldsymbol{\xi}\right)^\gamma \leq \int_\Omega \frac{\rho}{(J\rho)^\gamma}d\boldsymbol{x}. \qquad (6.105)$$

The differences between the two sides of (6.104) and (6.105) can be balanced by taking

$$\theta\left[\int_\Omega \rho\left(\sum_i (\nabla\xi_i)^T M^{-1}\nabla\xi_i\right)^{\frac{d\gamma}{2}} d\boldsymbol{x} - d^{\frac{d\gamma}{2}}\int_\Omega \frac{\rho}{(J\rho)^\gamma}d\boldsymbol{x}\right]$$
$$+ (1-\theta)d^{\frac{d\gamma}{2}}\left[\int_\Omega \frac{\rho}{(J\rho)^\gamma}d\boldsymbol{x} - \sigma^{1-\gamma}\left(\int_{\Omega_c} d\boldsymbol{\xi}\right)^\gamma\right] \qquad (6.106)$$

for a given value $\theta \in [0,1]$. Thus, we obtain the adaptation functional

$$I_{Hua}[\boldsymbol{\xi};\theta,\gamma] = \theta\left[\int_{\Omega}\rho\left(\sum_i(\nabla\xi_i)^T M^{-1}\nabla\xi_i\right)^{\frac{d\gamma}{2}} d\boldsymbol{x} - d^{\frac{d\gamma}{2}}\int_{\Omega}\frac{\rho}{(J\rho)^{\gamma}}d\boldsymbol{x}\right]$$
$$+ (1-\theta)d^{\frac{d\gamma}{2}}\int_{\Omega}\frac{\rho}{(J\rho)^{\gamma}}d\boldsymbol{x}, \tag{6.107}$$
$$= \theta\int_{\Omega}\rho\left(\sum_i(\nabla\xi_i)^T M^{-1}\nabla\xi_i\right)^{\frac{d\gamma}{2}} d\boldsymbol{x}$$
$$+(1-2\theta)d^{\frac{d\gamma}{2}}\int_{\Omega}\frac{\rho}{(J\rho)^{\gamma}}d\boldsymbol{x}, \tag{6.108}$$

for parameters $\theta \in [0,1]$ and $\gamma > 1$, where $\rho = \sqrt{\det(M)}$ and $J = \det(\boldsymbol{J})$. The first term of I_{Hua} in (6.107) corresponds to the alignment requirement and, in particular, reduces to a functional equivalent to I_{ali} in (6.101) when $\gamma = 1$.[2] The second term in the functional is the same as (6.103), which represents the equidistribution requirement. By design, the terms have the same dimension, and the balance between equidistribution and alignment is controlled by the dimensionless parameter θ.

The functional (6.108) can be shown to have invariance properties corresponding to those for the equidistribution and alignment conditions (cf. Theorems 4.1.2 and 4.3.2).

Theorem 6.4.1 *Functional (6.108) is invariant in the sense of functional equivalence under a scaling transformation of M and under rotation, translation, and dilation transformations of $\boldsymbol{\xi}$.*

Proof. Notice that (6.108) can be rewritten as

$$I_{Hua}[\boldsymbol{\xi};\theta,\gamma] = \theta\int_{\Omega}\rho\left(\mathrm{tr}\left(\boldsymbol{J}^{-1}M^{-1}\boldsymbol{J}^{-T}\right)\right)^{\frac{d\gamma}{2}} d\boldsymbol{x}$$
$$+(1-2\theta)d^{\frac{d\gamma}{2}}\int_{\Omega}\rho\,\det\left(\boldsymbol{J}^{-1}M^{-1}\boldsymbol{J}^{-T}\right)^{\frac{\gamma}{2}} d\boldsymbol{x}. \tag{6.109}$$

When M changes to cM for some positive constant c, ρ changes to $c^{\frac{d}{2}}\rho$. With the new monitor function, the functional takes the form

$$\tilde{I}[\boldsymbol{\xi}] = c^{\frac{d(1-\gamma)}{2}}\left[\theta\int_{\Omega}\rho\left(\mathrm{tr}\left(\boldsymbol{J}^{-1}M^{-1}\boldsymbol{J}^{-T}\right)\right)^{\frac{d\gamma}{2}} d\boldsymbol{x}\right.$$
$$\left.+(1-2\theta)d^{\frac{d\gamma}{2}}\int_{\Omega}\rho\,\det\left(\boldsymbol{J}^{-1}M^{-1}\boldsymbol{J}^{-T}\right)^{\frac{\gamma}{2}} d\boldsymbol{x}\right],$$

which is obviously equivalent to the original functional (6.108).

[2] For a given boundary condition, functionals are said to be equivalent if they have the same minimizers.

The invariance of (6.108) under a translation transformation of $\boldsymbol{\xi}$ is clear. Consider rotation and dilation transformations of $\boldsymbol{\xi}$, i.e.,

$$\boldsymbol{\xi} = cQ\tilde{\boldsymbol{\xi}}, \tag{6.110}$$

where c is a positive constant and Q is a constant orthogonal matrix. By the chain rule,

$$\boldsymbol{J} = \frac{\partial \boldsymbol{x}}{\partial \boldsymbol{\xi}} = \frac{\partial \boldsymbol{x}}{\partial \tilde{\boldsymbol{\xi}}} \frac{\partial \tilde{\boldsymbol{\xi}}}{\partial \boldsymbol{\xi}} = \tilde{\boldsymbol{J}}(cQ)^{-1},$$

which gives

$$\boldsymbol{J}^{-1} = cQ\tilde{\boldsymbol{J}}^{-1}. \tag{6.111}$$

Thus,

$$\begin{aligned} I[\boldsymbol{\xi}] &= c^{d\gamma}\Big[\ \theta \int_\Omega \rho \left(\mathrm{tr}\left(Q\tilde{\boldsymbol{J}}^{-1} M^{-1} \tilde{\boldsymbol{J}}^{-T} Q^T\right)\right)^{\frac{d\gamma}{2}} d\boldsymbol{x} \\ &\qquad +(1-2\theta) d^{\frac{d\gamma}{2}} \int_\Omega \rho \det\left(\tilde{Q}\boldsymbol{J}^{-1} M^{-1} \tilde{\boldsymbol{J}}^{-T} Q^T\right)^{\frac{\gamma}{2}} d\boldsymbol{x}\ \Big] \\ &= c^{d\gamma}\Big[\ \theta \int_\Omega \rho \left(\mathrm{tr}\left(\tilde{\boldsymbol{J}}^{-1} M^{-1} \tilde{\boldsymbol{J}}^{-T}\right)\right)^{\frac{d\gamma}{2}} d\boldsymbol{x} \\ &\qquad +(1-2\theta) d^{\frac{d\gamma}{2}} \int_\Omega \rho \det\left(\tilde{\boldsymbol{J}}^{-1} M^{-1} \tilde{\boldsymbol{J}}^{-T}\right)^{\frac{\gamma}{2}} d\boldsymbol{x}\ \Big] \\ &= c^{d\gamma} I[\tilde{\boldsymbol{\xi}}], \end{aligned}$$

and we conclude that (6.107), and therefore (6.108), is invariant under rotation and dilation transformations of $\boldsymbol{\xi}$. □

We now consider two special cases of (6.108). (i) For $\theta = 1/2$, only the first term remains, leaving the simpler functional

$$I_{Hua}[\boldsymbol{\xi}; 0.5, \gamma] = \frac{1}{2} \int_\Omega \rho \left(\sum_i (\nabla \xi_i)^T M^{-1} \nabla \xi_i\right)^{\frac{d\gamma}{2}} d\boldsymbol{x}. \tag{6.112}$$

This is in turn a special case of the functional in Example 6.2.1. When $d\gamma/2 \geq 1$, it is coercive and convex and has at least a minimizer in the admissible set $\mathscr{A}$ defined in (6.54) provided that $\mathscr{A}$ is nonempty. In practice, (6.112) may not sufficiently weight the equidistribution portion, necessitating use of (6.108) with a smaller value of θ than 1/2. (ii) In [176] $\theta = 0.1$ is used and found to work well for all problems tested. It is also recommended in [176] that $\gamma = 2$, in which case (6.108) becomes

$$I_{Hua}[\boldsymbol{\xi};\theta] := I_{Hua}[\boldsymbol{\xi};\theta,2] = \theta \int_\Omega \rho \left(\sum_i (\nabla \xi_i)^T M^{-1} \nabla \xi_i\right)^d d\boldsymbol{x} + (1-2\theta)d^d \int_\Omega \frac{\rho}{(J\rho)^2} d\boldsymbol{x}. \qquad (6.113)$$

This functional is of the form considered in Example 6.2.2 and is essentially the same as (6.72). It follows from that example that (6.113) is coercive and polyconvex for $\theta \in (0, \frac{1}{2})$, and thus the existence of its minimizers is guaranteed by Theorem 6.2.3.[3] The integrand of (6.113) can be written in the form (6.28), with

$$F(\boldsymbol{a}^1, \boldsymbol{a}^2, \boldsymbol{a}^3, J, \boldsymbol{x}) = F_1(\rho, \beta) + F_2(\rho, J) = \theta \rho \beta^d + (1-2\theta)d^d \frac{\rho}{(J\rho)^2}. \qquad (6.114)$$

Thus, the partial derivatives required for the MMPDEs (6.37) and (6.39) (with coefficient matrices in (6.34)) are given by

$$\begin{cases} \frac{\partial F}{\partial \boldsymbol{a}^i} = 2\frac{\partial F_1}{\partial \beta} M^{-1} \boldsymbol{a}^i = 2d\theta\rho\beta^{d-1} M^{-1} \boldsymbol{a}^i, \\ \frac{\partial F}{\partial J} = \frac{\partial F_2}{\partial J} = -2(1-2\theta)d^d \rho^{-1} J^{-3}, \\ \frac{\partial F_1}{\partial \beta} = d\theta\rho\beta^{d-1}, \quad \frac{\partial^2 F_1}{\partial \beta^2} = d(d-1)\theta\rho\beta^{d-2}, \quad \frac{\partial^2 F_1}{\partial \beta \partial \rho} = d\theta\beta^{d-1}, \\ \frac{\partial F_2}{\partial J} = -2(1-2\theta)d^d \rho^{-1} J^{-3}, \quad \frac{\partial^2 F_2}{\partial J^2} = 6(1-2\theta)d^d \rho^{-1} J^{-4}, \\ \frac{\partial^2 F_2}{\partial J \partial \rho} = 2(1-2\theta)d^d \rho^{-2} J^{-3}. \end{cases} \qquad (6.115)$$

We re-emphasize that minimizing a functional of the form (6.108) or (6.113) serves as a balance between the goals of satisfying the equidistribution and alignment conditions. The influence of the monitor function on the structure of the resulting (approximately) M-uniform mesh is clear: the shape and orientation of mesh cells are determined from M through (4.75) or (4.29) while the cell size is determined through (4.74) or (4.28). Moreover, as discussed in Chapter 5, the monitor function can be chosen with the goal of minimizing an interpolation error or an a posteriori estimate of the solution error.

6.4.4 Another mesh adaptation functional

In a similar way as above we can also develop a functional of the form (6.74) given in Example 6.2.3 based on equidistribution and alignment. To see this, we start with (6.108) or (6.109) using new dimensionless constant weights θ_1 and θ_3: for any given $\gamma > 1$,

---

[3] The existence of minimizer can also be proven for the more general functional form (6.108) for positive even integer γ but not other values of γ. This is because J^{-1} is not necessarily positive and the second term is not convex about J^{-1}.

$$\theta_1 \left[\rho \left(\text{tr} \left(\boldsymbol{J}^{-1} M^{-1} \boldsymbol{J}^{-T} \right) \right)^{\frac{d\gamma}{2}} - d^{\frac{d\gamma}{2}} \rho \det \left(\boldsymbol{J}^{-1} M^{-1} \boldsymbol{J}^{-T} \right)^{\frac{\gamma}{2}} \right]$$
$$+ \theta_3 d^{\frac{d\gamma}{2}} \rho \det \left(\boldsymbol{J}^{-1} M^{-1} \boldsymbol{J}^{-T} \right)^{\frac{\gamma}{2}}$$
$$= \theta_1 \rho \left(\text{tr} \left(\boldsymbol{J}^{-1} M^{-1} \boldsymbol{J}^{-T} \right) \right)^{\frac{d\gamma}{2}} + (\theta_3 - \theta_1) d^{\frac{d\gamma}{2}} \rho \det \left(\boldsymbol{J}^{-1} M^{-1} \boldsymbol{J}^{-T} \right)^{\frac{\gamma}{2}}. \quad (6.116)$$

These terms, based on the alignment and equidistribution conditions, also correspond to the first and third terms in (6.74), respectively (where $p = d\gamma$, $s = \gamma$, and $P = \boldsymbol{J}^{-1}$). For the second term, note that

$$\text{Cof}(PM^{-1}P^T) = \text{Cof}(\boldsymbol{J}^{-1} M^{-1} \boldsymbol{J}^{-T}) = (J\rho)^{-2} \boldsymbol{J}^T M \boldsymbol{J}.$$

Taking $q = d\gamma/(d-1)$ in (6.74), then $q \geq p/(p-1) = d\gamma/(d\gamma - 1)$ so the second condition of (6.75) is satisfied. It follows from the arithmetic-mean geometric-mean inequality that

$$\begin{aligned}
\left(\text{tr} \left(\text{Cof}(PM^{-1}P^T) \right) \right)^{\frac{q}{2}} &= (J\rho)^{-\frac{d\gamma}{d-1}} \left(\text{tr} \left(\boldsymbol{J}^T M \boldsymbol{J} \right) \right)^{\frac{d\gamma}{2(d-1)}} \\
&\geq (J\rho)^{-\frac{d\gamma}{d-1}} \left(d \det \left(\boldsymbol{J}^T M \boldsymbol{J} \right)^{\frac{1}{d}} \right)^{\frac{d\gamma}{2(d-1)}} \\
&= d^{\frac{d\gamma}{2(d-1)}} (J\rho)^{-\gamma} \\
&= d^{\frac{d\gamma}{2(d-1)}} \det \left(\boldsymbol{J}^{-1} M^{-1} \boldsymbol{J}^{-T} \right)^{\frac{\gamma}{2}}.
\end{aligned}$$

Thus, an integrand which corresponds to the alignment condition and has the form of the second term of (6.74) is

$$\theta_2 \left[d^{\frac{d\gamma(d-2)}{2(d-1)}} \rho (J\rho)^{-\frac{d\gamma}{d-1}} \left(\text{tr} \left(\boldsymbol{J}^T M \boldsymbol{J} \right) \right)^{\frac{d\gamma}{2(d-1)}} - d^{\frac{d\gamma}{2}} \det \left(\boldsymbol{J}^{-1} M^{-1} \boldsymbol{J}^{-T} \right)^{\frac{\gamma}{2}} \right], \quad (6.117)$$

where θ_2 is a dimensionless constant weight. Finally, for the last term of (6.74), we notice that for any $\nu > 0$,

$$\det \left(\boldsymbol{J}^{-1} M^{-1} \boldsymbol{J}^{-T} \right)^{-\frac{\nu}{2}} = (J\rho)^{\nu}.$$

From Theorem A.0.3,

$$\left(\frac{1}{\sigma} \int_{\Omega} \rho (J\rho)^{\nu} d\boldsymbol{x} \right)^{\frac{1}{\nu}} \geq \left(\frac{1}{\sigma} \int_{\Omega} \rho (J\rho)^{-1} d\boldsymbol{x} \right)^{-1} = \left(\frac{|\Omega_c|}{\sigma} \right)^{-1}, \quad (6.118)$$

with equality if and only if $J\rho =$ constant. Thus, an integrand which is both associated with the equidistribution condition and of the form of the last term of (6.74) is

$$\frac{\theta_4}{\sigma^{\gamma+\nu}} \rho (J\rho)^{\nu} = \frac{\theta_4}{\sigma^{\gamma+\nu}} \rho \det \left(\boldsymbol{J}^{-1} M^{-1} \boldsymbol{J}^{-T} \right)^{-\frac{\nu}{2}}, \quad (6.119)$$

where θ_4 is a dimensionless constant weight and the factor involving $\sigma = \int_\Omega \rho d\boldsymbol{x}$ has been added to make this integrand have the same dimension as the others. Adding the terms in (6.116), (6.117), and (6.119), we get

$$\begin{aligned} I[\boldsymbol{\xi};\theta_1,\theta_2.\theta_3,\theta_4] = \theta_1 & \int_\Omega \rho \left(\operatorname{tr}\left(\boldsymbol{J}^{-1}M^{-1}\boldsymbol{J}^{-T}\right)\right)^{\frac{d\gamma}{2}} d\boldsymbol{x} \\ & + \theta_2\, d^{\frac{d\gamma(d-2)}{2(d-1)}} \int_\Omega \frac{\rho}{(J\rho)^{\frac{d\gamma}{d-1}}} \left(\operatorname{tr}\left(\boldsymbol{J}^T M \boldsymbol{J}\right)\right)^{\frac{d\gamma}{2(d-1)}} d\boldsymbol{x} \\ & + (\theta_3 - \theta_1 - \theta_2)\, d^{\frac{d\gamma}{2}} \int_\Omega \rho \det\left(\boldsymbol{J}^{-1}M^{-1}\boldsymbol{J}^{-T}\right)^{\frac{\gamma}{2}} d\boldsymbol{x} \\ & + \frac{\theta_4}{\sigma^{\gamma+\nu}} \int_\Omega \rho \det\left(\boldsymbol{J}^{-1}M^{-1}\boldsymbol{J}^{-T}\right)^{-\frac{\nu}{2}} d\boldsymbol{x}, \end{aligned} \tag{6.120}$$

where $\gamma > 1$, $\nu > 0$, and θ_1, ..., θ_4 are dimensionless constant weights. Note that this functional can be rewritten as

$$\begin{aligned} & I[\boldsymbol{\xi};\theta_1,\theta_2.\theta_3,\theta_4] \\ & = \theta_1 \int_\Omega \rho \left(\sum_i (\nabla \xi_i)^T M^{-1} \nabla \xi_i\right)^{\frac{d\gamma}{2}} d\boldsymbol{x} \\ & + \theta_2\, d^{\frac{d\gamma(d-2)}{2(d-1)}} \int_\Omega \frac{\rho}{(J\rho)^{\frac{d\gamma}{d-1}}} \left(\sum_i \left(\frac{\partial \boldsymbol{x}}{\partial \xi_i}\right)^T M \frac{\partial \boldsymbol{x}}{\partial \xi_i}\right)^{\frac{d\gamma}{2(d-1)}} d\boldsymbol{x} \\ & + (\theta_3 - \theta_1 - \theta_2)\, d^{\frac{d\gamma}{2}} \int_\Omega \frac{\rho}{(J\rho)^\gamma} d\boldsymbol{x} + \frac{\theta_4}{\sigma^{\gamma+\nu}} \int_\Omega \rho\, (J\rho)^\nu d\boldsymbol{x}, \end{aligned} \tag{6.121}$$

It is easy to verify that this functional, like (6.108), is also invariant under a scaling transformation of M and rotation, translation, and dilation transformations of the computational coordinate $\boldsymbol{\xi}$. This functional is more complicated than (6.108), but this complexity may be worthwhile for some applications. Recall from Example 6.2.3 that the functional satisfies the conditions of Theorem 6.2.4 and has a minimizer in the admissible set $\mathscr{A}$ defined in (6.61) if $\mathscr{A}$ is nonempty and the positive weights satisfy

$$\theta_3 - \theta_1 - \theta_2 > 0. \tag{6.122}$$

Since the minimizer satisfies $J = \det(\boldsymbol{J}) > 0$ almost everywhere in Ω, this coordinate transformation is guaranteed to be nonsingular. Such theoretical guarantees can be rare to obtain for many variational methods, including (6.108), although no singular meshes have been observed numerically for this latter approach when the equations are properly discretized and solved.

6.4.5 Numerical examples

In this subsection we present four examples for which the functions or monitor functions are given explicitly. Numerical results are obtained with MMPDE5 in the form (6.39) for the functionals (6.101) and (6.113), which are discretized using finite differences in space and the (modified) backward Euler scheme (cf. (6.82) and (6.83)). The resulting linear systems are solved using the GMRES iterative method, preconditioned with the level-one fill-in ILU (incomplete LU) preconditioner [288]. The balancing function is taken as in (6.40), and $\tau = 1$ in the first two examples (for generating adaptive meshes for given functions), and $\tau = 10^{-2}$ for the third and fourth (where time-dependent monitor functions are used). The boundary point distribution is computed using a 1D MMPDE. The computational domain is taken as the unit square with a uniform mesh of size $J \times K$ (with $J = K$).

Example 6.4.1 In our first example an adaptive mesh is generated for the monitor function

$$M(x,y) = \rho(x,y)^2 \boldsymbol{v}\boldsymbol{v}^T + \boldsymbol{v}^{\perp}(\boldsymbol{v}^{\perp})^T, \quad \forall (x,y) \in \Omega = (0,1) \times (0,1) \qquad (6.123)$$

where $\boldsymbol{v} = (v_1, v_2)^T$ is a constant normal vector, $\boldsymbol{v}^{\perp} = (v_2, -v_1)^T$, and

$$\rho(x,y) = 1 + 100 \exp\left(-50 \left| y - \frac{1}{2} - \frac{1}{4}\sin(2\pi x)\right|\right). \qquad (6.124)$$

It is easy to see that this monitor function has the two eigenpairs $(\rho^2, \boldsymbol{v})$ and $(1, \boldsymbol{v}^{\perp})$.

Both functionals (6.101) and (6.113) are used for this example. Recall that (6.101) is associated with the alignment condition (4.75) (and also reduces to the harmonic mapping method discussed in §6.5.2) for this 2D case). The function ρ is chosen to illustrate mesh alignment with a monitor function. Indeed, it is expected that a minimizer of (6.101) satisfies (4.75) approximately in an integral sense. Noticing that $\rho(x,y)$ has a bump around the curve

$$y - \frac{1}{2} - \frac{1}{4}\sin(2\pi x) = 0, \qquad (6.125)$$

from Theorem 4.2.3 we anticipate that $h_{\boldsymbol{v}}(K)/h_{\boldsymbol{v}^{\perp}}(K) \approx 1/\rho$ for mesh elements in regions around the curves and $h_{\boldsymbol{v}}(K)/h_{\boldsymbol{v}^{\perp}}(K) \approx 1$ for elements in other regions, where $h_{\boldsymbol{v}}(K)$ and $h_{\boldsymbol{v}^{\perp}}(K)$ denote the lengths of element K along directions $\boldsymbol{v}$ and $\boldsymbol{v}^{\perp}$, respectively. Meshes obtained with $\boldsymbol{v} = (1,0)^T$, $(0,1)^T$, and $(1/\sqrt{2}, 1/\sqrt{2})^T$, shown in Figure 6.2(a), 6.3(a), and 6.4(a), respectively, do indeed show mesh alignment with the monitor function.. For example, in Figure 6.2(a) mesh elements around the curve (6.125) (depicted as a solid line) are thinner in the x-direction than in the y-direction, and other elements are much more regular. This is consistent with the above analysis since $\boldsymbol{v} = (0,1)^T$, $h_{\boldsymbol{v}}(K)/h_{\boldsymbol{v}^{\perp}}(K) = h_x(K)/h_y(K) \approx 1/\rho$ for elements around the curve, where $h_x(K)$ and $h_y(K)$ are the lengths of K in the x and y directions, respectively. However, one can also see from the figures that the mesh density is not necessarily higher around the curve. This is because functional (6.101) associated with alignment does not take into consideration equidistribution, so there is no guarantee that the mesh density is higher in regions where $\rho = \sqrt{\det(M)}$ is larger. To emphasize this, we have also plotted alongside these figures the meshes obtained with functional (6.113) (with $\theta = 0.1$). As expected, those meshes have a higher mesh density but somewhat weaker alignment around the curve. Interestingly, in the regions marked by "O" the mesh elements have much better alignment and concentration than in other regions along the curve. A common feature of these regions is that the eigenvalue ρ^2 has the greatest change (a bump) in the eigendirection $\boldsymbol{v}$. As we shall see, this is consistent with the qualitative analysis given in §6.5.1, even though the functional (6.113) cannot be cast in the form considered there ((6.139)).

This last observation suggests using a monitor function of the form

$$M = \frac{f(\psi)}{\|\nabla\psi\|^2}\nabla\psi(\nabla\psi)^T + \nabla\psi^{\perp}(\nabla\psi^{\perp})^T = I + \frac{f(\psi)-1}{\|\nabla\psi\|^2}\nabla\psi(\nabla\psi)^T \qquad (6.126)$$

to obtain a mesh with good alignment and concentration around a curve defined by $\psi(\boldsymbol{x}) = 0$.[4] Here, $\nabla\psi^{\perp}$ denotes a vector perpendicular to and of the same length as $\nabla\psi$, and $f(\psi)$ is a function of the distance from a given point to $\psi(\boldsymbol{x}) = 0$ that is greater than one and increases as the distance tends to zero. The monitor function (6.126) can also be used to concentrate mesh points in regions where an arbitrary function $\psi(\boldsymbol{x})$ has the largest gradient. In such a case, the function f can be defined as $f(\psi) = 1 + \|\nabla\psi\|^2$.

Using the modified monitor function (6.126) for $f = \rho$ in (6.124) and $\psi(x,y) = y - \frac{1}{2} - \frac{1}{4}\sin(2\pi x)$ in (6.125), we have

$$M = I + \frac{\rho^2 - 1}{\|\nabla\psi\|^2}\nabla\psi(\nabla\psi)^T. \qquad (6.127)$$

Meshes obtained using (6.101) and (6.113) with this monitor function are shown in Figure 6.5. Now mesh alignment and concentration improve for both functionals, and the mesh in Figure 6.5(b) obtained with (6.113) (with $\theta = 0.1$) shows both good alignment and concentration. □

For the remaining examples in this subsection, we only consider a functional having an equidistribution component, viz., (6.113) with $\theta = .1$, and solve by integrating MMPDE5 in (6.39) from $t = 0$ to $t = 1$, unless stated otherwise.

Example 6.4.2 The goal here is to generate an adaptive mesh for the function

[4] A similar form of the monitor function is also suggested for the generalized variable diffusion and harmonic mapping methods – see (6.148), (6.152), and (6.160).

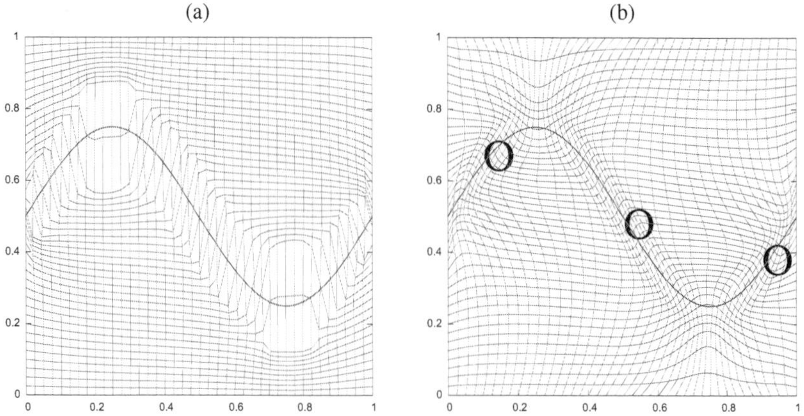

Fig. 6.2 Example 6.4.1. Adaptive meshes are generated using functionals (a) (6.101) and (b) (6.113) (with $\theta = 0.1$) for monitor function (6.123) with $\boldsymbol{v} = (1,0)^T$. The curve (6.125) is shown as a solid line.

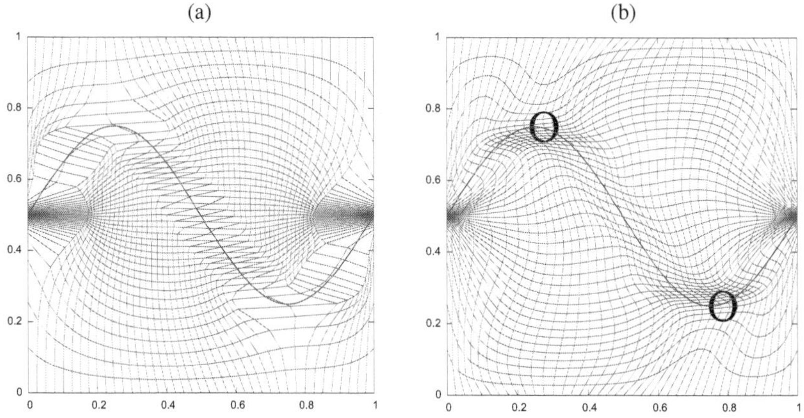

Fig. 6.3 Example 6.4.1. Adaptive meshes are generated using functionals (a) (6.101) and (b) (6.113) (with $\theta = 0.1$) for monitor function (6.123) with $\boldsymbol{v} = (0,1)^T$. The curve (6.125) is shown as a solid line.

$$u = \tanh\left(100\left((x-\frac{1}{2})^2 + (y-\frac{1}{2})^2 - \frac{1}{16}\right)\right) \tag{6.128}$$

for $(x,y) \in \Omega = (0,1) \times (0,1)$. Figure 6.6 shows that the adaptive meshes obtained with monitor functions (5.190) and (5.192) are nearly the same. This is because the function (6.128) exhibits no strong anisotropic behavior, and the monitor function (5.192) thus leads to an almost isotropic mesh. The balancing effect of θ between alignment and equidistribution can be seen from Figure 6.7 and 6.8 (a) and (b) where adaptive meshes and the maximum norm of the alignment and equidistribution qual-

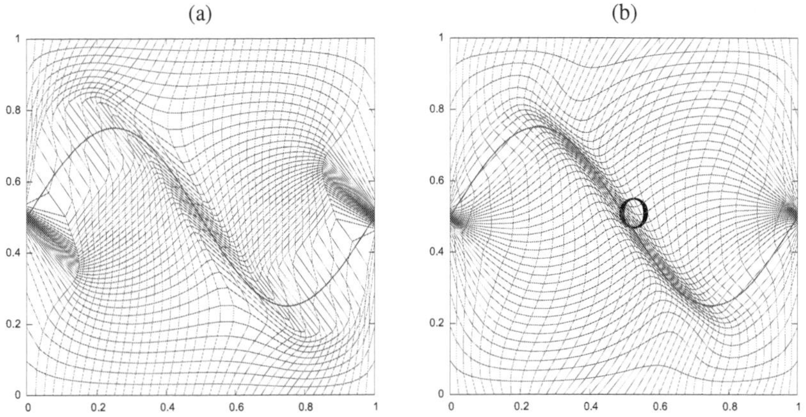

Fig. 6.4 Example 6.4.1. Adaptive meshes are generated using functionals (a) (6.101) and (b) (6.113) ($\theta = 0.1$) for monitor function (6.123) with $\boldsymbol{v} = (\frac{1}{\sqrt{2}}, \frac{1}{\sqrt{2}})^T$. The curve (6.125) is shown as a solid line.

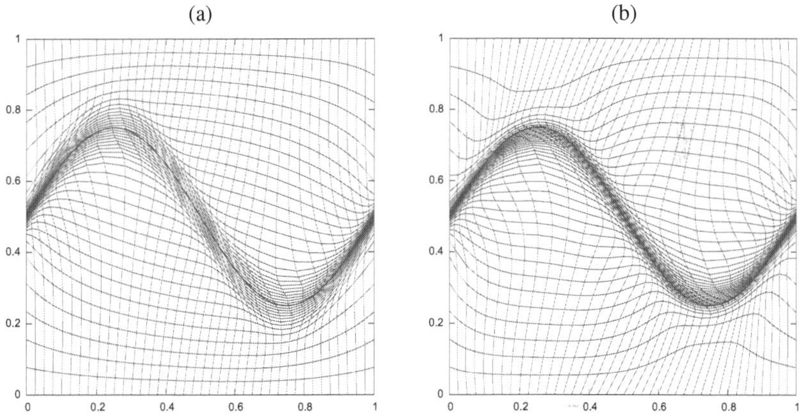

Fig. 6.5 Example 6.4.1. Adaptive meshes are generated using functionals (a) (6.101) and (b) (6.113) ($\theta = 0.1$) for monitor function (6.127). The curve (6.125) is shown as a solid line.

ity measures obtained with various values of θ are given. The convergence history plotted in Figure 6.8(c) shows that the linear interpolation error in the H^1 semi-norm decreases at a rate of first order as $J = K$ increases, confirming the analysis in §5.2. □

Example 6.4.3 In this example, an adaptive mesh is generated for the function

$$u = \tanh(100y) - \tanh(50(2x - 2y - 1)), \quad \forall (x, y) \in \Omega = (0, 1) \times (0, 1) \quad (6.129)$$

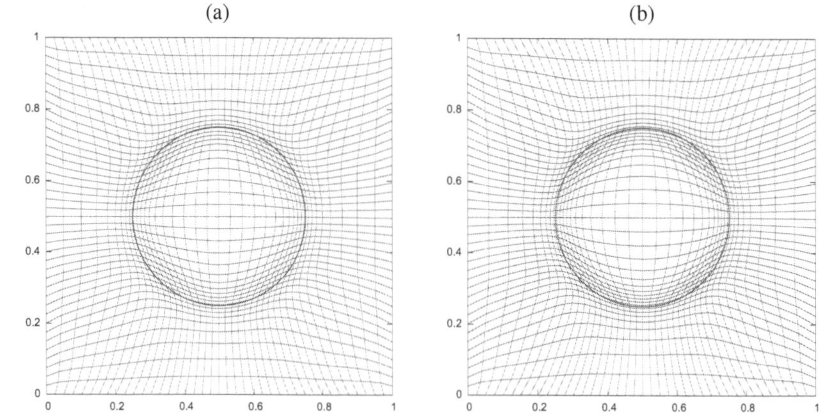

Fig. 6.6 Example 6.4.2. Adaptive meshes are obtained with functional (6.113) ($\theta = 0.1$) using (a) scalar monitor function (5.190) and (b) monitor function (5.192) (with parameters $p = q = 2$, $l = 2$, and $m = 1$). The solid circles in the plots indicate the position of the steep jump in the solution.

which models the interaction of a boundary layer and an oblique shock wave. Adaptive meshes obtained with monitor functions (5.190) and (5.192) are shown in Figure 6.9. Notice that the solution exhibits strong anisotropic behavior in the boundary and interior layers. Consequently, meshes obtained with monitor function (5.192) have a better alignment in the layer regions than those obtained with the scalar monitor function (5.190) and lead to a smaller interpolation errors (cf. Figure 6.10). Adaptive meshes obtained with different values of θ are shown in Figure 6.11. □

Example 6.4.4 Here, a moving mesh is generated for the monitor function $M = \rho(x,y,t)I$ with

$$\rho(x,y,t) = 1 + 10\exp\left(-50\left|\left(x - \frac{1}{2} - \frac{1}{4}\cos(2\pi t)\right)^2 + \left(y - \frac{1}{2} - \frac{1}{4}\sin(2\pi t)\right)^2 - \left(\frac{1}{10}\right)^2\right|\right) \tag{6.130}$$

for $(x,y) \in \Omega = (0,1) \times (0,1)$ and $t \in [0,1]$. An adaptive initial mesh is generated for $M = \rho(x,y,0)I$ using the same procedure as in the previous two examples. Then the moving mesh equation (6.39) is integrated with $\tau = 0.01$. The mesh of size 41×41 at $t = 0, 0.25, 0.5$, and 0.75 is shown in Figure 6.12. □

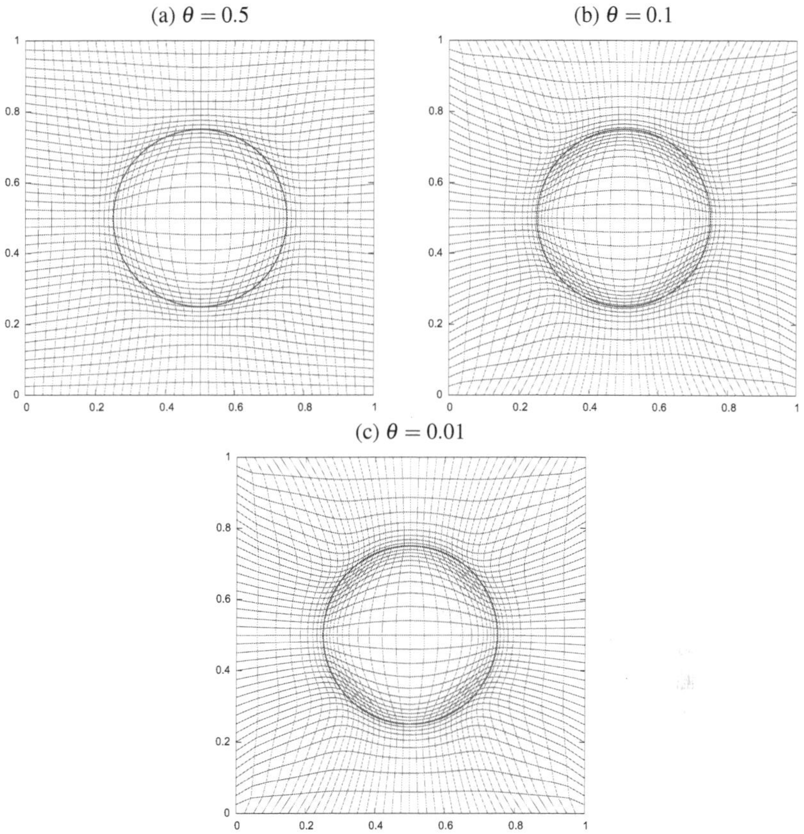

Fig. 6.7 Example 6.4.2. Adaptive meshes are obtained using the monitor function (5.192) with functional (6.113) for (a) $\theta = 0.5$, (b) $\theta = 0.1$, and (c) $\theta = 0.01$. The solid circles in the plots indicate the position of the steep jump in the solution.

6.5 Methods based on physical and geometric models

A number of variational methods have been developed based on other geometric and physical considerations. They are commonly designed to mimic a physical process or a geometric model, and most of them are not directly related to the equidistribution and alignment conditions or to an error analysis. As a result, it is often challenging to understand how the monitor function influences the behavior of a resulting mesh and thus how to decide when one choice of a monitor function is more appropriate than another.

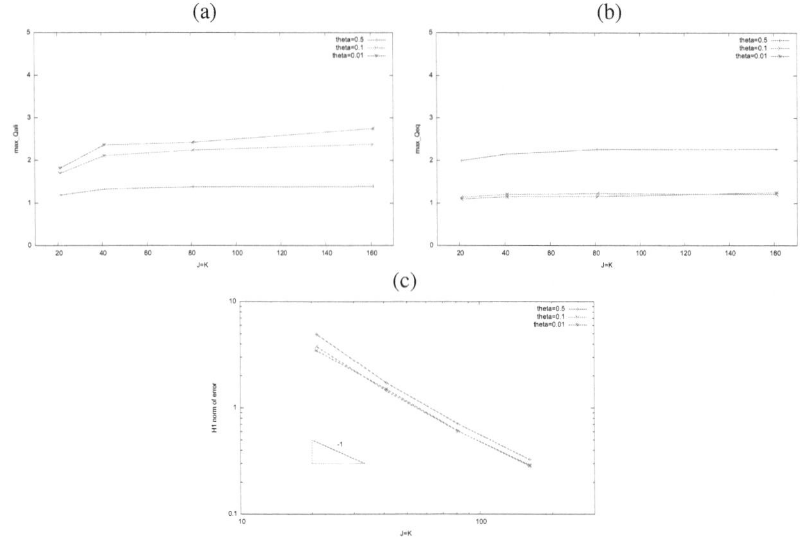

Fig. 6.8 Example 6.4.2. Numerical results are obtained with functional (6.113) and monitor function (5.192). (a) The maximum norm of the alignment quality measure; (b) the maximum norm of the equidistribution quality measure; (c) The H^1 semi-norm of linear interpolation error.

6.5.1 Variable diffusion methods

One of the earliest mesh generation methods is the variable diffusion method proposed by Winslow [342]. It employs the elliptic system of differential equations

$$-\nabla \cdot \left(\frac{1}{w} \nabla \xi_i \right) = 0, \quad i = 1, 2, ..., d \tag{6.131}$$

for generating an adaptive mesh, where $w = w(\boldsymbol{x}) > 0$ is a given weight function. This is a (steady-state) diffusion equation, mimicking a diffusion process where a heterogeneous diffusion coefficient results in an uneven concentration distribution. The goal here is to obtain a non-uniform distribution of constant coordinate lines (or surfaces in 3D) $\xi_i =$ constant, and thus to achieve mesh adaptivity using a weight function $w = w(\boldsymbol{x})$ which depends upon the physical solution.

The equation (6.131) is the Euler-Lagrange equation for the functional

$$I_{Win}[\boldsymbol{\xi}] = \frac{1}{2} \int_{\Omega} \frac{1}{w} \sum_i (\nabla \xi_i)^T \nabla \xi_i d\boldsymbol{x}, \tag{6.132}$$

which is of the form (6.11) for the special case (6.28) with

$$F_1(\rho, \beta) = \frac{1}{2} \sum_i (\nabla \xi_i)^T M^{-1} \nabla \xi_i = \frac{1}{2} \beta, \quad F_2(\rho, J) = 0, \tag{6.133}$$

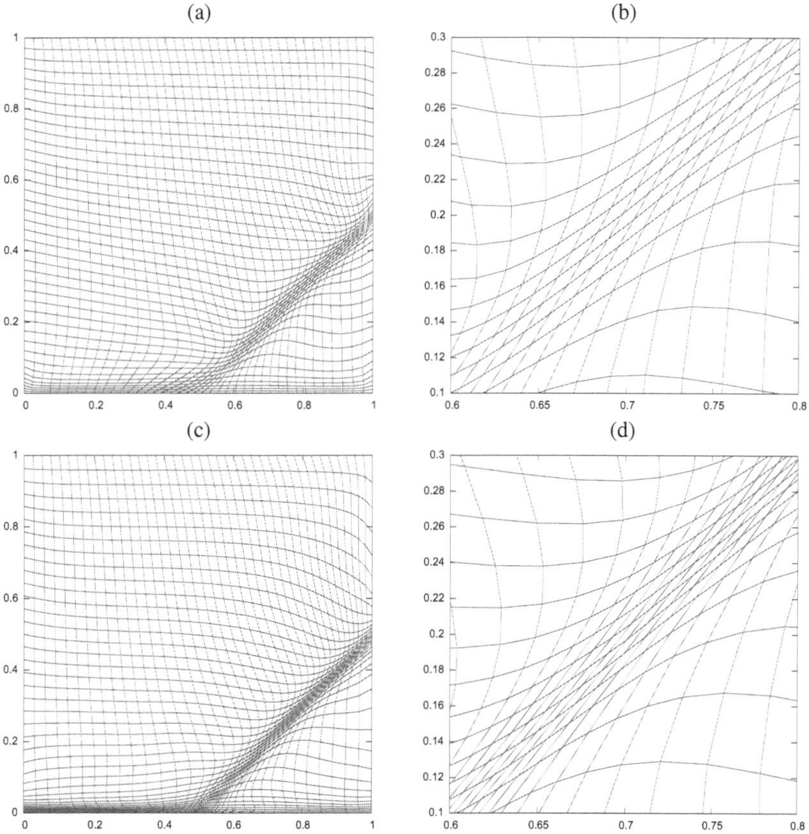

Fig. 6.9 Example 6.4.3. Adaptive meshes are obtained with functional (6.113) (with $\theta = 0.1$) and (a) scalar monitor function (5.190) and (c) monitor function (5.192) (with parameters $p = q = 2$, $l = 2$, and $m = 1$). Plots (b) and (d) are snapshots of the shock regions for (a) and (c), respectively.

and

$$M = w(\boldsymbol{x})I. \tag{6.134}$$

The MMPDEs (6.37) and (6.39) (with coefficient matrices in (6.34)) have a particularly simple form for this method. Specifically, (6.37) becomes

$$\dot{\boldsymbol{x}} = -\frac{1}{\tau p(\boldsymbol{x},t)} \sum_i \boldsymbol{a}_i \nabla \cdot \left(M^{-1} \boldsymbol{a}^i \right) \tag{6.135}$$

and the coefficients for (6.39) are

$$\begin{aligned} A_{ij} &= \left((\boldsymbol{a}^i)^T M^{-1} \boldsymbol{a}^j \right) I, \\ B_i &= I \sum_k \left((\boldsymbol{a}^k)^T \frac{\partial M^{-1}}{\partial \xi_k} \boldsymbol{a}^i \right). \end{aligned} \tag{6.136}$$

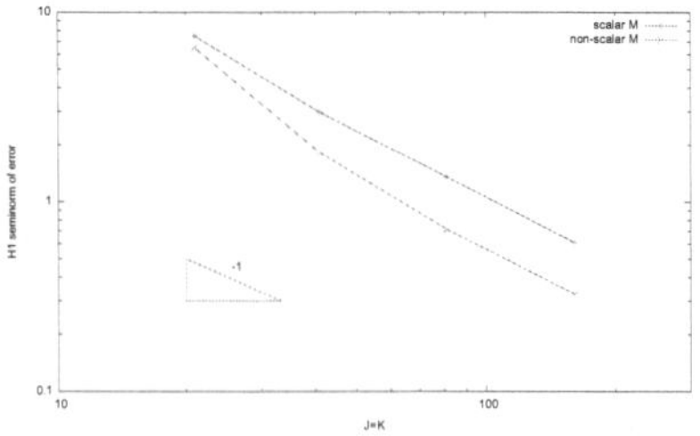

Fig. 6.10 Example 6.4.3. Convergence history of linear interpolation error in H^1 semi-norm is shown for adaptive meshes obtained with functional (6.113) (with $\theta = 0.1$) and scalar and non-scalar monitor functions (5.190) and (5.192).

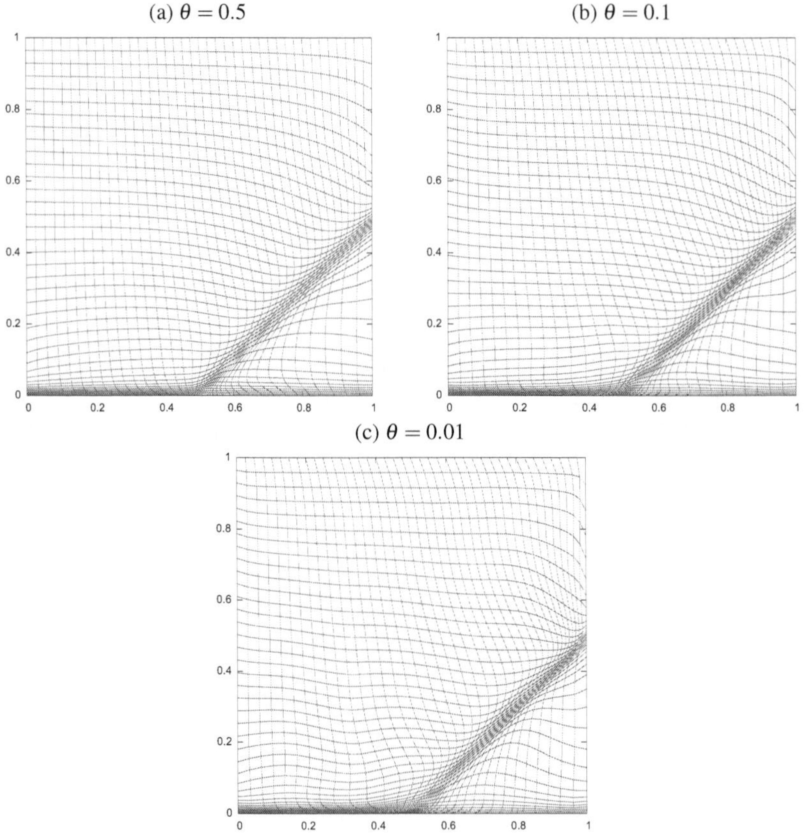

Fig. 6.11 Example 6.4.3. Adaptive meshes are obtained with functional (6.113) for (a) $\theta = 0.5$, (b) $\theta = 0.1$, and (c) $\theta = 0.01$. The monitor function (5.192) is used.

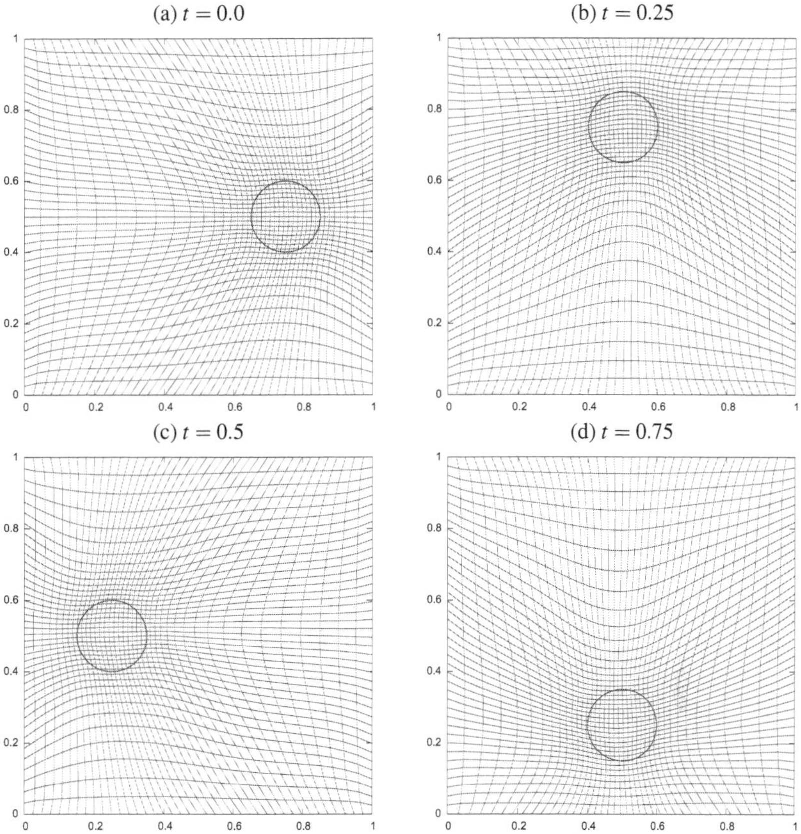

Fig. 6.12 Example 6.4.4. A moving mesh of size 41×41, obtained using MMPDE5 (6.39) (with $\tau = 0.01$) and functional (6.113) (with $\theta = 0.1$), is shown at different time instants.

It is interesting to contrast the functional $I_{Win}[\boldsymbol{\xi}]$ with the alignment functional $I_{ali}[\boldsymbol{\xi}]$ in (6.101) or the special case (6.112) of the functional (6.108) in the last section. Since it lacks the integration weight term ρ, (6.132) cannot be simply viewed as a functional associated with the alignment condition for a scalar-matrix monitor function $M = wI$. A qualitative analysis of the influence of the monitor function on the behavior of a resulting mesh is given below for a general functional (6.139), which includes (6.132) as a special case.

The construction of the functional $I_{Win}[\boldsymbol{\xi}]$ is motivated by physical considerations and has no explicit relation to an error analysis. As a consequence, there is no obvious optimal choice for the weight function $w = w(\boldsymbol{x})$. Generally speaking, w should be chosen to depend upon the physical solution $v = v(\boldsymbol{x})$ in such a way that it is large in regions where more mesh nodes are needed. A simple choice is

$$w = \sqrt{1 + |\nabla v|^2}. \tag{6.137}$$

A more sophisticated choice is the monitor function in (5.190) of §5.2,

$$w(\boldsymbol{x}) = \left[1 + \frac{1}{\alpha} \|D^l v\|_{l_p} \right]^{\frac{2q}{d+q(l-m)}}, \tag{6.138}$$

where α is defined in (5.190) and the physical meanings of l, m, p, and q are given in Table 5.1.

Recall that it is based on an isotropic bound on the interpolation error. Regardless of how the scalar function $w = w(\boldsymbol{x})$ is chosen, the mesh adaptation in (6.132) is isotropic since all directions are treated equally. It is thus unsuitable for applications where the physical solution exhibits a strong anisotropic behavior and a corresponding anisotropic mesh adaptation is desired. To enable this feature, a matrix-valued diffusion coefficient is used [82], giving the *generalized variable diffusion method*

$$I_{CHR}[\boldsymbol{\xi}] = \frac{1}{2} \int_{\Omega} \sum_i (\nabla \xi_i)^T M^{-1} \nabla \xi_i d\boldsymbol{x}, \tag{6.139}$$

where $M = M(\boldsymbol{x})$ is now a general symmetric, positive definite matrix. It is not difficult to see that this functional is invariant under a scaling transformation of M and/or rotation, translation, and dilation transformations of the computational coordinate $\boldsymbol{\xi}$. Moreover, it is coercive and uniformly convex, so the existence and uniqueness of its minimizer are guaranteed by Theorems 6.2.1 and 6.2.2.

The following qualitative analysis, adapted from [82], helps to provide an understanding of the influence of M on the mesh behavior and thereby to determine how to choose an appropriate monitor function M. The description is split into three parts.

(i) Influence of source terms on solution behavior for elliptic PDEs. Our basic tool for providing insight into the solution behavior for elliptic PDEs will be Green's function (e.g., see Courant and Hilbert [106] and Protter and Weinberger [276]). Using it, we derive a result which is used to gain an understanding of the influence of the monitor function on the mesh behavior.

Consider the differential operator

$$L = \sum_{ij} a_{ij}(\boldsymbol{x}) \frac{\partial^2}{\partial x_i \partial x_j} + \sum_i b_i(\boldsymbol{x}) \frac{\partial}{\partial x_i}, \quad \boldsymbol{x} \in \Omega \subset \mathbb{R}^d \tag{6.140}$$

where the coefficients $a_{ij} = a_{ji}$ and b_i are continuous in Ω and satisfy the uniform ellipticity condition: there exists a constant $\theta > 0$ such that

$$\sum_{ij} a_{ij}(\boldsymbol{x})\xi_i\xi_j \geq \theta|\boldsymbol{\xi}|^2, \quad \forall \boldsymbol{\xi} \in \mathbb{R}^d, \quad \forall \boldsymbol{x} \in \Omega.$$

If the boundary $\partial\Omega$ is sufficiently smooth, then Green's function for the differential operator L, denoted by $\mathscr{G} = \mathscr{G}(\boldsymbol{x},\boldsymbol{y})$, exists in Ω. Moreover, $\mathscr{G}$ has the following properties:

$$\begin{aligned}
&\mathscr{G}(\boldsymbol{x},\boldsymbol{y}) = \mathscr{G}(\boldsymbol{y},\boldsymbol{x}) > 0, \quad \forall \boldsymbol{x} \in \Omega, \forall \boldsymbol{y} \in \Omega;\\
&\mathscr{G}(\boldsymbol{x},\boldsymbol{y}) = 0, \quad \forall \boldsymbol{x} \in \partial\Omega, \forall \boldsymbol{y} \in \Omega;\\
&\mathscr{G}(\boldsymbol{x},\boldsymbol{y}) \to |\boldsymbol{x}-\boldsymbol{y}|^{2-d}, \quad \text{as } \boldsymbol{x} \to \boldsymbol{y} \text{ for } d > 2;\\
&\mathscr{G}(\boldsymbol{x},\boldsymbol{y}) \to -\log|\boldsymbol{x}-\boldsymbol{y}|, \quad \text{as } \boldsymbol{x} \to \boldsymbol{y} \text{ for } d = 2.
\end{aligned}$$

The solution $u(\boldsymbol{x})$ for the Dirichlet boundary value problem

$$L[u] = f \text{ in } \Omega, \quad u = h \text{ on } \partial\Omega, \tag{6.141}$$

can be expressed in terms of Green's function as

$$u(\boldsymbol{x}) = -\int_\Omega f(\boldsymbol{y})\mathscr{G}(\boldsymbol{x},\boldsymbol{y})d\boldsymbol{y} - \int_{\partial\Omega} h(\boldsymbol{y})\frac{\partial\mathscr{G}}{\partial\boldsymbol{n}}(\boldsymbol{x},\boldsymbol{y})dS, \tag{6.142}$$

where $\frac{\partial\mathscr{G}}{\partial\boldsymbol{n}}$ denotes the directional derivative of $\mathscr{G}$ along the outward normal $\boldsymbol{n}$ to $\partial\Omega$.

To examine the effect of f on the behavior of u, let v be the solution of the corresponding homogeneous problem

$$L[v] = 0 \text{ in } \Omega, \quad v = h \text{ on } \partial\Omega. \tag{6.143}$$

Clearly,

$$v(\boldsymbol{x}) = -\int_{\partial\Omega} h(\boldsymbol{y})\frac{\partial\mathscr{G}}{\partial\boldsymbol{n}}(\boldsymbol{x},\boldsymbol{y})dS, \tag{6.144}$$

and it follows from (6.142) and the positivity of $\mathscr{G}$ that

$$f \geq 0 \text{ in } \Omega \quad \Rightarrow \quad u \leq v \text{ in } \Omega.$$

This result is a direct conclusion of the well-known comparison theorem [106]. Its geometric meaning is that u-contour lines (or surfaces when $d > 2$) are shifted along the direction of increasing u from the "uniform, reference" v-contour lines. This can be seen by noticing that the inequality $u \leq v$ in Ω implies that for any constant c, the line $u(\boldsymbol{x}) = c$ lies in the region $v \geq c$. The shift of u-contour lines is illustrated in Figure 6.13.

The situation becomes more complicated when f changes sign in Ω. To see this, consider first an extreme (point charge) case with

$$f(\boldsymbol{x}) = \delta(\boldsymbol{x}-\boldsymbol{x}_1) - \delta(\boldsymbol{x}-\boldsymbol{x}_2), \tag{6.145}$$

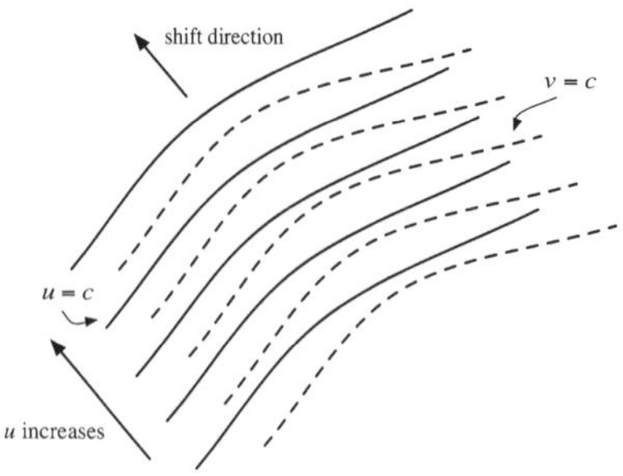

Fig. 6.13 A sketch of shifts of u-contour lines from reference v-contour lines for the case $f \geq 0$. (Cao et al. [82]. ©1999 Society for Industrial and Applied Mathematics. Reprinted with permission. All rights reserved.)

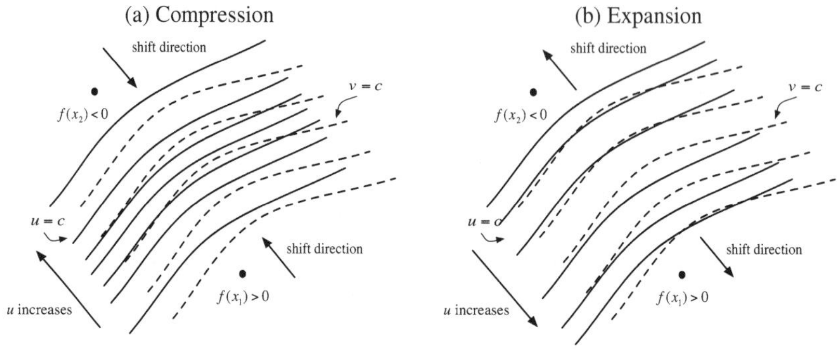

Fig. 6.14 Compression and expansion of u-contour lines caused by the two-point source term (6.145). Compression occurs when the increasing u direction coincides with the direction where f changes from positive to negative. (Cao et al. [82]. ©1999 Society for Industrial and Applied Mathematics. Reprinted with permission. All rights reserved.)

where $\boldsymbol{x}_1$ and $\boldsymbol{x}_2$ are two distinct points in the domain and δ is the Dirac delta function. The solution to the boundary value problem (6.141) is now

$$u(\boldsymbol{x}) = -\mathscr{G}(\boldsymbol{x},\boldsymbol{x}_1) + \mathscr{G}(\boldsymbol{x},\boldsymbol{x}_2) + v(\boldsymbol{x}).$$

Since $\mathscr{G}(\boldsymbol{x},\boldsymbol{y}) \to +\infty$ as $\boldsymbol{x} \to \boldsymbol{y}$, the behavior of u near $\boldsymbol{x}_1$ and $\boldsymbol{x}_2$ is dominated by $-\mathscr{G}(\boldsymbol{x},\boldsymbol{x}_1)$ and $\mathscr{G}(\boldsymbol{x},\boldsymbol{x}_2)$, respectively. Thus, u-contour lines shift in the increasing u direction near $\boldsymbol{x}_1$ and in the decreasing u direction near $\boldsymbol{x}_2$. If $\boldsymbol{x}_1$ and $\boldsymbol{x}_2$ are sufficiently close to each other, this will cause expansion or compression of u-contour lines in between. This behavior is illustrated in Figure 6.14.

The analysis of the point charge source function can be extended to a general function f. Because of the singularity of Green's function $\mathscr{G}(\boldsymbol{x},\boldsymbol{y})$ at $\boldsymbol{y} = \boldsymbol{x}$, the main contribution to the volume integral for the solution $u(\boldsymbol{x})$ of (6.142) comes from the values of f near $\boldsymbol{x}$. Thus, $(u - v)$ becomes negative in the region where f

is sufficiently large and positive. Geometrically, u-contour lines in the region will shift from the reference v-contour lines in the increasing u direction. Similarly, they will shift in the decreasing u direction where f is sufficiently large and negative. Consequently, *u-contour lines will be compressed or expanded in a region where f changes rapidly from positive to negative or from negative to positive. In particular, compression will occur when the increasing u direction coincides with the direction where f changes from positive to negative.*

The analysis also holds when f depends upon the unknown function, i.e., $f = f(\boldsymbol{x}, u)$. In this case, the solution $u = u(\boldsymbol{x})$ to the Dirichlet boundary value problem can still be expressed in terms of Green's function by

$$u(\boldsymbol{x}) = -\int_{\Omega} f(\boldsymbol{y}, u(\boldsymbol{y}))\mathscr{G}(\boldsymbol{x},\boldsymbol{y})d\boldsymbol{y} - \int_{\partial\Omega} h(\boldsymbol{y})\frac{\partial \mathscr{G}}{\partial \boldsymbol{n}}(\boldsymbol{x},\boldsymbol{y})dS. \tag{6.146}$$

Although this is a nonlinear equation, the analysis for the solution behavior given above remains valid.

(ii) Influence of monitor function on mesh behavior. The above result can be used to analyze the effect of the monitor function on the mesh behavior for the functional (6.139), with its Euler-Lagrange equation

$$\nabla \cdot (M^{-1}\nabla \xi_i) = 0, \quad i = 1, ..., d. \tag{6.147}$$

Taking the eigen-decomposition

$$M = [\boldsymbol{v}_1, ..., \boldsymbol{v}_d] \begin{bmatrix} \lambda_1 & & 0 \\ & \ddots & \\ 0 & & \lambda_d \end{bmatrix} \begin{bmatrix} (\boldsymbol{v}_1)^T \\ \vdots \\ (\boldsymbol{v}_d)^T \end{bmatrix} = \sum_i \lambda_i \boldsymbol{v}_i (\boldsymbol{v}_i)^T, \tag{6.148}$$

where λ_i and $\boldsymbol{v}_i$ are respectively the eigenvalues and normalized eigenvectors of M, (6.147) can be rewritten as

$$\sum_j \nabla \cdot \left(\frac{\boldsymbol{v}_j}{\lambda_j} \frac{\partial \xi_i}{\partial \boldsymbol{v}_j} \right) = 0, \quad i = 1, ..., d$$

where $(\partial/\partial \boldsymbol{v}_j)$ denotes a directional derivative along $\boldsymbol{v}_j$. After some algebraic manipulation, we obtain

$$\sum_j \frac{1}{\lambda_j} \frac{\partial^2 \xi_i}{\partial \boldsymbol{v}_j^2} = \sum_j \left(\frac{1}{\lambda_j} \frac{\partial \lambda_j}{\partial \boldsymbol{v}_j} - \nabla \cdot \boldsymbol{v}_j \right) \frac{1}{\lambda_j} \frac{\partial \xi_i}{\partial \boldsymbol{v}_j}, \quad i = 1, ..., d. \tag{6.149}$$

To see the effect of a change in λ_k $(k = 1, ..., d)$ on the behavior of ξ_i, after multiplying by λ_k (6.149) is a PDE of the form (6.141) with

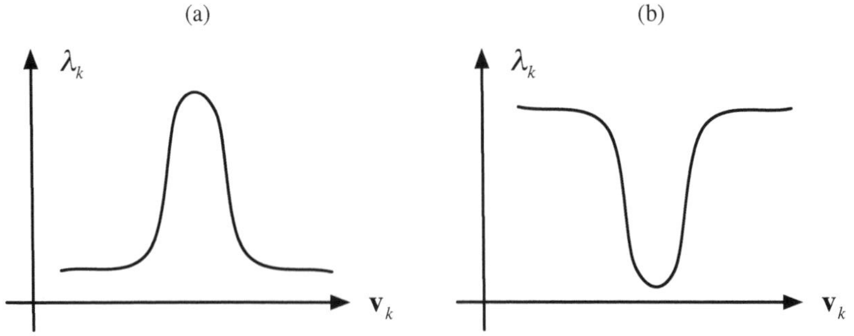

Fig. 6.15 A bump (a) or a dip (b) in λ_k along $\boldsymbol{v}_k$ direction.

$$L[\xi_i] \equiv \sum_j \frac{\lambda_k}{\lambda_j} \frac{\partial^2 \xi_i}{\partial \boldsymbol{v}_j^2} - \sum_{j \neq k} \left(\frac{1}{\lambda_j} \frac{\partial \lambda_j}{\partial \boldsymbol{v}_j} - \nabla \cdot \boldsymbol{v}_j \right) \frac{\lambda_k}{\lambda_j} \frac{\partial \xi_i}{\partial \boldsymbol{v}_j} + (\nabla \cdot \boldsymbol{v}_k) \frac{\partial \xi_i}{\partial \boldsymbol{v}_k}, \quad (6.150)$$

$$f(\boldsymbol{x}, \xi_i) = \frac{1}{\lambda_k} \frac{\partial \lambda_k}{\partial \boldsymbol{v}_k} \frac{\partial \xi_i}{\partial \boldsymbol{v}_k}. \quad (6.151)$$

The operator L has a form as in (6.140) and is uniformly elliptic if M is chosen (as is done in practice) to have eigenvalues strictly positive and bounded from above by a constant on $\bar{\Omega}$. Then the analysis for the behavior of the solution to BVP (6.142) applies. Moreover, the special form of f implies that the direction in which f changes from positive to negative coincides with the direction of increasing ξ_i if $\partial \xi_i / \partial \boldsymbol{v}_k \neq 0$ and λ_k has a bump along the $\boldsymbol{v}_k$ direction (see Figure 6.15(a)). If on the other hand, f changes from negative to positive along the direction of increasing ξ_k, then there is a dip in λ_k (see Figure 6.15(b)). From (i), we can conclude that *a bump or a dip in λ_k along direction $\boldsymbol{v}_k$ will likely cause coordinate lines not parallel to $\boldsymbol{v}_k$ to respectively compress or expand in the $\boldsymbol{v}_k$ direction.*

Note that this result is qualitative and does not explicitly specify where compression or expansion of coordinate lines occurs. Moreover, the compression or expansion is relative motion compared with that for the reference coordinate lines, which are themselves the solutions of the homogeneous BVP and generally non-uniform. The non-uniformity provides the mesh adaptivity. It depends upon several factors, including the boundary correspondence between the computational and physical domains, the deviation in ellipticity of the differential operator (6.150) from the Laplace operator, which often yields a uniform mesh, and the first order derivative terms in (6.150), which represent spatial changes in the eigenfunctions and other eigenvalues. A decrease in the ratios λ_k/λ_j, $j \neq k$ reduces the effects of the changes in λ_j and $\boldsymbol{v}_j$ ($j = 1, \ldots, d$, $j \neq k$) on mesh adaptation. With the exception of the boundary correspondence, the effects of these factors are complicated and difficult to analyze – see [82] for some illustrative examples.

Table 6.1 Values for p in (6.153) and corresponding monitor functions and functionals.

p	Monitor Function (6.147)	Functional (6.139)
-1	$M = \frac{\tilde{M}}{\sqrt{\det(\tilde{M})}}$ with $\tilde{M} = I + \nabla u (\nabla u)^T$	(6.156) in §6.5.2, Harmonic mapping
0	$M = I + \nabla u (\nabla u)^T$, arclength	(6.139)
1	$M = I\sqrt{1+\|\nabla u\|^2}$, scalar-matrix	(6.132), Variable diffusion

The mesh behavior is more restricted for the functional (6.132), a version of the functional (6.139) with the scalar matrix $M = w(\boldsymbol{x})I$. In this case, the eigenvalue decomposition (6.148) is not unique since *any* set of d orthogonal vectors can form a complete eigenvector system for M. So while the conclusion that coordinate lines compress or expand in the direction where the eigenvalue $\lambda = w(\boldsymbol{x})$ changes most significantly remains valid, the effect is isotropic in the sense that it applies equally for all directions with respect to changes in λ.

(iii) Choice of monitor function for general functional (6.139). The above analysis, although inconclusive, does offer some understanding of the influence of M on the mesh behavior for the functional (6.139) and thereby provides guidance on how to choose an appropriate monitor function. It is suggested in [82] that the choice of M involves interpreting its effects through the eigen-decomposition (6.148). The arclength-type monitor function serves as an insightful example: Given a physical solution $u = u(\boldsymbol{x})$, if

$$\begin{aligned} \boldsymbol{v}_1 &= \nabla u / |\nabla u|, \quad \boldsymbol{v}_2, \ldots, \boldsymbol{v}_d \text{ are orthogonal complements of } \boldsymbol{v}_1, \\ \lambda_1 &= \sqrt{1 + |\nabla u|^2}, \quad \lambda_2, \ldots, \lambda_d \text{ are functions of } \lambda_1, \end{aligned} \tag{6.152}$$

then coordinate lines compress or expand in the ∇u direction whenever the gradient changes significantly. In two dimensions, a special choice for the other eigenvalue is

$$\lambda_2 = \lambda_1^p \tag{6.153}$$

for some integer p. Several values for p and corresponding monitor functions and functionals are listed in Table 6.1.

Example 6.5.1 This example is similar to Example 6.4.1 except that the functional (6.139) is used here. Meshes obtained are shown in Figure 6.16. They are comparable to those in Figure 6.2(b)–6.5(b), indicating that functional (6.139) has a similar behavior as functional (6.113) (with $\theta = 0.1$) although it is unclear analytically how the former is related to the alignment and equidistribution conditions. At a more detailed level, one can see that the meshes in Figure 6.16 (b) and (c) mis-

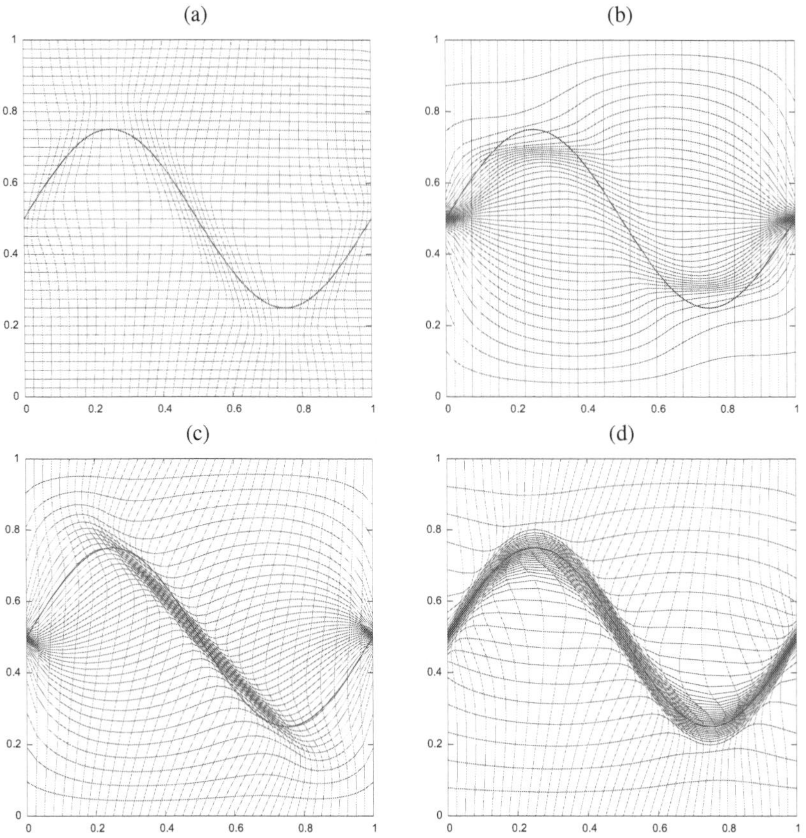

Fig. 6.16 Example 6.5.1. Numerical results are obtained with functional (6.139) and (a) monitor function (6.123) with $\boldsymbol{v} = (1,0)^T$, (b) monitor function (6.123) with $\boldsymbol{v} = (0,1)^T$, (c) monitor function (6.123) with $\boldsymbol{v} = (1/\sqrt{2}, 1/\sqrt{2})^T$, and (d) monitor function (6.127). The curve (6.125) is shown as a solid line.

concentrate in regions near points (0.25, 0.75), (0.75, 0.25) and (0.35, 0.7), (0.65, 0.3) whereas those in Figure 6.3(b) and 6.4(b) do not. This suggests that equidistribution plays a stronger role in (6.113) (with $\theta = 0.1$) than in (6.139).

The meshes in Figure 6.16 are consistent with the discussion for Example 6.4.1 and the qualitative analysis given in this section. □

Example 6.5.2 This example is similar to Example 6.4.2 except that we use the functional (6.139) for adaptive mesh generation. As in Example 6.4.2, the scalar monitor function (5.190) leads to nearly the same results as for the non-scalar monitor function (5.192), although the results with (5.190) are not presented here. One can see from the numerical results shown in Figure 6.17(a) that the mesh points are concentrated correctly around the solid circle, indicating the location of the steep

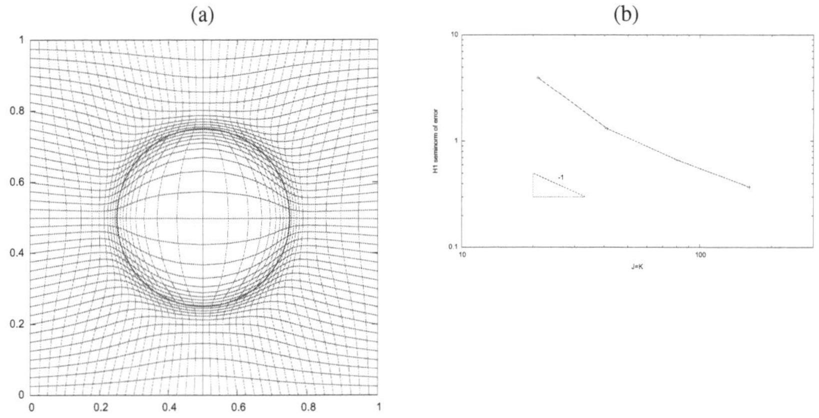

Fig. 6.17 Example 6.5.2. Numerical results are obtained with functional (6.139) and monitor function (5.192). (a) An adaptive mesh of size 41×41 and (b) The H^1 semi-norm of linear interpolation error. The solid circle indicates the location of steep jumps in the function.

jump of the function. The convergence history of the H^1 semi-norm of the linear interpolation error, shown in Figure 6.17(b), is comparable to that in Figure 6.8(c) for functional (6.113). □

Example 6.5.3 In this example, an adaptive mesh is generated for the function (6.129) of Example 6.4.3 using functional (6.139). Figure 6.18 shows adaptive meshes obtained with scalar and non-scalar monitor functions (5.190) and (5.192). The convergence history of linear interpolation error is shown in Figure 6.19. □

Example 6.5.4 Again using (6.139) in this example, a moving mesh is generated for the time-dependent monitor function $M = \rho(x,y,t)I$, with $\rho(x,y,t)$ given in (6.130) of Example 6.4.4 and using $\tau = 0.01$ for MMPDE5. Figure 6.20 shows a moving mesh of size 41×41 at different time instants. □

6.5.2 Harmonic mapping methods

Harmonic mappings have sometimes been used to determine the coordinate transformation for generating adaptive mesh generation. To shed light on strengths and weaknesses of this approach, in this subsection we give a brief description of harmonic mappings and related harmonic analysis (see references below for more details).

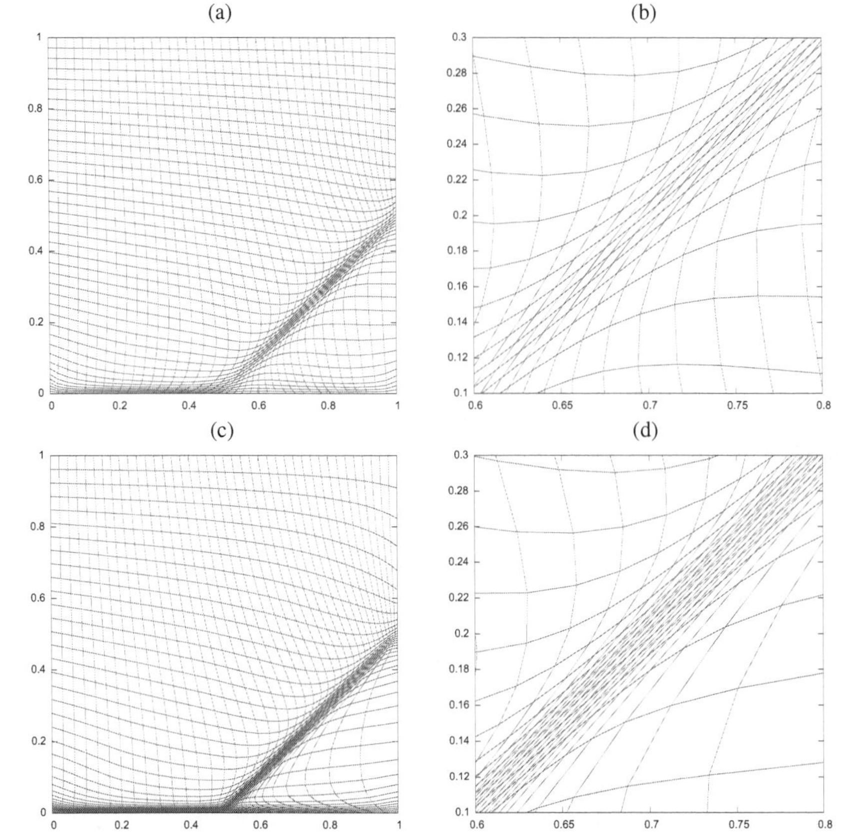

Fig. 6.18 Example 6.5.3. Adaptive meshes are obtained with functional (6.139) and (a) scalar monitor function (5.190) and (c) monitor function (5.192) (with parameters $p = q = 2$, $l = 2$, and $m = 1$). Plots (b) and (d) are snapshots of (a) and (c), respectively.

If $\mathscr{X}$ and $\mathscr{Y}$ are compact Riemannian manifolds with metric tensors (g_{ij}) and $(h_{\alpha\beta})$ in the local coordinates $\boldsymbol{x} = (x_1, ..., x_d)$ and $\boldsymbol{\xi} = (\xi_1, ..., \xi_d)$, respectively, then the *energy integral for a mapping* $\boldsymbol{\xi} = \boldsymbol{\xi}(\boldsymbol{x}) : \mathscr{X} \to \mathscr{Y}$ is defined as

$$E[\boldsymbol{\xi}] = \int_{\mathscr{X}} \sqrt{g} \sum_{i,j,\alpha,\beta} g^{ij} h_{\alpha\beta} \frac{\partial \xi_\alpha}{\partial x_i} \frac{\partial \xi_\beta}{\partial x_j} d\boldsymbol{x}, \tag{6.154}$$

where $g = \det(G)$, $G = (g_{ij})$, and $G^{-1} = (g^{ij})$. A mapping $\boldsymbol{\xi} = \boldsymbol{\xi}(\boldsymbol{x})$ is called *harmonic* if it is a minimum for $E[\boldsymbol{\xi}]$. Thus, it solves the Euler-Lagrange equation for the energy, which can be shown to be

$$\sum_{ij} \frac{\partial}{\partial \xi_i} \left(\sqrt{g} g^{ij} \frac{\partial \xi_\alpha}{\partial x_j} \right) + \sqrt{g} \sum_{ij\beta\gamma} g^{ij} \Gamma^{\alpha}_{\beta\gamma}(\mathscr{Y}) \frac{\partial \xi_\beta}{\partial x_i} \frac{\partial \xi_\gamma}{\partial x_j} = 0, \tag{6.155}$$

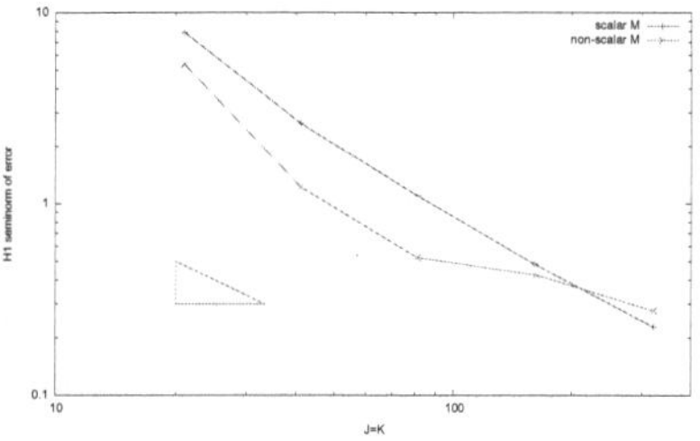

Fig. 6.19 Example 6.5.3. The H^1 semi-norm of linear interpolation error is plotted as function of $J = K$ for adaptive meshes obtained with functional (6.139) and scalar and non-scalar monitor functions (5.190) and (5.192).

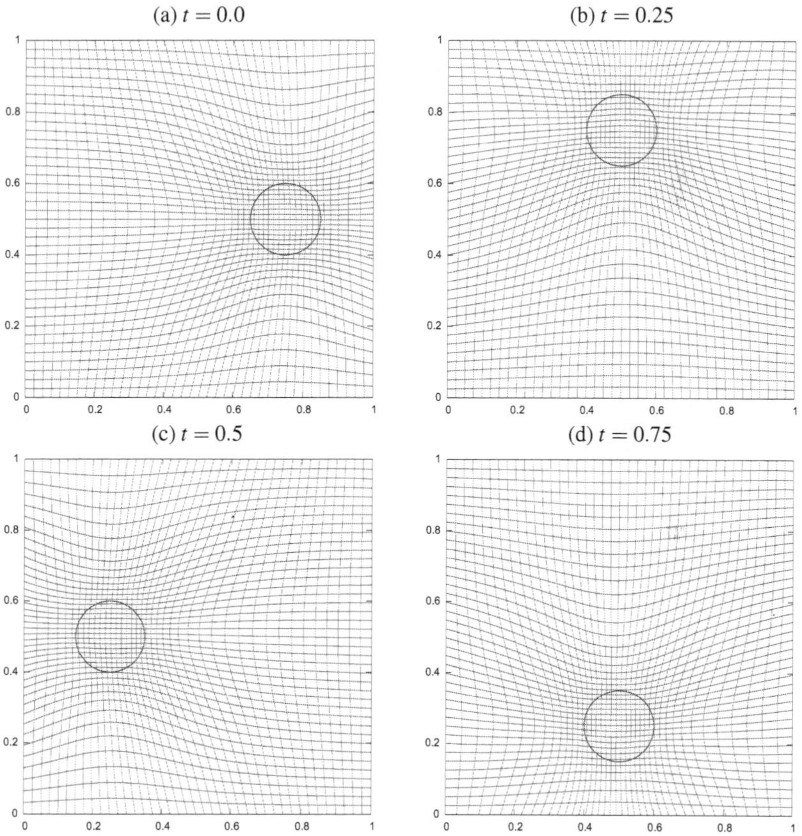

Fig. 6.20 Example 6.5.4. A moving mesh of size 41×41 is obtained with MMPDE5 (6.39) (with $\tau = 0.01$) and functional (6.139) at different time instants.

where $\Gamma^{\alpha}_{\beta\gamma}(\mathscr{Y})$ is the Christoffel symbol of the second kind,

$$\Gamma^{\alpha}_{\beta\gamma}(\mathscr{Y}) = \frac{1}{2}\sum_{\lambda} h^{\alpha\lambda}\left(\frac{\partial h_{\lambda\beta}}{\partial \xi_{\gamma}} + \frac{\partial h_{\lambda\gamma}}{\partial \xi_{\beta}} - \frac{\partial h_{\beta\gamma}}{\partial \xi_{\lambda}}\right).$$

Equation (6.155) is nonlinear in general. For a given boundary correspondence, a solution exists when $\mathscr{Y}$ has nonpositive Riemannian curvature and a convex boundary, and invertibility of the solution is guaranteed when $\dim(\mathscr{X}) = \dim(\mathscr{Y}) = 2$ – see Hamilton [166] and Schoen and Yau [294]. The reader is also referred to Liao [231] and Liao and Smale [233] for discussion on singularities of 3D harmonic mappings.

Dvinsky uses a harmonic mapping for mesh adaptation in [129]. Taking $\mathscr{X} = \Omega$ and $\mathscr{Y} = \Omega_c$, and choosing Ω_c to have a Euclidean geometry (i.e., $h_{\alpha\beta} = \delta_{\alpha\beta}$, giving zero curvature) and a convex boundary, then the harmonic mapping is potentially suitable for mesh generation since existence and invertibility of $\boldsymbol{\xi} = \boldsymbol{\xi}(\boldsymbol{x}) : \Omega \to \Omega_c$ are guaranteed. (Recall that such general theoretical guarantees of invertibility can be uncommon in the field of mesh generation.) More importantly, there is no restriction imposed on the metric tensor $G = (g_{ij})$, which serves as the monitor function $M(\boldsymbol{x})$ for controlling mesh concentration. Thus, the energy of the harmonic mapping becomes

$$I_{Dvi}[\boldsymbol{\xi}] = \int_{\Omega} \rho \sum_{i,j,\alpha} m^{ij} \frac{\partial \xi_{\alpha}}{\partial x_i}\frac{\partial \xi_{\alpha}}{\partial x_j} d\boldsymbol{x} = \int_{\Omega} \rho \sum_{i} (\nabla \xi_i)^T M^{-1} \nabla \xi_i d\boldsymbol{x}, \tag{6.156}$$

where $\rho = \sqrt{\det(M)}$. This can be cast in the form of (6.11) for the special case (6.28) with

$$F(\boldsymbol{a}^1, \boldsymbol{a}^2, \boldsymbol{a}^3, J, \boldsymbol{x}) = F_1(\rho, \beta) + F_2(\rho, J) = \rho\beta. \tag{6.157}$$

The partial derivatives needed for MMPDEs (6.37) and (6.39) (with coefficient matrices given in (6.34)) are

$$\begin{aligned}
&\frac{\partial F}{\partial \boldsymbol{a}^i} = 2\frac{\partial F_1}{\partial \beta} M^{-1}\boldsymbol{a}^i = 2\rho M^{-1}\boldsymbol{a}^i,\\
&\frac{\partial F}{\partial J} = 0,\\
&\frac{\partial F_1}{\partial \beta} = \rho, \quad \frac{\partial^2 F_1}{\partial \beta^2} = 0, \quad \frac{\partial^2 F_1}{\partial \beta \partial \rho} = 0,\\
&\frac{\partial F_2}{\partial J} = 0, \quad \frac{\partial^2 F_2}{\partial J^2} = 0, \quad \frac{\partial^2 F_2}{\partial J \partial \rho} = 0.
\end{aligned} \tag{6.158}$$

Perhaps the primary advantage of the harmonic mapping minimizing the energy integral is to provide as smooth a mapping as possible subject to the adaptivity constraints imposed by a given monitor function. The price paid for this smoothness is the inability of the mapping to provide equidistribution, as discussed below.

To understand the influence of M on the mesh behavior, note first that in two dimensions ($d = 2$) the functional (6.156) coincides with the alignment functional (6.101) and thus can be interpreted as a functional addressing the mesh alignment condition (4.75) but not the equidistribution condition (4.74). From Theorem 4.2.4, the shape and orientation of a physical element at point $\boldsymbol{x}$ are determined by $M(\boldsymbol{x})$ through (4.75) in such a way that the principal axes of the circumscribed ellipse of the element are formed by the eigenvectors of M while the shape is governed by $a_i/a_j = \sqrt{\lambda_j/\lambda_i}$, where λ_l and a_l ($l = i, j$) denote respectively the eigenvalues of M and the semi-length of the principal axes along the corresponding eigenvectors. However, since the functional (6.156) does not take the equidistribution condition (4.74) into consideration, the resulting mesh cannot be guaranteed to be dense (in terms of size of mesh cells) in regions where the mesh density function $\rho = \sqrt{\det(M)(\boldsymbol{x})}$ is large. This is illustrated in the examples below. Moreover, if the monitor function is chosen to be the scalar matrix $M(x,y) = w(x,y)I$, then $a_i/a_j = 1$ and the circumscribed ellipses of mesh elements become circles, so the resulting mesh is independent of weight function $w = w(x,y)$.[5] That is, *no* mesh adaptivity occurs for the harmonic mapping in this case.

The effect of the monitor function on mesh alignment for arbitrary dimension d can be investigated using our analysis of the general functional (6.139). If M is the monitor function used in (6.156), then it can be written in the form (6.139) with monitor function

$$\tilde{M} = \frac{M}{\rho} = \frac{M}{\sqrt{\det(M)}} = \sum_i \sqrt{\frac{\lambda_i}{\prod_{j\neq i}\lambda_j}}\, \boldsymbol{v}_i(\boldsymbol{v}_i)^T, \tag{6.159}$$

where the λ_i's and $\boldsymbol{v}_i$'s are the respective eigenvalues and eigenvectors of M. From the analysis in §6.5.1, a bump in $\sqrt{\lambda_i/\prod_{j\neq i}\lambda_j}$ along $\boldsymbol{v}_i$ causes compression in the $\boldsymbol{v}_i$ direction of the coordinate lines which are not parallel to $\boldsymbol{v}_i$. This allows us to interpret the mesh behavior for a variety of choices of the monitor function M which have been proposed in the past.

For example, Dvinsky [129] uses

$$M = I + \frac{f(\psi)}{\|\nabla\psi\|^2}\nabla\psi(\nabla\psi)^T, \tag{6.160}$$

where $\psi(\boldsymbol{x})$ is defined such that the values of $\boldsymbol{x}$ satisfying $\psi(\boldsymbol{x}) = 0$ characterize the attraction locations and $f(\psi)$ is a function of the distance from a given point to $\psi(\boldsymbol{x}) = 0$ that increases as the distance tends to zero. It is easy to verify that the eigenvalues of M are $\lambda_1 = 1 + f(\psi)$ and $\lambda_2 = \cdots = \lambda_d = 1$, and the corresponding mutually orthogonal eigenvectors are $\boldsymbol{v}_1 = \nabla\psi/\|\nabla\psi\|$ and $\boldsymbol{v}_2, \ldots, \boldsymbol{v}_d$. Thus, coordi-

[5] The last fact can also be verified directly for the functional (6.156).

nate lines are likely compressed – or mesh cells have a short length scale in the $\nabla\psi$ direction – whenever $f(\psi)$ has a bump. If the computed solution of a physical PDE is $u = u(\boldsymbol{x})$, then one can choose $\psi(\boldsymbol{x}) = u(\boldsymbol{x})$ and $f(\psi) = \|\nabla u\|^2$, giving

$$M = I + \nabla u (\nabla u)^T \tag{6.161}$$

in (6.160). Other choices of M are the monitor functions defined in §5.2.5.

It is easy to verify that the functional I_{Dvi} in (6.156) has the same invariance properties as stated in Theorem 6.4.1 for functional (6.108). Moreover, (6.156) is coercive and uniformly convex, so the existence and uniqueness of its minimizer are guaranteed by Theorems 6.2.1 and 6.2.2. (Recall that in 2D the invertibility of the minimizer is also guaranteed from the basic theory of harmonic mappings if $\partial\Omega_c$ is convex.)

Ivanenko and Charakhch'yan [201, 96] propose an interesting harmonic mapping method called the variational barrier method for use in mesh generation and adaptation. Its functional, defined as the harmonic energy written on the surface of the control function $u = u(\boldsymbol{x})$, takes the form

$$I_{IC}[\boldsymbol{x}] = \int_{\Omega_c} \operatorname{tr}\left(\boldsymbol{J}^{-1} M^{-1} \boldsymbol{J}^{-T}\right) \sqrt{\det(\boldsymbol{J}^T M \boldsymbol{J})} d\boldsymbol{\xi}, \tag{6.162}$$

where M is the monitor function defined in (6.161). In 2D this functional is

$$I_{IC,2D}[x,y] = \int_{\Omega_c} \frac{\operatorname{tr}\left(\boldsymbol{J}^T M \boldsymbol{J}\right)}{\det(\boldsymbol{J}^T M \boldsymbol{J})^{\frac{1}{2}}} d\boldsymbol{\xi}, \tag{6.163}$$

so the integrand is basically the ratio of the left-hand side to the right-hand side of the alignment condition (4.75). Thus, the functionals (6.163) and (6.162) can be interpreted as functionals addressing the mesh alignment condition (4.75) but not the equidistribution condition (4.74).

The idea of the variational barrier method is generalized by Azarenok [18, 19] to allow mesh adaptation on the computational mesh itself. Without including this control, the functional is given by

$$I_{Aza}[\boldsymbol{x}] = \int_{\Omega_c} \frac{\left(\operatorname{tr}\left(\boldsymbol{J}^T M \boldsymbol{J}\right)\right)^{\frac{d}{2}}}{\sqrt{\det(\boldsymbol{J}^T M \boldsymbol{J})}} d\boldsymbol{\xi}. \tag{6.164}$$

Obviously, this functional addresses the mesh alignment condition (4.75).

Example 6.5.5 Functional (6.156) is simply (6.101) in 2D, and meshes obtained with this functional in Example 6.4.1 are given in Figure 6.2(a)–6.5(a). □

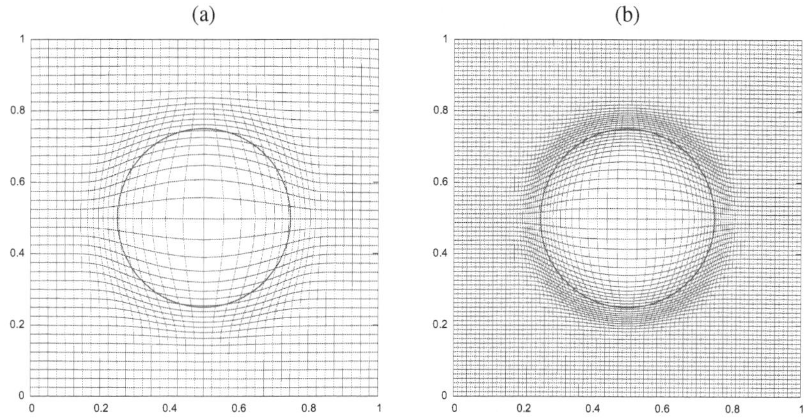

Fig. 6.21 Example 6.5.6. Adaptive meshes are obtained with functional (6.156) and monitor function (5.192). The size of the meshes is (a) 41×41 and (b) 81×81. The solid circles indicate the location of steep jumps in the function.

Example 6.5.6 This example is the same as Example 6.4.2 except that we now use the functional (6.156) for adaptive mesh generation. Since (6.156) gives no mesh adaptivity for scalar monitor functions, we use only the non-scalar monitor function (5.192). Adaptive meshes are shown in Figure 6.21. One can see that the mesh points are not correctly concentrated around the steep jump of the function. This confirms the analysis showing that functional (6.156) does not take the equidistribution condition (4.74) into consideration, leaving no guarantee that the resulting mesh is dense in regions where the mesh density function $\rho = \sqrt{\det(M)(\boldsymbol{x})}$ is large. This is also reflected by the mesh quality measures shown in Figure 6.22(a), where the alignment measure being nearly 1 and equidistribution measure being relatively large indicate that the alignment condition (4.75) is closely satisfied by the meshes but not the equidistribution condition (4.74). Figure 6.22(b) shows that the error with the resulting meshes is also larger than those for meshes obtained with functionals (6.113) and (6.139) (cf. Figure 6.8(c) and 6.17(c)). □

Example 6.5.7 Here, a 41×41 adaptive mesh is generated for function (6.129) of Example 6.4.3 using functional (6.156). The mesh and the linear interpolation error as a function of $J = K$ are shown in Figure 6.23. Unlike in the previous example, the mesh points concentrate correctly around the boundary and oblique layers. □

Example 6.5.8 In this example, a 41×41 moving mesh is generated for the time-dependent monitor function

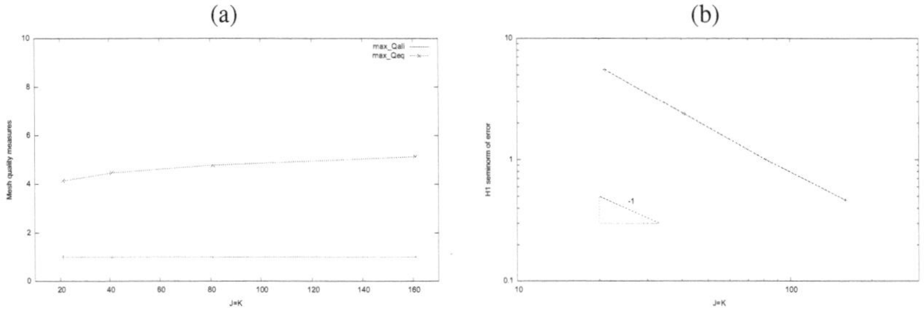

Fig. 6.22 Example 6.5.6. Numerical results are obtained with functional (6.156) and monitor function (5.192). (a) The maximum norm of the alignment and equidistribution quality measures and (b) The H^1 semi-norm of linear interpolation error.

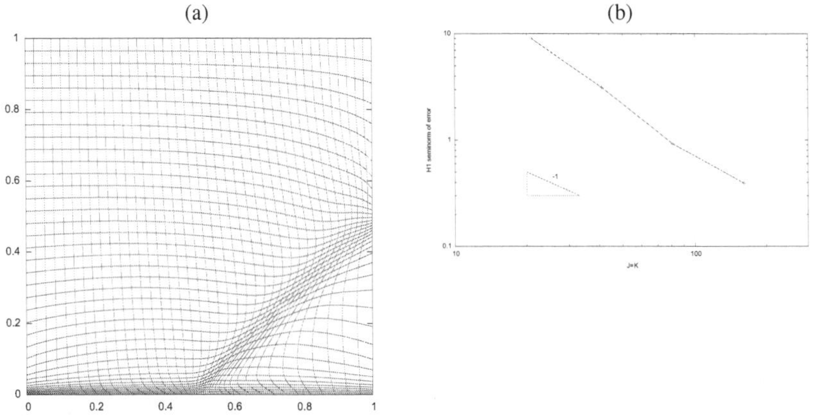

Fig. 6.23 Example 6.5.7. Numerical results are obtained with functional (6.156) and monitor function (5.192). (a) An adaptive mesh of size 41×41 and (b) The H^1 semi-norm of linear interpolation error.

$$
\begin{aligned}
M(x,y,t) &= \frac{\rho^2}{\|\nabla\rho\|^2}\nabla\rho(\nabla\rho)^T + \frac{1}{\|\nabla\rho\|^2}\nabla\rho^{\perp}(\nabla\rho^{\perp})^T \\
&= \frac{1}{\|\nabla\rho\|^2}\left[\nabla\rho,\nabla\rho^{\perp}\right]\begin{bmatrix}\rho^2 & 0\\ 0 & 1\end{bmatrix}\left[\nabla\rho,\nabla\rho^{\perp}\right]^T,
\end{aligned}
\tag{6.165}
$$

where $\rho(x,y,t)$ is given in (6.130) of Example 6.4.4 and $\nabla\rho^{\perp} = (\frac{\partial\rho}{\partial y}, -\frac{\partial\rho}{\partial x})^T$. Recall that using the scalar monitor function $M = \rho I$ with the harmonic mapping functional would lead to no mesh adaptation. The form of (6.165) is motivated from a qualitative analysis for the generalized variable diffusion functional (6.139) in §6.5.1 (see (6.126), (6.148) and (6.152)), so we investigate whether or not the coordinate lines not parallel to $\nabla\rho$ would compress around the circle

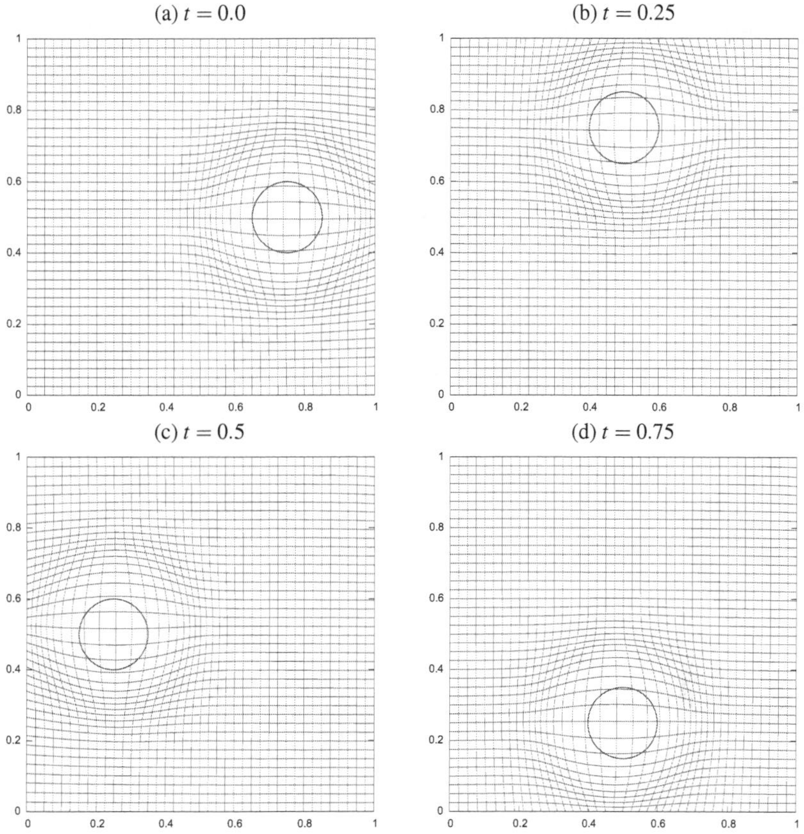

Fig. 6.24 Example 6.5.8. A moving mesh of size 41×41 is obtained using moving mesh equation (6.39) (with $\tau = 0.01$) and functional (6.156) at different time instants.

$$\left(x - \frac{1}{2} - \frac{1}{4}\cos(2\pi t)\right)^2 + \left(y - \frac{1}{2} - \frac{1}{4}\sin(2\pi t)\right)^2 = \left(\frac{1}{10}\right)^2.$$

Unfortunately, the mesh points do not properly concentrate around the circle for the harmonic mapping (6.156). The resulting moving mesh is shown in Figure 6.24 for several time instants. □

6.5.3 Hybrid methods and directional control

Motivated by Winslow's work [341] and the desire to use elliptic PDEs to generate adaptive meshes, Brackbill and Saltzman [58] propose a method which balances mesh adaptation with mesh smoothing and orthogonality. Their functional is given by

$$I_{BS}[\boldsymbol{\xi}] = \theta_a I_a[\boldsymbol{\xi}] + \theta_s I_s[\boldsymbol{\xi}] + \theta_o I_o[\boldsymbol{\xi}], \tag{6.166}$$

where θ_a, θ_s, and θ_o are weights and

$$I_a[\boldsymbol{\xi}] = \int_{\Omega} w(\boldsymbol{x}) J d\boldsymbol{x}, \tag{6.167}$$

$$I_s[\boldsymbol{\xi}] = \int_{\Omega} \sum_i (\nabla \xi_i)^T \nabla \xi_i d\boldsymbol{x}, \tag{6.168}$$

$$I_o[\boldsymbol{\xi}] = \int_{\Omega} \sum_{i \neq j} \left((\nabla \xi_i)^T \nabla \xi_j\right)^2 d\boldsymbol{x}. \tag{6.169}$$

The role of the functional I_a in controlling mesh concentration can be seen from its relation to the equidistribution condition (4.74) with $\rho = \sqrt{w(\boldsymbol{x})}$ and (6.118) with $\nu = 1$: minimizing $I_a[\boldsymbol{\xi}] = \int_{\Omega} J w(\boldsymbol{x}) d\boldsymbol{x} = \int_{\Omega} J \rho^2 d\boldsymbol{x}$ helps to approximately satisfy the equdistribution condition

$$J \sqrt{w(\boldsymbol{x})} = \frac{\sigma}{|\Omega_c|}, \tag{6.170}$$

where $\sigma = \int_{\Omega} \rho d\boldsymbol{x} = \int_{\Omega} \sqrt{w(\boldsymbol{x})} d\boldsymbol{x}$. The functional I_s aids in controlling mesh smoothness since its minimizer satisfies Laplace's equation

$$\Delta \xi_i = 0, \quad i = 1, \ldots, d. \tag{6.171}$$

Finally, a minimizer of I_o satisfies the orthogonality condition

$$(\nabla \xi_i)^T \nabla \xi_j = 0, \quad \forall i \neq j. \tag{6.172}$$

Since mesh adaptation in (6.166) is controlled using the scalar weight function $w = w(\boldsymbol{x})$, most natural choices of it tend to produce isotropic meshes. Consequently, with the functional I_a being associated with the equidistribution condition (6.170) for $\sqrt{w(\boldsymbol{x})} = \rho = \sqrt{\det(M)}$, a reasonable choice of the monitor function based on the isotropic error bound (5.190) is

$$w(\boldsymbol{x}) = \left[1 + \frac{1}{\alpha} \|D^l v\|_{l_p}\right]^{\frac{2dq}{d+q(l-m)}}, \tag{6.173}$$

where α is given in (5.190) and the meanings of l, m, p, and q are given in Table 5.1.

The functional (6.166) can be cast in the form (6.11) with

$$F(\boldsymbol{a}^1, \boldsymbol{a}^2, \boldsymbol{a}^3, J, \boldsymbol{x}) = \theta_a w(\boldsymbol{x}) J + \theta_s \sum_i (\boldsymbol{a}^i)^T \boldsymbol{a}^i + \theta_o \sum_{i \neq j} \left((\boldsymbol{a}^i)^T \boldsymbol{a}^j\right)^2. \tag{6.174}$$

The derivatives required for MMPDEs (6.37) and (6.39) (with coefficient matrices in (6.27)) are

$$
\begin{aligned}
\frac{\partial F}{\partial \boldsymbol{a}^i} &= 2\theta_s \boldsymbol{a}^i + 2\theta_o \sum_{j\neq i} \left((\boldsymbol{a}^i)^T \boldsymbol{a}^j \right) \boldsymbol{a}^j, \\
\frac{\partial F}{\partial J} &= \theta_a w(\boldsymbol{x}), \\
\frac{\partial^2 F}{\partial \boldsymbol{a}^s \partial \boldsymbol{a}^k} &= \begin{cases} 2\theta_s I + 2\theta_o \sum_{j\neq s} \boldsymbol{a}^j (\boldsymbol{a}^j)^T, & \text{for } s = k \\ 2\theta_o \left((\boldsymbol{a}^s)^T \boldsymbol{a}^k \right) I + 2\theta_o \boldsymbol{a}^k (\boldsymbol{a}^s)^T, & \text{for } s \neq k \end{cases} \\
\frac{\partial^2 F}{\partial J \partial \boldsymbol{x}} &= \theta_a \nabla w(\boldsymbol{x}), \\
\frac{\partial^2 F}{\partial \boldsymbol{a}^k \partial J} &= \frac{\partial^2 F}{\partial J \partial \boldsymbol{a}^k} = 0, \quad \frac{\partial^2 F}{\partial \boldsymbol{a}^i \partial \boldsymbol{x}} = 0, \quad \frac{\partial^2 F}{\partial J^2} = 0.
\end{aligned}
\tag{6.175}
$$

While the functional (6.166) provides a mechanism for an explicit control of mesh concentration, smoothness, and orthogonality, the terms characterizing these properties unfortunately have different dimensions, so the weights θ_a, θ_s, and θ_o cannot be dimensionless. As a consequence, they usually need to be re-chosen for each new application [102]. Finding suitable choices can be quite difficult, with substantial fine tuning required to determine the right balance between the terms. Another difficulty is that $I_o[\boldsymbol{\xi}]$ is not convex. The implication of this is that the adaptation functional $I_{BS}[\boldsymbol{\xi}]$ is not polyconvex, and the existence of its minimizer is not guaranteed.

An interesting variant of (6.167) is given by Steinberg and Roache [306], who propose minimizing

$$
I_{SR}[\boldsymbol{\xi}] = \int_{\Omega_c} J^2 d\boldsymbol{\xi}, \tag{6.176}
$$

with the intent of keeping cell volumes uniform across the domain. They come by this functional by observing that this unconstrained minimization problem is mathematically equivalent to the minimization of (6.176) subject to the global implicit constraint

$$
\int_{\Omega_c} J d\boldsymbol{\xi} = |\Omega|, \tag{6.177}
$$

and that the problem

$$
\min_{x_1,\ldots,x_m} \sum_{i=1}^{m} w_i x_i^2, \qquad \text{subject to } \sum_{i=1}^{m} x_i = 1 \tag{6.178}
$$

has the equidistributing solution $w_1 x_1 = \cdots = w_m x_m$. This physical meaning for the functional (6.176) can be easily explained formally using Theorem A.0.3. Since

$$
\int_{\Omega_c} J^2 d\boldsymbol{\xi} = \int_{\Omega} J\boldsymbol{x}
$$

is the functional I_a with $w(\boldsymbol{x}) = 1$, the equidistribution condition (6.170) is simply

$$J = \frac{|\Omega|}{|\Omega_c|}. \tag{6.179}$$

In other words, minimization of the functional (6.176) does indeed tend to make cell volumes uniform across the domain as desired.

A hybrid method with directional control. Recognizing the importance of having dimensionally homogeneous terms in an adaptation functional, Brackbill [57] presents a hybrid method which combines the variable diffusion functional I_{Win} with a directional control functional. Specifically, the functional is

$$I_{Bra}[\boldsymbol{\xi}] = (1-\theta)I_{Win}[\boldsymbol{\xi}] + \theta I_d[\boldsymbol{\xi}], \tag{6.180}$$

where $\theta \in [0,1)$ is a dimensionless parameter, and the dimensionally homogeneous terms are I_{Win} defined in (6.132) and the directional control functional

$$\begin{aligned} I_d[\boldsymbol{\xi}] &= \int_\Omega \frac{1}{w} \sum_i \|\boldsymbol{u}_i \times \nabla \xi_i\|^2 \, d\boldsymbol{x} \\ &= \int_\Omega \sum_i (\nabla \xi_i)^T \left[\frac{1}{w} \left(\|\boldsymbol{u}_i\|^2 I - \boldsymbol{u}_i (\boldsymbol{u}_i)^T \right) \right] \nabla \xi_i d\boldsymbol{x}. \end{aligned} \tag{6.181}$$

Here, $\boldsymbol{u}_i$ $(i = 1, ..., d)$ are prescribed vector fields, typically chosen on physical grounds as directions of preferred mesh alignment, and minimizing I_d increases alignment of the normals $\nabla \xi_i$ of the mesh surfaces of constant ξ_i with the vectors $\boldsymbol{u}_i$ $(i = 1, ..., d)$. Mesh orthogonality is increased if these vectors are chosen to be orthogonal to each other. Note as well that I_d can formally be given an energy formulation (6.156).

Two slightly different functionals for directional control,

$$I_{GE}[\boldsymbol{\xi}] = \int_\Omega \sum_i \|\boldsymbol{u}_i \times \nabla \xi_i\|^2 J d\boldsymbol{x} \tag{6.182}$$

and

$$\tilde{I}_{GE}[\boldsymbol{\xi}] = \int_\Omega \sum_i \|\boldsymbol{u}_i \times \nabla \xi_i\|^2 \, d\boldsymbol{x} \tag{6.183}$$

have been suggested by Guannakopoulos and Engel [160], who demonstrate numerically the ability of the functionals to cause coordinate line alignment with a prescribed vector field.

6.5.4 Jacobian-weighted methods

A Jacobian-weighted method [211, 214] minimizes an integral form of the Jacobian matrix of the coordinate transformation in the least squares sense. Its functional is

$$I_{Knu}[\boldsymbol{\xi}] = \int_\Omega \|\boldsymbol{J}^{-1} - S\|_F^2 d\boldsymbol{x}, \tag{6.184}$$

where $S = S(\boldsymbol{x})$ is a prescribed matrix field and $\|\cdot\|_F$ is the Frobenius matrix norm. This functional is coercive and uniformly convex, so the existence and uniqueness of a minimizer are guaranteed by Theorems 6.2.1 and 6.2.2. Denoting the row-vectors of S by $\boldsymbol{s}_i^T$, i.e., $S^T = [\boldsymbol{s}_1, ..., \boldsymbol{s}_d]$, (6.184) can be rewritten as

$$I_{Knu}[\boldsymbol{\xi}] = \int_\Omega \sum_i |\nabla \xi_i - \boldsymbol{s}_i|^2 d\boldsymbol{x} = \int_\Omega \sum_i (\boldsymbol{a}^i - \boldsymbol{s}_i)^T (\boldsymbol{a}^i - \boldsymbol{s}_i) d\boldsymbol{x}. \tag{6.185}$$

Thus, I_{Knu} can also be interpreted as a directional control functional for alignment of the normals of the mesh surfaces $\xi_i = constant$ with the prescribed vectors $\boldsymbol{s}_i$ $(i = 1, ..., d)$. The weight matrix S has an unambiguous geometric meaning – the reference (or target) inverse Jacobian matrix – so like the monitor function in I_{Hua} of §6.4, its effect on the mesh behavior is fairly clear.

The functional (6.185) is in the form (6.11) with

$$F(\boldsymbol{a}^1, \boldsymbol{a}^2, \boldsymbol{a}^3, J, \boldsymbol{x}) = \sum_i (\boldsymbol{a}^i - \boldsymbol{s}_i)^T (\boldsymbol{a}^i - \boldsymbol{s}_i),$$

with corresponding partial derivatives

$$\begin{aligned} &\frac{\partial F}{\partial \boldsymbol{a}^i} = 2(\boldsymbol{a}^i - \boldsymbol{s}_i), \quad \frac{\partial F}{\partial J} = \frac{\partial^2 F}{\partial J^2} = 0; \\ &\frac{\partial^2 F}{\partial \boldsymbol{a}^s \partial \boldsymbol{a}^k} = \delta_{sk} I, \quad \frac{\partial^2 F}{\partial \boldsymbol{a}^s \partial J} = \frac{\partial^2 F}{\partial J \partial \boldsymbol{a}^s} = 0, \\ &\frac{\partial^2 F}{\partial J \partial \boldsymbol{x}} = 0, \quad \frac{\partial^2 F}{\partial \boldsymbol{a}^i \partial \boldsymbol{x}} = -2\nabla \boldsymbol{s}_i = -2 \begin{bmatrix} (\nabla (\boldsymbol{s}_i)_1)^T \\ \vdots \\ (\nabla (\boldsymbol{s}_i)_d)^T \end{bmatrix}. \end{aligned} \tag{6.186}$$

The Euler-Lagrange equation for I_{Knu} is the simple elliptic system

$$\nabla^2 \xi_i = \nabla \cdot \boldsymbol{s}_i, \quad i = 1, ..., d, \tag{6.187}$$

as can be seen from (6.13). As a consequence, the resulting MMPDEs have a particularly straightforward form. For instance, MMPDE (6.37) becomes

$$\dot{\boldsymbol{x}} = -\frac{2}{\tau p(\boldsymbol{x}, t)} \sum_i \boldsymbol{a}_i \nabla \cdot (\boldsymbol{a}^i - \boldsymbol{s}_i), \tag{6.188}$$

while from (6.27) the coefficient matrices in MMPDE (6.39) are

$$A_{i,j} = 2I(\boldsymbol{a}^i)^T \boldsymbol{a}^j, \qquad B_i = -2\sum_k (\boldsymbol{a}^k)^T \nabla \boldsymbol{s}_i \boldsymbol{a}_k. \tag{6.189}$$

There can be significant consequences for methods which directly control the Jacobian matrix of the coordinate transformation instead of using a metric control as done by I_{Hua} of §6.4. From the analysis of §4.2, specifying the Jacobian matrix gives full control of the shape, size, and orientation of the physical mesh elements, but such control requires making appropriate choices for both the weight matrix $S = S(\boldsymbol{x})$ and the correspondence between $\delta\Omega$ and $\delta\Omega_c$. As a result, the functional (6.184) is not invariant under a scaling transformation of S and rotation and dilation transformations of $\boldsymbol{\xi}$. For example, under the rotation transformation given in (6.110) with $c = 1$, the functional (6.184) becomes

$$\begin{aligned} I_{Knu}[\boldsymbol{\xi}] &= \int_\Omega \|Q\tilde{\boldsymbol{J}}^{-1} - S\|_F^2 d\boldsymbol{x} \\ &= \int_\Omega \|Q(\tilde{\boldsymbol{J}}^{-1} - Q^T S)\|_F^2 d\boldsymbol{x} \\ &= \int_\Omega \|\tilde{\boldsymbol{J}}^{-1} - Q^T S\|_F^2 d\boldsymbol{x}. \end{aligned} \tag{6.190}$$

The right-hand side is neither equal to nor equivalent to $I_{Knu}[\tilde{\boldsymbol{\xi}}]$, so I_{Knu} is not invariant under (6.110). Equation (6.190) also shows that with the new computational coordinate system, $Q^T S$ instead of S, should be used instead of S if the same mesh adaptivity is to be maintained. In other words, the choice of S depends upon the choice of the computational coordinate system (and in the unstructured mesh case on the choice of the reference element), and this adds a layer of difficulty for making a proper choice of S.

Another difficulty is choosing S consistent with the boundary correspondence. More explicitly, specifying the Jacobian matrix for a mapping by continuity also specifies the boundary correspondence; therefore, prescribing any inconsistent boundary conditions cannot result in the mapping minimizing the functional satisfying $\boldsymbol{J} \approx S^{-1}$ near the boundary. We illustrate this below for a very simple example.

Example 6.5.9 Consider the case of a 2D mapping from the unit square onto itself, and choose $S = I$. If the boundary correspondence

$$\begin{cases} x(\xi,0) = \xi, \quad x(\xi,1) = \xi, \quad x(0,\eta) = 0, \quad x(1,\eta) = 1, \\ y(\xi,0) = 0, \quad y(\xi,1) = 0, \quad y(0,\eta) = \eta, \quad y(1,\eta) = \eta, \end{cases} \tag{6.191}$$

is used, then the minimizer $\boldsymbol{J} = S^{-1}$ for the functional I_{Knu} (6.184) is

$$x_\xi = 1, \quad x_\eta = 0, \quad y_\xi = 0, \quad y_\eta = 1, \tag{6.192}$$

and the coordinate transformation is $x(\xi,\eta) = \xi, y(\xi,\eta) = \eta$. However, for other boundary correspondences, even simple ones such as

$$\begin{cases} x(\xi,0) = 0, & x(\xi,1) = 1, \quad x(0,\eta) = \eta, \quad x(1,\eta) = \eta, \\ y(\xi,0) = \xi, & y(\xi,1) = \xi, \quad y(0,\eta) = 0, \quad y(1,\eta) = 1, \end{cases} \tag{6.193}$$

the minimizer of (6.184) does not satisfy $\boldsymbol{J} \approx S^{-1}$ near the boundary.

In contrast, when the metric control of §4.2 is employed, the orthogonal matrix U and diagonal matrix Σ in the SVD $\boldsymbol{J} = U\Sigma V^T$ are specified, but the orthogonal matrix V is free of influence from the monitor function and is determined by the boundary correspondence. This is illustrated for the example here if one takes $M = I$ for the functional in (6.113). It is easily verified that a minimizer subject to the boundary correspondence (6.191) is $x(\xi,\eta) = \xi, y(\xi,\eta) = \eta$ and subject to (6.193) is $x(\xi,\eta) = \eta, y(\xi,\eta) = \xi$. □

In [214], Knupp suggests choosing the weight matrix

$$S^{-1} = U\Sigma, \tag{6.194}$$

where the orthogonal matrix $U = [\boldsymbol{u}_1, ..., \boldsymbol{u}_d]$ and diagonal matrix $\Sigma = \text{diag}(\sigma_1, ..., \sigma_d)$ are used for directional and length control, respectively. In this case, (6.194) corresponds to the SVD $\boldsymbol{J} = U\Sigma V^T$ with $V = I$, so in 2D,

$$S^{-1} = \begin{bmatrix} \cos\theta & -\sin\theta \\ \sin\theta & \cos\theta \end{bmatrix} \begin{bmatrix} \sigma_1 & 0 \\ 0 & \sigma_2 \end{bmatrix}, \tag{6.195}$$

For the so-called reference Jacobian method, Knupp, Margolin, and Shashkov [212] compute a so-called Lagrangian grid (cf. §7.1.1) and corresponding reference Jacobian matrix $\boldsymbol{J}_{ref}$, which can be viewed as S^{-1}, and the functional

$$I_{KMS}[\boldsymbol{x}] = \int_{\Omega_c} \frac{\|\boldsymbol{J} - \boldsymbol{J}_{ref}\|_F^2 \det(\boldsymbol{J}_{ref})}{J} d\boldsymbol{\xi} \tag{6.196}$$

is then minimized directly. The goal is to obtain a smooth unfolded mesh which maintains most features of the Lagrangian grid. The functional (6.196), though similar to (6.184), uses the Jacobian matrix $\boldsymbol{J}$ and reference Jacobian matrix $\boldsymbol{J}_{ref}$ instead of their inverses and contains the explicit barrier factor $1/J$ to prevent $\boldsymbol{J}$ from becoming singular.

6.5.5 Methods based on mechanical models

Several methods have been developed from the basic principles of continuum mechanics. For these methods, the coordinate transformation $\boldsymbol{x} = \boldsymbol{x}(\boldsymbol{\xi})$ is viewed as the deformation map from a reference cell in Ω_c to a general cell in Ω, and a mesh generation functional is defined as the integral of a function of invariants of the deformation tensor $C = \boldsymbol{J}^T\boldsymbol{J}$. From basic linear algebra, a $d \times d$ matrix C has d invariants (e.g., see Golub and van Loan [157]), denoted by $\mathscr{I}_1$, ..., $\mathscr{I}_d$, and in 2D and 3D these are

$$\begin{cases} \mathscr{I}_1 = \mathrm{tr}(C) = \sum\limits_i \left(\frac{\partial x}{\partial \xi_i}\right)^T \frac{\partial x}{\partial \xi_i}, \\ \mathscr{I}_{d-1} = \mathrm{tr}\,\mathrm{Cof}(C) = \det(C)\mathrm{tr}(C^{-1}) = J^2 \sum\limits_i (\nabla \xi_i)^T \nabla \xi_i, \\ \mathscr{I}_d = \det(C) = J^2. \end{cases} \tag{6.197}$$

Here, we have used the fact that the cofactor matrix of C, $\mathrm{Cof}(C)$, is related to the inverse matrix by $\mathrm{Cof}(C) = \det(C)C^{-T}$. The functional has a general form

$$I[\boldsymbol{x}] = \int_{\Omega_c} \sigma(\boldsymbol{x}, \mathscr{I}_1, \ldots, \mathscr{I}_d) d\boldsymbol{\xi}, \tag{6.198}$$

where σ is a function of $\boldsymbol{x}$ and these invariants. The functional is written in terms of the unknown mapping $\boldsymbol{x} = \boldsymbol{x}(\boldsymbol{\xi})$ since the original development of the methods is based on mechanical models. Transformed into a functional in terms of the inverse mapping, this is written as

$$I[\boldsymbol{\xi}] = \int_{\Omega} \frac{\sigma(\boldsymbol{x}, \mathscr{I}_1, \ldots, \mathscr{I}_d)}{J} d\boldsymbol{x}, \tag{6.199}$$

with the understanding that when necessary the roles of dependent and independent variables are interchanged in the invariants. This integrand is called the stored energy function in continuum mechanics. As for other methods in this chapter, the optimal mapping is chosen to be a minimizer of this energy integral.

Jacquotte [202, 203] utilizes several axioms and principles from elasticity theory in deriving such a mesh generation functional. They include the following:

(i) *Hyperelasticity.* The stored energy function depends only upon $\boldsymbol{x}$ and the deformation gradient $\boldsymbol{J} = \partial \boldsymbol{x}/\partial \boldsymbol{\xi}$: $\sigma = \sigma(\boldsymbol{x}, \boldsymbol{J})$.

(ii) *Homogeneity.* The stored energy function is independent of the position of the cell: $\sigma = \sigma(\boldsymbol{J})$.

(iii) *The axiom of frame indifference.* The stored energy function is invariant under rotation of the $\boldsymbol{x}$ coordinate system: $\sigma(Q\boldsymbol{J}) = \sigma(\boldsymbol{J})$ for all orthogonal matrices Q.

(iv) *Isotropy property.* The stored energy function is independent of the orientation of the reference system (or invariant under rotation of the $\boldsymbol{\xi}$ coordinate system): $\sigma(\boldsymbol{J}Q) = \sigma(\boldsymbol{J})$ for all orthogonal matrices Q.

These requirements lead to the general form for the stored energy function σ in (6.198). A special choice arising in a 3D nonlinear elasticity model is

$$\sigma_{Jac,3d}(\mathscr{I}_1, \mathscr{I}_2, \mathscr{I}_3) \equiv c_1(\mathscr{I}_1 - \mathscr{I}_3 - 2) + c_2(\mathscr{I}_2 - 2\mathscr{I}_3 - 1) + k(J-1)^2, \quad (6.200)$$

where c_1, c_2, and k are constants satisfying the condition

$$k > 4(c_1 + c_2)/3 > 0, \quad (6.201)$$

which is needed to guarantee positiveness of the volume term and convexity of the stored energy function in a neighborhood of the identity mapping. The function (6.200) has been normalized so that it vanishes for the identity mapping, i.e., $\sigma_{Jac}(3,3,1) = 0$, so that the minimization of the functional $I[\boldsymbol{x}]$ in the mesh generation problem can give the solution $\boldsymbol{x}(\boldsymbol{\xi}) = \boldsymbol{\xi}$. For mesh adaptation, it takes the modified form

$$\begin{aligned}\sigma_{Jac,w}(\boldsymbol{x}, \mathscr{I}_1, \mathscr{I}_2, \mathscr{I}_3) \equiv\ & c_1(\mathscr{I}_1 - \mathscr{I}_3 - 2) + c_2(\mathscr{I}_2 - 2\mathscr{I}_3 - 1) \\ & + k(wJ-1)^2, \end{aligned} \quad (6.202)$$

where $w = w(\boldsymbol{x})$ is a weight function used for controlling mesh concentration. Now $\sigma_{Jac,w}$ does not satisfy the above homogeneity condition, consistent with the fact that an adaptive mesh is generally neither uniform nor satisfied such a condition. Minimization of the integral of the last term yields the solution $wJ = 1$, which is of course the equidistribution condition (4.74) with scalar monitor function $M = wI$.

In 2D, the stored energy function is

$$\begin{aligned}\sigma_{Jac,2d}(\mathscr{I}_1, \mathscr{I}_2) &= c(\mathscr{I}_1 - 2J) + (k-c)(J-1)^2 \\ &= c((x_\xi - y_\eta)^2 + (x_\eta + y_\xi)^2) + (k-c)(J-1)^2 \\ &= c(x_\xi^2 + x_\eta^2 + y_\xi^2 + y_\eta^2) + (k-c)(J-1)^2, \end{aligned} \quad (6.203)$$

and the convexity and positiveness conditions become $k > c > 0$.

A similar argument is used by de Almeida [113] to propose the general form $\sigma = \sigma(\boldsymbol{x}, \mathscr{I}_1, ..., \mathscr{I}_d)$ for the stored energy function. Several special choices for σ are

$$\begin{cases} \sigma_1 = \frac{\mathscr{I}_1}{2} = \frac{1}{2}\sum\limits_i \left(\frac{\partial \boldsymbol{x}}{\partial \xi_i}\right)^T \frac{\partial \boldsymbol{x}}{\partial \xi_i}, \\ \sigma_2 = \frac{\mathscr{I}_1}{2\sqrt{\mathscr{I}_d}} = \frac{1}{2J}\sum\limits_i \left(\frac{\partial \boldsymbol{x}}{\partial \xi_i}\right)^T \frac{\partial \boldsymbol{x}}{\partial \xi_i}, \\ \sigma_3 = \frac{\mathscr{I}_{d-1}}{2\sqrt{\mathscr{I}_d}} = \frac{J}{2}\sum\limits_i (\nabla \xi_i)^T \nabla \xi_i, \\ \sigma_4 = \sigma_3^2. \end{cases} \tag{6.204}$$

The functional associated with σ_1, called the unweighted length functional by Knupp and Steinberg [213], may result in a folded mesh for non-convex domains. The function σ_3 leads to the variable diffusion functional (6.139) or the harmonic mapping functional (6.156) with $M = I$. The function σ_4 is designed to have a stronger response to mapping singularity. In 2D, it is essentially the functional proposed by Liao [235],

$$\begin{aligned} I_{Liao}[x,y] &= \int_{\Omega_c} \left(\frac{\mathrm{tr}(C)}{\sqrt{\det(C)}} \right)^2 d\xi d\eta \\ &= \int_{\Omega_c} \left(\frac{x_\xi^2 + x_\eta^2 + y_\xi^2 + y_\eta^2}{J} \right)^2 d\xi d\eta. \end{aligned} \tag{6.205}$$

Interestingly, other methods not directly related to a mechanical model can also be cast in the form (6.198) or (6.199). These include the variable diffusion and harmonic mapping methods (corresponding to σ_3) and the functionals (6.108) and (6.121) (with $M = I$) based on the equidistribution and alignment conditions. From (6.197), it is easy to show that the latter two correspond respectively to the stored energy functions

$$\sigma_{Hua} = \theta \sqrt{\mathscr{I}_d} \left(\frac{\mathscr{I}_{d-1}}{\mathscr{I}_d} \right)^{\frac{d\gamma}{2}} + (1 - 2\theta) d^{\frac{d\gamma}{2}} (\mathscr{I}_d)^{\frac{1}{2} - \frac{\gamma}{2}} \tag{6.206}$$

and

$$\begin{aligned} \sigma = {} & \theta_1 \sqrt{\mathscr{I}_d} \left(\frac{\mathscr{I}_{d-1}}{\mathscr{I}_d} \right)^{\frac{d\gamma}{2}} + \theta_2 d^{\frac{d\gamma(d-2)}{2(d-1)}} (\mathscr{I}_1)^{\frac{d\gamma}{2(d-1)}} (\mathscr{I}_d)^{\frac{1}{2} - \frac{d\gamma}{2(d-1)}} \\ & + (\theta_3 - \theta_1 - \theta_2) d^{\frac{d\gamma}{2}} (\mathscr{I}_d)^{\frac{1}{2} - \frac{\gamma}{2}} + \frac{\theta_4}{\left(\int\limits_{\Omega_c} \sqrt{\mathscr{I}_d} d\boldsymbol{\xi} \right)^{\gamma + \nu}} (\mathscr{I}_d)^{\frac{1+\nu}{2}}. \end{aligned} \tag{6.207}$$

Note that in the current context of mesh generation (with $M = I$), $\int_{\Omega_c} \sqrt{\mathscr{I}_d} d\boldsymbol{\xi} = |\Omega|$. As we shall see later, for mesh adaptation we can use a solution-dependent deformation tensor $C_M = \boldsymbol{J}^T M \boldsymbol{J}$, In which case $\mathscr{I}_d = (J\rho)^2$ and $\int_{\Omega_c} \sqrt{\mathscr{I}_d} d\boldsymbol{\xi} = \int_\Omega \rho d\boldsymbol{x}$.

Liseikin [238] suggests using dimensionless mesh conformity measures for mesh generation. Two examples are

$$\sigma_{Lis,1} = \left(\frac{\mathscr{I}_1}{\mathscr{I}_d^{1/d}} \right)^{\gamma}, \quad \gamma > 0, \tag{6.208}$$

$$\sigma_{Lis,2} = \left(\frac{\mathscr{I}_{d-1}}{\mathscr{I}_d^{1-1/d}} \right)^{\gamma}, \quad \gamma > 0. \tag{6.209}$$

Note that $\sigma_{Lis,1}$ is simply a power of the geometric quality measure Q_{geo} (cf. (4.101)). To interpret $\sigma_{Lis,2}$, note from (6.197) and (4.102) that

$$\frac{\mathscr{I}_{d-1}}{\mathscr{I}_d^{1-1/d}} = \mathscr{I}_d^{1/d} \operatorname{tr}(C^{-1}) = d\, \hat{Q}_{geo}^{2(d-1)/d}(\boldsymbol{x}).$$

Since both Q_{geo} and $\hat{Q}_{geo}$ are equivalent to $\mu_{\max}/\mu_{\min}$ in the sense of (4.103) and (4.104), where $\mu_{\max}$ and $\mu_{\min}$ are the respective maximum and minimum singular values of $\boldsymbol{J}$, one can conclude that $\sigma_{Lis,1}$, $\sigma_{Lis,2}$, Q_{geo}, and $\hat{Q}_{geo}$ are all basically equivalent to the ratio $\mu_{\max}/\mu_{\min}$.

Branets and Carey [60] propose using

$$\sigma_{BC} = (1-\theta) \frac{\left(\frac{1}{d}\mathscr{I}_1\right)^{d/2}}{\sqrt{\mathscr{I}_d}} + \frac{\theta}{2} \left(\frac{\sqrt{\mathscr{I}_d}}{\alpha} + \frac{\alpha}{\sqrt{\mathscr{I}_d}} \right), \tag{6.210}$$

where $\theta \in [0,1)$ is a parameter balancing the two terms and α is defined as

$$\alpha = \frac{\int_{\Omega_c} \sqrt{\mathscr{I}_d} d\boldsymbol{\xi}}{\int_{\Omega_c} d\boldsymbol{\xi}}. \tag{6.211}$$

The first term in (6.210) is equivalent to the geometric quality measure Q_{geo} and regularizes the shape of mesh cells while the second term regularizes the size of mesh cells. Note the close similarity of (6.210) to (6.206) and (6.207), with all of them having terms which address the shape and size of mesh cells.

A simple technique to modify mesh generation methods to adaptive methods. Few of the methods discussed in this section have mesh adaptation directly incorporated into their formulations. Even for those which do, such as Jacquotte's method with (6.202) and Winslow's variable diffusion method, only a scalar weight function which gives isotropic mesh control is introduced. However, there is a simple but effective way to incorporate solution adaptation into the mesh generation formulations. For a typical such method, one introduces corresponding adaptive methods by making the following substitutions:

$$\begin{array}{ccl} \text{mesh generation} & \longrightarrow & \text{mesh adaptation} \\ \boldsymbol{J}^T\boldsymbol{J} & \longrightarrow & C_M = \boldsymbol{J}^T M \boldsymbol{J} \text{ (solution-dependent)} \\ J = \sqrt{\det(\boldsymbol{J}^T\boldsymbol{J})} & \longrightarrow & \sqrt{\det(\boldsymbol{J}^T M \boldsymbol{J})} \end{array} \tag{6.212}$$

where $M = M(\boldsymbol{x})$ is of course the monitor function depending upon the physical solution. This is the basic approach taken in [175] to develop the adaptive methods in §6.4.

6.5.6 *Methods based on Monge-Ampère equation / Monge-Kantorovich optimal transport problem*

These types of methods are different from the previous variational methods in that they seek a coordinate transformation as the solution of a constraint optimization problem instead of the minimizer of a functional. They can be derived in two different but mathematically equivalent ways, one based on the Monge-Ampère (MA) equation and the other based on the Monge-Kantorovich (MK) optimal transport problem. The MA equation and the MK problem have significant applications in differential geometry, image process, and several other fields in science; e.g., see Evans [138]. As adaptive mesh generation methods, they both have the attractive feature of satisfying the equidistribution condition (4.74) by construction. Indeed, the MA equation can be interpreted as specifying the Jacobian for a coordinate transformation via this equidistribution condition, while the MK problem seeks a coordinate transformation closest to the identity mapping among all coordinate transformations satisfying (4.74).

Derivation based on Monge-Ampère equation. Given a mesh density function $\rho = \rho(\boldsymbol{x})$ (or $\rho = \rho(\boldsymbol{x},t)$ for time-dependent problems), the coordinate transformation $\boldsymbol{x} = \boldsymbol{x}(\boldsymbol{\xi}) : \Omega_c \to \Omega$ needed for mesh generation is determined through the specification of its Jacobian via the equidistribution condition (4.74), i.e.,

$$\det\left(\frac{\partial \boldsymbol{x}}{\partial \boldsymbol{\xi}}\right) = \frac{\sigma}{|\Omega_c|\,\rho(\boldsymbol{x})}, \tag{6.213}$$

where $\sigma = \int_\Omega \rho(\boldsymbol{x}) d\boldsymbol{x}$. As we know, this equation is insufficient to uniquely specify $\boldsymbol{x} = \boldsymbol{x}(\boldsymbol{\xi})$, and here one requires it to also be irrotational, i.e.,

$$\nabla_\xi \times \boldsymbol{x} = 0. \tag{6.214}$$

From the Helmholtz decomposition theorem (see §7.1.1), this is equivalent to requiring that

$$\boldsymbol{x}(\boldsymbol{\xi}) = \nabla_\xi \phi \tag{6.215}$$

for a potential function ϕ. Substituting (6.215) into (6.213) leads to the Monge-Ampère (MA) equation

$$\det(H(\phi)) = \frac{\sigma}{|\Omega_c|\,\rho(\nabla_{\boldsymbol{\xi}}\phi)}, \tag{6.216}$$

where $H(\phi) = \nabla^2_{\boldsymbol{\xi}}\phi$ is the Hessian matrix of ϕ with respect to the independent variable $\boldsymbol{\xi}$. In 2D this second-order nonlinear PDE is

$$\det\begin{bmatrix}\phi_{\xi\xi} & \phi_{\xi\eta}\\ \phi_{\eta\xi} & \phi_{\eta\eta}\end{bmatrix} = \frac{\sigma}{|\Omega_c|\,\rho(\phi_\xi,\phi_\eta)} \tag{6.217}$$

or

$$\phi_{\xi\xi}\phi_{\eta\eta} - \phi^2_{\xi\eta} = \frac{\sigma}{|\Omega_c|\,\rho(\phi_\xi,\phi_\eta)}. \tag{6.218}$$

Boundary conditions for the MA equation are obtained by requiring that $\boldsymbol{x}(\boldsymbol{\xi})$ maps $\partial\Omega_c$ to $\partial\Omega$, i.e.,

$$\nabla_{\boldsymbol{\xi}}\phi(\partial\Omega_c) = \partial\Omega. \tag{6.219}$$

For example, consider a 2D case for which the closed unit square is mapped to itself. The boundary correspondence can be taken as

$$\begin{cases} x(0,\eta)=0, \quad x(1,\eta)=1, & \text{for the west and east sides, resp.}\\ y(\xi,0)=0, \quad y(\xi,1)=1, & \text{for the south and north sides, resp.}\end{cases} \tag{6.220}$$

The corresponding boundary condition for the MA equation are then

$$\begin{cases} \phi_\xi(0,\eta)=0, \quad \phi_\xi(1,\eta)=1, & \text{for west and east sides, resp.}\\ \phi_\eta(\xi,0)=0, \quad \phi_\eta(\xi,1)=1, & \text{for south and north sides, resp.}\end{cases} \tag{6.221}$$

The boundary value problem (6.216) and (6.219) has a unique solution (up to an additive constant) when Ω_c and Ω are convex; e.g., see Brenier [61]. The MA equation has several invariance properties. It is trivial to see that it is invariant under a scaling transformation of ρ and a translation transformation of $\boldsymbol{\xi}$. It is not difficult to see that (6.216) is also invariant under a rotation transformation $\boldsymbol{\xi} \to Q\tilde{\boldsymbol{\xi}}$ for an orthogonal matrix Q if ρ satisfies $\rho(Q\boldsymbol{x}) = \rho(\boldsymbol{x})$.

Derivation based on Monge-Kantorovich optimal transport. The original mass transport problem, proposed by Monge [259] in the eighteenth century, asks how best to move a pile of soil to a fill with the least amount of work (e.g., see Evans [138]). Mathematically, this can be stated as follows: Given two positive density functions ρ_0 and ρ_1 of equal mass, find $\boldsymbol{x}$ that transfers the density ρ_0 to ρ_1 and minimizes the cost

$$C(\boldsymbol{x}) = \int_{\mathbb{R}^d} \|\boldsymbol{x}(\boldsymbol{\xi}) - \boldsymbol{\xi}\|^p \rho_0(\boldsymbol{\xi}) d\boldsymbol{\xi} \quad p \geq 1. \tag{6.222}$$

This implies that $\boldsymbol{x}$ satisfies the **Mass Transport Problem** (of ρ_0 to ρ_1) if

$$\int_{\boldsymbol{x}^{-1}(A)} \rho_0(\boldsymbol{\xi}) d\boldsymbol{\xi} = \int_A \rho_1(\boldsymbol{x}) d\boldsymbol{x} \quad \forall A \subset \mathbb{R}^d. \tag{6.223}$$

It is important to observe that the Jacobian equation for this mass transport problem is simply a generalized equidistribution condition

$$\rho_1(\boldsymbol{x}) J(\boldsymbol{\xi}) = \rho_0. \tag{6.224}$$

For $p = 2$, this optimization problem has a unique solution which can be expressed as the gradient of a convex potential function; e.g., see Brenier [61] and Caffarelli [74, 75].

In mesh generation terms, we take $\rho_0 \equiv 1$ (corresponding to a uniform mesh in Ω_c), and reformulate (6.222) and (6.224) as the constraint optimization problem

$$\begin{cases} \min I[\boldsymbol{x}] \equiv \frac{1}{2} \int_{\Omega_c} \|\boldsymbol{x}(\boldsymbol{\xi}) - \boldsymbol{\xi}\|^2 d\boldsymbol{\xi} \\ \text{subject to} \quad \rho(\boldsymbol{x}(\boldsymbol{\xi})) J(\boldsymbol{\xi}) = \frac{\sigma}{|\Omega_c|}, \quad \forall \boldsymbol{\xi} \in \Omega_c \end{cases} \tag{6.225}$$

where $I[\boldsymbol{x}]$ represents the total amount of work, the minimization is over all mappings $\boldsymbol{x} = \boldsymbol{x}(\boldsymbol{\xi}) : \Omega_c \to \Omega$, and J is the Jacobian. That is, the minimization problem seeks a coordinate transformation closest to the identity mapping among all coordinate transformations satisfying the equidistribution condition (4.74). Inserting the gradient of this potential function into the constraint (equidistribution) condition, one obtains the MA equation (6.216).

To show more formally this connection between the MK problem and the MA equation, we give a derivation by solving the constraint problem using the Lagrange multipliers method [61]. Letting $\lambda = \lambda(\boldsymbol{\xi})$ be the Lagrange multiplier, we have the new functional as

$$\tilde{I}[\boldsymbol{x}, \lambda] = \int_{\Omega_c} \left[\frac{1}{2} \|\boldsymbol{x} - \boldsymbol{\xi}\|^2 + \lambda \left(\rho(\boldsymbol{x}) J - \frac{\sigma}{|\Omega_c|} \right) \right] d\boldsymbol{\xi}. \tag{6.226}$$

The variation of this functional has the form

$$\delta \tilde{I} = \int_{\Omega_c} \left[(\boldsymbol{x} - \boldsymbol{\xi}) \cdot \delta \boldsymbol{x} + \left(\rho(\boldsymbol{x}) J - \frac{\sigma}{|\Omega_c|} \right) \delta \lambda + \lambda J \nabla \rho \cdot \delta \boldsymbol{x} + \lambda \rho \delta J \right] d\boldsymbol{\xi}.$$

Since

$$J = \boldsymbol{a}_1 \cdot (\boldsymbol{a}_2 \times \boldsymbol{a}_3),$$

we have

$$\delta J = \sum_{i=1}^{3} \delta \boldsymbol{a}_i \cdot (\boldsymbol{a}_j \times \boldsymbol{a}_k) = \sum_{i=1}^{3} \frac{\partial \delta \boldsymbol{x}}{\partial \xi_i} \cdot (\boldsymbol{a}_j \times \boldsymbol{a}_k) = \sum_{i=1}^{3} \frac{\partial \delta \boldsymbol{x}}{\partial \xi_i} \cdot (J\boldsymbol{a}^i) \qquad (i,j,k) \text{ cyclic}$$

Using $\delta \boldsymbol{x} = 0$ on $\partial \Omega_c$, we can rewrite $\delta \tilde{I}$ as

$$\delta \tilde{I} = \int_{\Omega_c} \Big[\, (\boldsymbol{x} - \boldsymbol{\xi}) \cdot \delta \boldsymbol{x} + \left(\rho(\boldsymbol{x}) J - \frac{\sigma}{|\Omega_c|} \right) \delta \lambda + \lambda J \nabla \rho \cdot \delta \boldsymbol{x} \\ - \sum_{i=1}^{3} \frac{\partial}{\partial \xi_i} (\lambda \rho J \boldsymbol{a}^i) \cdot \delta \boldsymbol{x} \, \Big] \, d\boldsymbol{\xi}.$$

The Euler-Lagrange equations are thus

$$\boldsymbol{x} - \boldsymbol{\xi} + \lambda J \nabla \rho - \sum_{i=1}^{3} \frac{\partial}{\partial \xi_i} (\lambda \rho J \boldsymbol{a}^i) = 0, \tag{6.227}$$

$$\rho(\boldsymbol{x}) J - \frac{\sigma}{|\Omega_c|} = 0. \tag{6.228}$$

From identities (3.8) and (3.10), (6.227) reduces to

$$\boldsymbol{x} - \boldsymbol{\xi} = J \rho \sum_{i=1}^{3} \boldsymbol{a}^i \frac{\partial \lambda}{\partial \xi_i}.$$

Combined with (6.228), it can be further simplified to

$$\boldsymbol{x} - \boldsymbol{\xi} = \frac{\sigma}{|\Omega_c|} \sum_{i=1}^{3} \boldsymbol{a}^i \frac{\partial \lambda}{\partial \xi_i}.$$

Multiplying this equation by $\boldsymbol{a}_k$ and using the orthogonality condition (3.7), we get

$$\frac{\sigma}{|\Omega_c|} \frac{\partial \lambda}{\partial \xi_k} = (\boldsymbol{x} - \boldsymbol{\xi}) \cdot \boldsymbol{a}_k = (\boldsymbol{x} - \boldsymbol{\xi}) \cdot \frac{\partial \boldsymbol{x}}{\partial \xi_k}.$$

From this it follows that

$$\begin{aligned} \frac{\partial}{\partial \xi_k} \frac{1}{2} \|\boldsymbol{x} - \boldsymbol{\xi}\|^2 &= (\boldsymbol{x} - \boldsymbol{\xi}) \cdot \left(\frac{\partial \boldsymbol{x}}{\partial \xi_k} - \frac{\partial \boldsymbol{\xi}}{\partial \xi_k} \right) \\ &= (\boldsymbol{x} - \boldsymbol{\xi}) \cdot \frac{\partial \boldsymbol{x}}{\partial \xi_k} - (x_k - \xi_k) \\ &= \frac{\sigma}{|\Omega_c|} \frac{\partial \lambda}{\partial \xi_k} - (x_k - \xi_k), \end{aligned}$$

which can be rewritten as

$$x_k = \xi_k + \frac{\sigma}{|\Omega_c|} \frac{\partial \lambda}{\partial \xi_k} - \frac{\partial}{\partial \xi_k} \frac{1}{2} \|\boldsymbol{x} - \boldsymbol{\xi}\|^2,$$

or in a vector form,

$$x = \nabla_\xi \left(\frac{\sigma}{|\Omega_c|}\lambda + \frac{1}{2}\|\boldsymbol{\xi}\|^2 - \frac{1}{2}\|x - \boldsymbol{\xi}\|^2 \right). \tag{6.229}$$

Equation (6.229) is exactly in the form (6.215) with

$$\phi = \frac{\sigma}{|\Omega_c|}\lambda + \frac{1}{2}\|\boldsymbol{\xi}\|^2 - \frac{1}{2}\|x - \boldsymbol{\xi}\|^2,$$

implying that the solution of the minimization problem (6.225) can be expressed as the gradient of a potential function. The Monge-Ampère equation (6.216) is then obtained as before by inserting $x(\boldsymbol{\xi}) = \nabla_{\boldsymbol{\xi}}\phi$ into (6.228).

The parabolic Monge-Ampère (PMA) equation. Motivated by MMPDEs (§2.3 and §6.1.2), Budd and Williams [71, 72] study a so-called parabolic Monge-Ampère equation for generating adaptive moving meshes for time-dependent problems. Specifically, they use

$$\tau\left(I - \frac{1}{\beta^2}\Delta_\xi\right)\phi_t = \left[\det(H(\phi))\frac{\rho(\nabla_\xi\phi)}{\sigma}\right]^{\frac{1}{d}}, \tag{6.230}$$

where $\tau > 0$ and $\beta > 0$ are user-prescribed parameters (and d is the spatial dimension). The parameter τ plays a similar role as before in the MMPDEs for adjusting the response time of mesh movement to changes in ρ, while β controls the degree of spatial smoothness of the mesh (cf. (2.128)). Note that the scaling power $1/d$ is necessary for both sides of (6.230) to have the same dimension about ϕ when ρ is constant.

Sulman et al. [308, 309] use an alternative form of the PMA for mesh adaptation,

$$\phi_t = \log\left(\det(H(\phi))\frac{\rho(\nabla_\xi\phi)}{\sigma}\right). \tag{6.231}$$

The properties of this equation such as global existence and uniqueness of the solution to the initial-boundary value problem and convergence to steady state can be seen in [309].

Equations (6.230) and (6.231), together with (6.215), can be used to either generate an adaptive mesh for a given time-independent mesh density function ρ or an adaptive moving mesh for a time-dependent problem. In the first case, the mesh can be obtained by solving (6.230) or (6.231) to steady state with the initial condition

$$\phi(\boldsymbol{\xi}, 0) = \frac{1}{2}\|\boldsymbol{\xi}\|^2, \tag{6.232}$$

which corresponds to the identity mapping or a uniform mesh.

Numerical solution procedure and examples. The MA equation (6.216) and PMA equation (6.230) can be discretized in space using finite differences and finite elements; e.g., see Budd and Williams [71], Delzanno et al. [117], Budd et al. [68], Obermann [269] and references therein. Both equations are highly nonlinear, with nonlinearities coming from the Hessian of ϕ and the mesh density function ρ (itself a nonlinear function of the gradient of ϕ in general). They can be solved by a direct iterative method such as Newton's iteration [117] or a relaxation method [71]. Indeed, the PMA equation can be used as a relaxation method (with t as the continuation parameter) for solving the MA equation. Since the MA and PMA equations are closely tied to the equidistribution condition (4.74), the mesh density function can be chosen as in (5.190) when they are used for the numerical solution of a PDE problem.

Example 6.5.10 In this example a moving mesh is generated for the mesh density function

$$\rho(x,y,t) = 1+5\exp\left(-50\left|\left(x-\frac{1}{2}-\frac{1}{4}\cos(2\pi t)\right)^2 + \left(y-\frac{1}{2}-\frac{1}{2}\sin(2\pi t)\right)^2 - \left(\frac{1}{10}\right)^2\right|\right). \tag{6.233}$$

A mesh of size 30×30 obtained with the PMA equation is shown in Figure 6.25 for several time instants. In the computation, $\tau = 0.01$, $\Omega = \Omega_c = (0,1)\times(0,1)$, and a uniform mesh is used for Ω_c. The PMA equation is discretized using finite differences in space and integrated using an explicit Runge-Kutta formula in time. It can be seen that the mesh has correct concentration (due to equidistribution) and good regularity. Notice also that the mesh lines are orthogonal near the boundary due to the Neumann boundary condition (6.221). □

Example 6.5.11 This example is similar to the previous one except that the mesh density function is

$$\rho(x,y,t) = 1+5\exp\left(-50\left|y-\frac{1}{2}-\frac{1}{4}\sin(2\pi x)\sin(2\pi t)\right|\right). \tag{6.234}$$

A mesh of size 30×30 obtained with the PMA equation is shown in Figure 6.26 for several time instants. □

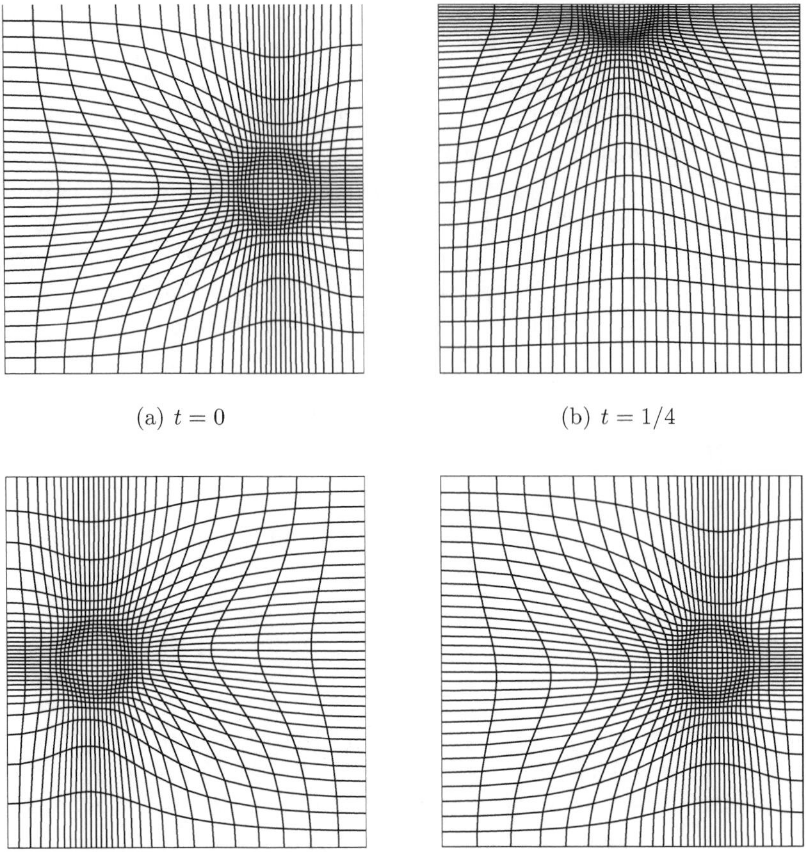

(a) $t = 0$ (b) $t = 1/4$

(c) $t = 1/2$ (d) $t = 10$

Fig. 6.25 Example 6.5.10. A moving mesh is obtained with the PMA equation for the given mesh density function (6.233). Reprinted from Budd et al. [68], with permission from Cambridge University Press.

6.5.7 Summary

Some major features of the methods discussed in §6.4 and §6.5 are summarized in Table 6.2. One question, which is not listed in the table but important to the success of those methods, is if the monitor function or equivalent can be chosen based on error control or minimization. The answer for this question is yes for some of the methods but not so obvious for the others; see the detailed discussion in this and previous sections.

Table 6.2 Summary of methods in §6.4 and §6.5.

Method		Related to		Invariant Transformations		Other Properties
Functional (or Mesh Eq.)	Motivation	Equid.[a]	Align.[b]	$M/\rho/w/S/\boldsymbol{J}_{ref}$	$\boldsymbol{\xi}$	
$I_{Hua}[\boldsymbol{\xi};\theta,\gamma]$ (6.108)	Equid. & align.	Yes	Yes	Scaling	R[c],T[d], D[e]	$\theta=0.5, d\gamma\geq 2$: Coercive & convex $\theta<0.5,\gamma=2$: Coercive & polyconvex
$I[\boldsymbol{\xi};\theta_1,\theta_2,\theta_3,\theta_4]$ (6.121)	Equid. & align.	Yes	Yes	Scaling	R,T,D	$\theta_3-\theta_1-\theta_2>0,\gamma>1$: Coercive, polyconvex, & $\det(\boldsymbol{J})>0$
$I_{Win}[\boldsymbol{\xi}]$, (6.132) $I_{CHR}[\boldsymbol{\xi}]$, (6.139)	Variable diffusion	Unknown	Unknown	Scaling	R,T,D	Coercive & convex
$I_{Dvi}[\boldsymbol{\xi}]$, (6.156)	Harmonic map	No	Yes	Scaling	R,T,D	Coercive & convex
$I_{IC}[\boldsymbol{\xi}]$, (6.162) $I_{Aza}[\boldsymbol{\xi}]$, (6.164)	Variational barrier	No	Yes	Scaling	R,T,D	Coercive & convex
$I_{BS}[\boldsymbol{\xi}]$, (6.166)	Hybrid control	Yes	No	Scaling	R,T,D	
$I_{Bra}[\boldsymbol{\xi}]$, (6.180)	Directional control	Unknown	Yes	Scaling	T,D	Coercive & convex
$I_{Knu}[\boldsymbol{\xi}]$, (6.184)	Jacobian-weighted	Yes	Yes		T	Coercive & convex
$I_{KMS}[\boldsymbol{\xi}]$, (6.196)	Reference Jacobian	Yes	Yes		T	Coercive & convex
$\sigma_{Jac,w}$, (6.202)	Mechanical model	Yes	No		T	
MA Eq. (6.213) PMA Eq. (6.230) PMA Eq. (6.231)	Jacobian specification, equid.	Yes	Unknown		R,T	Elliptic/Parabolic

[a] Equidistribution.

[b] Alignment.

[c] Rotation.

[d] Translation.

[e] Dilation.

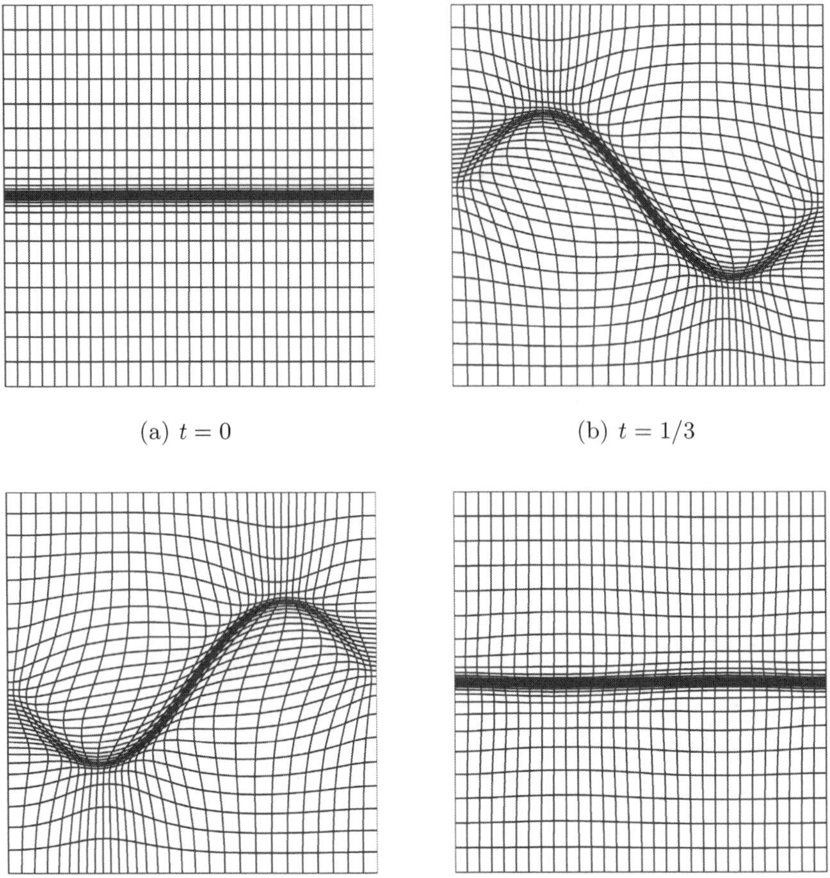

(a) $t = 0$ (b) $t = 1/3$

(c) $t = 2/3$ (d) $t = 1$

Fig. 6.26 Example 6.5.11. A moving mesh is obtained with the PMA equation for the given mesh density function (6.234). Reprinted from Budd et al. [68], with permission from Cambridge University Press.

6.6 Examples of applications

In this section numerical results are presented for several examples selected because they have been frequently used as test problems for moving mesh methods and have relatively simple formulations. The reader is referred to the biographical notes in §6.7 for references to other applications.

Example 6.6.1 In this example we solve the initial-boundary value problem for the 2D Burgers' equation

$$u_t = 0.005\Delta u - uu_x - uu_y \tag{6.235}$$

for $(x,y) \in \Omega = (0,1)\times(0,1)$ and $t \in (0.25, 1.25]$. The Dirichlet boundary condition and the initial condition are chosen such that the problem admits the exact solution

$$u(x,y,t) = \frac{1}{1+\exp\left(\frac{x+y-t}{0.01}\right)}, \tag{6.236}$$

which is a straight-line wave (u is constant along line $x+y=c$) moving in the direction $\theta = 45^o$.

MMPDE (6.39) (with $\tau = 0.1$), combined with functional I_{Hua} in (6.113) (with $\theta = 0.1$) and monitor function (5.192) (with $m = 1$), is used for adaptive moving mesh generation. Both the MMPDE and physical PDE are discretized in space by finite differences on a rectangular mesh in the computational domain (cf. §6.3.1). The extended system is then integrated alternately with quasi-Lagrange treatment of mesh movement (cf. §2.6 and §3.2). A fixed time step $\Delta t = 0.001$ is used in the integration.

An adaptive moving mesh of size 41×41 and a contour plot of the computed solution at several time instants are shown in Figure 6.27. A mesh obtained with the two-mesh strategy (cf. §3.4) (with a mesh of size 21×21 being moved by the MMPDE) is shown in Figure 6.28. This mesh is slightly less concentrated than that in Figure 6.27 and also leads to a larger error. However, the required computation time is significantly reduced with the two-mesh strategy; also see [175]. □

Example 6.6.2 (Combustion model) This model problem in combustion theory is given by

$$\begin{cases} u_t = \Delta u - \frac{R}{\alpha\delta} u e^{\delta(1-1/T)}, \\ L\,T_t = \Delta T + \frac{R}{\delta} u e^{\delta(1-1/T)}, \end{cases} \quad (x,y)\in\Omega \equiv (-1,1)\times(-1,1),\ t>0 \tag{6.237}$$

subject to the boundary and initial conditions

$$\begin{cases} u(x,y,t) = 1,\ T(x,y,t) = 1, & (x,y)\in\partial\Omega \\ u(x,y,0) = 1,\ T(x,y,0) = 1, & (x,y)\in\Omega. \end{cases} \tag{6.238}$$

Here, L, α, δ, and R are physical parameters and the variables u and T denote the concentration and temperature of a chemical which is undergoing a one-step reaction in the domain Ω. For small times the temperature gradually increases from unity, with a "hot spot" forming at the origin. At a finite time, ignition occurs, causing the temperature at the origin to increase rapidly to approximately $1+\alpha$. A flame

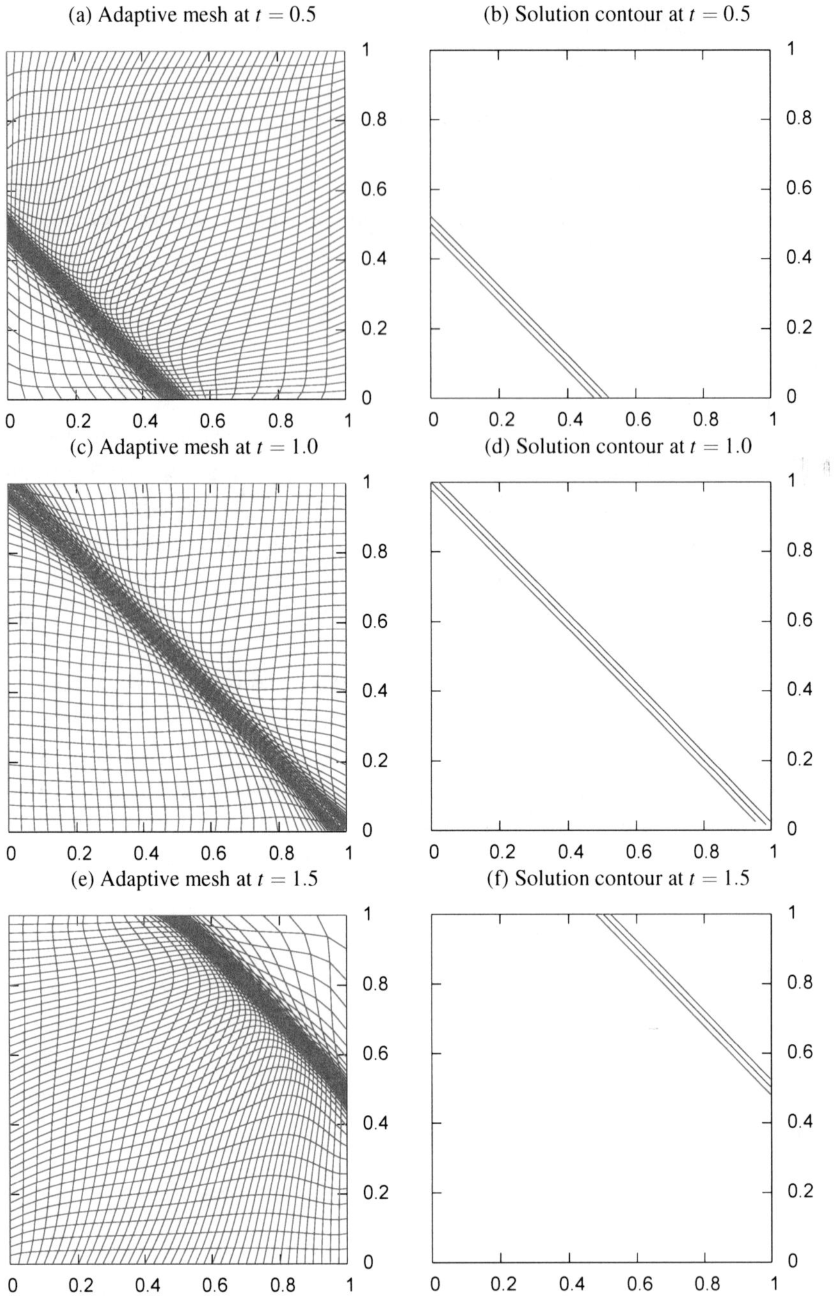

Fig. 6.27 Example 6.6.1. Adaptive 41×41 mesh and contour plots of the solution (for $u = 0.1$, 0.5, and 0.9) at different time instants. Error is $\int_{0.25}^{1.5} \|u^h - u\|_{L^2(\Omega)} dt = 7.54 \times 10^{-4}$.

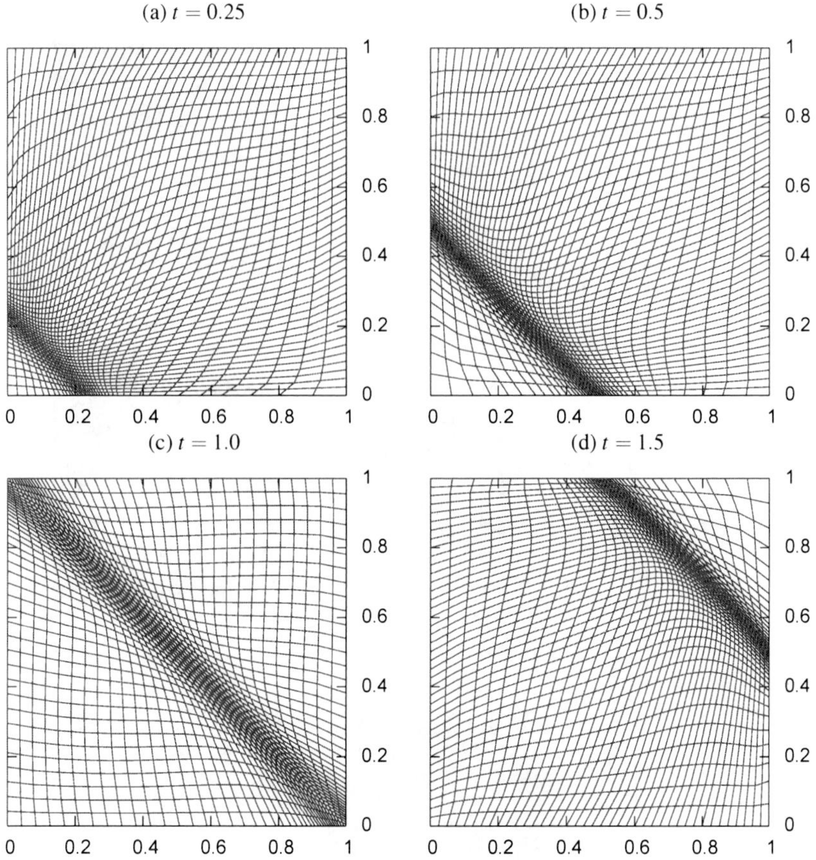

Fig. 6.28 Example 6.6.1. Adaptive 41×41 mesh obtained using the two-mesh technique, with a 21×21 mesh being moved by the MMPDE. Error $\int_{0.25}^{1.5} \|u^h - u\|_{L^2(\Omega)} dt = 1.16 \times 10^{-3}$.

front then forms and propagates toward the boundary of the domain at a high speed. The degree of difficulty of the problem is determined by the value of δ. Following Moore and Flaherty [260], the physical parameters are chosen as $L = 0.9$, $\alpha = 1$, $\delta = 20$, and $R = 5$.

A moving finite difference solution is shown in Figure 6.29, where MMPDE (6.39) (with $\tau = 0.01$), the functional I_{Hua} in (6.113) (with $\theta = 0.1$), and the monitor function (5.192) (with $m = 1$) are used in the computation.

The initial-boundary value problem (6.237) and (6.238) is also solved on a J-shape domain using a moving mesh finite element method by Cao et al. [81]. Figure 6.30 shows the computed mesh and temperature T using MMPDE (6.39), the functional I_{CHR} in (6.139), and the monitor function $M = \sqrt{1 + \frac{1}{2}\|\nabla T\|^2} I$. □

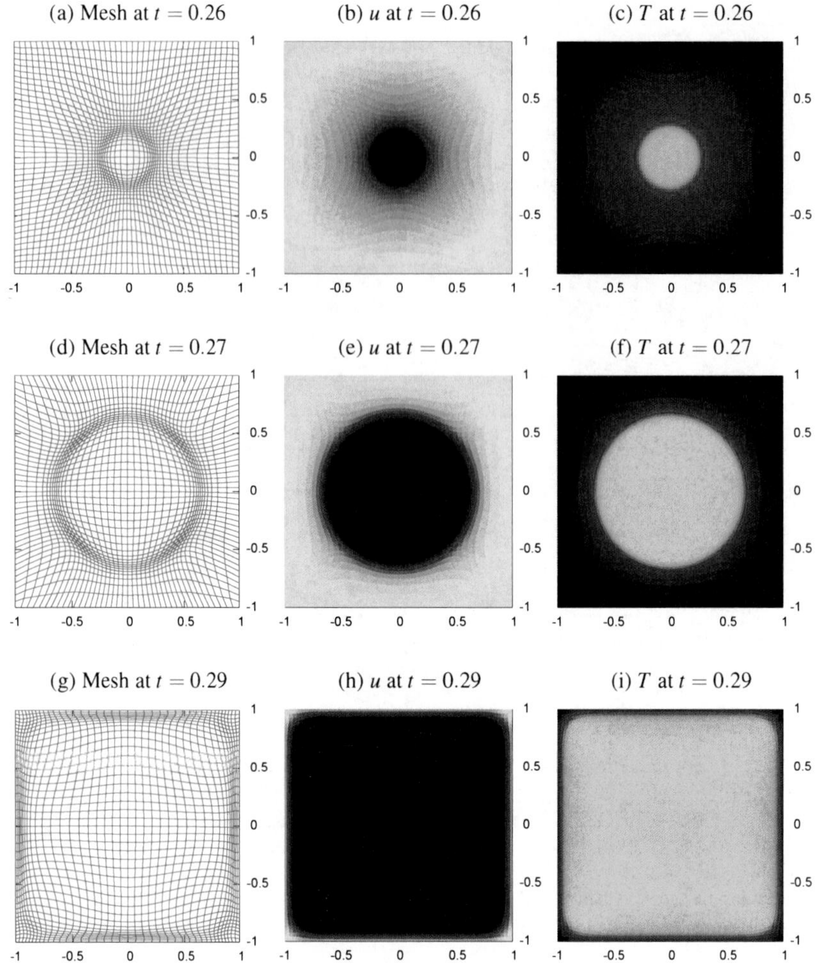

Fig. 6.29 Example 6.6.2. Adaptive 41×41 mesh, and corresponding computed solution u (where yellow represents 1.0 and black represents 0) and T (where yellow represents 2.2 and black represents 1), are shown at various time instants.

Example 6.6.3 (Convection in porous medium) This model describes the buoyancy-driven horizontal spreading of heat and chemical species through a fluid saturated porous medium. From Darcy's law and the homogeneous porous medium assumption, conservation of mass, momentum, energy, and the constituent give rise to the following system [268]:

$$\begin{cases} -\nabla^2 \psi &= Ra\left(\frac{\partial T}{\partial x} + N\frac{\partial C}{\partial x}\right), \\ \frac{\partial T}{\partial t} + \frac{\partial T}{\partial x}\frac{\partial \psi}{\partial y} - \frac{\partial T}{\partial y}\frac{\partial \psi}{\partial x} &= \nabla^2 T, \\ \frac{\phi}{\sigma}\frac{\partial C}{\partial t} + \frac{\partial C}{\partial x}\frac{\partial \psi}{\partial y} - \frac{\partial C}{\partial y}\frac{\partial \psi}{\partial x} &= \frac{1}{Le}\nabla^2 C, \end{cases} \tag{6.239}$$

where ψ is the stream function of the flow, T the temperature, C the concentration of the constituent, Ra the Darcy-modified Rayleigh number, N the buoyancy ratio, Le the Lewis number, ϕ the porosity ratio, and σ the heat capacity ratio. The physical domain is shown in Figure 6.31. The initial and boundary conditions are

$$\psi|_{t=0} = 0, \qquad T|_{t=0} = C|_{t=0} = \begin{cases} 1, & \text{for } x \le \frac{1}{2}, \\ 0, & \text{for } x > \frac{1}{2}, \end{cases} \tag{6.240}$$

and

$$\psi|_{\partial\Omega} = 0, \qquad \left.\frac{\partial T}{\partial \boldsymbol{n}}\right|_{\partial\Omega} = \left.\frac{\partial C}{\partial \boldsymbol{n}}\right|_{\partial\Omega} = 0, \qquad \text{for } t > 0 \tag{6.241}$$

where $\boldsymbol{n}$ denotes the unit outward normal to the boundary $\partial\Omega$.

The model with parameters $Ra = 1000$, $N = 0$, $Le = 1$, and $\frac{\phi}{\sigma} = 1$ is simulated using a moving mesh finite element method by Cao et al. [81]. The solution domain is initially partitioned almost uniformly into 3833 elements. MMPDE (6.39), functional (6.139), and the monitor function $M = \sqrt{1 + |\nabla T|^2}\, I$ are used for mesh movement. From the results shown in Figure 6.32, it is clear that the mesh adapts well to the temperature and successfully follows the motion of the thin layer of large temperature and concentration variation. As the variation in temperature and concentration is gradually smoothed out by diffusion, the mesh becomes more uniform. □

Example 6.6.4 (Coupling of groundwater flow and NAPL transport.) This example is a model problem for multiphase flow and transport in a groundwater environment [251, 252, 287]. It models the process of nonaqueous phase liquids (NAPLs) dissolved into the aqueous phase. The governing equations are

$$\frac{\partial \theta_n}{\partial t} = -\frac{k_{na}\,(C_a^* - C_a)}{\rho_n}, \tag{6.242}$$

$$\frac{\partial(\theta_a C_a)}{\partial t} = \nabla \cdot (D\nabla C_a - q_a C_a) + k_{na}\,(C_a^* - C_a), \tag{6.243}$$

$$\nabla \cdot \left(\frac{k k_{ra}}{\mu}\,(\nabla p_a - \rho_a g \nabla x)\right) + \left(\frac{1}{\rho_a} - \frac{1}{\rho_n}\right) k_{na}(C_a^* - C_a) = 0, \tag{6.244}$$

where the subscripts "a" and "n" represent the aqueous and nonaqueous phases, respectively, the superscript "$*$" indicates an equilibrium condition with the companion phase involved in the mass transfer, $\theta = \theta(t,x,y)$ is the volumetric fraction, $C_a = C_a(t,x,y)$ is the concentration of the NAPL dissolved in water, ρ is density,

k_{na} is a mass transfer coefficient representing a mass transfer process referenced to a loss by the nonaqueous phase and a gain by the aqueous phase, q_a is the water flux, D is the dispersivity, $p_a = p_a(t,x,y)$ is the aqueous phase pressure, k is permeability, k_{ra} is relative permeability, μ is viscosity, g is the gravity constant, and x is the vertical coordinate. Also, $\theta_n + \theta_a = n$, where the porosity n is taken to be constant here, and the water flux is given by Darcy's law,

$$q_a = -\frac{k k_{ra}}{\mu}\left(\nabla p_a - \rho_a g \nabla x\right).$$

The equations (6.242) and (6.243) are for the volumetric fraction of NAPL or NAPL content and the NAPL dissolved in water, respectively, while (6.244) is for the aqueous phase pressure.

Huang and Zhan [194] consider a physical scenario where the aqueous phase is being flushed from the left boundary and the dissolved NAPL is being eluted from the right boundary. Mathematically, the left and right boundary conditions are a specified flux for the aqueous phase, while the top and bottom boundary conditions are no-flow. The initial residual NAPL saturation and other variables are homogeneous. This is a typical laboratory condition, except that a perturbation in the residual NAPL saturation near the left boundary, where a portion of the boundary is NAPL free, indicates that clean water is flushing in. In the numerical simulation, the values of the physical parameters are taken as those given in Table 15.1 of [252], which are reasonably representative of conditions encountered in two-dimensional physical experiments. MMPDE (6.39), the functional (6.139), and the arc-length monitor function $M = I + \nabla C_a \nabla C_a^T$ are used for mesh movement. A moving finite difference result is shown in Figure 6.33. □

Example 6.6.5 (Stefan problems) The 2D enthalpy formulation of Stefan (phase change) problems can be written in the general form

$$\frac{\partial u}{\partial t} = \Delta\theta + f(x,y,t), \quad (x,y) \in \Omega \tag{6.245}$$

where θ is the temperature, $u(\theta)$ is the enthalpy, and $f(x,y,t)$ represents any body heating or cooling sources [109]. If the substance with specific heat constants c_1 and c_2 undergoes a phase change at the temperature $\theta = \theta_m$, then the enthalpy relationship to the temperature is

$$u(\theta) = \begin{cases} u(\theta_m^-) + c_1(\theta - \theta_m), & \theta < \theta_m \\ u(\theta_m^-) + \lambda + c_2(\theta - \theta_m), & \theta \geq \theta_m \end{cases} \tag{6.246}$$

where λ is the latent heat and $u(\theta_m^-)$ is the enthalpy at θ_m^- (before jump). In numerical simulation, this function is often replaced with a continuously differentiable, regularized function. A regularization proposed by Egolf and Manz [131] takes the form

$$u(\theta) = \begin{cases} u(\theta_m^-) + c_1(\theta - \theta_m) + \frac{\lambda}{2} exp\left(-\frac{|\theta - \theta_m|}{\varepsilon^-}\right), & \theta < \theta_m \\ u(\theta_m^-) + \lambda + c_2(\theta - \theta_m) - \frac{\lambda}{2} exp\left(-\frac{|\theta - \theta_m|}{\varepsilon^+}\right), & \theta \geq \theta_m \end{cases} \tag{6.247}$$

where

$$\varepsilon^- = \frac{\lambda}{2(c_2 - c_1)} \left(\frac{(c_2 - c_1)\varepsilon}{\lambda} + 1 - \sqrt{1 + \left(\frac{(c_2 - c_1)\varepsilon}{\lambda}\right)^2} \right),$$

$$\varepsilon^+ = \frac{\lambda}{2(c_2 - c_1)} \left(\frac{(c_2 - c_1)\varepsilon}{\lambda} - 1 + \sqrt{1 + \left(\frac{(c_2 - c_1)\varepsilon}{\lambda}\right)^2} \right),$$

and $\varepsilon > 0$ is the regularization parameter. Several physical scenarios have been simulated numerically by Beckett et al. [47] using a moving mesh finite element method based on MMPDE (6.39) (with $\tau = 0.1$), functional (6.139), and the scalar monitor function (5.232). Figure 6.34 shows an adaptive mesh and the corresponding computed interface at various time instants for an oscillating source problem where $\Omega = (-1,1) \times (-1,1)$, the initial condition is $\theta(x,y,0) = y/10$, the boundary conditions are $\theta(x,y,t) = y/10$ for the three sides with $y > -1$ and a homogeneous Neumann condition on the bottom side $y = -1$, and the oscillating heat source is given by

$$f(x,y,t) = \cos(\frac{t}{5}) \max\left(0, 3.125 - 50((x+\frac{1}{5})^2 + (y+\frac{1}{2})^2)\right)$$
$$+ \sin(\frac{t}{5}) \max\left(0, 3.125 - 50((x+\frac{1}{5})^2 + (y-\frac{1}{2})^2)\right).$$

□

6.7 Biographical notes

The variational approach and related elliptic PDE mesh generators have received considerable attention from scientists and engineers in the past. The approach is particularly straightforward for finite difference computations, where its implementation is simple and requires little expertise with data structures to implement the mesh refinement. Variational methods have typically been used for generating structured meshes, but they can also be used for generating unstructured meshes – e.g., see [81]. Interestingly, many of the smoothing methods used for improving mesh quality in unstructured mesh refinement can be viewed as variational methods or their variants and, vice versa, variational methods can be used for improving mesh quality in unstructured mesh refinement – e.g., see Bank and Smith [36] and Chen [98].

A large number of variational and elliptic PDE-based methods have been developed. In addition to those mentioned in §6.4 and §6.5, the papers [341, 324, 306, 235, 163, 217, 280, 95], among others, deserve special attention. The reader is also referred to the books [92, 213, 238, 325] and references therein for further reading in this topic.

Moving mesh methods of variational type have been successfully used for solving a large range of application problems. In addition to the problems discussed in §6.6, other examples include *fluid dynamics* (Yanenko et al. [345], Tan et al. [313], Tang [318]), *groundwater flow and multi-phase flow* (Huang and Zhan [194] and Wang [336]), *problems with blow-up solutions* (Budd et al. [67], Budd and Williams [71], Ceniceros and Hou [95], and Ren and Wang [280]), *chemotaxis systems* (Budd et al. [66]), *reactive flow and reaction diffusion systems* (Zegeling and Kok [351] and Azarenok and Tang [21]), *the nonlinear Schrödinger equation* (Ceniceros [94], Budd et al. [63], Ren and Wang [280]), *phase change problems* (Lynch [243], Lang [224], Feng et al. [142], Tan et al. [314], Yu et al. [346], Di et al. [120, 118], Beckett et al. [47], Mackenzie and Mekwi [246], Tan et al. [312]), *shear layer calculations* (Di et al. [119], Tang [318]), *gas dynamics* (Azarenok et al. [200, 17, 20]), *hyperbolic conservation laws* (Azarenok [17], Tang [318]), Stockie et al. [307], Tang and Tang [316, 317], Tang [315]), *problems with high vorticity* (Ceniceros and Hou [95]), *magneto-hydrodynamics* (Zegeling et al. [348, 350, 333, 349], Tan [311], and Han and Tang [167] Van Dam [332]), *meteorological problems* (Budd and Piggott [69]), *crystal growth* (Li [230], Wang et al. [337]), and *combustion problems* (Lang et al. [223]).

6.8 Exercises

1. Show that functional $I[\xi] = \int_0^1 \frac{1}{\rho(x)} \left(\frac{d\xi}{dx}\right)^2 dx$ can be transformed into $\hat{I}[x] = \int_0^1 \left(\rho(x) \frac{dx}{d\xi}\right)^{-1} d\xi$ by interchanging the roles of dependent and independent variables, assuming that the coordinate transformation $x = x(\xi) : [0,1] \to [0,1]$ has the inverse $\xi = \xi(x) : [0,1] \to [0,1]$. Verify that $I[\xi]$ and $\hat{I}[x]$ have Euler-Lagrange equations that are mathematically equivalent.
2. Show that (6.6) is the Euler-Lagrange equation of (6.7).
3. Derive (6.8) from (6.6).

4. Find an equivalent functional of (6.7) expressed in terms of $x = x(\xi, \eta)$ and $y = y(\xi, \eta)$. Compare your result with functional (6.10). Explain why your functional is less likely to produce a folded mesh.
5. Complete the derivation for (6.14).
6. Derive (6.46) from (6.45).
7. Show that functional (6.7) is coercive and uniformly convex.
8. Use Lemma 6.2.1 to show that the functional

$$I[\xi,\eta] = \int_{\Omega} \left(\left(\frac{\partial \xi}{\partial x}\right)^2 + \left(\frac{\partial \xi}{\partial y}\right)^2 + \left(\frac{\partial \eta}{\partial x}\right)^2 + \left(\frac{\partial \xi}{\partial x}\right)^2 \right)^{\gamma} dxdy,$$

where $\gamma \geq 1$ is a constant, is convex.
9. Show that the functional

$$I[\xi,\eta] = \int_{\Omega} \Big[\left(\frac{\partial \xi}{\partial x}\right)^2 + \left(\frac{\partial \xi}{\partial y}\right)^2 + \left(\frac{\partial \eta}{\partial x}\right)^2 + \left(\frac{\partial \xi}{\partial x}\right)^2 + \left(\frac{\partial \xi}{\partial x}\frac{\partial \eta}{\partial y} - \frac{\partial \xi}{\partial y}\frac{\partial \eta}{\partial x}\right)^2 \Big] dxdy$$

is not convex but polyconvex.
10. Give a 2D functional of the form (6.74).
11. Derive (6.98) and (6.99).
12. Use dimensional analysis to show that the terms in functional (6.108) are dimensionally homogeneous.
13. Confirm the validity of the results in Table 6.1.
14. Show that the functional (6.156) will result in a mesh without adaptivity in 2D when a scalar-matrix monitor function is used.
15. Show that functionals (6.167) – (6.169) have different dimensions.
16. Show the equality (6.181).
17. For some of the mesh generation methods in §6.5, use (6.212) to find the corresponding adaptive mesh methods, and analyze their capability to satisfy the mesh equidistribution and alignment conditions.

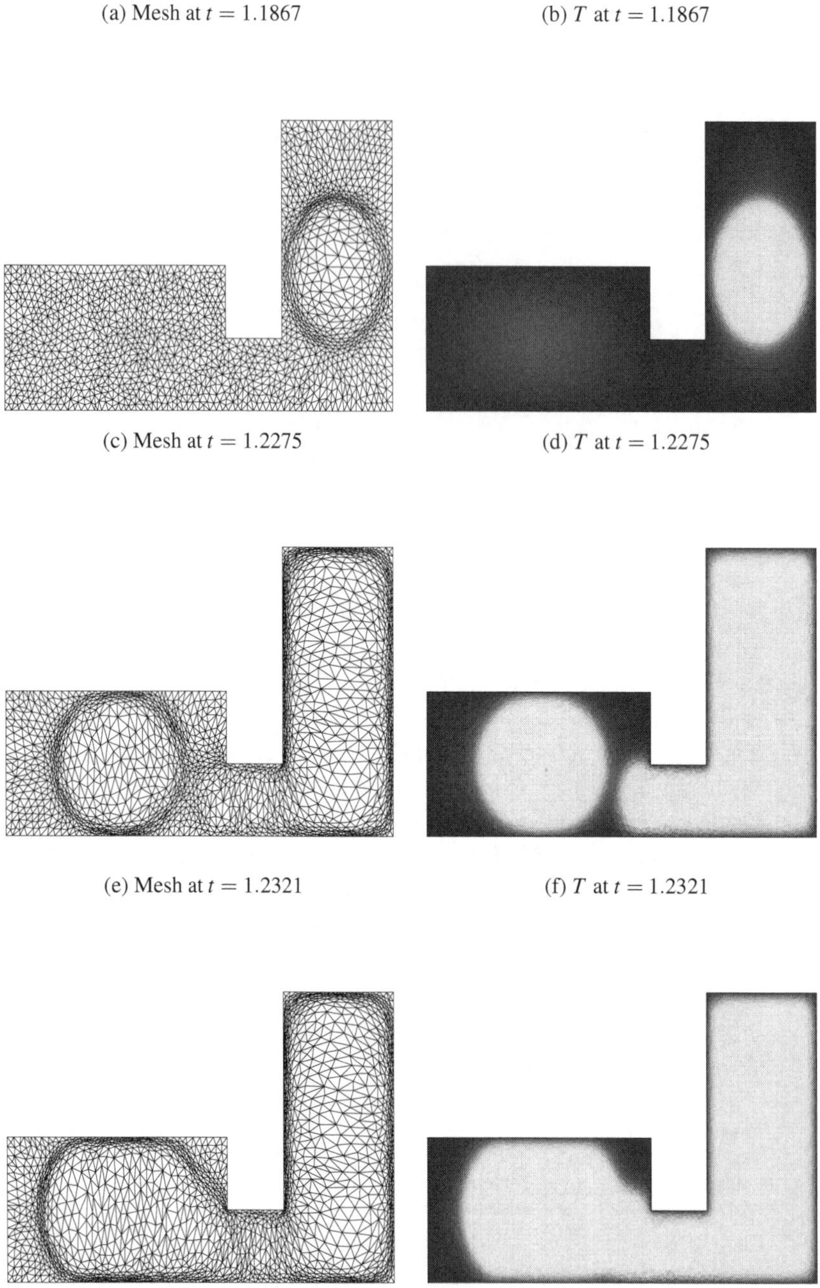
(a) Mesh at $t = 1.1867$
(b) T at $t = 1.1867$
(c) Mesh at $t = 1.2275$
(d) T at $t = 1.2275$
(e) Mesh at $t = 1.2321$
(f) T at $t = 1.2321$

Fig. 6.30 Example 6.6.2. Adaptive mesh and corresponding solution T (where white represents 2.2 and black represents 1), obtained by a moving mesh finite element method, are shown at various time instants. Reprinted from Cao et al. [81], with permission from Elsevier.

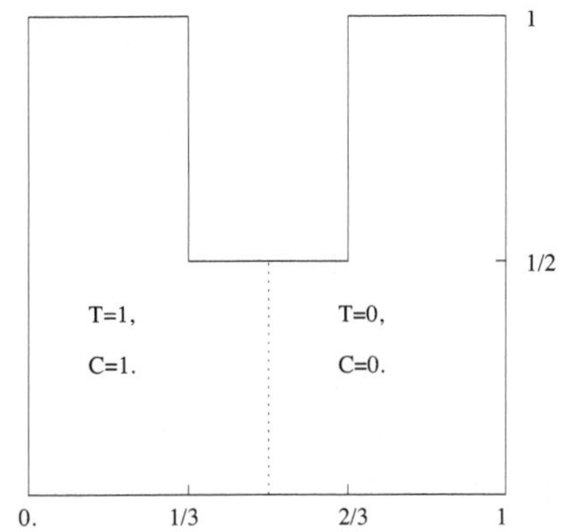

Fig. 6.31 Physical domain for Example 6.6.3.

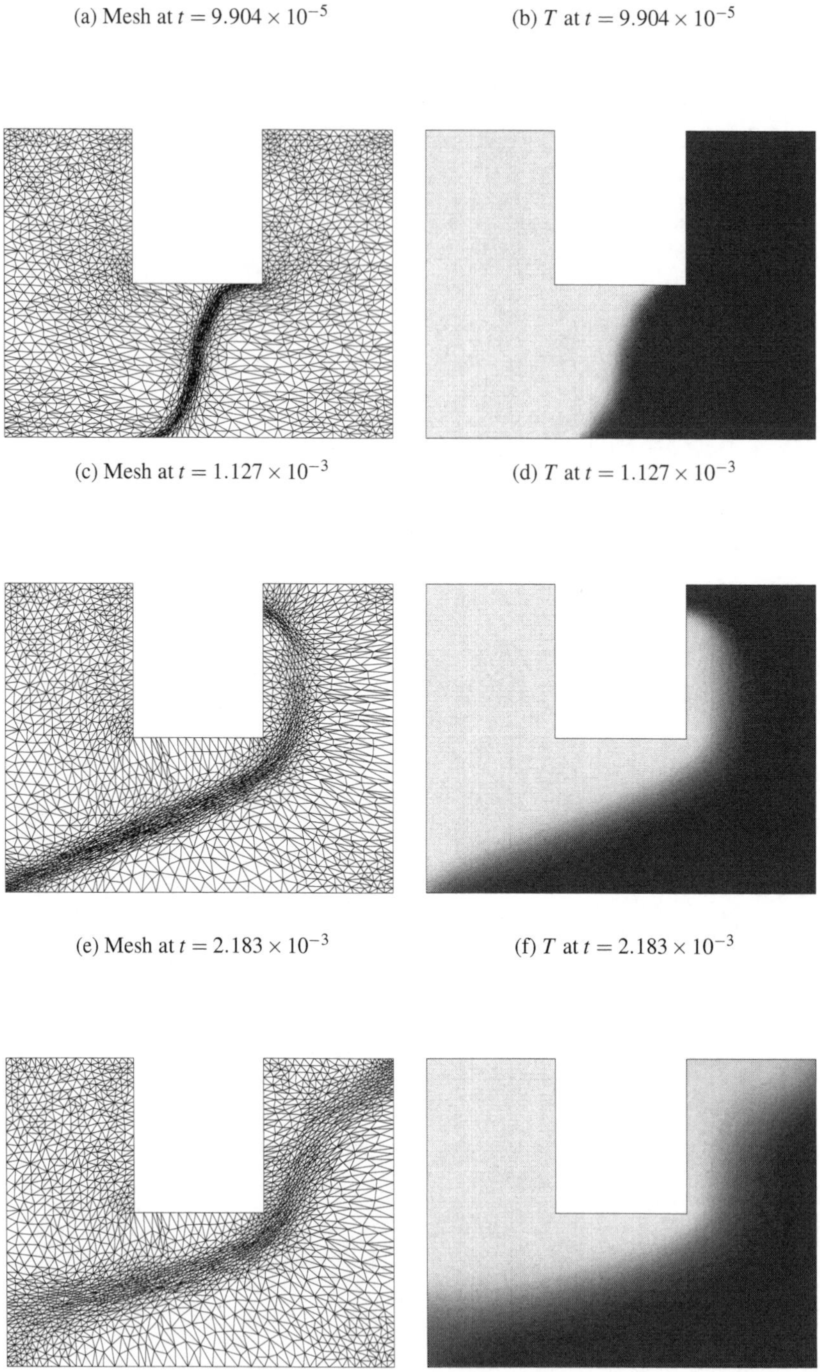
(a) Mesh at $t = 9.904 \times 10^{-5}$
(b) T at $t = 9.904 \times 10^{-5}$
(c) Mesh at $t = 1.127 \times 10^{-3}$
(d) T at $t = 1.127 \times 10^{-3}$
(e) Mesh at $t = 2.183 \times 10^{-3}$
(f) T at $t = 2.183 \times 10^{-3}$

Fig. 6.32 Example 6.6.3. Adaptive mesh and corresponding solution T (where white represents 1 and black represents 0), obtained by a moving mesh finite element method, are shown at various time instants. Reprinted from Cao et al. [81], with permission from Elsevier.

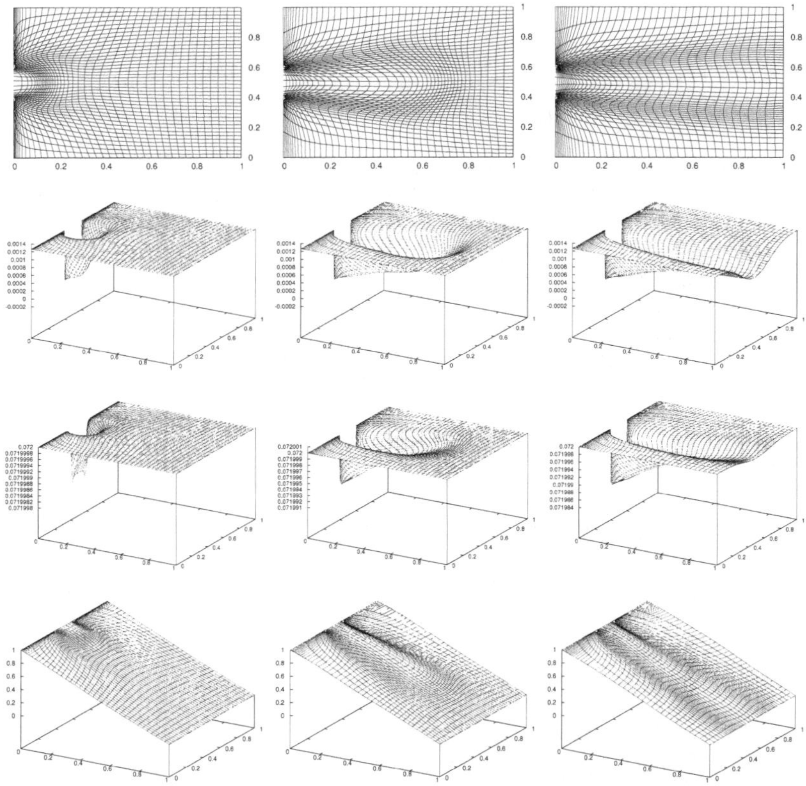

Fig. 6.33 Example 6.6.4. Adaptive moving mesh (first row), NAPL (second row), dissolved NAPL in water (third row), and aqueous phase pressure (fourth row) for the NAPL-flow coupling problem. The first, second, and third columns correspond to time instants $t = 1.821 \times 10^{-2}$, 9.426×10^{-2}, and 1.714×10^{-1}, respectively. Reprinted from Huang and Zhan [194], with permission from the American Mathematical Society.

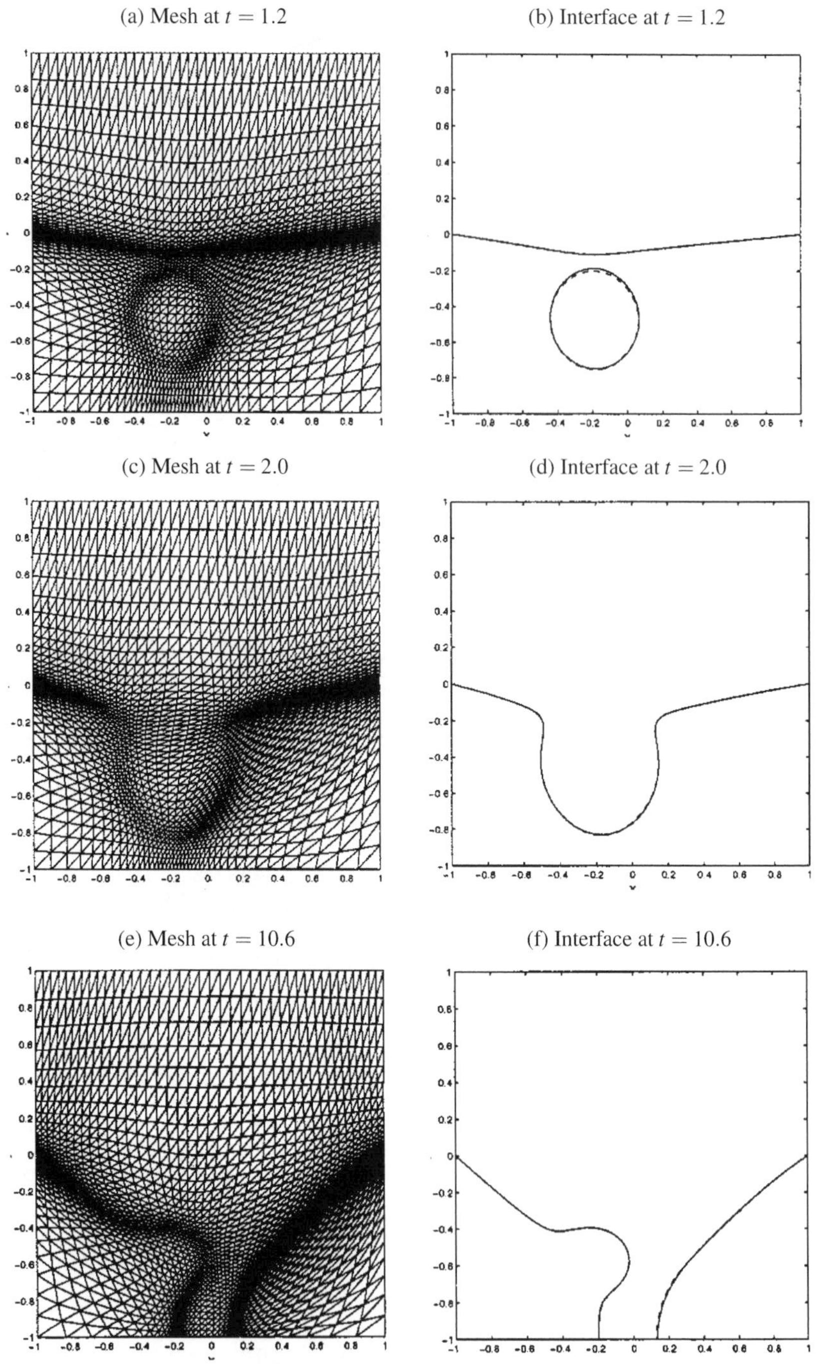

Fig. 6.34 Example 6.6.5. Adaptive mesh and interface prediction. Reprinted from Beckett et al. [47], with permission from Elsevier.

Chapter 7
Velocity-Based Adaptive Methods

In this chapter we discuss *velocity-based* adaptive moving mesh methods. Although the classification of methods as being either velocity-based or location-based can at times be somewhat artificial, the former are generally characterized by the fact that their formulations directly target the mesh velocity, with the subsequent mesh points determined by integrating the velocity field. Some of these methods are motivated by the Lagrangian method in computational fluid dynamics (e.g., see Fletcher [148] or §7.1.1) and some others are based on minimizing a quantity related to error. A fortuitous property of the Lagrangian methods is that it is well-suited to maintaining sharper material interfaces since convection terms are eliminated from the governing equations. A disadvantage is that the meshes have a tendency to tangle and lose spatial resolution of the solution. Unfortunately, the Lagrangian-like moving mesh methods also inherit this disadvantage of Lagrangian methods, and major effort has gone into the development of these methods so as to avoid mesh tangling and/or regain spatial accuracy.

7.1 Methods based on geometric conservation law

In this section we discuss three methods based on the Geometric Conservation Law (GCL) [322]: the GCL method developed in [84], the deformation map method developed by Liao and coworkers [53, 78, 236, 295], and a moving mesh finite element method developed by Baines and coworkers [31, 340]. Like the methods based on the Monge-Ampère equation described in §6.5.6, these GCL-based methods also enforce the equidistribution condition (4.74). However, they are formulated directly in terms of mesh velocity and as a consequence share many of the advantages and disadvantages of the Lagrangian method in computational fluid dynamics [148].

7.1.1 GCL method

The GCL method [84] is based on the Geometric Conservation Law (3.21) and the equidistribution condition (4.74). Dividing (4.74) by σ and differentiating the resulting equation with respect to t (with $\boldsymbol{\xi}$ being fixed), we obtain

$$J\frac{\partial}{\partial t}\left(\frac{\rho}{\sigma}\right)+J\nabla\left(\frac{\rho}{\sigma}\right)\cdot\dot{x}+\dot{J}\left(\frac{\rho}{\sigma}\right)=0.$$

Combining (3.21) with the above equation gives rise to

$$\frac{\partial}{\partial t}\left(\frac{\rho}{\sigma}\right)+\nabla\cdot\left(\frac{\rho}{\sigma}\dot{x}\right)=0. \tag{7.1}$$

This is basically the continuity equation in fluid dynamics, with "flow velocity" $\dot{\boldsymbol{x}}$ and "fluid density" ρ/σ. It can also be considered as a condition for determining the divergence of mesh speed $\dot{\boldsymbol{x}}$. Thus, the equidistribution condition (4.74) is directly used to determine the divergence of $\dot{\boldsymbol{x}}$.

Equation (7.1) can be derived in a slightly different way. Taking $u=\rho/\sigma$ in (3.24) (a form of GCL), we have

$$\frac{d}{dt}\int_{A(t)}\frac{\rho}{\sigma}d\boldsymbol{x}=\int_{A(t)}\left(\frac{\partial}{\partial t}\left(\frac{\rho}{\sigma}\right)+\nabla\cdot\left(\frac{\rho}{\sigma}\dot{\boldsymbol{x}}\right)\right)d\boldsymbol{x}.$$

Then (7.1) follows from the facts that $A(t)$ is arbitrary and that, due to the equidistribution condition (4.74),

$$\frac{d}{dt}\int_{A(t)}\frac{\rho}{\sigma}d\boldsymbol{x}=\frac{d}{dt}\int_{A_c}\frac{J\rho}{\sigma}d\boldsymbol{\xi}=\frac{d}{dt}\left(\frac{|A_c|}{|\Omega_c|}\right)=0. \tag{7.2}$$

The single equation (7.1) is insufficient to determine the vector field $\dot{\boldsymbol{x}}$, and the motivation for finding supplementary conditions on $\dot{\boldsymbol{x}}$ is provided by the Helmholtz decomposition theorem for vectors, which states that a continuous and differentiable vector field can be decomposed into the orthogonal sum of a gradient of a scalar field and the curl of a vector field. Thus, $\dot{\boldsymbol{x}}$ can be determined by specifying both its divergence through (7.1) and its curl. We require $\dot{\boldsymbol{x}}$ to satisfy the general curl condition

$$\nabla\times w(\dot{\boldsymbol{x}}-\boldsymbol{v}_{ref})=0, \tag{7.3}$$

where $w>0$ is a weight function and $\boldsymbol{v}_{ref}$ is a user-specified reference vector field. Different choices for w and $\boldsymbol{v}_{ref}$ lead to different curl conditions on $\dot{\boldsymbol{x}}$, and these choices are discussed later.

Requirement (7.3) implies that there exists a potential function ϕ such that

$$w(\dot{\boldsymbol{x}}-\boldsymbol{v}_{ref})=\nabla\phi$$

or

$$\dot{\boldsymbol{x}} = \frac{1}{w}\nabla\phi + \boldsymbol{v}_{ref}. \tag{7.4}$$

Inserting this into (7.1) leads to

$$\nabla \cdot \left(\frac{\rho}{w\sigma}\nabla\phi\right) = -\frac{\partial}{\partial t}\left(\frac{\rho}{\sigma}\right) - \nabla \cdot \left(\frac{\rho}{\sigma}\boldsymbol{v}_{ref}\right) \qquad \text{in } \Omega. \tag{7.5}$$

The boundary condition on ϕ can be obtained by requiring that the mesh points do not move out the domain, i.e., $\dot{\boldsymbol{x}}\cdot\boldsymbol{n} = 0$ where $\boldsymbol{n}$ denotes the outward normal to $\partial\Omega$. From (7.4), this gives the boundary condition

$$\frac{\partial\phi}{\partial n} = -w\boldsymbol{v}_{ref}\cdot\boldsymbol{n} \qquad \text{on } \partial\Omega. \tag{7.6}$$

To summarize, the potential function ϕ is determined by solving the elliptic equation (7.5) subject to the Neumann boundary condition (7.6). Once ϕ is known, the mesh location can be obtained by integrating (7.4).

The basic GCL method can also be cast in a least squares formulation [84].

Relation to Lagrangian methods in fluid dynamics. In fluid dynamics there exist two basic ways to represent the flow field, the *Eulerian representation* and the *Lagrangian representation* (e.g., see Batchelor [38]). In the Eulerian representation, fluid motion is observed at a fixed spatial location $\boldsymbol{x}$ and the flow velocity is described by a function in the form of $\boldsymbol{v}_{flow}(\boldsymbol{x},t)$. Governing equations expressed in terms of the independent variables $\boldsymbol{x}$ and t are typically solved numerically on a fixed mesh. Numerical methods based on the Eulerian representation are called Eulerian methods. While Eulerian meshes trivially avoid mesh tangling, solutions are diffusive so it is difficult to maintain sharp material interfaces.

In the alternative Lagrangian representation, fluid motion is observed by following individual fluid particles as they move through space and time. The flow velocity is described by a function of the form of $\hat{\boldsymbol{v}}_{flow}(\boldsymbol{\xi},t)$, where $\boldsymbol{\xi}$ is a vector field used to label fluid particles and is often chosen as the position of the particles at an initial time t_0. The position of any particle $\boldsymbol{\xi}$, $\boldsymbol{x} = \boldsymbol{x}(\boldsymbol{\xi},t)$, is then determined by

$$\dot{\boldsymbol{x}} = \hat{\boldsymbol{v}}_{flow}(\boldsymbol{\xi},t), \tag{7.7}$$

subject to the initial condition

$$\boldsymbol{x}(\boldsymbol{\xi},t_0) = \boldsymbol{\xi}. \tag{7.8}$$

Since both $\boldsymbol{v}_{flow}(\boldsymbol{x},t)$ and $\hat{\boldsymbol{v}}_{flow}(\boldsymbol{\xi},t)$ describe the same velocity field, it follows from the definition of $\boldsymbol{x}(\boldsymbol{\xi},t)$ that

$$\boldsymbol{v}_{flow}(\boldsymbol{x}(\boldsymbol{\xi},t),t) = \hat{\boldsymbol{v}}_{flow}(\boldsymbol{\xi},t).$$

Inserting this into (7.7) gives

$$\dot{\boldsymbol{x}} = \boldsymbol{v}_{flow}(\boldsymbol{x},t), \tag{7.9}$$

which simply states that the velocity of a particle at a time t is equal to the velocity at its location $\boldsymbol{x}$. A particular advantage of the Lagrangian representation is that the advective terms in the governing equations vanish identically. Lagrangian methods – numerical methods based on the Lagrangian representation – are less diffusive than Eulerian ones and well-suited to maintaining sharp material interfaces. Moreover, the non-singularity of the coordinate transformation from the Lagrangian coordinates to the Euler coordinates is guaranteed by the incompressibility condition of the fluid. However, multidimensional Lagrangian meshes, generated by solving (7.9), are often too skewed to be used directly in the numerical solution of PDEs. This inadequacy has been a driving force behind the development of hybrid methods such as particle methods [174], methods of characteristics [139], and Arbitrary Lagrangian-Eulerian (ALE) methods [173, 249].

There is a close relation between the GCL method and Lagrangian methods, and in fact, the GCL method can be regarded as a generalization of these methods. To see this, choose the control vector field $\boldsymbol{v}_{ref}$ in (7.4) to be the flow velocity $\boldsymbol{v}_{flow}$. Taking ρ = constant as a special case (no adaption), for incompressible fluid flow where $\nabla \cdot \boldsymbol{v}_{flow} = 0$ we have ϕ = constant, and (7.4) reduces to (7.9).

Relation to the L^2 Monge-Kantorovich problem. Benamou and Brenier [48] introduce a so-called fluid dynamics formulation of the L^2 Monge-Kantorovich problem (cf. §6.5.6). For given density functions $\rho_0(\boldsymbol{x})$ and $\rho_1(\boldsymbol{x})$, the formulation involves minimizing a cost function, i.e.,

$$\min_{\boldsymbol{v},\rho/\sigma} \int_0^1 \int_\Omega \frac{\rho(\boldsymbol{x},t)}{2\sigma} \|\boldsymbol{v}(\boldsymbol{x},t)\|^2 d\boldsymbol{x}dt, \tag{7.10}$$

subject to the constraints $\rho(\boldsymbol{x},0) = \rho_0(\boldsymbol{x})$, $\rho(\boldsymbol{x},1) = \rho_1(\boldsymbol{x})$, and (7.1), which is now

$$\frac{\partial}{\partial t}\left(\frac{\rho}{\sigma}\right) + \nabla \cdot \left(\frac{\rho}{\sigma}\boldsymbol{v}\right) = 0. \tag{7.11}$$

Using the Lagrange multiplier method we can obtain the Euler-Lagrange equations for the above constraint optimization problem. They include equation (7.11) and

$$\frac{1}{2}\|\boldsymbol{v}\|^2 - \frac{\partial \phi}{\partial t} - \nabla\phi \cdot \boldsymbol{v} = 0, \tag{7.12}$$

$$\boldsymbol{v} - \nabla\phi = 0, \tag{7.13}$$

where the potential function $\phi = \phi(\boldsymbol{x},t)$ is the Lagrange multiplier. Inserting (7.13) into (7.11) and (7.12) gives

$$\frac{\partial}{\partial t}\left(\frac{\rho}{\sigma}\right) + \nabla \cdot \left(\frac{\rho}{\sigma}\nabla\phi\right) = 0, \tag{7.14}$$

$$\frac{\partial \phi}{\partial t} + \frac{1}{2}\|\nabla\phi\|^2 = 0. \tag{7.15}$$

Note that this system cannot be integrated as ODEs due to the constraints $\rho(\boldsymbol{x},0) = \rho_0(\boldsymbol{x})$ and $\rho(\boldsymbol{x},1) = \rho_1(\boldsymbol{x})$. An augmented Lagrangian method is proposed for solving this system in [48]. The formulation and the solution method are used by Sulman et al. [308] for the purpose of mesh adaptation.

Note that (7.14) corresponds to (7.5) with $w = 1$ and $\boldsymbol{v}_{ref} = 0$. In this sense, the fluid dynamics formulation of the L^2 Monge-Kantorovich problem can be viewed as a special example of the GCL formulation where ρ connects ρ_0 at $t = 0$ and ρ_1 at $t = 1$ in a special way. Moreover, the optimization of (7.10) with respect to $\boldsymbol{v}$ only (and subject to (7.11)) will result precisely in a GCL formulation.

Choice of w, $\boldsymbol{v}_{ref}$, and ρ. We consider here the two obvious choices for the weight function $w = \rho/\sigma$ and $w = 1$. The former corresponds to the deformation map method (see §7.1.2 below) and generally does not result in an irrotational mesh velocity field. The latter results in an irrotational mesh velocity field when $\boldsymbol{v}_{ref} = 0$ (cf. (7.3)). While numerous other options are possible, their mathematical and/or physical significance is unclear.

The choice of the control vector field $\boldsymbol{v}_{ref}$ is generally problem dependent. For fluid dynamics problems, a good choice of it can be the flow velocity, since this is likely to reduce the magnitude of the convection term. However, when mesh adaption is allowed (i.e., ρ is not constant), grid movement due to the mesh adaption may increase the convection term and make $\boldsymbol{v}_{ref}$ more difficult to choose. Generally speaking, when physical intuition for choosing $\boldsymbol{v}_{ref}$ is unavailable, the best option is likely to be to simply choose $\boldsymbol{v}_{ref} = \boldsymbol{0}$.

The choice of the adaptation function ρ should generally be based on the equidistribution condition (4.74). Consequently, it is natural to define it to be $\rho = \sqrt{\det(M)}$ with the monitor function determined as in Chapter 5. It can also be chosen based on other considerations, such as the scaling invariance argument used in [31].

Example 7.1.1 In this example we generate a moving adaptive mesh for a given mesh density function

$$\rho(x,y,t) = 1 + 5t\exp(-50|(x-\frac{1}{2})^2 + (y-\frac{1}{2})^2 - (\frac{1}{4})^2|) \tag{7.16}$$

for $t \in [0,1]$ and $(x,y) \in (0,1)\times(0,1)$. A mesh at $t = 1$ generated with the GCL method (with $w = 1$ and $\boldsymbol{v}_{ref} = \boldsymbol{0}$) is shown in Figure 7.1. □

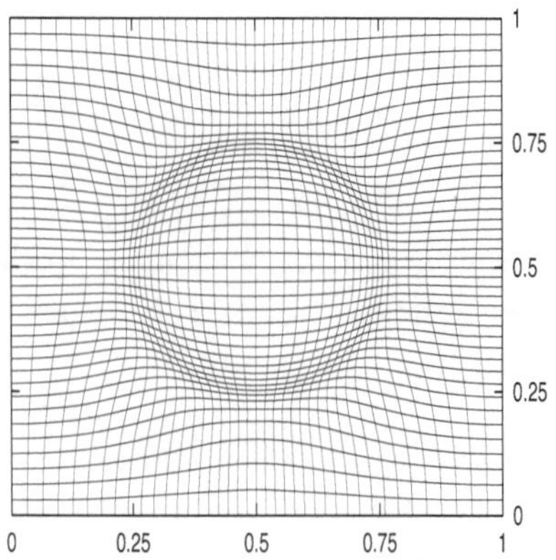

Fig. 7.1 Example 7.1.1. A mesh at $t = 1$ is generated using the GCL method with $w = 1$ and $\mathbf{v}_{ref} = \mathbf{0}$. (Cao et al. [84].

Example 7.1.2 Here a moving mesh is generated for the time mesh density function

$$\rho(x,y,t) = \begin{cases} 1+100(t+0.1)\exp(-50|(x-\frac{1}{2})^2+(y-\frac{1}{2})^2-0.09|), \ \forall t \in (-0.1,0) \\ 1+10\exp(-50|(x-\frac{1}{2}-t)^2+(y-\frac{1}{2})^2-0.09|), \ \forall t \geq 0. \end{cases} \tag{7.17}$$

This function is defined using two time phases, one from $t = -0.1$ to $t = 0$ and the other for $t > 0$. The purpose of the first phase is to produce an adaptive mesh for $t = 0$, starting from a uniform mesh at $t = -0.1$. For $t > 0$, the function simulates a circular peak which moves right at speed 1 and eventually leaves the domain while maintaining its shape.

A moving mesh obtained for $w = 1$ is shown in Figure 7.2. As expected, the mesh points are concentrated around the circle $(x-\frac{1}{2})^2+(y-\frac{1}{2})^2 = 0.3^2$ at the beginning $(t = 0)$ and then follow the movement of the circular peak. It is interesting to note that the mesh is not very smooth. This appears to be an inherent feature of the GCL method because the locations of the mesh points are not governed by either an elliptic or parabolic PDE. □

Example 7.1.3 For this example a moving mesh is generated for the mesh density function

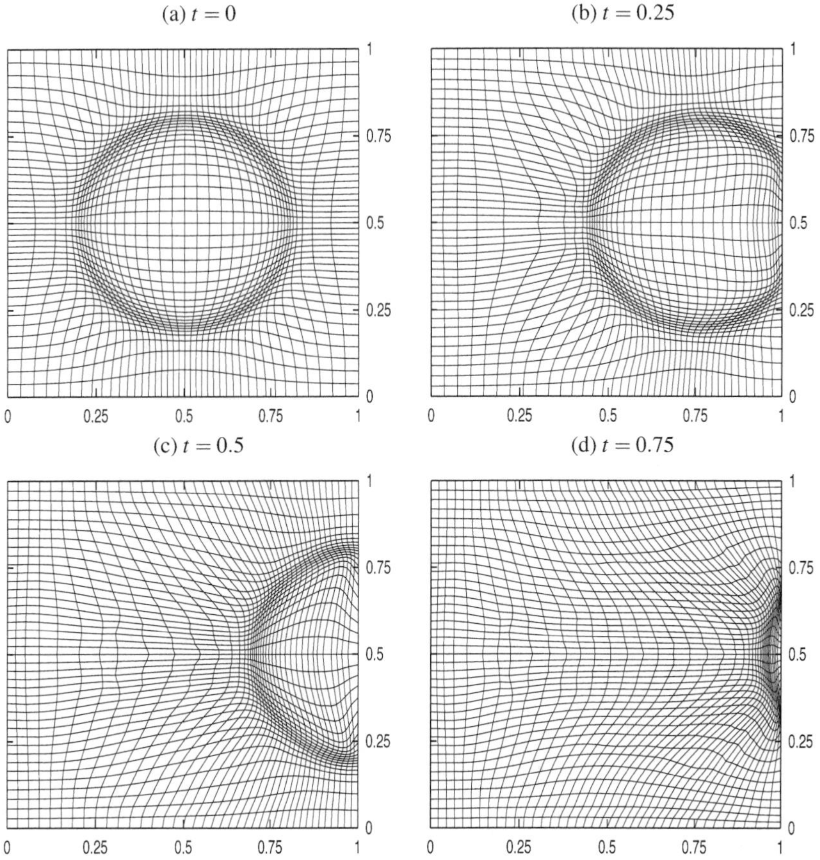

Fig. 7.2 Example 7.1.2. A moving mesh is generated using the GCL method with $w = 1$ and $\mathbf{v}_{ref} = \mathbf{0}$. (Cao et al. [84]. ©2002 Society for Industrial and Applied Mathematics. Reprinted with permission. All rights reserved.)

$$\rho(x,y,t) = \begin{cases} 1+50(0.1+t)\exp(-50|(x-\frac{3}{4})^2+(y-\frac{1}{2})^2-.01|), \\ \qquad \text{for } -0.1 < t < 0 \\ 1+5\exp(-50|(x-\frac{1}{2}-\frac{1}{4}\cos(2\pi t))^2+(y-\frac{1}{2}-\frac{1}{4}\sin(2\pi t))^2-.01|), \\ \qquad \text{for } t \geq 0. \end{cases} \tag{7.18}$$

The largest values of ρ occur around a small circle which rotates about the point $(\frac{1}{2}, \frac{1}{2})$.

This is a difficult test problem for many moving mesh methods, especially for ones using a Lagrangian representation. For such methods, as the concentration of mesh points follows the small rotating circle, if some of the boundary mesh points stay fixed (as in this case where the four corner points are fixed), then the mesh can

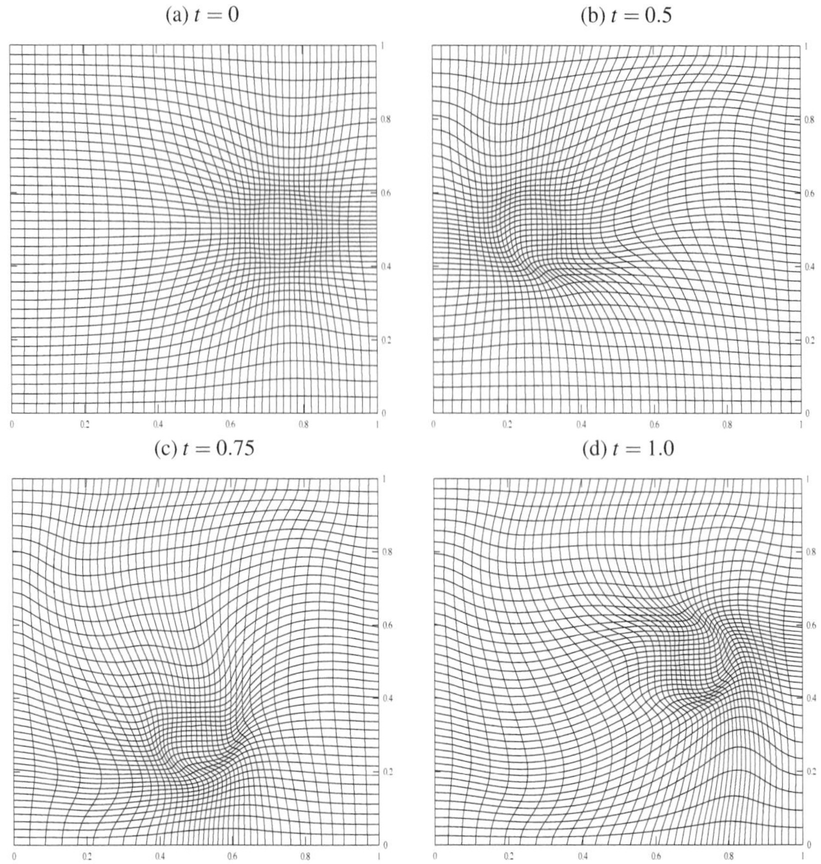

Fig. 7.3 Example 7.1.3. A moving mesh is generated using the GCL method with $w = 1$ and $\boldsymbol{v}_{ref} = \boldsymbol{0}$. (Cao et al. [84]. ©2002 Society for Industrial and Applied Mathematics. Reprinted with permission. All rights reserved.)

be expected to become more and more skewed. Indeed, the current GCL method suffers from this difficulty. Although the GCL condition guarantees non-singularity of the Jacobian of the coordinate transformation in the continuous case, it does not prevent the mesh from becoming increasingly skewed. Points of such a highly skewed mesh can have a tendency to tangle each other numerically. This skewness is illustrated in Figure 7.3 for the case $w = 1$. □

7.1.2 Deformation map method

The deformation map is introduced by Moser [110, 261] to study volume elements of a compact Riemannian manifold when proving the existence of a C^1 diffeomorphism with specified Jacobian. The map is subsequently adapted by Liao and coworkers [53, 236, 295] to generate adaptive moving meshes. It takes the form

$$\begin{cases} \dot{\boldsymbol{x}} = \frac{1}{\rho(\boldsymbol{x},t)} \nabla \phi(\boldsymbol{x},t), & \text{in } \Omega \\ \Delta \phi = -\frac{\partial \rho}{\partial t}, & \text{in } \Omega \\ \frac{\partial \phi}{\partial n} = 0, & \text{on } \partial \Omega. \end{cases} \tag{7.19}$$

It is easy to see that (7.19) corresponds to the GCL method (7.4) and (7.5) for the case $\boldsymbol{v}_{ref} = 0$ and $w = \rho$. Note that the mesh velocity is generally not irrotational in this case. Indeed, from (7.3), we have

$$\nabla \times \dot{\boldsymbol{x}} = -\frac{1}{\rho} \nabla \rho \times \dot{\boldsymbol{x}}. \tag{7.20}$$

Figure 7.4 shows a moving mesh obtained with the deformation map method (or GCL method with $w = \rho$) for Example 7.1.2. It is apparent that the moving mesh is considerably different from the one in Figure 7.2 (generated with the GCL method with $w = 1$). In our experience, the moving meshes generated with $w = 1$ are usually less skewed than those obtained with $w = \rho$. Although it is difficult to predict the precise influence caused by the choice $w = \rho$, it appears that the choice $w = 1$, which produces an irrotational mesh velocity, will generally produce better behaved adaptive meshes.

7.1.3 Static version

For many situations, one needs to generate a fixed mesh having a specified mesh topology and a prescribed mesh density distribution $\tilde{\rho}(\boldsymbol{x})$. An example is to produce an initial adaptive mesh which evenly distributes the interpolation error of some initial data (cf. Example 7.1.1). In such cases, we may define a time-dependent mesh density function by

$$\rho(\boldsymbol{x},t) = (1-t) + t\,\tilde{\rho}(\boldsymbol{x}), \qquad 0 \le t \le 1 \tag{7.21}$$

and then use continuation, integrating the mesh system (7.4) and (7.5) from $t = 0$ to 1. A variant of this procedure can also be used to create an adaptive mesh at each time level for a time-dependent problem [241].

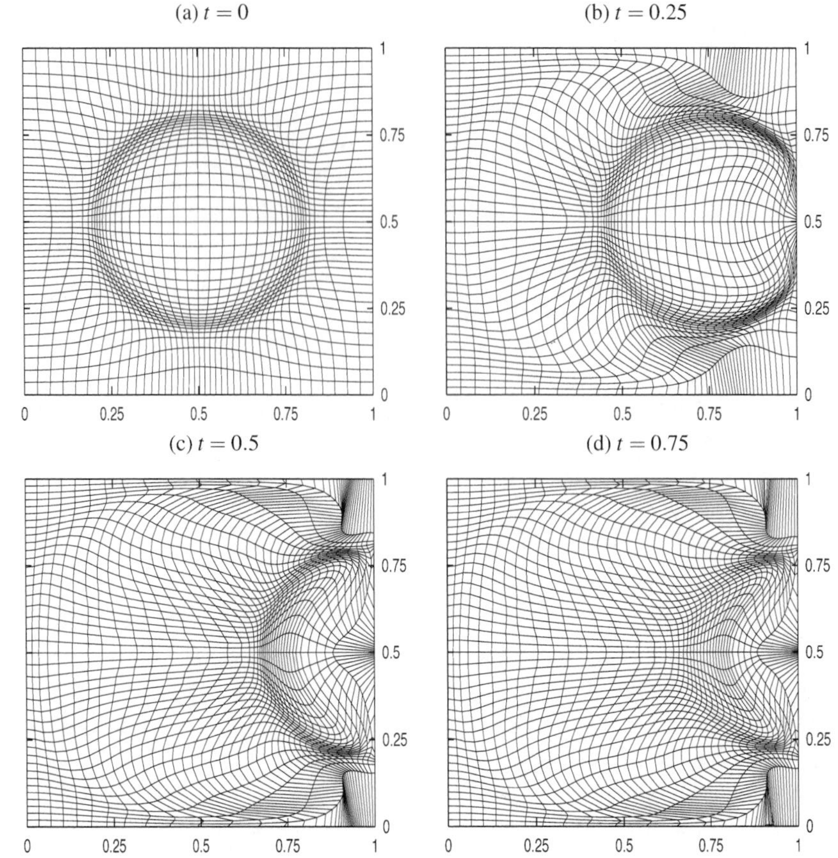

Fig. 7.4 Example 7.1.2. A moving mesh is generated using the GCL method with $w = \rho$ and $\boldsymbol{v}_{ref} = \boldsymbol{0}$. (Cao et al. [84]. ©2002 Society for Industrial and Applied Mathematics. Reprinted with permission. All rights reserved.)

7.1.4 A moving mesh finite element method based on GCL

Baines et al. [32, 340, 31, 33] have developed a moving finite element method bearing a strong relation to the GCL method of §7.1.1 and the deformation method of §7.1.2. To describe this method, consider a physical differential equation of the form

$$u_t = \mathscr{L}u, \quad \text{in } \Omega \tag{7.22}$$

where $\mathscr{L}$ is a differential operator, and for simplicity assume that the Dirichlet boundary condition

$$u = g, \quad \text{on } \partial\Omega \tag{7.23}$$

is given. A mesh density function is chosen of the form $\rho = \rho(u, \nabla u)$.

Taking $w = 1$ in (7.5) and noticing that σ is a function of t, we have

$$\nabla \cdot (\rho \nabla \phi) = \frac{\rho}{\sigma}\frac{d\sigma}{dt} - \left(\frac{\partial \rho}{\partial u} + \frac{\partial \rho}{\partial \nabla u} \cdot \nabla \right) u_t - \nabla \cdot (\rho \boldsymbol{v}_{ref}).$$

Inserting (7.22) into the above equation, multiplying it by a test function $v \in H^1(\Omega)$, integrating over Ω, and performing integration by parts, (7.5) can be rewritten in the weak form

$$\begin{aligned}\int_{\Omega(t)} \rho \nabla \phi \cdot \nabla v d\boldsymbol{x} &= -\frac{1}{\sigma}\frac{d\sigma}{dt}\int_{\Omega(t)} v\rho d\boldsymbol{x} \\ &+ \int_{\Omega(t)} v \left(\frac{\partial \rho}{\partial u} + \frac{\partial \rho}{\partial \nabla u} \cdot \nabla \right) \mathscr{L} u d\boldsymbol{x} - \int_{\Omega(t)} \rho \boldsymbol{v}_{ref} \cdot \nabla v d\boldsymbol{x} \\ &+ \int_{\partial\Omega(t)} v\rho \frac{\partial \phi}{\partial n} dS + \int_{\partial\Omega(t)} v\rho \boldsymbol{v}_{ref} \cdot \boldsymbol{n} dS, \qquad \forall v \in H^1(\Omega) \end{aligned} \quad (7.24)$$

where $\boldsymbol{n}$ is the unit outward normal to the boundary $\partial\Omega$ and for generality we consider the situation where Ω can change with time. The weak formulation of (7.4) is

$$\int_{\Omega(t)} v(\dot{\boldsymbol{x}} - \nabla \phi - \boldsymbol{v}_{ref}) d\boldsymbol{x} = 0, \quad \forall v \in H^1(\Omega). \quad (7.25)$$

Taking $u = v\rho/\sigma$ in (3.24) we have

$$\begin{aligned}\frac{d}{dt}\int_{\Omega(t)} \frac{v\rho}{\sigma} d\boldsymbol{x} &= \int_{\Omega(t)} \left(\frac{\partial}{\partial t}\left(\frac{v\rho}{\sigma}\right) + \nabla \cdot \left(\frac{v\rho\dot{\boldsymbol{x}}}{\sigma}\right)\right) d\boldsymbol{x} \\ &= \int_{\Omega(t)} \frac{\rho}{\sigma}(v_t + \nabla v \cdot \dot{\boldsymbol{x}}),\end{aligned}$$

where (7.1) has been used in the last step. Thus, for any v satisfying

$$v_t + \nabla v \cdot \dot{\boldsymbol{x}} = 0, \quad (7.26)$$

we get

$$\frac{d}{dt}\int_{\Omega(t)} \frac{v\rho}{\sigma} d\boldsymbol{x} = 0$$

or

$$\int_{\Omega(t)} v\rho d\boldsymbol{x} = \sigma(t) \frac{\int_{\Omega(0)} v(\boldsymbol{x},0)\rho(\boldsymbol{x},0) d\boldsymbol{x}}{\sigma(0)} \quad (7.27)$$

for the general time-dependent domain. If we take $v = 1$ in (7.27), then differentiating both sides with respect to t and using the Leibniz rule (3.25), we get

$$\frac{d\sigma}{dt} = \int_{\Omega(t)} \left(\frac{\partial \rho}{\partial u} + \frac{\partial \rho}{\partial \nabla u} \cdot \nabla \right) \mathscr{L} u d\boldsymbol{x} + \int_{\partial\Omega(t)} \rho \dot{\boldsymbol{x}} \cdot \boldsymbol{n} dS. \quad (7.28)$$

We can now describe the moving mesh FEM (Finite Element Method), for which equations (7.27) and (7.28) are used to update the physical solution and parameter σ, respectively. Denote the moving mesh by $\mathscr{T}_h(t)$; see §3.3.3. Let the corresponding linear basis functions be $\phi_j(\boldsymbol{x},t)$, $i = 1,...,N_{vi},...,N_v$, where the last $(N_v - N_{vi})$ functions are associated with boundary vertices. Then the linear finite element approximations are

$$\begin{cases} u^h(\boldsymbol{x},t) = \sum_{i=1}^{N_v} u_j(t)\phi_j(\boldsymbol{x},t), \quad \phi^h(\boldsymbol{x},t) = \sum_{i=1}^{N_v} \Phi_j(t)\phi_j(\boldsymbol{x},t), \\ \dot{\boldsymbol{x}}^h(\boldsymbol{x},t) = \sum_{i=1}^{N_v} \dot{\boldsymbol{x}}_j(t)\phi_j(\boldsymbol{x},t), \end{cases} \tag{7.29}$$

where $\Phi_j(t) \approx \phi(\boldsymbol{x}_j(t),t)$ (which is not to be confused with the basis function $\phi_j(\boldsymbol{x},t)$). Define

$$c_j = \frac{\int_{\Omega(0)} \phi_j(\boldsymbol{x},0)\rho(u^h(\boldsymbol{x},0),\nabla u^h(\boldsymbol{x},0))d\boldsymbol{x}}{\sigma(0)}, \quad j = 1,...,N_{vi}. \tag{7.30}$$

The method involves the following sequence of steps to update $U = \{u_j(t)\}$, $\Phi = \{\Phi_j(t)\}$, $\boldsymbol{X} = \{\boldsymbol{x}_j\}$, and $\sigma(t)$:

(i) Given $\boldsymbol{X}$ and $\sigma(t)$, update U by solving

$$\int_{\Omega(t)} \phi_j(\boldsymbol{x},t)\rho(u^h(\boldsymbol{x},t),\nabla u^h(\boldsymbol{x},t))d\boldsymbol{x} = \sigma(t)c_j, \quad j = 1,...,N_{vi} \tag{7.31}$$

subject to the Dirichlet boundary condition (7.23) at the boundary vertices. Equation (7.31) is obtained by taking $v = \phi_j(\boldsymbol{x},t)$ in (7.27). Note that this choice of v is permissible since ϕ_j satisfies (7.26) (see (3.97)).

(ii) Given U, solve discrete versions of (7.24) and (7.28) for Φ and σ of the form

$$\begin{aligned} \int_{\Omega(t)} \rho^h \nabla\phi^h \cdot \nabla\phi_j d\boldsymbol{x} = & -c_j \frac{d\sigma}{dt} + \int_{\Omega(t)} \phi_j \left(\frac{\partial\rho^h}{\partial u} + \frac{\partial\rho^h}{\partial\nabla u}\cdot\nabla\right)\mathscr{L}u^h d\boldsymbol{x} \\ & - \int_{\Omega(t)} \rho^h \boldsymbol{v}_{ref}\cdot\nabla\phi_j d\boldsymbol{x} + \int_{\partial\Omega(t)} \phi_j\rho^h \frac{\partial\phi^h}{\partial n} dS \\ & + \int_{\partial\Omega(t)} \phi_j\rho^h \boldsymbol{v}_{ref}\cdot\boldsymbol{n} dS, \qquad i = 1,...,N_{vi} \end{aligned} \tag{7.32}$$

$$\frac{d\sigma}{dt} = \int_{\Omega(t)} \left(\frac{\partial\rho^h}{\partial u} + \frac{\partial\rho^h}{\partial\nabla u}\cdot\nabla\right)\mathscr{L}u^h d\boldsymbol{x} + \int_{\partial\Omega(t)} \rho^h \dot{\boldsymbol{x}}^h\cdot\boldsymbol{n} dS, \tag{7.33}$$

where (7.31) has been used. Equation (7.32) is subject to a proper boundary condition for ϕ^h.

(iii) Given Φ, update $\boldsymbol{X}$ by solving the discrete form of (7.25)

$$\int_{\Omega(t)} \phi_j(\dot{\boldsymbol{x}}^h - \nabla\phi^h - \boldsymbol{v}_{ref})d\boldsymbol{x} = 0, \quad i = 1,...,N_v. \tag{7.34}$$

The above algorithm imposes the Dirichlet boundary condition in a strong sense and unlike the continuous problem does not generally conserve the mass. For example, consider the porous medium equation,

$$u_t = \nabla(u^m \nabla u), \tag{7.35}$$

where m is a positive integer. It is known that (7.35) admits a family of compact support self-similar solutions with moving boundaries on which $u = 0$. Taking the moving compact support as $\Omega(t)$, it is easy to show that the total mass, $\int_{\Omega(t)} u d\boldsymbol{x}$, is conserved for (7.35), so a reasonable choice for the density function is $\rho = u$. Unfortunately, the total mass is not generally conserved for the numerical solution since $\{\phi_j\}_{j=1}^{N_{vi}}$ in (7.31) do not form a partition of unity (i.e., $\sum_{j=1}^{N_{vi}} \phi_j \neq 1$ for all $\boldsymbol{x} \in \Omega(t)$). A remedy [196] is to use modified interior basis functions $\tilde{\phi}_j$ satisfying

$$\sum_{j=1}^{N_{vi}} \tilde{\phi}_j = \sum_{j=1}^{N_v} \phi_j = 1. \tag{7.36}$$

Another way is to use weak imposition of the Dirichlet boundary condition. This can be done [32] by computing u_j at both interior and boundary vertices through (7.31) (i.e., using $j = 1, ..., N_v$ instead of $j = 1, ..., N_{vi}$) and using these conditions in evaluating the integrals in (7.32) and (7.33).

For the porous medium equation (7.35), the mesh density function can also be defined based on scaling invariance; see [31] for details.

Example 7.1.4 Consider the porous medium equation (7.35). In d dimensions ($d = 1$ or 2) it admits a radially symmetric self-similar solution of the form [264]

$$u(r,t) = \begin{cases} \frac{1}{\lambda^d}\left(1 - \left(\frac{r}{r_0\lambda}\right)^2\right)^{\frac{1}{m}}, & \text{for } r \le r_0\lambda \\ 0, & \text{for } r > r_0\lambda \end{cases} \tag{7.37}$$

where r is the radial coordinate and

$$\lambda = \left(\frac{t}{t_0}\right)^{\frac{1}{2+dm}}, \quad t_0 = \frac{r_0^2 m}{2(2+dm)}.$$

A solution obtained with the moving mesh FEM (with weak imposition of the Dirichlet boundary condition) is shown in Figure 7.5. □

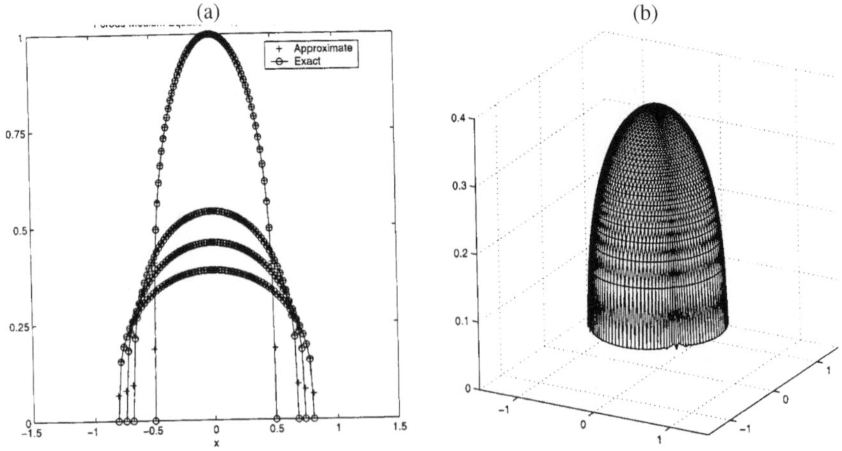

Fig. 7.5 Example 7.1.4. Shown are (a) slices of the exact and approximate solutions along $y = 0$ for $m = 3$ at time instances $t = 0, 0.5, 1.0$, and 2.0 and (b) the approximate solution surface at $t = 2$. Reprinted from Baines et al. [32], with permission from Elsevier.

7.2 MFE – moving finite element method

The original moving finite element method (MFE) developed by Miller and Miller [258] and Miller [253] (also see Baines [29], Carlson and Miller [88, 89], and Miller [256] and references therein) generates a moving mesh through the mesh velocity. Consider a time-dependent physical PDE of the form (7.22) and denote the corresponding coordinate transformation by $\boldsymbol{x} = \boldsymbol{x}(\boldsymbol{\xi}, t) : \Omega_c \to \Omega$. From (3.19) we can rewrite (7.22) as

$$\dot{u} - \nabla u \cdot \dot{\boldsymbol{x}} = \mathscr{L}u, \tag{7.38}$$

where $\dot{u} = (\partial/\partial t)u(\boldsymbol{x}(\boldsymbol{\xi}, t), t)$ and the mesh velocity $\dot{\boldsymbol{x}} = (\partial/\partial t)\boldsymbol{x}(\boldsymbol{\xi}, t)$. In continuous form the MFE can be viewed as determining the solution $u(\boldsymbol{x}(\boldsymbol{\xi}, t), t)$ and the mesh $\boldsymbol{x} = \boldsymbol{x}(\boldsymbol{\xi}, t)$ by minimizing a weighted L^2-norm of the residual of (7.38) with respect to $\dot{u}$ and $\dot{\boldsymbol{x}}$, viz.,

$$\min_{\dot{u}, \dot{\boldsymbol{x}}} \int_{\Omega} (\dot{u} - \nabla u \cdot \dot{\boldsymbol{x}} - \mathscr{L}u)^2 w \, d\boldsymbol{x}, \tag{7.39}$$

where w is a weight function. The classical version of MFE uses $w = 1$ [258, 253] and the gradient weighted MFE (GWMFE) [88, 89, 255] uses $w = 1/(1 + |\nabla u|^2)$. A nice feature of the MFE (and GWMFE) is that the mesh attempts to follow a path corresponding to the smallest weighted L^2-norm of the residual of the discrete equations. In particular, for (diffusion-domianted) parabolic PDEs whose solutions tend to a steady-state, then the mesh will tend to be a steady-state and become a locally optimal mesh that produces the least error among all meshes with the same connectivity [207, 253]. Note that the last property does not hold for hyperbolic equations [88] although some variants of the MFE, such as the stabilized MFE of [256] and the least squares MFE of [257], are designed to maintain the property for some types of hyperbolic problems. The difficulty with the MFE (and GWMFE) is that the mesh equations resulting from the minimization of functional (7.39) can become degenerate, and its numerical computation requires careful regularization. This can be seen from the Euler-Lagrange equations for (7.39), i.e.,

$$\dot{u} - \nabla u \cdot \dot{\boldsymbol{x}} - \mathscr{L} u = 0, \tag{7.40}$$

$$(\dot{u} - \nabla u \cdot \dot{\boldsymbol{x}} - \mathscr{L} u) \nabla u = 0. \tag{7.41}$$

The first equation is simply (7.38). However, (7.40) and (7.41) are clearly not independent and form a degenerate system for $\dot{u}$ and $\dot{\boldsymbol{x}}$. It is thus not surprising that the MFE equations can sometimes become indeterminate since they are derived from a discretization of (7.39) (see (7.45) below).

To consider the actual implementation of the MFE, denote the mesh by $\mathscr{T}_h(t) = \{K(t)\}$ and the corresponding linear finite element space by $\mathscr{S}^h(t)$ (cf. (3.92)). For a basis $\{\phi_j(\boldsymbol{x},t)\}$ of $\mathscr{S}^h$, the approximate solution u in $\mathscr{S}^h$ has the representation

$$u^h(\boldsymbol{x},t) = \sum_j u_j(t)\phi_j(\boldsymbol{x},t). \tag{7.42}$$

From (3.98),

$$\frac{\partial}{\partial t} u^h(\boldsymbol{x},t) = \sum_j \dot{u}_j \phi_j(\boldsymbol{x},t) - \nabla u^h \cdot \Pi_1 \dot{x}, \tag{7.43}$$

where $\Pi_1 \dot{x}$ is the piecewise linear mesh velocity satisfying

$$\Pi_1 \dot{x} = \sum_j \dot{\boldsymbol{x}}_j(t)\phi_j(\boldsymbol{x},t) \tag{7.44}$$

and $\dot{\boldsymbol{x}}_j$'s are the mesh velocities. Replacing u by u^h in (7.22) and using (7.43), we obtain the minimization problem

$$\min_{\dot{u}_j, \dot{\boldsymbol{x}}_j} \int_\Omega w \left(\sum_j \dot{u}_j \phi_j(\boldsymbol{x},t) - \nabla u^h \cdot \Pi_1 \dot{x} - \mathscr{L} u^h \right)^2 d\boldsymbol{x}, \tag{7.45}$$

which resembles (7.39) by viewing $\sum_j \dot{u}_j \phi_j(\boldsymbol{x},t)$ as $\dot{u}^h$. To avoid the possible indeterminateness of a solution to the discrete equations, a penalty term is added to (7.45). Using the so-called internodal viscosity form proposed by Miller [254], one obtains the modified minimization problem

$$\min_{\dot{u}_j, \dot{\boldsymbol{x}}_j} \Bigg\{ \int_\Omega w \left(\sum_j \dot{u}_j \phi_j(\boldsymbol{x},t) - \nabla u^h \cdot \Pi_1 \dot{x} - \mathscr{L} u^h \right)^2 d\boldsymbol{x}$$

$$+\varepsilon^2 \int_\Omega \left(\|\nabla \sum_j \dot{u}_j \phi_j\|^2 + \|\nabla \sum_j \dot{\boldsymbol{x}}_j \phi_j\|_F^2 \right) d\boldsymbol{x}, \Bigg\} \qquad (7.46)$$

where ε^2 is an appropriately chosen small internodal viscosity coefficient. The MFE discrete equations can be readily obtained as

$$\sum_j \dot{u}_j \int_\Omega w \phi_j \phi_k d\boldsymbol{x} - \sum_j \int_\Omega w (\dot{\boldsymbol{x}}_j \cdot \nabla u^h) \phi_j \phi_k d\boldsymbol{x}$$

$$+ \varepsilon^2 \sum_j \dot{u}_j \int_\Omega (\nabla \phi_j \cdot \nabla \phi_k) d\boldsymbol{x} = \int_\Omega w \phi_k \mathscr{L} u^h d\boldsymbol{x}, \qquad (7.47)$$

$$\sum_j \dot{u}_j \int_\Omega w \phi_j \phi_k \nabla u^h d\boldsymbol{x} - \sum_j \int_\Omega w (\dot{\boldsymbol{x}}_j \cdot \nabla u^h) \phi_j \phi_k \nabla u^h d\boldsymbol{x}$$

$$- \varepsilon^2 \sum_j \dot{\boldsymbol{x}}_j \int_\Omega (\nabla \phi_j \cdot \nabla \phi_k) d\boldsymbol{x} = \int_\Omega w \phi_k \nabla u^h \mathscr{L} u^h d\boldsymbol{x}. \qquad (7.48)$$

This system, rewritten in abstract form as

$$F(\dot{U}, \dot{\boldsymbol{X}}, U, \boldsymbol{X}, t) = 0, \qquad (7.49)$$

is solved for the physical solution $U(t) = \{u_j(t)\}$ and the mesh locations $\boldsymbol{X}(t) = \{\boldsymbol{x}_j\}$.

When the differential operator $\mathscr{L}$ involves second order derivatives, the integrals on the right-hand sides of (7.47) and (7.48) are not well defined since $u^h \in H^1(\Omega)$ is not in the domain of $\mathscr{L}$. In such case, they can be interpreted and calculated in terms of mollification [258], i.e., piecewise linear functions are smoothed with a mollification radius $\delta > 0$, the integrals are calculated, and then the limit is taken as $\delta \to 0$. These limiting integrals turn out to be independent of the particular mollification used. The details of the mollification method and examples can be found in [88, 89]. An alternative interpretation involves distributions, or formal integration by parts [262] (cf. (5.195)), although it may not be applicable for some situations [254].

To conclude this section it should be pointed out that the so-called moving best fit (MBF) method developed by Baines [30, 28] deserves special attention. It aims to avoid the inherent singularities of the MFE and thus the need to use regularization. The reader is referred to Baines [29] for a detailed description and discussion of the method.

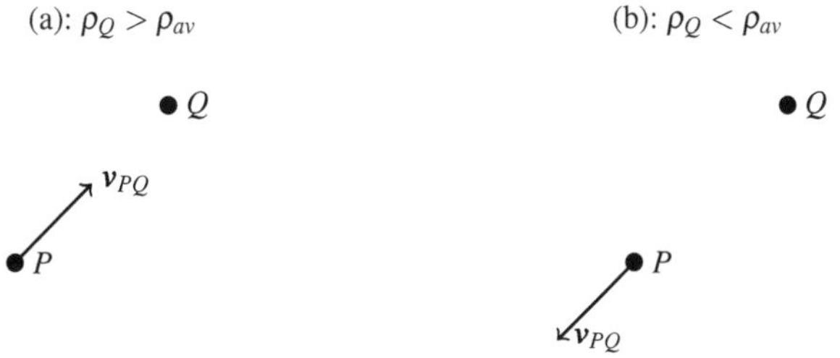

Fig. 7.6 Illustration of mesh velocity.

7.3 Other approaches

7.3.1 Method based on attraction-repulsion

Let $\rho(\boldsymbol{x})$ be a (strictly positive) density function and ρ_{av} be a local average. The method proposed by Anderson and Rai [12] is based on the assumption that points where ρ is larger than ρ_{av} should attract other points, and points where ρ is smaller than ρ_{av} should repel them. For any two points, P and Q, Figure 7.6 illustrates the corresponding mesh velocity. More precisely, the mesh velocity at P due to error at Q is written as

$$\boldsymbol{v}_{PQ} = K \frac{\rho_Q - \rho_{av}}{||\boldsymbol{x}_Q - \boldsymbol{x}_P||^{\gamma+1}} (\boldsymbol{x}_Q - \boldsymbol{x}_P), \tag{7.50}$$

where K and γ are two positive parameters. The distance between the two points is used so that the greater the distance, the less they influence each other.

Expression (7.50) is used in [12] to define the global mesh velocity field in one dimension and in two dimensions in a tensor product manner. The global mesh velocity at vertex i is expressed as

$$\dot{\boldsymbol{x}}_i = -\boldsymbol{J}_i \boldsymbol{\xi}_t = K \boldsymbol{J}_i \sum_{j \neq i} \frac{\rho_j - \rho_{av}}{||\boldsymbol{x}_j - \boldsymbol{x}_i||^{\gamma+1}} (\boldsymbol{x}_j - \boldsymbol{x}_i), \tag{7.51}$$

where the summation is over all mesh points. Selection of K and γ is discussed for the 1D situation in [12], but it is unclear how the choice is extended to multi-dimensions. The numerical results also show the method to be more successful in 1D than in multi-dimensions.

7.3.2 Methods based on spring models

With this type of method the mesh is viewed as a spring system for which each edge has a given spring stiffness constant (see Figure 7.7). For example, Gnoffo [155, 156] defines the stiffness constant for the edge connecting points i and j as a function of the gradient of some unknown function u,

$$k_j(\boldsymbol{x}_i) = 1 + c\frac{|u(\boldsymbol{x}_j) - u(\boldsymbol{x}_i)|}{\|\boldsymbol{x}_j - \boldsymbol{x}_i\|}, \tag{7.52}$$

where c is a positive constant. By Hooke's law the force related to this edge is given by $\boldsymbol{F}_{ji} = (\boldsymbol{x}_j - \boldsymbol{x}_i)k_j(\boldsymbol{x}_i)$. The mesh equations are then obtained by requiring that the spring forces at each point are at equilibrium, viz.,

$$\sum_{j \neq i} (\boldsymbol{x}_j - \boldsymbol{x}_i) k_j(\boldsymbol{x}_i) = 0, \tag{7.53}$$

or

$$\boldsymbol{x}_i = \frac{\sum_{j \neq i} \boldsymbol{x}_j k_j(\boldsymbol{x}_i)}{\sum_{j \neq i} k_j(\boldsymbol{x}_i)}, \tag{7.54}$$

where the summation is over all neighboring points of i. A Gauss-Seidel iteration can be used to update the position. Indeed, we have

$$\boldsymbol{x}_i^{new} = \frac{\sum_{j \neq i} \boldsymbol{x}_j k_j(\boldsymbol{x}_i^{old})}{\sum_{j \neq i} k_j(\boldsymbol{x}_i^{old})}, \tag{7.55}$$

where the $\boldsymbol{x}_j$'s on the right-hand side take the latest available approximations.

Note that (7.54) can be interpreted as a weighted averaging or a weighted Laplacian smoothing. Consequently, (7.53) can be viewed as a discretization on a uniform computational mesh of Laplace's equation

$$\nabla \cdot (w(\boldsymbol{x}) \nabla x_l) = 0, \quad l = 1, \ldots, d \tag{7.56}$$

for some variable diffusion function $w(\boldsymbol{x})$, where x_k's are the components of the coordinate transformation $\boldsymbol{x} = \boldsymbol{x}(\boldsymbol{\xi})$.

Habashi et al. [162] define the stiffness constant as

$$k_j(\boldsymbol{x}_i) = \frac{d_M(\boldsymbol{x}_j - \boldsymbol{x}_i)}{\|\boldsymbol{x}_j - \boldsymbol{x}_i\|}, \tag{7.57}$$

where $d_M(\boldsymbol{x}_j - \boldsymbol{x}_i)$ denotes the metric distance between points i and j. They use the slightly different Gauss-Seidel update strategy

$$\boldsymbol{x}_i^{new} = \boldsymbol{x}_i^{old} + \omega \frac{\sum_{j \neq i} (\boldsymbol{x}_j - \boldsymbol{x}_i^{old}) k_j(\boldsymbol{x}_i^{old})}{\sum_{j \neq i} k_j(\boldsymbol{x}_i^{old})}, \tag{7.58}$$

where $\omega > 0$ is a relaxation factor. Using the analogy of (7.55) to (7.53), we can loosely interpret (7.58) as a Euler-type discretization of the differential equation

$$\dot{x}_l = \nabla \cdot (w(\boldsymbol{x}) \nabla x_l), \quad l = 1, ..., d. \tag{7.59}$$

It can be generalized to including an anisotropic diffusion matrix $M(\boldsymbol{x})$, and in the moving mesh framework

$$\tau \dot{x}_l = \frac{1}{p(\boldsymbol{x})} \nabla \cdot (M(\boldsymbol{x}) \nabla x_l), \quad l = 1, ..., d \tag{7.60}$$

where $\tau > 0$ is a parameter and $p(\boldsymbol{x})$ is a positive function. Note that both (7.59) and (7.60) are parabolic equations which can be considered as gradient flow equations for the mesh $\boldsymbol{x} = \boldsymbol{x}(\boldsymbol{\xi})$; cf. (6.35). As a consequence, a natural way of choosing τ, $p(\boldsymbol{x})$, $w(\boldsymbol{x})$, and $M(\boldsymbol{x})$ would be as done for the MMPDEs in Chapter 6.

Tomita et al. [328] define

$$k_j(\boldsymbol{x}_i) = c \frac{\|\boldsymbol{x}_j - \boldsymbol{x}_i\| - \bar{d}}{\|\boldsymbol{x}_j - \boldsymbol{x}_i\|} \tag{7.61}$$

and update the position of the vertices through Newton's second law, i.e.,

$$m \frac{d^2 \boldsymbol{x}_i}{dt^2} = c \sum_{j \neq i} (\boldsymbol{x}_j - \boldsymbol{x}_i) k_j(\boldsymbol{x}_i) - \gamma \frac{d\boldsymbol{x}_i}{dt}, \tag{7.62}$$

where m is the mass of each point, $\gamma > 0$ is a damping coefficient, $c > 0$ is the spring constant, and $\bar{d} > 0$ is the natural spring length. The choice of $\bar{d}$ needs careful tuning: as it becomes longer, the ratio of maximum mesh interval to minimum becomes closer to one. If the natural spring length is larger than a critical value, however, the spring dynamic system does not have a stable equilibrium. The method is used in [328] to successfully generate a more homogeneous mesh system than the standard icosahedral one.

Interestingly, we can combine (7.52) or (7.57) with (7.62) for mesh adaptation. In continuous form, this leads to the hyperbolic mesh equation

$$m \frac{\partial^2 x_l}{\partial t^2} = \frac{1}{p(\boldsymbol{x})} \nabla \cdot (M(\boldsymbol{x}) \nabla x_l) - \gamma \frac{\partial x_l}{\partial t}, \quad l = 1, ..., d. \tag{7.63}$$

This type of hyperbolic mesh equation is very different from mesh equations of parabolic type, and its relative advantages are unclear.

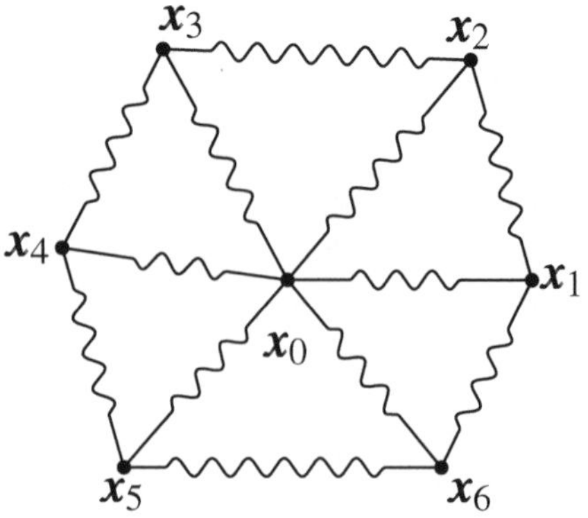

Fig. 7.7 A mesh is viewed as a spring system.

7.3.3 Methods based on minimizing convection terms

Motivated by Lagrangian methods in computational fluid dynamics, a number of moving mesh approaches have been developed by minimizing convective terms in governing equations. In this subsection we describe such a method (Petzold [274] and Hyman and Larrouturou [198]) that is based on minimizing the time rate of change of the solution and defines the mesh velocity in terms of the time and space derivatives of the solution. With the method, the solution is changing slowly in this moving mesh reference frame, and large time steps are taken without sacrificing accuracy. However, the mesh points may quickly tangle and fail to adequately resolve the solution. When this happens, it is suggested (e.g., see [199]) that a static rezone step be used to regain spatial accuracy, viz., the domain is re-meshed and the solution interpolated from the old to the new one.

We illustrate the method for the differential equation (7.22), which takes the form (7.38) in the computational coordinate. The mesh velocity is chosen to minimize the variation of u and x in the new coordinates $\boldsymbol{\xi}$ and t, i.e.,

$$\min_{\dot{\boldsymbol{x}}} \left(\dot{u}^2 + \alpha \|\dot{\boldsymbol{x}}\|^2\right) = \min_{\dot{\boldsymbol{x}}} \left((\nabla u \cdot \dot{\boldsymbol{x}} + \mathscr{L} u)^2 + \alpha \|\dot{\boldsymbol{x}}\|^2\right), \tag{7.64}$$

where $\alpha > 0$ is a parameter. This leads to

$$(\alpha + \nabla u \nabla u^T)\dot{\boldsymbol{x}} = -(\mathscr{L} u)\nabla u. \tag{7.65}$$

There is no mechanism in the above equation to avoid tangling of mesh points. Petzold [274] suggests adding a diffusion-like term, giving

$$(\alpha + \nabla u \nabla u^T)\dot{\boldsymbol{x}} = \lambda \Delta \dot{\boldsymbol{x}} - (\mathscr{L} u)\nabla u, \tag{7.66}$$

where $\lambda > 0$ is a parameter. It is found in 1D that with these smoothed mesh velocities the mesh points are unlikely to cross in a time step and the added term α is important even when mesh points are deleted and moved apart using the static rezoning strategy after every time step.

A weighted minimization is considered in [199] to obtain mesh equations which are invariant under scaling and translation of u and $\boldsymbol{x}$. The above-described method has been implemented mainly for 1D problems (e.g., see [199]), and it would be interesting to investigate their potential in multidimensions.

7.4 Exercises

1. Show that the Euler-Lagrange equations of the constrained optimization problem (7.10) are given by (7.11)–(7.13).
2. Derive (7.20).
3. Complete the derivation of (7.24).
4. Verify (7.37) is a solution of the porous medium equation (7.35).
5. Show that the Euler-Lagrange equations of (7.39) are given by (7.40) and (7.41).
6. Derive the continuous version of (7.46).
7. An ordinary differential equation corresponding to (7.63) takes the form

$$m\frac{d^2x}{dt^2} = -cx - \gamma\frac{dx}{dt},$$

   where m, c, and γ are positive constants. Show that this represents a dissipative system, and $x(t) \to 0$ as $t \to \infty$.
8. Derive (7.65) from (7.64).

Appendix A
Sobolev spaces

Throughout this book, Ω denotes a simply connected, open, bounded domain in the d-dimensional space $\mathbb{R}^d$, $\overline{\Omega}$ is its closure, and $|\Omega|$ denotes its length, area, or volume for $d = 1, 2$, or 3, respectively. The coordinates on Ω are denoted by $\boldsymbol{x} = (x_1, ..., x_d)^T$. Given a multi-index $\alpha = (\alpha_1, \alpha_2, ..., \alpha_d)$ of non-negative integers, let $|\alpha| = \alpha_1 + \cdots + \alpha_d$ and

$$D^\alpha u = \frac{\partial^{|\alpha|} u}{\partial x_1^{\alpha_1} \cdots \partial x_d^{\alpha_d}}.$$

An l-th order partial derivative is also denoted by

$$D^{(i_1,...,i_l)} u = \frac{\partial^l u}{\partial x_{i_1} \cdots \partial x_{i_l}},$$

where $(i_1, ..., i_l)$ is an integer vector of l components with $1 \le i_1, ..., i_l \le d$. The set of all l-th order partial derivatives is denoted by either $D^{|\alpha|} u$ or $D^l u$ with $l = |\alpha|$. Following the convention, we will denote D^1 by the gradient operator ∇, i.e., $\nabla = D^1$.

Lebesgue space $L^p(\Omega)$ $(1 \le p < \infty)$ is the vector space of the functions $u : \Omega \to \mathbb{R}$ for which $|u|^p$ is Lebesgue integrable on Ω, so $\int_\Omega |u(\boldsymbol{x})|^p d\boldsymbol{x} < \infty$. It is a Banach space with respect to the norm

$$\|u\|_{L^p(\Omega)} = \left(\int_\Omega |u(\boldsymbol{x})|^p d\boldsymbol{x} \right)^{\frac{1}{p}}.$$

$L^\infty(\Omega)$ is the Banach space of the measurable functions which are defined on Ω and bounded outside a set of measure zero. It is equipped with the norm

$$\|u\|_{L^\infty(\Omega)} = \text{ess} \sup_{\boldsymbol{x} \in \Omega} |u(\boldsymbol{x})|.$$

We list several fundamental inequalities associated with Lebesgue spaces in the following.

Theorem A.0.1 (Minkowski's inequality or the triangle inequality) *For* $1 \le p \le \infty$,

$$\|u+v\|_{L^p(\Omega)} \le \|u\|_{L^p(\Omega)} + \|v\|_{L^p(\Omega)} \quad \forall u, v \in L^p(\Omega). \tag{A.1}$$

Theorem A.0.2 (Hölder's inequality) *Assume* $1 \le p, q \le \infty$, $\frac{1}{p} + \frac{1}{q} = 1$. *Then,*

$$\int_\Omega |uv| d\boldsymbol{x} \le \|u\|_{L^p(\Omega)} \|v\|_{L^q(\Omega)} \quad \forall u \in L^p(\Omega),\ \forall v \in L^q(\Omega). \tag{A.2}$$

The Lebesgue spaces have the embedding property

$$L^1(\Omega) \hookleftarrow L^2(\Omega) \hookleftarrow \cdots \hookleftarrow L^\infty(\Omega).$$

This is an immediate result of Hölder's inequality. Indeed, for any two integers $1 \le q < p \le \infty$ and any function $u \in L^p(\Omega)$, we have

$$\|u\|_{L^q(\Omega)} = \left(\int_\Omega |u|^q \cdot 1 d\boldsymbol{x} \right)^{\frac{1}{q}} \le \left(\||u|^q\|_{L^{\frac{p}{q}}(\Omega)} \|1\|_{L^{\frac{p}{p-q}}(\Omega)} \right)^{\frac{1}{q}} = |\Omega|^{\frac{1}{q}-\frac{1}{p}} \|u\|_{L^p(\Omega)}.$$

The following theorem is a generalization of Hölder's inequality because of the weight function term and general exponent. It is used repeatedly in this book, particularly when dealing with the equidistribution principle in Chapter 6. The interested reader is referred to Hardy et al. [168] for its proof.

Theorem A.0.3 *Given a weight function* $w(\boldsymbol{x}) \ge 0$ *with* $\int_\Omega w d\boldsymbol{x} = 1$, *for an arbitrary function* f *and real number* r *define*

$$M_r(f) = \left(\int_\Omega w |f|^r d\boldsymbol{x} \right)^{\frac{1}{r}},$$

with the limits $M_0(f) = \exp(\int_\Omega w \log |f| d\boldsymbol{x})$ *(geometric mean),* $M_{+\infty} = \max |f|$, *and* $M_{-\infty} = \min |f|$. *Then*

$$M_r(f) < M_s(f) \tag{A.3}$$

for $-\infty \le r < s \le +\infty$ *unless (a)* $M_r(f) = M_s(f) = +\infty$, *which can happen only if* $r \ge 0$, *or (b)* $M_r(f) = M_s(f) = 0$, *which can happen only if* $s \le 0$, *or (c)* $f \equiv$ *constant.*

Sobolev spaces deal with function derivatives. Define $C_0^\infty(\Omega) \equiv \bigcap_{m=0}^\infty C_0^m(\Omega)$, where $C^m(\Omega)$ denotes the vector space of m-times differentiable functions on Ω, $C_0^m(\Omega) = \{u \in C^m(\Omega) \mid \text{supp}(u) \subset \Omega\}$, and $\text{supp}(u) = \overline{\{\boldsymbol{x} \mid \boldsymbol{x} \in \Omega, u(\boldsymbol{x}) \ne 0\}}$. For a multi-index $\alpha = (\alpha_1, ..., \alpha_d)$ and for a Lebesgue integrable function u on Ω, if there is a Lebesgue integrable function v_α on Ω which satisfies the condition

$$\int_{\Omega} u D^{\alpha} \psi d\mathbf{x} = (-1)^{|\alpha|} \int_{\Omega} v_{\alpha} \psi d\mathbf{x}, \quad \forall \psi \in C_0^{\infty}(\Omega),$$

then v_α is said to be a distributional derivative or a generalized derivative of u of order $|\alpha|$, and it is denoted by $D^{\alpha}u = v_{\alpha}$.

For a given integer $m \geq 0$ and real number $p \in [1,\infty]$, the Sobolev space $W^{m,p}(\Omega)$ is defined as the vector space of the functions $u \in L^p(\Omega)$ such that for each multi-index α with $|\alpha| \leq m$, the distributional derivative $D^{\alpha}u$ belongs to $L^p(\Omega)$. With the norm

$$\|u\|_{W^{m,p}(\Omega)} = \left(\sum_{|\alpha| \leq m} \int_{\Omega} |D^{\alpha} u|^p d\mathbf{x} \right)^{\frac{1}{p}},$$

it is a Banach space. The semi-norm $|u|_{W^{m,p}(\Omega)}$ is defined by

$$|u|_{W^{m,p}(\Omega)} = \left(\sum_{|\alpha| = m} \int_{\Omega} |D^{\alpha} u|^p d\mathbf{x} \right)^{\frac{1}{p}} = \left(\int_{\Omega} \|D^m u\|_{l_p}^p d\mathbf{x} \right)^{\frac{1}{p}},$$

where $\|\cdot\|_{l_p}$ denotes the l_p matrix norm. The scaled semi-norm $\langle u \rangle_{W^{m,p}(\Omega)}$ is defined by

$$\langle u \rangle_{W^{m,p}(\Omega)} = \left(\frac{1}{|\Omega|} \sum_{|\alpha| = m} \int_{\Omega} |D^{\alpha} u|^p d\mathbf{x} \right)^{\frac{1}{p}} = \left(\frac{1}{|\Omega|} \int_{\Omega} \|D^m u\|_{l_p}^p d\mathbf{x} \right)^{\frac{1}{p}}. \tag{A.4}$$

Note that $L^p(\Omega) = W^{0,p}(\Omega)$, and for the special case $p = 2$ we denote $H^m(\Omega) = W^{m,2}(\Omega)$.

The trace operator plays an important role in studying the existence and uniqueness of solutions to BVPs of PDEs. Loosely speaking, the trace operation defines the notion of restriction of a function on the domain boundary. Consider a bounded open domain $\Omega \subset \mathbb{R}^d$ with Lipschitz (continuous) boundary $\partial\Omega$. Then the trace operator T is defined as

$$Tu = u|_{\partial\Omega}, \quad \forall u \in C^1(\bar{\Omega}). \tag{A.5}$$

Some important properties of the trace operator are given in the following lemmas. The interested reader is referred to, e.g., Grisvard [159] for their proofs.

Theorem A.0.4 *Any bounded open convex subset of $\mathbb{R}^d$ has a Lipschitz boundary.*

Theorem A.0.5 (The trace theorem) *Let Ω be a bounded open subset of $\mathbb{R}^d$ with a Lipschitz boundary $\partial\Omega$. Then for any number $1 < p < \infty$, the trace operator T, which is defined on $C^1(\bar{\Omega})$, has a unique continuous extension as an operator from $W^{1,p}(\Omega)$ onto $W^{1-\frac{1}{p},p}(\partial\Omega)$.*

Theorem A.0.6 *Let Ω be a bounded open subset of $\mathbb{R}^d$ with a Lipschitz boundary $\partial\Omega$. Then for any number $1 < p < \infty$, there exists a constant C such that*

$$\int_{\partial D} |Tu|^p dS \leq C \left[\varepsilon^{1-\frac{1}{p}} \int_D |\nabla u|^p d\mathbf{x} + \varepsilon^{-\frac{1}{p}} \int_D |u|^p d\mathbf{x} \right] \tag{A.6}$$

for all $u \in W^{1,p}(\Omega)$ and $\varepsilon \in (0,1)$.

For simplicity, the trace of $u \in W^{1,p}(\Omega)$ is denoted by $u|_{\partial\Omega}$ when this can be done without causing any confusion.

Poincaré-type inequalities which relate the integral of a function to that of its gradient are also frequently used throughout the book. There are two types of these inequalities, those concerned with functions having vanishing trace and those with the difference between functions and their average. The proofs of the following two theorems can be found, e.g., in Evans [137] and Chen and Wu [101].

Theorem A.0.7 (Poincaré's inequality; $u|_{\partial\Omega} = 0$) *Let Ω be a bounded, open subset of $\mathbb{R}^d$. For $1 \leq q \leq p < \infty$,*

$$\|u\|_{L^q(\Omega)} \leq C\|\nabla u\|_{L^p(\Omega)}, \quad \forall u \in W_0^{1,p}(\Omega) \tag{A.7}$$

where $W_0^{1,p}(\Omega) = \{u \in W^{1,p}(\Omega) : \; u|_{\partial\Omega} = 0\}$ and C is a constant depending only on d, p, q, and Ω.

Theorem A.0.8 (Poincaré's inequality; u_Ω) *Let Ω be a bounded, connected, open subset of $\mathbb{R}^d$ with Lipschitz boundary $\partial\Omega$. For $1 \leq q \leq p < \infty$,*

$$\|u - u_\Omega\|_{L^q(\Omega)} \leq C\|\nabla u\|_{L^p(\Omega)}, \quad \forall u \in W^{1,p}(\Omega) \tag{A.8}$$

where C is a constant depending only on d, p, q, and Ω and u_Ω is the average of u on Ω, i.e.,

$$u_\Omega = \frac{1}{|\Omega|} \int_\Omega u d\mathbf{x}. \tag{A.9}$$

It is thoretically interesting and also practically important to find an explicit expression for the constant C in Poincaré's inequalities. But this is available only for special domains such as convex ones; see Payne and Weinberger [272], Bebendorf [40], Acosta and Durán [2], and Chua and Wheeden [103]. The proofs of the following two theorems can be found in [103].

Theorem A.0.9 (Poincaré's inequality for a convex domain; u_Ω) *Let $\Omega \subset \mathbb{R}^d$ be a bounded convex domain with diameter h_Ω. Then, for any $1 \leq q \leq p < \infty$,*

$$\|u - u_\Omega\|_{L^q(\Omega)} \leq c_{p,q} h_\Omega |\Omega|^{\frac{1}{q}-\frac{1}{p}} \|\nabla u\|_{L^p(\Omega)}, \quad \forall u \in W^{1,p}(\Omega) \tag{A.10}$$

where $c_{p,q}$ is a constant depending only on p and q which has values or bounds as follows:

(1) If $1 < q \le p < \infty$,

$$c_{p,q} \le q^{\frac{1}{q}} \left(\frac{p}{p-1} \right)^{\frac{1}{p}-\frac{1}{q}} 2^{\frac{p-1}{p}}.$$

(2) If $q = 1 < p < \infty$,

$$\left(\int_0^1 (x(1-x))^{\frac{p}{p-1}} dx \right)^{\frac{p-1}{p}} \le c_{p,1} \le 2 \left(\int_0^1 (x(1-x))^{\frac{p}{p-1}} dx \right)^{\frac{p-1}{p}}.$$

(3) If $p = q = 1$, $c_{1,1} = \frac{1}{2}$.
(4) If $p = q = 2$, $c_{2,2} = \frac{1}{\pi}$.

Theorem A.0.10 (Poincaré's inequality for a convex domain; u_Ω) *Let $\Omega \subset \mathbb{R}^d$ be a bounded convex domain contained in a parallelepiped defined by*

$$\mathscr{P} = \left\{ \boldsymbol{x}_0 + \sum_{i=1}^d \beta_i \boldsymbol{v}_i : 0 \le \beta_i \le h_i,\ i = 1, ..., d \right\} \tag{A.11}$$

for some positive numbers h_i, $i = 1, ..., d$ and d linearly independent unit vectors $\boldsymbol{v}_i$, $i = 1, ..., d$. Then, for any $1 \le q \le p < \infty$,

$$\|u - u_\Omega\|_{L^q(\Omega)} \le c_{p,q} |\Omega|^{\frac{1}{q}-\frac{1}{p}} \left\| \sum_{i=1}^d h_i |\nabla u \cdot \boldsymbol{v}_i| \right\|_{L^p(\Omega)}, \quad \forall u \in W^{1,p}(\Omega) \tag{A.12}$$

where $c_{p.q}$ is the same constant as in (A.10).

It is interesting to point out that the result of Theorem A.0.9 is isotropic while that of Theorem A.0.10 is anisotropic in nature (since it takes into consideration the directional derivatives — see Chapter 5). Moreover, as an application example of Theorem A.0.10 in finite element computation, we can take Ω as the generic element K of an affine family $\{\mathscr{T}_h\}$ with the equilateral, unitary-volume reference element $\hat{K}$. Let the affine mapping from $\hat{K}$ to K be F_K and the SVD decomposition of the Jacobian matrix F_K' be

$$F_K' = U \Sigma V^T = [\boldsymbol{u}_1, ..., \boldsymbol{u}_d] \begin{bmatrix} \sigma_1 & \cdots & 0 \\ \vdots & \ddots & \vdots \\ 0 & \cdots & \sigma_d \end{bmatrix} V^T, \tag{A.13}$$

where U and V are orthogonal matrices and $\sigma_i > 0$, $i = 1, ..., d$. The circumscribed ellipsoid (denoted by $\mathscr{E}$) of K is described by equation (4.44). It is obvious that K is contained in the parallelepiped

$$\mathscr{P}_K = \left\{ \boldsymbol{x}_K + \sum_{i=1}^{d} \beta_i \boldsymbol{u}_i \, : \, 0 \le \beta_i \le \frac{\hat{h}\sigma_i}{2}, \; i = 1, ..., d \right\}, \tag{A.14}$$

where $\boldsymbol{x}_K$ is the center of K. Then,

$$\begin{aligned} \sum_{i=1}^{d} \frac{\hat{h}\sigma_i}{2} |\nabla u \cdot \boldsymbol{u}_i| &\le \frac{\hat{h} d^{\frac{1}{2}}}{2} \left[\sum_{i=1}^{d} \sigma_i^2 |\nabla u \cdot \boldsymbol{u}_i|^2 \right]^{\frac{1}{2}} \\ &= \frac{\hat{h} d^{\frac{1}{2}}}{2} \left[\mathrm{tr}((F_K')^T \nabla u \nabla u^T F_K') \right]^{\frac{1}{2}}, \end{aligned}$$

where the last step can be verified directly using (A.13). Combining this with Theorem A.0.10, we obtain, for $1 \le q \le p < \infty$ and $u \in W^{1,p}(K)$,

$$\begin{aligned} &\|u - u_K\|_{L^q(K)} \\ &\le \frac{\hat{h} d^{\frac{1}{2}} c_{p,q}}{2} |K|^{\frac{1}{q} - \frac{1}{p}} \left(\int_K \left[\mathrm{tr}((F_K')^T \nabla u(\boldsymbol{x}) \nabla u^T(\boldsymbol{x}) F_K') \right]^{\frac{p}{2}} d\boldsymbol{x} \right)^{\frac{1}{p}}. \end{aligned} \tag{A.15}$$

This is in the same form as (5.42) ($m = 0$) except that the current estimate works only for the averaging approximation and the constant $c_{p,q}$ has an almost explicit expression.

Appendix B
Arithmetic-mean geometric-mean inequality and Jensen's inequality

The following two inequalities are used frequently throughout the book. The first is a generalization of the well-known arithmetic-mean geometric-mean inequality and the second is Jensen's inequality.

Theorem B.0.11 (Generalized arithmetic-mean geometric-mean inequality.) *Let $w_1, \ldots, w_m$ be m weights satisfying $w_i > 0$ and $\sum_i w_i = 1$. Then, for any positive numbers $a_1, \ldots, a_m$,*

$$\left(\sum_{i=1}^{m} w_i a_i^s\right)^{\frac{1}{s}} \leq \left(\sum_{i=1}^{m} w_i a_i^t\right)^{\frac{1}{t}}$$

for any numbers $-\infty \leq s < t \leq \infty$, with equality iff (if and only if) $a_1 = \cdots = a_m$. Here, by convention

$$\left(\sum_{i=1}^{m} w_i a_i^s\right)^{\frac{1}{s}} = \begin{cases} \prod_{i=1}^{m} a_i^{w_i} & \text{for } s = 0, \\ \max_{1\leq i\leq m} a_i & \text{for } s = \infty, \\ \min_{1\leq i\leq m} a_i & \text{for } s = -\infty. \end{cases}$$

This theorem reduces to the well known arithmetic-mean geometric-mean inequality when $s = 0$, $t = 1$, and $w_1 = \cdots = w_m = \frac{1}{m}$. A refined version of the arithmetic-mean geometric-mean inequality [218] is

$$\frac{1}{m(m-1)} \sum_{i<j} \left(\sqrt{a_i} - \sqrt{a_j}\right)^2 \leq \frac{1}{m} \sum_i a_i - \left(\prod_i a_i\right)^{\frac{1}{m}} \leq \frac{1}{m} \sum_{i<j} \left(\sqrt{a_i} - \sqrt{a_j}\right)^2. \quad \text{(B.1)}$$

Theorem B.0.12 (Jensen's inequality.) *For any m positive numbers $a_1, \ldots, a_m$, the inequality*

$$\left(\sum_{i=1}^{m} a_i^s\right)^{\frac{1}{s}} \geq \left(\sum_{i=1}^{m} a_i^t\right)^{\frac{1}{t}}$$

holds for any numbers s and t with $0 < s \leq t$.

The following corollary can be easily proven using Theorems B.0.11 and B.0.12.

Corollary B.0.1 *For any real number* $t > 0$,

$$|a+b|^t \leq c_t(|a|^t + |b|^t), \quad \forall a,b \in \mathbb{R} \tag{B.2}$$

$$|a|^t - c_t|b|^t \leq c_t|a-b|^t, \quad \forall a,b \in \mathbb{R} \tag{B.3}$$

where

$$c_t = 2^{\max\{t-1,0\}}. \tag{B.4}$$

Moreover, if $t \in (0,1)$, *then it holds that*

$$\left||a|^t - |b|^t\right| \leq |a-b|^t \quad \forall a,b \in \mathbb{R}. \tag{B.5}$$

References

1. B. Achchab, S. Achchab, and A. Agouzal. Some remarks about the hierarchical a posteriori error estimate. *Numer. Meth. P. D. E.*, 20:919–932, 2004.
2. G. Acosta and R. G. Durán. An optimal Poincaré inequality in L^1 for convex domains. *Proc. Amer. Math. Soc.*, 132:195–202 (electronic), 2004.
3. G. Acosta, R. G. Duran, and J. D. Rossi. An adaptive time step procedure for a parabolic problem with blow-up. *Computing*, 68:343–373, 2002.
4. R. A. Adams. *Sobolev Spaces*. Academic Press, New York, 1975.
5. S. Adjerid and J. E. Flaherty. A moving finite element method with error estimation and refinement for one-dimensional time dependent partial differential equations. *SIAM J. Numer. Anal.*, 23:778–795, 1986.
6. S. Adjerid and J. E. Flaherty. A moving-mesh finite element method with local refinement for parabolic partial differential equations. *Comput. Meth. Appl. Mech. Engrg.*, 55:3–26, 1986.
7. A. Agouzal, K. Lipnikov, and Y. Vassilevski. Generation of quasi-optimal meshes based on a posteriori error estimates. In *Proceedings, 16th International Meshing Roundtable*, pages 139–148, Sandia National Laboratories, Albuquerque, NM, 2008. Sandia Report 98-2250.
8. M. Ainsworth and J. T. Oden. *A posteriori error estimation in finite element analysis*. Pure and Applied Mathematics (New York). Wiley-Interscience [John Wiley & Sons], New York, 2000.
9. D. Ait-Ali-Yahia, W. G. Habashi, A. Tam, M. G. Vallet, and M. Fortin. A directionally-adaptive finite element method for high-speed flows. Technical Report 96-2553, AIAA Paper, 1996.
10. D. A. Anderson. Adaptive mesh schemes based on grid speeds. Technical Report 83-1931, AIAA Paper, 1983.
11. D. A. Anderson. Application of adaptive grids to transient problems. In I. Babuška, J. Chandra, and J. E. Flaherty, editors, *Adaptive Computational Methods for Partial Differential Equations*, pages 208–223, SIAM, Philadelphia, 1983.
12. D. A. Anderson and M. M. Rai. The use of solution adaptive grids in solving partial differential equations. *Appl. Math. Comput.*, 10/11:317–338, 1982. Numerical grid generation (Nashville, Tenn., 1982).
13. D. A. Anderson and M. M. Rai. The use of solution adaptive grids in solving partial differential equations. In J. F. Thompson, editor, *Numerical Grid Generation*, pages 317–338, North-Holland, Amsterdam, 1983.
14. J. D. Anderson. *Computational Fluid Dynamics: The Basics with Applications*. McGraw-Hill, Inc., New York, 1995.

15. T. Apel, M. Berzins, P. K. Jimack, G. Kunert, A. Plaks, I. Tsukerman, and M. Walkley. Mesh shape and anisotropic elements: theory and practice. In J. R. Whiteman, editor, *The Mathematics of Finite Elements and Applications X*, pages 367–376, Elsevier, Oxford, 2000. MAFELAP 1999 (Uxbridge).
16. U. Ascher, R. M. M. Mattheij, and R. D. Russell. *Numerical Solution of Boundary Value Problems for Ordinary Differential Equations*. Prentice-Hall, Englewood Cliffs, NJ, 1988.
17. B. N. Azarenok. Variational barrier method of adaptive grid generation in hyperbolic problems of gas dynamics. *SIAM J. Numer. Anal.*, 40:651–682 (electronic), 2002.
18. B. N. Azarenok. A variational hexahedral grid generator with control metric. *J. Comput. Phys.*, 218:720–747, 2006.
19. B. N. Azarenok. A method of constructing adaptive hexahedral moving grids. *J. Comput. Phys.*, 226:1102–1121, 2007.
20. B. N. Azarenok, S. A. Ivanenko, and T. Tang. Adaptive mesh redistribution method based on Godunov's scheme. *Commun. Math. Sci.*, 1:152–179, 2003.
21. B. N. Azarenok and T. Tang. Second-order Godunov-type scheme for reactive flow calculations on moving meshes. *J. Comput. Phys.*, 206:48–80, 2005.
22. I. Babuška and A. K. Aziz. On the angle condition in the finite element method. *SIAM J. Numer. Anal.*, 13:214–226, 1976.
23. I. Babuška, J. Chandra, and J. E. Flaherty, editors. *Adaptive Computational Methods for Partial Differential Equations*, SIAM, Philadelphia, 1983.
24. I. Babuška and W. C. Rheinboldt. A-posteriori error estimates for the finite element method. *Int. J. Numer. Meth. Engrg.*, 12:1597–1615, 1978.
25. I. Babuška and W. C. Rheinboldt. Analysis of optimal finite-element meshes in R^1. *Math. Comput.*, 33:435–463, 1979.
26. I. Babuška and T. Strouboulis. *The Finite Element Method and Its Reliability*. Oxford Science Publication, New York, 2001. Numer. Math. Sci. Comput.
27. S. B. Baden, N. P. Chrisochoides, D. B. Gannon, and M. L. Norman, editors. *Structured Adaptive Mesh Refinement (SAMR) Grid Methods*, The IMA Volumes in Mathematics and its Applications, Springer, New York, Berlin, 2000. Vol. 117.
28. M. J. Baines. Algorithms for optimal discontinuous piecewise linear and constant L_2 fits to continuous functions with adjustable nodes in one and two dimensions. *Math. Comp.*, 62:645–669, 1994.
29. M. J. Baines. *Moving Finite Elements*. Oxford University Press, Oxford, 1994.
30. M. J. Baines. On the relationship between the moving finite-element procedure and best piecewise L_2 fits with adjustable nodes. *Numer. Methods Partial Differential Equations*, 10:191–203, 1994.
31. M. J. Baines, M. E. Hubbard, and P. K. Jimack. A moving mesh finite element algorithm for fluid flow problems with moving boundaries. *Internat. J. Numer. Methods Fluids*, 47:1077–1083, 2005. 8th ICFD Conference on Numerical Methods for Fluid Dynamics. Part 2.
32. M. J. Baines, M. E. Hubbard, and P. K. Jimack. A moving mesh finite element algorithm for the adaptive solution of time-dependent partial differential equations with moving boundaries. *Appl. Numer. Math.*, 54:450–469, 2005.
33. M. J. Baines, M. E. Hubbard, P. K. Jimack, and A. C. Jones. Scale-invariant moving finite elements for nonlinear partial differential equations in two dimensions. *Appl. Numer. Math.*, 56:230–252, 2006.
34. C. Bandle and H. Brunner. Blowup in diffusion equations: a survey. *J. Comput. Appl. Math.*, 97:3–22, 1998.
35. R. E. Bank and R. F. Santos. Analysis of some moving space-time finite element methods. *SIAM J. Numer. Anal.*, 30:1–18, 1993.

36. R. E. Bank and R. K. Smith. Mesh smoothing using a posteriori error estimates. *SIAM J. Numer. Anal.*, 34:979–997, 1997.
37. G. I. Baranblatt. *Dimensional Analysis.* Gordon & Breach, New York, 1989.
38. G. K. Batchelor. *An Introduction to Fluid Dynamics.* Cambridge University Press, Cambridge, 1973.
39. J. Baumgarte. Stabilization of constraints and integrals of motion in dynamical systems. *Comput. Meth. Appl. Mech. Engrg.*, 1:1–16, 1972.
40. M. Bebendorf. A note on the Poincaré inequality for convex domains. *Z. Anal. Anwendungen*, 22:751–756, 2003.
41. J. Bebernes and S. Bricher. Final time blowup profiles for semilinear parabolic equations via center manifold theory. *SIAM J. Math. Anal.*, 23:852–869, 1992.
42. G. Beckett and J. A. Mackenzie. Convergence analysis of finite difference approximations on equidistributed grids to a singularly perturbed boundary value problem. *Appl. Numer. Math.*, 35:87–109, 2000.
43. G. Beckett and J. A. Mackenzie. On a uniformly accurate finite difference approximation of a singularly perturbed reaction-diffusion problem using grid equidistribution. *J. Comput. Appl. Math.*, 131:381–405, 2001.
44. G. Beckett and J. A. Mackenzie. Uniformly convergent high order finite element solutions of a singularly perturbed reaction-diffusion equation using mesh equidistribution. *Appl. Numer. Math.*, 39:31–45, 2001.
45. G. Beckett, J. A. Mackenzie, A. Ramage, and D. M. Sloan. On the numerical solution of one-dimensional PDEs using adaptive methods based on equidistribution. *J. Comput. Phy.*, 167:372–392, 2001.
46. G. Beckett, J. A. Mackenzie, A. Ramage, and D. M. Sloan. Computational solution of two-dimensional unsteady PDEs using moving mesh methods. *J. Comput. Phys.*, 182:478–495, 2002.
47. G. Beckett, J. A. Mackenzie, and M. L. Robertson. A moving mesh finite element method for the solution of two-dimensional Stefan problems. *J. Comput. Phys.*, 168:500–518, 2001.
48. J.-D. Benamou and Y. Brenier. A computational fluid mechanics solution to the Monge-Kantorovich mass transfer problem. *Numer. Math.*, 84:375–393, 2000.
49. M. Berger and R. V. Kohn. A rescaling algorithm for the numerical calculation of blowing-up solutions. *Comm. Pure Appl. Math*, 41:841–863, 1988.
50. M. Berzins. A solution-based triangular and tetrahedral mesh quality indicator. *SIAM J. Sci. Comput.*, 19:2051–2060, 1998.
51. J. G. Blom, J. M. Sanz-Serna, and J. G. Verwer. On simple moving grid methods for one-dimensional evolutionary partial differential equations. *J. Comput. Phys.*, 74:191–213, 1988.
52. J. G. Blom and J. G. Verwer. On the use of the arclength and curvature monitor in a moving grid method which is based on the method of lines. Technical Report NM-N8902, CWI, Amsterdam, 1989.
53. P. Bochev, G. Liao, and G. d. Pena. Analysis and computation of adaptive moving grids by deformation. *Numer. Meth. PDEs*, 12:489–506, 1996.
54. H. Borouchaki, P. L. George, P. Hecht, P. Laug, and E. Saletl. Delaunay mesh generation governed by metric specification: Part I. Algorithms. *Fin. Elem. Anal. Des.*, 25:61–83, 1997.
55. H. Borouchaki, P. L. George, and B. Mohammadi. Delaunay mesh generation governed by metric specification: Part II. Applications. *Fin. Elem. Anal. Des.*, 25:85–109, 1997.
56. R. L. Bowers and J. R. Wilson. *Numerical Modeling in Applied Physics and Astrophysics.* Jones and Bartlett Publishers, Boston, 1991.
57. J. U. Brackbill. An adaptive grid with directional control. *J. Comput. Phys.*, 108:38–50, 1993.

58. J. U. Brackbill and J. S. Saltzman. Adaptive zoning for singular problems in two dimensions. *J. Comput. Phys.*, 46:342–368, 1982.
59. C. Brändle, F. Quirós, and J. D. Rossi. An adaptive numerical method to handle blow-up in a parabolic system. *Numer. Math.*, 102:39–59, 2005.
60. L. Branets and G. F. Carey. A local cell quality metric and variational grid smoothing algorithm. In *Proceedings, 12th International Meshing Roundtable*, Sandia National Laboratories, Albuquerque, NM, 2003.
61. Y. Brenier. Polar factorization and monotone rearrangement of vector-valued functions. *Comm. Pure Appl. Math.*, 44:375–417, 1991.
62. S. C. Brenner and L. R. Scott. *The Mathematical Theory of Finite Element Methods*. Springer-Verlag, New York, 1994.
63. C. Budd, S. Chen, and R. D. Russell. New self-similar solutions of the nonlinear Schrödinger equation with moving mesh methods. *J. Comput. Phys.*, 152:756–789, 1999.
64. C. Budd, G. Collins, W. Huang, and R. D. Russell. Self-similar numerical solutions of the porous medium equation using moving mesh methods. *Phil. Trans. R. Soc. Lond. A*, 357:1047–1078, 1999.
65. C. Budd, O. Koch, and E. Weinmüller. Computation of self-similar solution profiles for the nonlinear Schrödinger equation. *Computing*, 77:335–346, 2006.
66. C. J. Budd, R. Carretero-González, and R. D. Russell. Precise computations of chemotactic collapse using moving mesh methods. *J. Comput. Phys.*, 202:463–487, 2005.
67. C. J. Budd, W. Huang, and R. D. Russell. Moving mesh methods for problems with blow-up. *SIAM J. Sci. Comput.*, 17:305–327, 1996.
68. C. J. Budd, W. Huang, and R. D. Russell. Adaptivity with moving grids. *Acta Numerica*, 18:111–241, 2009.
69. C. J. Budd and M. D. Piggott. Geometric integration and its applications. In *Handbook of Numerical Analysis, Vol. XI*, Handb. Numer. Anal., XI, pages 35–139. North-Holland, Amsterdam, 2003.
70. C. J. Budd, V. Rottschäfer, and J. F. Williams. Multibump, blow-up, self-similar solutions of the complex Ginzburg-Landau equation. *SIAM J. Appl. Dyn. Syst.*, 4:649–678 (electronic), 2005.
71. C. J. Budd and J. F. Williams. Parabolic Monge-Ampère methods for blow-up problems in several spatial dimensions. *J. Phys. A Math. Gen.*, 39:5425–5444, 2006.
72. C. J. Budd and J. F. Williams. Moving mesh generation using the parabolic Monge-Ampère equation. *SIAM J. Sci. Comput.*, 31:3438–3465, 2009.
73. H. G. Burchard. Splines (with optimal knots) are better. *Appl. Anal.*, 3:309–319, 1974.
74. L. A. Caffarelli. The regularity of mappings with a convex potential. *J. Amer. Math. Soc.*, 5:99–104, 1992.
75. L. A. Caffarelli. Boundary regularity of maps with convex potentials II. *Ann. Math.*, 144:453–496, 1996.
76. G. Caginalp. Stefan and Hele-Shaw type models as asymptotic limits of the phase-field equations. *Phys. Rev. A (3)*, 39:5887–5896, 1989.
77. G. Caginalp and E. A. Socolovsky. Computation of sharp phase boundaries by spreading: The planar and spherically symmetric cases. *J. Comput. Phys.*, 95:85–100, 1991.
78. X. Cai, D. Fleitas, B. Jiang, and G. Liao. Adaptive grid generation based on the least-squares finite-element method. *Comput. Math. Appl.*, 48:1077–1085, 2004.
79. S. A. Canann, J. R. Tristano, and M. L. Staten. An approach to combined laplacian and optimization-based smoothing for triangular, quadrilateral, and quad-dominant meshes. In *Proceedings, 7th International Meshing Roundtable*, Sandia National Laboratories, Albuquerque, NM, 1998.

80. W. Cao. On the error of linear interpolation and the orientation, aspect ratio, and internal angles of a triangle. *SIAM J. Numer. Anal.*, 43:19–40 (electronic), 2005.
81. W. Cao, W. Huang, and R. D. Russell. An r-adaptive finite element method based upon moving mesh pdes. *J. Comp. Phys.*, 149:221–244, 1999.
82. W. Cao, W. Huang, and R. D. Russell. A study of monitor functions for two dimensional adaptive mesh generation. *SIAM J. Sci. Comput.*, 20:1978–1994, 1999.
83. W. Cao, W. Huang, and R. D. Russell. Comparison of two-dimensional r-adaptive finite element methods using various error indicators. *Math. Comput. Simulation*, 56:127–143, 2001.
84. W. Cao, W. Huang, and R. D. Russell. A moving mesh method based on the geometric conservation law. *SIAM J. Sci. Comput.*, 24:118–142, 2002.
85. W. Cao, W. Huang, and R. D. Russell. Approaches for generating moving adaptive meshes: location versus velocity. *Appl. Numer. Math.*, 47:121–138, 2003.
86. G. Carey. *Computational grids: Generation, adaptation, and solution strategies.* Taylor & Francis, Washington, DC, 1997.
87. G. F. Carey and H. T. Dinh. Grading functions and mesh redistribution. *SIAM J. Numer. Anal.*, 22:1028–1040, 1985.
88. N. Carlson and K. Miller. Design and application of a gradient-weighted moving finite element code, Part I, In 1-D. *SIAM J. Sci. Comput.*, 19:728–765, 1998.
89. N. Carlson and K. Miller. Design and application of a gradient-weighted moving finite element code. Part II In 2-D. *SIAM J. Sci. Comput.*, 19:766–798, 1998.
90. J. Castillo. Mathematical aspects of variational grid generation I. In J. Hauser and C. Taylor, editors, *Numerical Grid Generation in Computational Fluid Dynamics*, Pineridge Press Limited, Swansea, U.K., 1986.
91. J. E. Castillo. A discrete variational grid generation method. *SIAM J. Sci. Statist. Comput.*, 12:454–468, 1991.
92. J. E. Castillo, editor. *Mathematical Aspects of Numerical Grid Generation*, Frontiers in Applied Mathematics vol 8, Philadelphia, 1991. SIAM.
93. M. J. Castro-Díaz, F. Hecht, B. Mohammadi, and O. Pironneau. Anisotropic unstructured mesh adaption for flow simulations. *Int. J. Numer. Meth. Fluids*, 25:475–491, 1997.
94. H. D. Ceniceros. A semi-implicit moving mesh method for the focusing nonlinear Schrödinger equation. *Commun. Pure Appl. Anal.*, 1:1–18, 2002.
95. H. D. Ceniceros and T. Y. Hou. An efficient dynamically adaptive mesh for potentially singular solutions. *J. Comput. Phys.*, 172:609–639, 2001.
96. A. A. Charakhch'yan and S. A. Ivanenko. A variational form of the Winslow grid generator. *J. Comput. Phys.*, 136:385–398, 1997.
97. K. Chen. Error equidistribution and mesh adaptation. *SIAM J. Sci. Comput.*, 15:798–818, 1994.
98. L. Chen. Mesh smoothing schemes based on optimal Delaunay triangulations. In *Proceedings, 13th International Meshing Roundtable*, pages 109–120, Sandia National Laboratories, Albuquerque, NM, 2004. Sandia Report #2004-3765C.
99. L. Chen, P. Sun, and J. C. Xu. Optimal anisotropic meshes for minimizing interpolation errors in L^p-norm. *Math. Comput.*, 76:179–204, 2007.
100. L. Chen and J. C. Xu. Stability and accuracy of adapted finite element methods for singularly perturbed problems. *Numer. Math.*, 109:167–191, 2008.
101. Y.-Z. Chen and L.-C. Wu. *Second Order Elliptic Equations and Elliptic Systems.* American Mathematical Society, Providence, Rhode Island, 1998. Translations of Mathematical Mongraphs, Volume 174.
102. K. N. Christodoulou and L. E. Scriven. Discretization of free surface flows and other moving boundary problems. *J. Comput. Phys.*, 99:39–55, 1992.

103. S.-K. Chua and R. L. Wheeden. Estimates of best constants for weighted Poincaré inequalities on convex domains. *Proc. London Math. Soc. (3)*, 93:197–226, 2006.
104. P. G. Ciarlet. *The Finite Element Method for Elliptic Problems*. North-Holland, Amsterdam, 1978.
105. P. G. Ciarlet. *Mathematical elasticity. Vol. I*, volume 20 of *Studies in Mathematics and its Applications*. North-Holland Publishing Co., Amsterdam, 1988. Three-dimensional elasticity.
106. R. Courant and D. Hilbert. *Methods of Matheatical Physics*. Interscience Publishers, Inc., New York, 1953. Volumes I and II.
107. C. L. Cox and T. H. Payne. Mathematical modeling of unsaturated porous media flow and transport. In D. R. Shier and K. T. Wallenius, editors, *Applied Mathematical Modeling: A Multidisciplinary Approach*, pages 185–201, Chapman & Hall/CRC, London, Boca Raton, 2000.
108. J. M. Coyle, J. E. Flaherty, and R. Ludwig. On the stability of mesh equidistribution strategies for time-dependent partial differential equations. *J. Comput. Phys.*, 62:26–39, 1986.
109. J. Crank. *Free and Moving Boundary Problems*. Clarendon Press, New York, 1984.
110. B. Dacorogna and J. Moser. On a partial differential equation involving the Jacobian determinant. *Ann. Inst. Henri Poincare Analyse non lineaire*, 7:1–26, 1990.
111. E. F. D'Azevedo. Optimal triangular mesh generation by coordinate transformation. *SIAM J. Sci. Stat. Comput.*, 12:755–786, 1991.
112. E. F. D'Azevedo and R. B. Simpson. On optimal triangular meshes for minimizing the gradient error. *Numer. Math.*, 59:321–348, 1991.
113. V. F. de Almeida. Domain deformation mapping: application to variational mesh generation. *SIAM J. Sci. Comput.*, 20:1252–1275, 1999.
114. M. de Berg, M. van Kreveld, M. Overmars, and O. Schwarzkopf. *Computational Geometry*. Springer, Berlin, 2000.
115. C. de Boor. Good approximation by splines with variable knots. In A. Meir and A. Sharma, editors, *Spline Functions and Approximation Theory*, pages 57–73, Birkhäuser Verlag, Basel und Stuttgart, 1973.
116. C. de Boor. Good approximation by splines with variables knots II. In G. A. Watson, editor, *Lecture Notes in Mathematics 363*, pages 12–20, Springer-Verlag, Berlin, 1974. Conference on the Numerical Solution of Differential Equations, Dundee, Scotland, 1973.
117. G. L. Delzanno, L. Chacón, J. M. Finn, Y. Chung, and G. Lapenta. An optimal robust equidistribution method for two-dimensional grid adaptation based on Monge-Kantorovich optimization. *J. Comput. Phys.*, 227(23):9841–9864, 2008.
118. Y. Di, R. Li, and T. Tang. A general moving mesh framework in 3D and its application for simulating the mixture of multi-phase flows. *Commun. Comput. Phys.*, 3:582–602, 2008.
119. Y. Di, R. Li, T. Tang, and P. Zhang. Moving mesh finite element methods for the incompressible Navier-Stokes equations. *SIAM J. Sci. Comput.*, 26:1036–1056 (electronic), 2005.
120. Y. Di, R. Li, T. Tang, and P. Zhang. Level set calculations for incompressible two-phase flows on a dynamically adaptive grid. *J. Sci. Comput.*, 31:75–98, 2007.
121. D. S. Dodson. Optimal order approximation by polynomial spline functions. Technical report, Purdue University, 1972. Ph.D. thesis.
122. V. Dolejší. Anisotropic mesh adaptation for finite volume and finite element methods on triangular meshes. *Computing and Visualisation in Science*, 1:165–178, 1998.
123. J. Dompierre, P. Labbé, F. Guibault, and R. Camarero. Benchmarks for 3d unstructured tetrahedral mesh optimization. In *Proceedings, 7th International Meshing Roundtable*, pages 459–478, Sandia National Laboratories, Albuquerque, NM, 1996. Sandia Report 98-2250.
124. E. A. Dorfi and L. O'c Drury. Simple adaptive grids for 1-D initial value problems. *J. Comput. Phys.*, 69:175–195, 1987.

125. W. Dörfler. A robust adaptive strategy for the nonlinear Poisson equation. *Computing*, 55:289–304, 1995.
126. W. Dörfler and R. H. Nochetto. Small data oscillation implies the saturation assumption. *Numer. Math.*, 91:1–12, 2002.
127. T. Dupont. Mesh modification for evolution equations. *Math. Comput.*, 39:85–107, 1982.
128. T. F. Dupont and Y. Liu. Symmetric error estimates for moving mesh Galerkin methods for advection-diffusion equations. *SIAM J. Numer. Anal.*, 40:914–927 (electronic), 2002.
129. A. S. Dvinsky. Adaptive grid generation from harmonic maps on Riemannian manifolds. *J. Comput. Phys.*, 95:450–476, 1991.
130. H. A. Dwyer and B. R. Sanders. Numerical modeling of unsteady flame propagation. Technical Report SAND77-8275, Sandia National Laboratory, Livermore, 1978.
131. P. W. Egolf and H. Manz. Theory and modeling of phase change materials with and without mushy regions. *Int. J. Mass Heat Transfer*, 37:2917–2924, 1994.
132. P. R. Eiseman. Grid generation for fluid mechanics computation. *Ann. Rev. Fluid Mech.*, 17:487–522, 1985.
133. P. R. Eiseman. Adaptive grid generation. *Comput. Meth. Appl. Mech. Engrg.*, 64:321–376, 1987.
134. J. Emert and R. Nelson. Volume and surface area for polyhedra and polytopes. *Mathematics Magazine*, 70:365–371, 1997.
135. A. Ern and J. L. Guermond. *Theory and Practice of Finite Elements*. Springer-Verlag, New York, 2004.
136. J. W. Evans. Nerve axon equations: IV. The stable and the unstable impulse. *Indiana Univ. Math. J.*, 24:1169–1190, 1975.
137. L. C. Evans. *Partial Differential Equations*. American Mathematical Society, Providence, Rhode Island, 1998. Graduate Studies in Mathematics, Volume 19.
138. L. C. Evans. Partial differential equations and Monge-Kantorovich mass transfer. In *Current Developments in Mathematics, 1997 (Cambridge, MA)*, pages 65–126. Int. Press, Boston, MA, 1999.
139. R. E. Ewing and H. Wang. A summary of numerical methods for time-dependent advection-dominated partial differential equations. *J. Comput. Appl. Math.*, 128:423–445, 2001. Numerical analysis 2000, Vol. VII, Partial differential equations.
140. C. Farhat and P. Geuzaine. Design and analysis of robust ALE time-integrators for the solution of unsteady flow problems on moving grids. *Comput. Meth. Appl. Mech. Engrg.*, 193:4073–4095, 2004.
141. C. Farhat, P. Geuzaine, and C. Grandmont. The discrete geometric conservation law and the nonlinear stability of ALE schemes for the solution of flow problems on moving grids. *J. Comput. Phys.*, 174:669–694, 2001.
142. W. M. Feng, P. Yu, S. Y. Hu, Z. K. Liu, Q. Du, and L. Q. Chen. Spectral implementation of an adaptive moving mesh method for phase-field equations. *J. Comput. Phys.*, 220:498–510, 2006.
143. J. A. Ferreira. On the convergence on nonrectangular grids. *J. Comput. Appl. Math.*, 85:333–344, 1997.
144. B. H. Fiedler and R. J. Trapp. A fast dynamic grid adaption scheme for meteorological flows. *Mon. Weather Rev.*, 121:2879–2888, 1993.
145. B. A. Finlayson. *Numerical Methods for Problems with Moving Fronts*. Ravenna Park Publ., Inc., Seattle, 1992.
146. R. FitzHugh. Impulses and physiological states in theoretical models of nerve membrane. *Biophys. J.*, 1:445–466, 1961.

147. J. E. Flaherty, J. M. Coyle, R. Ludwig, and S. F. Davis. A new consistent spatial differencing scheme for the transonic full-potential equation. In I. Babuška, J. Chandra, and J. E. Flaherty, editors, *Adaptive Computational Methods for Partial Differential Equations*, pages 144–164, SIAM, Philadelphia, 1983.
148. C. A. J. Fletcher. *Computational Techniques for Fluid Dynamics, Volumes 1 and 2*. Springer-Verlag, Berlin, New York, 1988.
149. L. Formaggia and F. Nobile. A stability analysis for the arbitrary Lagrangian Eulerian formulation with finite elements. *East-West J. Numer. Math.*, 7:105–131, 1999.
150. P. Frey and P. L. George. *Mesh Generation: Application to Finite Elements*. Hermes Science, Oxford and Paris, 2000.
151. A. Gamliel and L. M. Abriola. A one-dimensional moving grid solution for the coupled nonlinear equations governing multi-phase flow in porous media. 1: Model development. *Int. J. Numer. Meth. Fluids*, 14:25–45, 1992.
152. A. Gamliel and L. M. Abriola. A one-dimensional moving grid solution for the coupled nonlinear equations governing multi-phase flow in porous media. 2: Example simulations and sensitivity analysis. *Int. J. Numer. Meth. Fluids*, 14:47–69, 1992.
153. I. M. Gelfand and S. V. Fomin. *Calculus of Variations*. Prentice-Hall, Inc., Englewood Cliffs, NJ, 1963.
154. P. L. George. Automatic mesh generation and finite element computation. In P. G. Ciarlet and J. L. Lions, editors, *Handbook of Numerical Analysis, Vol. IV*, pages 69–190, Elsevier Science B.V., New York, 1996.
155. P. A. Gnoffo. A vectorized, finite-volume, adaptive-grid algorithm for Navier-Stokes. *Appl. Math. Comput.*, 10/11:819–835, 1982. Numerical grid generation (Nashville, Tenn., 1982).
156. P. A. Gnoffo. A finite volume, adaptive grid algorithm applied to planetary entry flowfields. *AIME Journal*, 21:1249–1254, 1983.
157. G. H. Golub and C. F. van Loan. *Matrix Computation*. The Johns Hopkins University Press, Baltimore, Maryland, 1983.
158. J. B. Greenberg. A new self-adaptive grid method. *AIAA J.*, 23:317–320, 1985.
159. P. Grisvard. *Elliptic Problems in Nonsmooth Domains*. Pitman, Boston, London, 1985.
160. A. E. Guannakopoulos and A. J. Engel. Directional control in grid generation. *J. Comput. Phys.*, 74:422–439, 1988.
161. M. H. Gutknecht. Variants of BICGSTAB for matrices with complex spectrum. *SIAM J. Sci. Comput.*, 14:1020–1033, 1993.
162. W. G. Habashi, J. Dompierre, Y. Bourgault, D. Ait-Ali-Yahia, M. Fortin, and M.-G. Vallet. Anisotropic mesh adaptation: towards user-independent, mesh-independent and solver-independent CFD. Part I: General principles. *Int. J. Numer. Meth. Fluids*, 32:725–744, 2000.
163. R. Hagmeijer. Grid adaption based on modified anisotropic diffusion equations formulated in the parametric domain. *J. Comput. Phys.*, 115:169–183, 1994.
164. E. Hairer, S. P. Nørsett, and G. Wanner. *Solving Ordinary Differential Equations. I*, volume 8 of *Springer Series in Computational Mathematics*. Springer-Verlag, Berlin, second edition, 1993. Nonstiff problems.
165. E. Hairer and G. Wanner. *Solving Ordinary Differential Equations. II*, volume 14 of *Springer Series in Computational Mathematics*. Springer-Verlag, Berlin, second edition, 1996. Stiff and differential-algebraic problems.
166. R. Hamilton. *Harmonic Maps of Manifolds with Boundary*. Springer-Verlag, Berlin, 1975. Lecture Notes in Mathematics Vol. 471.
167. J. Han and H.-Z. Tang. An adaptive moving mesh method for two-dimensional ideal magnetohydrodynamics. *J. Comput. Phys.*, 220:791–812, 2007.
168. G. H. Hardy, J. E. Littlewood, and G. Pólya. *Inequalities*. Cambridge University Press, Cambridge, 1934.

169. D. F. Hawken, J. J. Gottlieb, and J. S. Hansen. Review of some adaptive node-movement techniques in finite element and finite difference solutions of pdes. *J. Comput. Phys.*, 95:254–302, 1991.
170. Y. He and W. Huang. A posteriori error analysis for finite element solution of elliptic differential equations using equidistributing meshes. 2009. (arXiv:0911.0065v1).
171. R. G. Hindman. Generalized coordinate forms of governing fluid equations and associated geometrically induced errors. *AIAA J.*, 20:1359–1367, 1982.
172. R. G. Hindman and J. Spencer. A new approach to truly adaptive grid generation. Technical Report 83-0450, AIAA Paper, 1983.
173. C. W. Hirt, A. A. Amsden, and J. L. Cook. An arbitrary Lagrangian-Eulerian computing method for all flow speeds. *J. Comput. Phys.*, 14:227–253, 1974.
174. R. W. Hockney and J. W. Eastwood. *Computer Simulation Using Particles*. McGraw-Hill Inc., New York, 1981.
175. W. Huang. Practical aspects of formulation and solution of moving mesh partial differential equations. *J. Comput. Phys.*, 171:753–775, 2001.
176. W. Huang. Variational mesh adaptation: isotropy and equidistribution. *J. Comput. Phys.*, 174:903–924, 2001.
177. W. Huang. Convergence analysis of finite element solution of one-dimensional singularly perturbed differential equations on equidistributing meshes. *International Journal of Numerical Analysis & Modeling*, 2:57–74, 2005.
178. W. Huang. Measuring mesh qualities and application to variational mesh adaptation. *SIAM J. Sci. Comput.*, 26:1643–1666, 2005.
179. W. Huang. Metric tensors for anisotropic mesh generation. *J. Comput. Phys.*, 204:633–665, 2005.
180. W. Huang. Mathematical principles of anisotropic mesh adaptation. *Comm. Comput. Phys.*, 1:276–310, 2006.
181. W. Huang. Anisotropic mesh adaptation and movement. In T. Tang and J. Xu, editors, *Adaptive Computations: Theory and Algorithms*, pages 68–158, Science Press, Beijing, 2007. Mathematics Monograph Series 6.
182. W. Huang, L. Kamenski, and J. Lang. A new anisotropic mesh adaptation method based upon hierarchical a posteriori error estimates. *J. Comput. Phys.*, 229:2179–2198, 2010.
183. W. Huang and X. P. Li. An anisotropic mesh adaptation method for the finite element solution of variational problems. *Fin. Elem. Anal. Des.*, 46:61–73, 2010.
184. W. Huang, J. Ma, and R. D. Russell. A study of MMPDE moving mesh methods for the numerical simulation of blowup in reaction diffusion equations. *J. Comput. Phys.*, 227:6532–6552, 2008.
185. W. Huang, Y. Ren, and R. D. Russell. Moving mesh methods based on moving mesh partial differential equations. *J. Comput. Phys.*, 113:279–290, 1994.
186. W. Huang, Y. Ren, and R. D. Russell. Moving mesh partial differential equations (MMPDEs) based upon the equidistribution principle. *SIAM J. Numer. Anal.*, 31:709–730, 1994.
187. W. Huang and R. D. Russell. A moving collocation method for the numerical solution of time dependent differential equations. *Appl. Numer. Math.*, 20:101–116, 1996.
188. W. Huang and R. D. Russell. Analysis of moving mesh partial differential equations with spatial smoothing. *SIAM J. Numer. Anal.*, 34:1106–1126, 1997.
189. W. Huang and R. D. Russell. A high dimensional moving mesh strategy. *Appl. Numer. Math.*, 26:63–76, 1997.
190. W. Huang and R. D. Russell. Moving mesh strategy based upon a gradient flow equation for two dimensional problems. *SIAM J. Sci. Comput.*, 20:998–1015, 1999.
191. W. Huang and R. D. Russell. Review of moving mesh methods for soving PDEs. *J. Comput. Appl. Math.*, 128:383–398, 2001.

192. W. Huang and D. M. Sloan. A simple adaptive grid method in two dimensions. *SIAM J. Sci. Comput.*, 15:776–797, 1994.
193. W. Huang and W. Sun. Variational mesh adaptation II: error estimates and monitor functions. *J. Comput. Phys.*, 184:619–648, 2003.
194. W. Huang and X. Zhan. Adaptive moving mesh modeling for two dimensional groundwater flow and transport. In *Recent Advances in Adaptive Computation (Hangzhou, 2004)*, volume 383 of *AMS Contemporary Mathematics*, pages 239–252. Amer. Math. Soc., Providence, RI, 2005.
195. W. Huang, L. Zheng, and X. Zhan. Adaptive moving mesh methods for simulating one-dimensional groundwater problems with sharp moving fronts. *Int. J. Numer. Meth. Engrg.*, 54:1579–1603, 2002.
196. M. E. Hubbard, M. J. Baines, and P. K. Jimack. Consistent Dirichlet boundary conditions for numerical solution of moving boundary problems. *Appl. Numer. Math.*, 59:1337–1353, 2009.
197. J. M. Hyman and B. Larrouturou. Dynamic rezone methods for partial differential equations in one space dimension. Technical Report LA-UR-86-1678, Los Alamos National Laboratory, Los Alamos, NM, 1986.
198. J. M. Hyman and B. Larrouturou. Dynamic rezone methods for partial differential equations in one space dimension. *Appl. Num. Math.*, 5:435–450, 1989.
199. J. M. Hyman, Shengtai Li, and L. R. Petzold. An adaptive moving mesh method with static rezoning for partial differential equations. *Comput. Math. Appl.*, 46:1511–1524, 2003.
200. S. A. Ivanenko and B. N. Azarenok. Application of moving adaptive grids for numerical solution of 2D nonstationary problems in gas dynamics. *Internat. J. Numer. Methods Fluids*, 39:1–22, 2002.
201. S. A. Ivanenko and A. A. Charakhch'yan. Curvilinear grids from convex quadrangles. *Zh. Vychisl. Mat. i Mat. Fiz.*, 28:503–514, 622, 1988.
202. O.-P. Jacquotte. A mechanical model for a new grid generation method in computational fluid dynamics. *Comput. Meth. Appl. Mech. Engrg.*, 66:323–338, 1988.
203. O.-P. Jacquotte. Generation, optimization and adaptation of multiblock grids around complex configurations in computational fluid dynamics. *Int. J. Numer. Meth. Engrg.*, 34:443–454, 1992.
204. O.-P. Jacquotte and G. Coussement. Structured mesh adaption: space accuracy and interpolation methods. *Comput. Meth. Appl. Mech. Engrg.*, 101:397–432, 1992. Reliability in computational mechanics (Kraków, 1991).
205. P. Jamet. Estimations de l'erreur pour des elements finis droits preque degeneres. *RAIRO Anal. Numer.*, 10:43–60, 1976.
206. X. Ji. Adaptive numerical integration based on coordinate transformations. Technical report, Department of Mathematics, University of Kansas, Lawrence, Kansas 66045, U. S. A., 2002. MA Thesis.
207. P. K. Jimack. On steady and large time solutions of the semi-discrete moving finite element equations for 1-D diffusion problems. *IMA J. Numer. Anal.*, 12:545–564, 1992.
208. J. Kautský and N. K. Nichols. Equidistributing meshes with constraints. *SIAM J. Sci. Statist. Comput.*, 1:499–511, 1980.
209. J. Kautský and N. K. Nichols. Smooth regrading of discretized data. *SIAM J. Sci. Statist. Comput.*, 3:145–159, 1982.
210. D. A. Knoll, R. B. Lowrie, and J. E. Morel. Numerical analysis of time integration errors for nonequilibrium radiation diffusion. *J. Comput. Phys.*, 226:1332–1347, 2007.
211. P. Knupp. Mesh generation using vector-fields. *J. Comput. Phys.*, 119:142–148, 1995.
212. P. Knupp, L. Margolin, and M. Shashkov. Reference jacobian optimization-based rezone strategies for arbitrary lagrangian eulerian methods. *J. Comput. Phys.*, 176:93–128, 2002.

213. P. Knupp and S. Steinberg. *Fundamentals of Grid Generation.* CRC Press, Boca Raton, 1994.
214. P. M. Knupp. Jacobian-weighted elliptic grid generation. *SIAM J. Sci. Comput.*, 17:1475–1490, 1996.
215. P. M. Knupp. Applications of mesh smoothing: copy, morph, and sweep on unstructured quadrilateral meshes. *Internat. J. Numer. Methods Engrg.*, 45:37–45, 1999.
216. P. M. Knupp. Algebraic mesh quality metrics. *SIAM J. Sci. Comput.*, 23:193–218 (electronic), 2001.
217. P. M. Knupp and N. Robidoux. A framework for variational grid generation: conditioning the Jacobian matrix with matrix norms. *SIAM J. Sci. Comput.*, 21:2029–2047, 2000.
218. H. Kober. On the arithmetic and geometric means and on Hölder's inequality. *Proc. Amer. Math. Soc.*, 9:452–459, 1958.
219. N. Kopteva and M. Stynes. A robust adaptive method for a quasi-linear one-dimensional convection-diffusion problem. *SIAM J. Numer. Anal.*, 39:1446–1467, 2001.
220. M. Krizek. On the maximum angle condition for linear tetrahedral elements. *SIAM J. Numer. Anal.*, 29:513–520, 1992.
221. G. Kunert. A posteriori error estimation for anisotropic tetrahedral and triangular finite element meshes. Technical report, TU Chemnitz, 1999. Ph. D. Thesis.
222. J. Lang. *Adaptive Multilevel Solution of Nonlinear Parabolic PDE Systems: Theory, Algorithm, and Applications*, volume 16 of *Lecture Notes in Computational Science and Engineering*. Springer-Verlag, Berlin, 2001.
223. J. Lang, W. Cao, W. Huang, and R. D. Russell. A two–dimensional moving finite element method with local refinement based on a posteriori error estimates. *Appl. Numer. Math.*, 46:75–94, 2003.
224. X. Lang. Moving mesh control volume solution for two dimensional phase change problems in enthalpy formulation. Technical report, Department of Mathematics, University of Kansas, 1997. Master's thesis.
225. G. Lapenta and L. Chacón. Cost-effectiveness of fully implicit moving mesh adaptation: A practical investigation in 1D. *J. Comput. Phys.*, 219:86–103, 2006.
226. Z. Lei, G. Liao, G. de la Pena, and D. Anderson. A moving grid algorithm based on deformation method. In B. K. Soni, J. F. Thompson, J. Hauser, and P. R. Eiseman, editors, *Proceedings of the 7th International Conference on Numerical Grid Generation in Computational Field Simulations*, 2001.
227. M. Lentini and V. Pereyra. An adaptive finite difference solver for nonlinear two-point boundary problems with mild boundary layers. *SIAM J. Numer. Anal.*, 14:94–111, 1977. Papers on the numerical solution of two-point boundary-value problems (NSF-CBMS Regional Res. Conf., Texas Tech Univ., Lubbock, Tex., 1975).
228. R. Li, T. Tang, and P. W. Zhang. Moving mesh methods in multiple dimensions based on harmonic maps. *J. Comput. Phys.*, 170:562–588, 2001.
229. R. Li, T. Tang, and P. W. Zhang. A moving mesh finite element algorithm for singular problems in two and three space dimensions. *J. Comput. Phys.*, 177:365–393, 2002.
230. Z. Li. Computation of crystalline microstructures with the mesh transformation method. In T. Tang and J. Xu, editors, *Adaptive Computations: Theory and Algorithms*, pages 211–241, Beijing, 2007. Science Press. Mathematics Monograph Series 6.
231. G. Liao. A study of regularity problem of harmonic maps. *Pacific J. Math.*, 131:291–302, 1988.
232. G. Liao, F. Liu, G. C. de la Pena, D. Peng, and S. Osher. Level-set-based deformation methods for adaptive grids. *J. Comput. Phys.*, 159:103–122, 2000.
233. G. Liao and N. Smale. Harmonic maps with nontrivial higher-dimensional singularities. In *Lecture Notes in Pure and Appl. Math. 144*, pages 79–89, Dekker, New York, 1993.

234. G. Liao and J. Xue. Moving meshes by the deformation method. *J. Comput. Appl. Math.*, 195:83–92, 2006.
235. G. J. Liao. Variational approach to grid generation. *Numer. Meth. P.D.E.*, 8:143–147, 1992.
236. G. J. Liao and D. Anderson. A new approach to grid generation. *Appl. Anal.*, 44:285–298, 1992.
237. T. Linß. *Layer-Adapted Meshes for Reaction-Convection-Diffusion Problems*. Springer-Verlag, Berlin, Heidelberg, 2010.
238. V. D. Liseikin. *Grid Generation Methods*. Springer, Berlin, 1999.
239. A. Liu and B. Joe. On the shape of tetrahedra from bisection. *Math. Comput.*, 63:141–154, 1994.
240. A. Liu and B. Joe. Relationship between tetrahedron quality measures. *BIT*, 34:268–287, 1994.
241. F. Liu, S. Ji, and G. Liao. An adaptive grid method and its application to steady Euler flow calculations. *SIAM J. Sci. Comput.*, 20:811–825, 1998.
242. Y. Liu, R. E. Bank, T. F. Dupont, S. Garcia, and R. F. Santos. Symmetric error estimates for moving mesh mixed methods for advection-diffusion equations. *SIAM J. Numer. Anal.*, 40:2270–2291 (electronic) (2003), 2002.
243. D. R. Lynch. Unified approach to simulation on deforming elements with application to phase change problems. *J. Comput. Phys.*, 47:387–411, 1982.
244. J. Mackenzie. Uniform convergence analysis of an upwind finite-differnce approximation of a convection-diffusion boundary value problem on an adaptive grid. *IMA J. Numer. Anal.*, 19:233–249, 1999.
245. J. A. Mackenzie and W. R. Mekwi. An analysis of stability and convergence of a finite-difference discretization of a model parabolic PDE in 1D using a moving mesh. *IMA J. Numer. Anal.*, 27:507–528, 2007.
246. J. A. Mackenzie and W. R. Mekwi. On the use of moving mesh methods to solve PDEs. In T. Tang and J. Xu, editors, *Adaptive Computations: Theory and Algorithms*, pages 242–278, Science Press, Beijing, 2007. Mathematics Monograph Series 6.
247. J. A. Mackenzie and M. L. Robertson. A moving mesh method for the solution of the one-dimensional phase-field equations. *J. Comput. Phys.*, 181:526–544, 2002.
248. N. K. Madsen. MOLAG: A method of lines adaptive grid interface for nonlinear partial differential equations. In B. Engquist and T. Smedsaas, editors, *PDE Software: Modules, Interfaces and Systems*, pages 207–224, North Holland, Amsterdam, 1984. Söderköping, Sweden, 22-26 August, 1983.
249. L. G. Margolin. Introduction to "an arbitrary Lagrangian-Eulerian computing method for all flow speeds". *J. Comput. Phys.*, 135:198–202, 1997.
250. D. Mihalas and B. W. Mihalas. *Foundations of Radiation Hydrodynamics*. Oxford University Press, New York, Oxford, 1984.
251. C. T. Miller, G. Christakos, P. T. Imhoff, J. F. McBride, J. A. Pedit, and J. A. Trangenstein. Multiphase flow and transport modeling in heterogeneous porous media: Challenges and approaches. *Adv. Water Resour.*, 21:77–120, 1998.
252. C. T. Miller, S. N. Gleyzer, and P. T. Imhoff. Numerical modeling of NAPL dissolution fingering in porous media. In H. M. Selim and L. Ma, editors, *Physical Nonequilibrium in Soils Modeling and Application*, Ann Arbor Press, Chelsea, Michigan, 1998.
253. K. Miller. Moving finite elements II. *SIAM J. Numer. Anal.*, 18:1033–1057, 1981.
254. K. Miller. A geometrical-mechanical interpretation of gradient-weighted moving finite elements. *SIAM J. Numer. Anal.*, 34:67–90, 1997.
255. K. Miller. Nonlinear Krylov and moving nodes in the method of lines. *J. Comput. Appl. Math.*, 183:275–287, 2005.

256. K. Miller. Stabilized moving finite elements for convection dominated problems. *J. Sci. Comput.*, 24:163–182, 2005.
257. K. Miller and M. J. Baines. Least squares moving finite elements. *IMA J. Numer. Anal.*, 21:621–642, 2001.
258. K. Miller and R. N. Miller. Moving finite elements I. *SIAM J. Numer. Anal.*, 18:1019–1032, 1981.
259. G. Monge. Mémoire sur la théorie des déblais et des remblais. In *Histoire de l'Académie Royale des Sciences de Paris*, pages 666–704. 1781.
260. P. K. Moore and J. E. Flaherty. Adaptive local overlapping grid methods for parabolic system in two space dimensions. *J. Comput. Phys.*, 98:54–63, 1992.
261. J. Moser. On the volume elements of a manifold. *Trans. AMS*, 120:286–294, 1965.
262. A. C. Mueller and G. F. Carey. Continuously deforming finite elements. *Internat. J. Numer. Methods Engrg.*, 21:2099–2126, 1985.
263. L. S. Mulholland, W. Huang, and D. M. Sloan. Pseudospectral solution of near-singular problems using numerical coordinate transformations based on adaptivity. *SIAM J. Sci. Comput.*, 19:1261–1298, 1998.
264. J. D. Murray. *Mathematical Biology, An Introduction*. Springer, Berlin, 2002. Third Edition.
265. J. Nagumo, S. Arimoto, and S. Yoshizawa. An active pulse transmission line simulating nerve axon. *Proc. Inst. Radio Engrg.*, 50:2061–2070, 1964.
266. T. Nakagawa. Blowing up of a finite difference solution to $u_t = u_{xx} + u^2$. *Appl. Math. Optim.*, 2:337–350, 1975/76.
267. T. Nakagawa and T. Ushijima. Finite element analysis of the semi-linear heat equation of blow-up type. In J. H. Miller, editor, *Topics in Numerical Analysis, III*, pages 275–291, Academic Press, London, New York, 1977.
268. D. A. Neid and A. Bejan. *Convection in Porous Media*. pringer-Verlag, New York, 1992.
269. Adam M. Oberman. Wide stencil finite difference schemes for the elliptic Monge-Ampère equation and functions of the eigenvalues of the Hessian. *Discrete Contin. Dyn. Syst. Ser. B*, 10:221–238, 2008.
270. C. V. Pao. *Nonlinear Parabolic and Elliptic Equations*. Plenum Press, New York, 1992.
271. V. N. Parthasarathy, C. M. Graichen, and A. F. Hathaway. A comparison of tetrahedron quality measures. *Fin. Elem. Anal. Des.*, 15:255–261, 1993.
272. L. E. Payne and H. F. Weinberger. An optimal Poincaré inequality for convex domains. *Arch. Rational Mech. Anal.*, 5:286–292 (1960), 1960.
273. V. Pereyra and E. G. Sewell. Mesh selection for discrete solution of boundary problems in ordinary differential equations. *Numer. Math.*, 23:261–268, 1975.
274. L. R. Petzold. Observations on an adaptive moving grid method for one-dimensional systems for partial differential equations. *Appl. Numer. Math.*, 3:347–360, 1987.
275. T. Plewa, T. Linde, and V. G. Weirs, editors. *Adaptive Mesh Refinement – Theory and Applications*, Lecture Notes in Computational Science and Engineering, Springer, New York, Berlin, 2005. Vol. 41.
276. M. H. Protter and H. F. Weinberger. *Maximum Principles in Differential Equations*. Prentice-Hall Inc., Englewood Cliffs, N. J., 1967.
277. J. D. Pryce. On the convergence of iterated remeshing. *IMA J. Numer. Anal.*, 9:315–335, 1989.
278. Y. Qiu and D. M. Sloan. Analysis of difference approximations to a singularly perturbed two-point boundary value problem on an adaptively generated grid. *J. Comput. Appl. Math.*, 101:1–25, 1999.
279. Y. Qiu, D. M. Sloan, and T. Tang. Numerical solution of a singularly perturbed two-point boundary value problem using equidistribution: analysis of convergence. *J. Comput. Appl. Math.*, 116:121–143, 2000.

280. W. Ren and X. Wang. An iterative grid redistribution method for singular problems in multiple dimensions. *J. Comput. Phys.*, 159:246–273, 2000.
281. Y. Ren. Theory and computation of moving mesh methods for solving time-dependent partial differential equations. Technical report, Department of Mathematics and Statistics, Simon Fraser University, 1992. Ph.D. Thesis.
282. Y. Ren and R. D. Russell. Moving mesh techniques based upon equidistribution and their stability. *SIAM J. Sci. Stat. Comput.*, 13:1265–1286, 1992.
283. R. J. Renka. Quadratic Shepard method for bivariate interpolation of scattered data. *ACM TOMS*, 14:149–150, 1988.
284. J. R. Rice. On the degree of convergence of nonlinear spline approximation. In I. J. Schoenberg, editor, *Approximations with Special Emphasis on Spline Functions*, pages 349–365, Academic Press, New York, London, 1969.
285. R. D. Russell and J. Christiansen. Adaptive mesh selection strategies for solving boundary value problems. *SIAM J. Numer. Anal.*, 15:59–80, 1978.
286. R. D. Russell, J. F. Williams, and X. Xu. Movcol4: A moving mesh code for fourth-order time-dependent partial differential equations. *SIAM J. Sci. Comput.*, 29:197–220, 2007.
287. T. F. Russell. Modeling of multiphase multicontaminant transport in the subsurface. *Rev. Geophys.*, 33:(Supplement 2) 1035–1047, 1995.
288. Y. Saad. *Iterative Methods for Sparse Linear Systems*. SIAM, Philadelphia, 2003. Second Edition.
289. Y. Saad and M. H. Schultz. GMRES: A generalized minimal residual algorithm for solving nonsymmetric linear systems. *SIAM J. Sci. Stat. Comput.*, 7:856–869, 1986.
290. J. Sacks and D. Ylvisaker. Designs for regression problems with corrected errors. *Ann. Math. Stat.*, 37:66–89, 1966.
291. J. Sacks and D. Ylvisaker. Designs for regression problems with corrected errors; many parameters. *Ann. Math. Stat.*, 39:49–69, 1968.
292. J. Sacks and D. Ylvisaker. Designs for regression problems with corrected errors III. *Ann. Math. Stat.*, 41:2057–2074, 1970.
293. C. D. Sarris. *Adaptive Mesh Refinement for Time-Domain Numerical Electromagnetics*. Morgan & Claypool Publishers, San Rafael, CA, 2007.
294. R. Schoen and S. T. Yau. On univalent harmonic maps between surfaces. *Inv. Math.*, 44:265–278, 1978.
295. B. Semper and G. Liao. A moving grid finite-element method using grid deformation. *Numer. Methods in PDEs*, 11:603–615, 1995.
296. L. F. Shampine. Solving $0 = F(t, y(t), y'(t))$ in Matlab. *J. Numer. Math.*, 10:291–310, 2002.
297. L. F. Shampine and M. W. Reichelt. The MATLAB ODE Suite. *SIAM J. Sci. Comput.*, 18:1–22, 1997.
298. L. F. Shampine, M. W. Reichelt, and J. A. Kierzenka. Solving index-1 DAEs in MATLAB and Simulink. *SIAM Review*, 41:538–552, 1999.
299. Z.-C. Shi, Z. Chen, T. Tang, and D. Yu, editors. *Recent Advances in Adaptive Computation*, Comtemporary Mathematics Volume 383, Providence, Rhode Island, 2005. American Mathematical Society. Proceedings of the International Conference on Recent Sdvances in Adaptive Computation, May 24-28, 2004, Zhejiang University, Hangzhou, China.
300. G. I. Shishkin. *Discrete Approximation of Singularly Perturbed Elliptic and Parabolic Equations*. Russian Academy of Sciences, Ural Section, Ekaterinburg, 1992 (in Russian).
301. C.-W. Shu. Discontinuous Galerkin methods: general approach and stability. In S. Bertoluzza, S. Falletta, G. Russo, and C.-W. Shu, editors, *Numerical Solutions of Partial Differential Equations*, pages 149–201, Birkhäuser, Basel, 2009. Advanced Courses in Mathematics CRM Barcelona.

302. C.-W. Shu. High order weighted essentially non-oscillatory schemes for convection dominated problems. *SIAM Review*, 51:82–126, 2009.
303. D. M. Sloan. A review of moving mesh methods for the numerical solution of partial differential equations. *Not. S. Afr. Math. Soc.*, 33:68–78, 2002.
304. G. A. Sod. A survey of several finite difference methods for systems of nonlinear hyperbolic conservation laws. *J. Comput. Phys.*, 27:1–31, 1978.
305. D. M. Y. Sommerville. *An Introduction to the Geometry of n Dimensions*. Methuen & Co. LTD., London, 1929.
306. S. Steinberg and P. J. Roache. Variational grid generation. *Numer. Meth. P.D.E.*, 2:71–96, 1986.
307. J. M. Stockie, J. A. Mackenzie, and R. D. Russell. A moving mesh method for one dimensional hyperbolic conservation law. *SIAM J. Sci. Comput.*, 22:1791–1813, 2000.
308. M. M. Sulman, J. F. Williams, and R. D. Russell. Optimal mass transport for higher dimensional adaptive grid generation. 2009 (submitted).
309. M. M. Sulman, J. F. Williams, R. D. Russell, and F. M. Beg. Volumic image registration methods based on solving the Monge-Ampère equation. 2009 (submitted).
310. J. L. Synge. *The Hypercircle in Mathematical Physics*. Cambridge University Press, Cambridge, 1957.
311. Z. Tan. Adaptive moving mesh methods for two-dimensional resistive magnetohydrodynamic PDE models. *Computers & Fluids*, 36:758–771, 2007.
312. Z. Tan, K. M. Lim, and B. C. Khoo. An adaptive mesh redistribution method for the incompressible mixture flows using phase-field model. *J. Comput. Phys.*, 225:1137–1158, 2007.
313. Z. Tan, K. M. Lim, and B. C. Khoo. An adaptive moving mesh method for two-dimensional incompressible viscous flows. *Commun. Comput. Phys.*, 3:679–703, 2008.
314. Z. Tan, T. Tang, and Z. Zhang. A simple moving mesh method for one- and two-dimensional phase-field equations. *J. Comput. Appl. Math.*, 190:252–269, 2006.
315. H.-Z. Tang. Solution of the shallow-water equations using an adaptive moving mesh method. *Internat. J. Numer. Methods Fluids*, 44:789–810, 2004.
316. H. Z. Tang and T. Tang. Adaptive mesh methods for one- and two-dimensional hyperbolic conservation laws. *SIAM J. Numer. Anal.*, 41:487–515 (electronic), 2003.
317. H.-Z. Tang and T. Tang. Multi-dimensional moving mesh methods for shock computations. In *Recent Advances in Scientific Computing and Partial Differential Equations (Hong Kong, 2002)*, volume 330 of *Contemp. Math.*, pages 169–183. Amer. Math. Soc., Providence, RI, 2003.
318. T. Tang. Moving mesh methods for computational fluid dynamics flow and transport. In *Recent Advances in Adaptive Computation (Hangzhou, 2004)*, volume 383 of *AMS Contemporary Mathematics*, pages 141–173. Amer. Math. Soc., Providence, RI, 2005.
319. T. Tang and J. Xu. *Adaptive Computations: Theory and Algorithms*. Science Press, Beijing, 2007. Mathematics Monograph Series 6.
320. T. W. Tee and Lloyd N. Trefethen. A rational spectral collocation method with adaptively transformed Chebyshev grid points. *SIAM J. Sci. Comput.*, 28:1798–1811 (electronic), 2006.
321. J. W. Thomas. *Numerical Partial Differential Equations, Conservation Laws and Elliptic Equations*. Springer, New York, Berlin, Heidelberg, 1999.
322. P. D. Thomas and C. K. Lombard. Geometric conservation law and its application to flow computations on moving grids. *AIAA J.*, 17:1030–1037, 1979.
323. J. F. Thompson. A survey of dynamically-adaptive grids in the numerical solution of partial differential equations. *Appl. Numer. Math.*, 1:3–27, 1985.
324. J. F. Thompson, F. C. Thames, and C. W. Mastin. Automatic numerical grid generation of body fitted curvilinear coordinate system of field containing any number of arbitrary two dimensional bodies. *J. Comput. Phys.*, 15:299–319, 1974.

325. J. F. Thompson, Z. A. Warsi, and C. W. Mastin. *Numerical Grid Generation: Foundations and Applications*. North-Holland, New York, 1985.
326. J. F. Thompson, Z. U. A. Warsi, and C. W. Mastin. Boundary-fitted coordinate systems for numerical solution of partial differential equations - a review. *J. Comput. Phys.*, 47:1–108, 1982.
327. J. F. Thompson and N. P. Weatherill. Structured and unstructured grid generation. *Critical Reviews Biomed. Engrg.*, 20:73–120, 1992.
328. H. Tomita, M. Satoh, and K. Goto. An optimization of the icosahedral grid modified by spring dynamics. *J. Comput. Phys.*, 183:307–331, 2002.
329. Y. Tourigny. The optimisation of the mesh in first-order systems least-squares methods. *J. Sci. Comput.*, 24:219–245, 2005.
330. Y. Tourigny and M. J. Baines. Analysis of an algorithm for generating locally optimal meshes for L_2 approximation by discontinuous piecewise polynomials. *Math. Comp.*, 66:623–650, 1997.
331. Y. Tourigny and F. Hülsemann. A new moving mesh algorithm for the finite element solution of variational problems. *SIAM J. Numer. Anal.*, 35:1416–1438 (electronic), 1998.
332. A. van Dam. Go with the flow: moving meshes and solution monitoring for compressible flow simulation. Technical report, Utrecht University, 2009. Ph.D. thesis.
333. A. van Dam and P. A. Zegeling. A robust moving mesh finite volume method applied to 1D hyperbolic conservation laws from magnetohydrodynamics. *J. Comput. Phys.*, 216:526–546, 2006.
334. H. A. van der Vorst. BI-CGSTAB: a fast and smoothly converging variant of BI-CG for the solution of nonsymmetric linear systems. *SIAM J. Sci. Stat. Comput.*, 13:631–644, 1992.
335. R. Verfürth. *A Review of A-Posteriori Error Estimation and Adaptive Mesh Refinement Techniques*. John Wiley and Teubner, Germany, 1996. Advances in Numerical Mathematics.
336. C. Wang, H.-Z. Tang, and Tiegang Liu. An adaptive ghost fluid finite volume method for compressible gas-water simulations. *J. Comput. Phys.*, 227:6385–6409, 2008.
337. H. Wang, R. Li, and T. Tang. Efficient computation of dendritic growth with r-adaptive finite element methods. *J. Comput. Phys.*, 227:5984–6000, 2008.
338. L. Wang and J. Shen. Error analysis for mapped Jacobi spectral methods. *J. Sci. Comput.*, 24:183–218, 2005.
339. A. J. Wathen and M. J. Baines. On the structure of the moving finite-element equations. *IMA J. Numer. Anal.*, 5:161–182, 1985.
340. B. V. Wells, M. J. Baines, and P. Glaister. Generation of arbitrary Lagrangian-Eulerian (ALE) velocities, based on monitor functions, for the solution of compressible fluid equations. *Internat. J. Numer. Methods Fluids*, 47:1375–1381, 2005. 8th ICFD Conference on Numerical Methods for Fluid Dynamics. Part 2.
341. A. Winslow. Numerical solution of the quasi-linear Poisson equation in a nonuniform triangle mesh. *J. Comput. Phys.*, 2:149–172, 1967.
342. A. M. Winslow. Adaptive mesh zoning by the equipotential method. Technical Report UCID-19062, Lawrence Livemore Laboratory, 1981 (unpublished).
343. X. Xu, W. Huang, R. D. Russell, and J. F. Williams. Convergence of de Boor's algorithm for generation of equidistributing meshes. *IMA J. Numer. Anal.*, 2010 (to appear).
344. A. Yamada, K. Inoue, T. Itoh, and K. Shimada. An approach for generating meshes similar to a reference mesh. In *Proceedings, 9th International Meshing Roundtable*, pages 101–107, Sandia National Laboratories, Albuquerque, NM, 2000.
345. N. N. Yanenko, E. A. Kroshko, V. V. Liseikin, V. M. Fomin, V. P. Shapeev, and Yu. A. Shitov. Methods for the construction of moving grids for problems of fluid dynamics with big deformations. In A. I. van de Vooren and P. J. Zandbergen, editors, *Proceedings of the*

Fifth International Conference on Numerical Methods in Fluid Dynamics, pages 454–459, Berlin, 1976. Springer. Lecture Notes in Physics Vol. 59.

346. P. Yu, L. Q. Chen, and Q. Du. Applications of moving mesh methods to the Fourier spectral approximations of phase-field equations. In *Recent Advances in Computational Sciences*, pages 80–99. World Sci. Publ., Hackensack, NJ, 2008.
347. P. A. Zegeling. *Moving-Grid Methods for Time-Dependent Partial Differential Equations*. CWI Tract 94, Stichting Mathematisch Centrum, Centrum voor Wiskunde en Informatica, Amsterdam, 1993.
348. P. A. Zegeling. On resistive MHD models with adaptive moving meshes. *J. Sci. Comput.*, 24:263–284, 2005.
349. P. A. Zegeling. Theory and application of adaptive moving grid methods. In T. Tang and J. Xu, editors, *Adaptive Computations: Theory and Algorithms*, pages 279–332, Science Press, Beijing, 2007. Mathematics Monograph Series 6.
350. P. A. Zegeling, W. D. de Boer, and H. Z. Tang. Robust and efficient adaptive moving mesh solution of the 2-D Euler equations. In *Recent Advances in Adaptive Computation*, volume 383 of *Contemp. Math.*, pages 375–386. Amer. Math. Soc., Providence, RI, 2005.
351. P. A. Zegeling and H. P. Kok. Adaptive moving mesh computations for reaction-diffusion systems. *J. Comput. Appl. Math.*, 168:519–528, 2004.
352. P. A. Zegeling, J. G. Verwer, and J. C. H. Van Eijkeren. Application of a moving grid method to a class of 1d brine transport problems in porous media. *Int. J. Numer. Meth. Fluids*, 15:175–191, 1992.
353. Z. Zhang and A. Naga. A new finite element gradient recovery method: Superconvergence property. *SIAM J. Sci. Comput.*, 26:1192–1213, 2005.
354. O. C. Zienkiewicz and J. Z. Zhu. The superconvergence patch recovery and a posteriori error estimates. Part 1: The recovery technique. *Int. J. Numer. Methods Engrg.*, 33:1331–1364, 1992.
355. O. C. Zienkiewicz and J. Z. Zhu. The superconvergence patch recovery and a posteriori error estimates. Part 2: Error estimates and adaptivity. *Int. J. Numer. Methods Engrg.*, 33:1365–1382, 1992.
356. M. Zlámal. On the finite element method. *Numer. Math.*, 12:394–409, 1968.

Nomenclature

Abbreviations:

1D	One dimensions
2D	Two dimensions
3D	Three dimensions
DAE	Differential-algebraic equation
GCL	Geometric conservation law
MMPDE	Moving mesh PDE
ODE	Ordinary differential equation
PDE	Partial differential equation
SDIRK	Singly diagonally implicit Runge-Kutta scheme
SVD	Singular value decomposition

Notation:

d	dimension of physical domain Ω
Ω	A open, bounded physical domain in $\mathbb{R}^d$
$\lvert\Omega\rvert$	Volume of Ω
$\boldsymbol{x} = (x_1, ..., x_d)^T$	Physical coordinates
(x, y)	Physical coordinates in 2D
Ω_c	computational domain in $\mathbb{R}^d$
$\boldsymbol{\xi} = (\xi_1, ..., \xi_d)^T$	Computational coordinates
(ξ, η)	Computational coordinates in 2D
$\mathscr{T}_h$	A triangulation or a mesh for the physical domain Ω
N, N_v, N_{vi}	Numbers of the elements, vertices, and interior vertices of $\mathscr{T}_h$, respectively
$\mathscr{T}_h^c$	A triangulation or a mesh for the computational domain Ω_c
K	Generic element

$\hat{K}$	Reference or master element
h_K, H_K	Diameter and in-diameter of element K, respectively
F_K	Mapping from $\hat{K}$ to K; often is invertible and affine
F'_K	Jacobian matrix of mapping $F_K : \hat{K} \to K$
$\boldsymbol{J}$	Jacobian matrix of maping $\boldsymbol{x} = \boldsymbol{x}(\boldsymbol{\xi})$, i.e., $\boldsymbol{J} = \frac{\partial \boldsymbol{x}}{\partial \boldsymbol{\xi}}$
J	The Jacobian (determinant), i.e., $J = \det(\boldsymbol{J})$
X_t	Linear interpolant of nodal mesh speeds
$\hat{u}$	Function defined on $\hat{K}$; $\hat{u} = u \circ F_K$
u^h	Computational solution
$\mathscr{S}^h$	A finite element space defined on a triangulation $\mathscr{T}_h$; finite dimensional
$H(u)$	Hessian of function u
$\|H(u)\|$	$= \sqrt{H(u)^2}$
$P_k(K)$	Set of polynomials of degree no more than k defined on K
Π_k	Interpolation operator which is invariant for all polynomials of degree no more than k
$\Pi_{k,K}$	Restriction of Π_k on K
$W^{m,p}(\Omega)$	Sobolev space
$H^m(\Omega)$	$= W^{m,2}(\Omega)$
$L^p(\Omega)$	Lebesgue space, i.e., $L^p(\Omega) = W^{0,p}(\Omega)$
$\|\cdot\|_{W^{m,p}(\Omega)}$, $\vert\cdot\vert_{W^{m,p}(\Omega)}$	Norm and semi-norm of $W^{m,p}(\Omega)$
$\langle\cdot\rangle_{W^{m,p}(\Omega)}$	Scaled semi-norm of $W^{m,p}(\Omega)$, i.e., $\langle\cdot\rangle_{W^{m,p}(\Omega)} = \vert\Omega\vert^{-\frac{1}{p}} \vert\cdot\vert_{W^{m,p}(\Omega)}$
$\|\cdot\|$	l_2 norm for vectors or matrices
$\|\cdot\|_F$	Frobenius norm for matrices
$\mathrm{tr}(\cdot)$, $\det(\cdot)$	Trace and determinant of matrices
$\mathbb{R}_+^{3\times 3}$	$\{P \in \mathbb{R}^{3\times 3};\ \det(P) > 0\}$
M	Monitor function or metric tensor
M_K	An average of M on element K or a monitor function defined on K
ρ	Mesh density (specification) function; $\rho = \sqrt{\det(M)}$
ρ_K	Mesh density (specification) function defined on K; $\rho_K = \sqrt{\det(M_K)}$

Index